Lecture Notes
in Control and Information Sciences 241

Editor: M. Thoma

Springer-Verlag London Ltd.

Yutaka Yamamoto and Shinji Hara (Eds)

Learning, Control and Hybrid Systems

Festschrift in honor of Bruce Allen Francis and Mathukumalli Vidyasagar on the occasion of their 50th birthdays

 Springer

Series Advisory Board

Editors

Yutaka Yamamoto, PhD
Department of Applied Analysis and Complex Dynamical Systems
Graduate School of Informatics, Kyoto University, Kyoto 606-8501, Japan

Shinji Hara, PhD
Department of Computational Intelligence and Systems Science
Tokyo Institute of Technology, 4259 Nagatchuta-cho, Midori-ku,
Yokohama 226-8502, Japan

ISBN 978-1-85233-076-7

British Library Cataloguing in Publication Data
Learning, control and hybrid systems. - (Lecture notes in control and information sciences ; 241)
 1. Hybrid computers 2. Intelligent control systems 3. Machine learning
 I. Hara, Shinji II. Yamamoto, Yutaka
 004.1'9
ISBN 978-1-85233-076-7

Library of Congress Cataloging-in-Publication Data
Learning, control and hybrid systems / Yutaka Yamamoto and Shinji Hara, eds.
   p.   cm. -- (Lecture notes in control and information sciences ; 241)
 Includes bibliographical references.
 ISBN 978-1-85233-076-7   ISBN 978-1-84628-533-2 (eBook)
 DOI 10.1007/978-1-84628-533-2

 1. Automatic control. 2. Signal processing--Digital techniques.
 I. Yamamoto, Yutaka. II. Hara, Shinji, 1952- . III. Series.
 TJ213.S45543   1998   98-38066
 629.8--dc21   CIP

Typesetting: Camera ready by authors

69/3830-543210 Printed on acid-free paper

Dedicated to

Bruce Allen Francis and Mathukumalli Vidyasagar

on the occasion of their 50th birthdays

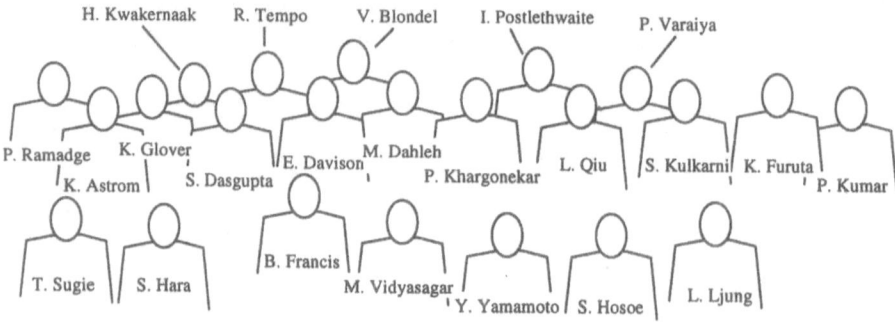

Preface

Two prominent control/system theorists Mathukumalli Vidyasagar and Bruce Allen Francis both turned 50 years of age in 1997, on September 29, and October 8, respectively, only about one week apart. To celebrate this memorable occasion, their friends got together at Bangalore (India) and had a workshop from January 4 through 8, 1998, under the same title of the present volume. This book grew out of this workshop.

Mathukumalli Vidyasagar was born in Guntur, Andhra Pradesh, India on September 29, 1947. He moved to the United States at the age of 13 along with his father (a renowned mathematician). He received the B.S., M.S., and Ph. D. degrees, all in Electrical Engineering, from the University of Wisconsin, in 1965, 1967, and 1969, respectively. Since then, he has taught at various universities in the U. S. and Canada, but at around the age of 40, he decided to return to his country. Since June 1989, he has been the Director of the Centre for Artificial Intelligence and Robotics (CAIR) in Bangalore, India, and has shown great leadership in guiding this institute. Among his honors in recognition of his research activities are the Distinguished Service Citation from his Alma Mater (The University of Wisconsin). He is also a Fellow of IEEE and the Third World Academy of Sciences as well as the Indian Academy of Sciences, the Indian National Science Academy and the Indian National Academy of Engineering.

He is the author or co-author of seven successful books (among them is the "Control System Synthesis" by MIT Press) and more than one hundred and twenty papers in archival journals. He is renowned for his fundamental contributions in nonlinear stability theory, robust control, especially, fractional representation approach, graph topology. In recent years, he has shown interest in various fields including robotics, artificial intelligence, neural networks, and learning theory.

Bruce Allen Francis was born in Toronto, Canada, on October 8, 1947. He obtained his B. A. Sc. and M. Eng. degrees in Mechanical Engineering and his Ph. D. degree in Electrical Engineering from the University of Toronto in 1969, 1970, and 1975, respectively. Since then, he has taught at various institutions, including the University of Cambridge and McGill University. He is presently a Professor in the ECE Department of the University of Toronto. He is a Fellow of the IEEE since 1988, and a co-recipient of two Outstanding Paper Awards for papers appearing in the IEEE Transactions on Automatic Control in 1984 and 1990, and also a co-recipient (with Doyle, Glover, and Khargonekar) of the 1991 IEEE W.R.G. Baker Prize Award (for the most outstanding paper reporting original work in the IEEE Transactions, Journals, Magazines, and Proceedings).

His contributions in linear systems, robust control, sampled-data systems are very fundamental. Of particular importance are the derivation of the internal model principle for servomechanism control, derivation of the first general solution to the H_∞ control problem, and sampled-data H^∞ control. He is recently interested in widening the scope of application of control theory; this attempt includes his new approach to digital signal processing. He has authored three books; among them is the very successful "A Course in H_∞ Control Theory" from Springer-Verlag, which greatly popularized this theory.

Bruce and Sagar had lots of crossover in between. It is less known that Bruce first started his career as Sagar's Ph. D. student in 1971-2, and later they were colleagues at University of Waterloo during 1982-4. Needless to say, they have several joint papers together.

On the personal sides of the editors, it has been always pleasure meeting and discussing with Bruce and Sagar. We are sure that the participants and the authors of this volume will fully agree. We would like to indulge ourselves a little in recalling some past experiences on our personal acquaintances.

Yamamoto (abbr. YY hereafter) got acquainted with Sagar in 1981 when he was visiting Japan as a JSPS fellow. On one day he gave Sagar and his wife Shakunthala a round-trip ride between Kyoto and Osaka Universities when Sagar had to give a lecture at Osaka. That was when YY got to learn about the graph topology and its relation with fractional representations. Since then Sagar visited Japan several times; also, Hara (abbr. SH hereafter) had a one-year visit with Sagar in 1988. Through candid conversations SH introduced Sagar to learning Japanese, which turned out to be quite a hobby for Sagar later. In 1994, Sagar again visited Tokyo Institute of Technology as a Nippon-Steel Visiting Chair Professor. He gave several lectures in Japanese, with Japanese LaTeX slides. It was a lot of fun for YY to help Sagar with Japanese through very late night telephone conversations. (Unfortunately, his wife and daughter had to return to India earlier than planned, and he was left alone as a single in Tokyo. As a result, he would often come back to office late at night to work.)

During his visit to Canada (in 1989), SH also had a visit with Bruce, and that was his first exposure to sampled-data H^∞ control. Much earlier in 1981, Bruce had attended YY's paper presentation in IFAC World Congress in Kyoto, but this somehow did not turn into a close exchange of opinions. Later around 1984, SH and YY started their joint work on repetitive control, and the crucial factor here is a generalization to an infinite-dimensional context of the internal model principle—one of the key contributions in servomechanism control by Bruce (along with M. Wonham). Then much later in summer of 1989, Bruce visited Japan as a JSPS fellow, and he gave a series of lectures on sampled-data control theory. At that time, YY was also contemplating a new model for sampled-data systems, and the series of lectures that Bruce gave greatly stimulated him. It soon (in the summer) culminated into a new function space model which is now also known under the name of lifting.

In commemorating their superb contributions to system and control theory, the primary objective of the Bangalore workshop, entitled "Learning, Control and Hybrid Systems" was to enlighten the current state of the art and future directions in systems and control theory. Reflecting their wide range of contributions and the variety of attendees, the contributions include many research areas. This volume is herewith dedicated to them.

The present volume consists of four parts, each entitled: "Learning and Computational Issues," "Hybrid Systems," "Modeling, Identification and Estimation," and "Robust Control." Each part reflects the current status of the field or a new approach in system theory.

The first part contains five papers related to learning and computational issues. The recent advances in robust control re-united various methods in operations research and control theory. Varied design and analysis problems are re-formulated as numerical optimization problems. On the other hand, it is gradually recognized that many control problems are computationally hard. For example, the computation of the structured singular value (μ analysis) is known to be NP hard. While this does not necessarily mean that such problems are genuinely intractable problems, it also shows that they may be computationally intractable if one imposes naive solutions. One possible direction is to relax the requirement of a "hard bound" in performance or in computational procedure. The first paper by Sagar himself discusses this issue based on a property in statistical learning theory, i.e., uniform convergence of empirical means (UCEM) property. He shows that the UCEM property assures the existence of an efficient randomized algorithm and proves that it holds in several robust control problems using recent results in VC-dimension theory. The second paper by Khargonekar and Tikku is focused on concrete randomized algorithms for robust stability analysis and robust controller synthesis. They provide several numerical examples to show the effectiveness of the proposed randomized algorithms. Blondel and Tsitsiklis investigate the decidability and the NP-hardness of stability and null-controllability tests for three classes of discrete-time nonlinear systems. The fourth paper by Hara and Yamada discusses the computational complexity of a fairly general class of non-convex optimization problems, named Matrix Product Eigenvalue Problem (MPEP), in robust control synthesis. Then the paper by Georgiou and Tannenbaum introduces a new technique using the Groebner bases for the computation of switching surfaces in time optimal control. The method is expected to be effective given the complexity of the task.

Part B is centered around hybrid systems. Needless to say, control systems are usually placed in hybrid environments. Purely linear control system structures hardly occur in practice. They are usually placed under higher-level logic controls, switching mechanisms that are to take care of some unpredicted operating condition changes, etc. Controllers are also implemented in a sampled-data setting, where the mixture of continuous and discrete timings occur simultaneously. In the past, control theorists often ignored such a hybrid nature of control environments, but it has been gradually recognized that absorbing such hybrid natures into the framework of control is a vital factor in making the control engineering more amenable to reality. This part collects varied contributions in such attempts.

The first paper by Bruce himself addresses the issue of sampled-data/digital control and gives an overview of his new approach toward digital signal processing via this new theory. The emphasis here is on the capability that the new, modern sampled-data theory, initiated by his leadership in the late 80's, can handle the aliasing problem in a unified, optimization framework, such as the H^∞ control theory. The second paper by Yamamoto and Khargonekar discusses basically the same issue; the emphasis is placed more on the continuous-time behavior. Various design examples are illustrated to show the superiority of the modern H^∞ design. The paper by Dasgupta also discusses a problem of digital signal processing. He shows a solution to the minimization of quantization error by bit allocation in biorthogonal, maximally decimated uniform filter banks. The fourth paper in this part by Hocherman-Frommer, Kulkarni and Ramadge discusses switching control. The switching occurs based on output prediction errors, and sufficient conditions are derived under which the closed-loop system remains exponentially stable while maintaining good tracking properties. The contribution by Åström discusses a task level control by taking an inverted pendulum as an laboratory example. There can be many different tasks that can be imposed on an inverted pendulum, for example, swing-up of the pendulum, catching a swinging pendulum, etc. The paper discusses various issues in such control problems. The next paper by Furuta and Pan addresses the issue of the variable structure control of sampled-data systems, and the proposed discrete-time controller quadratically stabilizes the system. Yurkovich's paper gives a perspective on fuzzy control. The role of heuristics, need of high-level decisions, etc. are examined from the viewpoint of conventional control engineers.

Part C includes various contributions related to modeling. The importance of modeling adequate for control cannot be over-emphasized, especially now robust control theory has reached some maturity. The first paper by Ljung discusses various issues in identification for control, e.g., frequency ranges on which the model should fit, iterative design, model validation etc. It gives a compact overview as well as directions for future study. The second paper by Dahleh discusses the identification of complex systems; here the issue is choosing a finitely parametrized family of models instead of just one. The third paper by Bai, Tempo and Ye discusses the issue of system parameter estimation, but from the new viewpoint of the analytic center approach, reflecting the recent advances in interior point algorithms. The estimation problem can be formulated as a convex programming problem, and the authors propose the method of analytic center. In contrast, the paper by Kulkarni and Posner give a nonparametric output prediction problem for nonlinear, discrete-time systems. Under some mild assumptions, the algorithm is shown to converge. Once we obtained a model, its simplification is often of crucial interest, for while the model should be accurate enough to capture the essential features of a plant, it should also be simple enough to be effective for system design. The paper by Glover, Goddard and Chu surveys recent results in the model reduction for uncertain, parameter-varying nonlinear systems. Various methods based on balancing are examined. The paper by Mitter and Sahai is focused on the role of the notion of information in control. They review Witsenhausen's notions of information pattern and show that there exists a family of nonlinear quantizing control laws which can be superior to the best linear one. The final paper in this part by Yanagisawa and Kimura addresses a control problem in a quantum mechanical context. In minimizing a variance, the authors are interestingly led to a least square optimal control problem.

Robust control has been the key word in control theory for at least the past decade and a half. Part D contains eight contributions in this subject in varied aspects—theory and applications.

The papers by Kwakernaak and Hosoe both deal with H^∞ control, the former dealing with descriptor systems, and the latter with emphasis on a unified approach. Spectral factorizations play crucial roles in both papers. The paper by Postlethwaite, Smerlas and Walker shows the first H^∞ controller that worked on a real helicopter. The difficulties encountered are in high levels of uncertainty and cross-axes coupling, but the test shows impressive results. The paper by Chen and Sugie proposes new parameter dependent multipliers to yield less conservative bounds for μ (the structured singular value). The next paper by Davison discusses the recent issues on the concern of the robustness of optimal controller in the presence of uncertainty both in the plant and the controller. The paper shows that a certain design approach can indeed possess such robustness. The next paper by Qiu and Chen discusses the time-domain performance limitations of feedback control—as opposed to usual frequency domain results. The paper by Honda, Suzuki and Sakamoto introduces quadratic constraints and discusses loopshaping in the open loop. The final paper by Gorbet, Morris and Wang deals with the problem of hysteresis. This nonlinear behavior appears in many devices such as shape memory alloys. The authors employ the Preisach model, and show that it can be placed in the standard dynamical system framework.

The papers are based on the presentations given in the Bangalore workshop, but we have also included some papers of the authors who could not attend the workshop for a variety of reasons. On the other hand, we could not unfortunately include some papers that were presented at the workshop.

Acknowledgments

The workshop was intense, academic, pleasant and very successful, both from scientific and social points of view. It pointed to various new directions, and serious exchanges of views and ideas were observed throughout. We are grateful to CAIR for undertaking the effort of organizing the workshop and for the smooth operation of the meeting. We thank Hideaki Ishii for the help in formatting the texts, and also Nicholas Pinfield and Alison Picken at Springer-Verlag for their valuable editorial assistance.

Yutaka Yamamoto

Shinji Hara

Kyoto and Nagatsuta, 1998

Contents

C. Modeling, Identification and Estimation

D. Robust Control

Part A

Learning and Computational Issues

Randomized Algorithms for Robust Controller Synthesis Using Statistical Learning Theory

M. Vidyasagar*

Abstract

By now it is known that several problems in the robustness analysis and synthesis of control systems are NP-complete or NP-hard. These negative results force us to modify our notion of "solving" a given problem. If we cannot solve a problem *exactly* because it is NP-hard, then we must settle for solving it *approximately*. If we cannot solve *all* instances of a problem, we must settle for solving "almost all" instances of a problem. An approach that is recently gaining popularity is that of using *randomized* algorithms. The notion of a randomized algorithm as defined here is somewhat different from that in the computer science literature, and enlarges the class of problems that can be efficiently solved. We begin with the premise that many problems in robustness analysis and synthesis can be formulated as the minimization of an objective function with respect to the controller parameters. It is argued that, in order to assess the performance of a controller as the plant varies over a prespecified family, it is better to use the *average* performance of the controller as the objective function to be minimized, rather than its *worst-case* performance, as the worst-case objective function usually leads to rather conservative designs. Then it is shown that a property from statistical learning theory known as uniform convergence of empirical means (UCEM) plays an important role in allowing us to construct efficient randomized algorithms for a wide variety of controller synthesis problems. In particular, whenever the UCEM property holds, there exists an efficient (i.e., polynomial-time) randomized algorithm. Using very recent results in VC-dimension theory, it is shown that the UCEM property holds in several problems such as robust stabilization and weighted H_2/H_∞-norm minimization. Hence it is possible to solve such problems efficiently using randomized algorithms.

*Centre for Artificial Intelligence and Robotics, Raj Bhavan Circle, High Grounds, Bangalore 560 001, India. E-Mail: sagar@cair.ernet.in

1 Introduction

During recent years it has been shown that several problems in the robustness analysis and synthesis of control systems are either NP-complete or NP-hard. See [16, 15, 1, 2] for some examples of such NP-hard problems. In the face of these and other negative results, one is forced to make some compromises in the notion of a "solving" a problem. An approach that is recently gaining popularity is the use of *randomized* algorithms, which are not required to work "all" of the time, only "most" of the time. Specifically, the probability that the algorithm fails can be made arbitrarily small (but of course not exactly equal to zero). In return for this compromise, one hopes that the algorithm is *efficient*, i.e., runs in polynomial-time. The idea of using randomization to solve control problems is suggested, among other places, in [17], [13]. In [10], [19], randomized algorithms are developed for a few problems such as: (i) determining whether a given controller stabilizes every plant in a structured perturbation model, (ii) determining whether there exists a controller of a specified order that stabilizes a given fixed plant, and so on.

The objective of the present paper is to demonstrate that some recent advances in statistical learning theory can be used to develop *efficient* randomized algorithms for a wide variety of controller analysis/synthesis problems. For such problems, it is shown that whenever a property known as uniform convergence of empirical means (UCEM) holds, there exists an efficient randomized algorithm for an associated function minimization problem. Some specific problems that fall within this framework include: robust stabilization of a family of plants by a fixed controller belonging to a specified family of controllers, and the minimization of weighted H_2/H_∞-norms. Note that what is developed here is a *broad framework* that can accommodate a wide variety of problems. The few specific problems solved here are meant only to illustrate the power and breadth of this framework. No doubt other researchers would be able to solve many more problems using the same general approach.

2 Paradigm of Controller Synthesis Problem

Suppose one is given a family of plants $\{G(x), \ x \in X\}$ parametrized by x, and a family of controllers $\{K(y), \ y \in Y\}$ parametrized by y. The objective is to find a *a single fixed controller* $K(y_0), y_0 \in Y$ that performs reasonably well for almost all plants $G(x)$. By choosing an appropriate performance index, many problems in controller synthesis can be covered by the above statement. The objective of this section is to put forward an abstract problem formulation that makes the above statement quite precise, and which forms the "universe of discourse" for the remainder of the paper. In particular, it is argued that, to avoid overly conservative designs, the performance of a controller should be taken as its *average* performance as the plant varies over a prespecified family, and not its *worst-case* performance.

Suppose $\psi(\cdot, \cdot)$ is a given cost function. Thus $\psi(G, K)$ is a measure of the performance of the system when the plant is G and the controller is K. The phrase "cost function" implies that lower values of ψ are preferred. For instance, if the objective is merely to choose a stabilizing controller, then one could define

$$\psi(G, K) := \begin{cases} 1, & \text{if the pair } (G, K) \text{ is unstable,} \\ 0, & \text{if the pair } (G, K) \text{ is stable.} \end{cases} \tag{2.1}$$

As a second example, one could choose

$$\psi(G, K) := \begin{cases} 1, & \text{if the pair } (G, K) \text{ is unstable,} \\ J(G, K)/[1 + J(G, K)], & \text{if the pair } (G, K) \text{ is stable,} \end{cases} \tag{2.2}$$

where

$$J(G, K) = \| W(I + GK)^{-1} \|_2$$

if one wishes to study filtering problems, and

$$J(G, K) := \| W(I + GK)^{-1} \|_\infty$$

if one wishes to study problems of optimal rejection of disturbances. Note that W is a given weighting matrix. Two points should be noted in the above definition: (i) The usual weighted H_2 or H_∞-norm denoted by $J(G, K)$ takes values in $[0, \infty)$. For purely technical reasons that will become clear later, this cost function is rescaled by defining $\psi = J/(1 + J)$, so that $\psi(G, K)$ takes values in $[0, 1]$. (ii) To guard against the possibility that $W(I + GK)^{-1}$ belongs to H_2 even though the pair (G, K) is unstable,[1] the cost function $\psi(G, K)$ is explicitly defined to be 1 (corresponding to $J = \infty$), if the pair (G, K) is unstable.

The preceding discussion pertains only to quantifying the performance of a *single plant-controller pair*. However, in problems of robust stabilization and robust performance, the cost function should reflect the performance of a *fixed* controller for a *variety* of plants. Since $G = G(x)$ and $K = K(y)$, let us define

$$g(x, y) := \psi[G(x), K(y)].$$

Note that g depends on both the plant parameter $x \in X$ and the controller parameter $y \in Y$. As such, g maps $X \times Y$ into $[0, 1]$. The aim is to define an objective function of y alone that quantifies the performance of the controller $K(y)$, so that by minimizing this objective function with respect to y one could find an "optimal" controller.

As a first attempt, one could choose

$$h(y) := \sup_{x \in X} g(x, y) = \sup_{x \in X} \psi[G(x), K(y)]. \tag{2.3}$$

---

[1]For example, consider the case where $W = 1/(s+1)^2$ and $(1 + GK)^{-1}$ is improper and behaves as $O(s)$ as $s \to \infty$, but has all of its poles in the open left half-plane.

Thus $h(y)$ measures the *worst-case* performance of a controller $K(y)$ as the plant varies over $\{G(x), \ x \in X\}$. For instance, if one chooses $\psi(\cdot, \cdot)$ as in (2.1), then $h(y) = 0$ if and only if the controller $K(y)$ stabilizes *every single* plant in $\{G(x), \ x \in X\}$. If $K(y)$ fails to stabilize even a single plant, then $h(y) = 1$. Thus minimizing the present choice of $h(\cdot)$ corresponds to solving the robust (or simultaneous) stabilization problem. Similarly, if $\psi(G, K)$ is chosen as in (2.2), then minimizing the associated $h(\cdot)$ corresponds to achieving the best possible *guaranteed* performance with robust stabilization.

It is widely believed that methods such as H_∞-norm minimization for achieving robust stabilization, and μ-synthesis for achieving guaranteed performance and robust stabilization, lead to overly conservative designs. Much of the conservatism of the designs can be attributed to the worst-case nature of the associated cost function. It seems much more reasonable to settle for controllers that work satisfactorily "most of the time." One way to capture this intuitive idea in a mathematical framework is to introduce a probability measure P on the set X, that reflects one's prior belief on the way that the "true" plant $G(x)$ is distributed in the set of possible plants $\{G(x), \ x \in X\}$. For instance, in a problem of robust stabilization, G_0 can be a nominal, or most likely, plant model, and the probability measure P can be "peaked" around G_0. The more confident one is about the nominal plant model G_0, the more sharply peaked the probability measure P can be. Once the probability measure P is chosen, the objective function to be minimized can be defined as

$$f(y) := E_P[g(x, y)] = E_P[\psi(G(x), K(y))]. \qquad (2.4)$$

Thus $f(y)$ is the *expected or average* performance of the controller $K(y)$ when the plant is distributed according to the probability measure P. The expected value type of objective function captures the intuitive idea that a controller can occasionally be permitted to perform poorly for some plant conditions, provided these plant conditions are not too likely to occur.

While the worst-case objective function defined in (2.3) is easy to understand and to interpret, the interpretation of the expected-value type of objective function defined in (2.4) needs a little elaboration. Suppose $\psi(\cdot, \cdot)$ is defined as in (2.1). Then $f(y)$ is the measure (or "volume") of the subset of $\{G(x), \ x \in X\}$ that *fails* to be stabilized by the controller $K(y)$. Alternatively, one can assert with confidence $1 - f(y)$ that the controller $K(y)$ stabilizes a plant $G(x)$ selected at random from $\{G(x), \ x \in X\}$ according to the probability measure P. More generally, suppose ψ (or equivalently g) assumes values in $[0, \infty)$ (and not just $[0, 1]$ as assumed elsewhere). Then a routine computation shows that, for each $\gamma > f(y)$, we have

$$P\{x \in X : \psi[G(x), K(y)] \geq \gamma\} = P\{x \in X : g(x, y) \geq \gamma\} \leq f(y)/\gamma.$$

Hence, given any $\gamma > f(y)$, it can be asserted with confidence $1 - f(y)/\gamma$ that $\psi[G(x), K(y)] \leq \gamma$ for a plant $G(x)$ chosen at random from $\{G(x), \ x \in X\}$ according to the probability measure P.

3 Various Types of "Near" Minima

Suppose Y is a given set, $f : Y \to \Re$ is a given function, and that it is desired to minimize $f(y)$ with respect to y. There are many problems, such as those mentioned in Section 1, in which finding the minimum value

$$f^* := \inf_{y \in Y} f(y)$$

is NP-hard. More precisely, given a number f_0, it is NP-hard to determine whether or not $f_0 \geq f^*$. In such cases, one has to be content with "nearly" minimizing $f(\cdot)$. The objective of this section is to introduce three different definitions of "near minima."

Definition 1 *Suppose $f : Y \to \Re$ and that $\epsilon > 0$ is a given number. A number $f_0 \in \Re$ is said to be a* **Type 1 near minimum** *of $f(\cdot)$ to accuracy ϵ, or an* **approximate near minimum** *of $f(\cdot)$ to accuracy ϵ, if*

$$\inf_{y \in Y} f(y) - \epsilon \leq f_0 \leq \inf_{y \in Y} f(y) + \epsilon, \tag{3.1}$$

or equivalently

$$\left| f_0 - \inf_{y \in Y} f(y) \right| \leq \epsilon.$$

Definition 2 *Suppose $f : Y \to \Re$, that Q is a given probability measure on Y, and that $\alpha > 0$ is a given number. A number $f_0 \in \Re$ is said to be a* **Type 2 near minimum** *of $f(\cdot)$ to level α, or a* **probable near minimum** *of $f(\cdot)$ to level α, if $f_0 \geq f^*$, and in addition*

$$Q\{y \in Y : f(y) < f_0\} \leq \alpha.$$

The notion of a probable near minimum can be interpreted as follows: f_0 is a probable near minimum of $f(\cdot)$ to level α if there is an "exceptional set" S with $Q(S) \leq \alpha$ such that

$$\inf_{y \in Y} f(y) \leq f_0 \leq \inf_{y \in Y \setminus S} f(y). \tag{3.2}$$

In other words, f_0 is bracketed by the infimum of $f(\cdot)$ over all of Y, and the infimum of $f(\cdot)$ over "nearly" all of Y.

It is important to note that even if f_0 is a probable minimum of $f(\cdot)$ to some level α, however small α might be, the difference $f_0 - f^*$ could be arbitrarily large, or even infinite. In fact, it is possible for a finite number to be a probable near minimum of a function that is unbounded from below. Thus a probable near minimum to level α can be interpreted as follows: If one person gives the number f_0 to the adversary, and challenges the adversary to "beat" f_0 by producing a $y \in Y$ such that $f(y) < f_0$, and if the adversary tries to produce a suitable y by choosing $y \in Y$ at random according to the probability measure Q, then his/her chances of winning are no more than α. However, if the adversary does succeed in producing a y such that $f(y) < f_0$, then the difference $f_0 - f(y)$ could be arbitrarily large.

Definition 3 *Suppose $f : Y \to \Re$, that Q is a given probability measure on Y, and that $\epsilon, \alpha > 0$ are given numbers. A number $f_0 \in \Re$ is said to be a* **Type 3 near minimum** *of $f(\cdot)$ to accuracy ϵ and level α, or a* **probably approximate near minimum** *of $f(\cdot)$ to accuracy ϵ and level α, if $f_0 \geq f^* - \epsilon$, and in addition*

$$Q\{y \in Y : f(y) < f_0 - \epsilon\} \leq \alpha.$$

Another way of saying this is that there exists an "exceptional set" $S \subseteq Y$ with $Q(S) \leq \alpha$ such that

$$\inf_{y \in Y} f(y) - \epsilon \leq f_0 \leq \inf_{y \in Y \setminus S} f(y) + \epsilon. \tag{3.3}$$

A comparison of (3.1), (3.2) and (3.3) brings out clearly the relationships between the various types of near minima.

4   A General Approach to Randomized Algorithms

In this section, a general approach is outlined that could be used to develop randomized algorithms for minimizing an objective function of the type (2.4). Subsequent sections contain a study of some specific situations in which this general approach could be profitably applied.

4.1   The UCEM Property

Let us return to the specific problem of minimizing the type of objective function introduced in (2.4), namely

$$f(y) = E_P[g(x,y)].$$

In general, evaluating an expected value *exactly* is not an easy task, since an expected value is just an integral with respect to some measure. However, it is possible to *approximate* an expected value to arbitrarily small error, as follows: Let us for the time being ignore the y variable, and suppose $a : X \to [0,1]$ is a measurable function. Then

$$E_P(a) = \int_X a(x) \, P(dx).$$

To approximate $E_P(a)$, one generates i.i.d. samples $x_1, \ldots, x_m \in X$ distributed according to P, and defines

$$\hat{E}(a; \mathbf{x}) := \frac{1}{m} \sum_{j=1}^{m} a(x_j).$$

The number $\hat{E}(a; \mathbf{x})$ is referred to as the *empirical mean* of the function $a(\cdot)$ based on the multisample $\mathbf{x} := [x_1 \ldots x_m]^t \in X^m$. Note that $\hat{E}(a; \mathbf{x})$ is itself a random variable on the product space X^m. Now one can ask: How good an approximation is $\hat{E}(a; \mathbf{x})$ to the true mean $E_P(a)$? As estimate is given by a well-known bound known as **Hoeffding's inequality** [7], which states that for each $\epsilon > 0$, we have

$$P^m\{\mathbf{x} \in X^m : |\hat{E}(a; \mathbf{x}) - E_P(a)| > \epsilon\} \leq 2 \exp(-2m\epsilon^2).$$

All the material presented thus far in this subsection is standard and classical. Now we come to some recent ideas. Suppose $\mathcal{A}$ is a *family* of measurable functions mapping X into $[0, 1]$. Note that $\mathcal{A}$ need *not* be a finite family. For each function $a \in \mathcal{A}$, one can form an empirical mean $\hat{E}(a; \mathbf{x})$ using a multisample $\mathbf{x} \in X^m$, as described above. Now let us define

$$q(m, \epsilon; \mathcal{A}) := P^m\{\mathbf{x} \in X^m : \sup_{a \in \mathcal{A}} |\hat{E}(a; \mathbf{x}) - E_P(a)| > \epsilon\},$$

or equivalently,

$$q(m, \epsilon; \mathcal{A}) := P^m\{\mathbf{x} \in X^m : \exists a \in \mathcal{A} \text{ s.t. } \hat{E}(a; \mathbf{x}) - E_P(a)| > \epsilon\}.$$

Then, after m i.i.d. samples have been drawn and an empirical mean $\hat{E}(a; \mathbf{x})$ is computed for each function $a \in \mathcal{A}$, it can be said with confidence $1 - q(m, \epsilon; \mathcal{A})$ that *every single* empirical mean $\hat{E}(a; \mathbf{x})$ is within ϵ of the corresponding true $E_P(a)$. The family $\mathcal{A}$ is said to have the property of **uniform convergence of empirical means (UCEM)** if $q(m, \epsilon; \mathcal{A}) \to 0$ as $m \to \infty$.

Note that if the family $\mathcal{A}$ is *finite*, then by repeated application of Hoeffding's inequality it follows that

$$q(m, \epsilon; \mathcal{A}) \leq 2|\mathcal{A}| \exp(-2m\epsilon^2). \tag{4.1}$$

Hence every finite family of functions has the UCEM property. However, an infinite family need not have the UCEM property. During the last twenty five years or so, many researchers have studied this property. A standard reference for some of the early results is [20], while a recent and thorough treatment can be found in [23]. In particular, Section 3.1 of [23] contains a detailed discussion of the UCEM property as well as several examples.

4.2   An Approach to Finding Approximate Near Minima with High Confidence

The notion of UCEM (uniform convergence of empirical means) introduced in the preceding subsection suggests the following approach to finding an approximate near minimum of the objective function (2.4). Let us introduce the notation

$$g_y(x) := g(x, y), \ \forall x \in X, \ \forall y \in Y.$$

Thus for each $y \in Y$, the function $g_y(\cdot)$ maps X into $[0, 1]$. Now define the associated family of functions $\mathcal{G} := \{g_y(\cdot), \ y \in Y\}$. Suppose now that $\mathbf{x} := [x_1 \ldots x_m]^t \in X^m$ is a collection of i.i.d. samples. For each function $g_y(\cdot) \in \mathcal{G}$, one can define its empirical mean based on the multisample $\mathbf{x}$ as

$$\hat{E}(g_y; \mathbf{x}) := \frac{1}{m} \sum_{j=1}^{m} g_y(x_i) = \frac{1}{m} \sum_{j=1}^{m} g(x_j, y), \ y \in Y.$$

As before, let

$$q(m, \epsilon; \mathcal{G}) := P^m \{\mathbf{x} \in X^m : \ \sup_{g_y \in \mathcal{G}} |\hat{E}(g_y; \mathbf{x}) - E_P(g_y)| > \epsilon\}. \qquad (4.2)$$

Observe that an equivalent way of writing $q(m, \epsilon; \mathcal{G})$ is as follows:

$$q(m, \epsilon; \mathcal{G}) := P^m \{\mathbf{x} \in X^m : \ \sup_{y \in Y} |\hat{E}(g_y; \mathbf{x}) - f(y)| > \epsilon\}.$$

The family $\mathcal{G}$ has the UCEM property if and only if $q(m, \epsilon; \mathcal{G}) \to 0$ as $m \to \infty$ for each $\epsilon > 0$.

Suppose the family $\mathcal{G}$ does indeed have the UCEM property, and consider the following approach: Given $\epsilon, \delta > 0$, choose the integer m large enough that

$$q(m, \epsilon; \mathcal{G}) < \delta.$$

Then, using one's favourite algorithm, find a value of $y \in Y$ that *exactly* minimizes the function $\hat{E}(g_y; \mathbf{x})$. In other words, choose $y_0 \in Y$ such that

$$\hat{E}(g_{y_0}; \mathbf{x}) = \min_{y \in Y} \hat{E}(g_y; \mathbf{x}).$$

Then it can be said with confidence $1 - \delta$ that y is an *approximate* near minimizer of the *original* objective function $f(\cdot)$ to accuracy ϵ.

The claim made above is easy to establish. Once the integer m is chosen large enough that $q(m, \epsilon; \mathcal{G}) < \delta$, it can be said with confidence $1 - \delta$ that

$$|f(y) - \hat{E}(g_y; \mathbf{x})| \leq \epsilon, \ \forall y \in Y.$$

In other words, the function $\hat{E}(g.; \mathbf{x})$ is a *uniformly close* approximation to the original objective function $f(\cdot)$. Hence it readily follows that an *exact* minimizer of $\hat{E}(g.; \mathbf{x})$ is also an *approximate* near minimizer of $f(\cdot)$ to accuracy ϵ.

Note that the above approach is an example of a *randomized* algorithm, in the sense that there is a nonzero probability (namely, $q(m, \epsilon; \mathcal{G})$) that the algorithm may fail to produce an approximate near minimum of $f(\cdot)$. By increasing the integer m of x-samples used in computing the empirical mean $\hat{E}(g_y; \mathbf{x})$, this failure probability can be made *arbitrarily small*, but it can never be made exactly equal to zero.

4.3 A Universal Algorithm for Finding Probable Near Minima

Suppose $h : Y \to \Re$, Q is a probability measure on Y, and that h is measurable with respect to the σ-algebra that underlies Q. Then the following algorithm produces, with arbitrarily high confidence, a probable near minimum of $h(\cdot)$ to a specified level.

Algorithm 1 *Given*

- *A probability measure Q on Y,*

- *A measurable function $h : Y \to \Re$,*

- *A level parameter $\alpha \in (0, 1)$, and*

- *A confidence parameter $\delta \in (0, 1)$.*

Choose an integer n such that

$$(1 - \alpha)^n \leq \delta, \text{ or equivalently } n \geq \frac{\lg(1/\delta)}{\lg[1/(1 - \alpha)]}. \tag{4.3}$$

Generate independent identically distributed (i.i.d.) samples $y_1, \ldots, y_n \in Y$ distributed according to Q. Define

$$\bar{h} := \min_{1 \leq i \leq n} h(y_i).$$

Then it can be said with confidence at least $1 - \delta$ that $\bar{h}$ is a probable near minimum of $h(\cdot)$ to level α.

This claim is proved in [23], Lemma 11.1, p. 357.

4.4   An Algorithm for Finding Probably Approximate Near Minima

The ideas in the preceding two subsections can be combined to produce a randomized algorithm for finding a probably approximate (or Type 3) near minimum of an objective function $f(\cdot)$ of the form (2.4). Actually two distinct algorithms are presented. The first is "universal," while the second algorithm is applicable only to situations where an associated family of functions has the UCEM property. The sample complexity estimates for the first "universal" algorithm are the best possible, whereas there is considerable scope for improving the sample complexity estimates of the second algorithm.

Suppose real parameters $\epsilon, \alpha, \delta > 0$ are given; the objective is to develop a randomized algorithm that constructs a probably approximate (Type 3) near minimum of

$$f(y) := E_P[g(x,y)]$$

to accuracy ϵ and level α, with confidence $1 - \delta$. In other words, the probability that the randomized algorithms fails to find a probably approximate near minimum to accuracy ϵ and level α must be at most δ.

Algorithm 2 *Given*

- *Sets X, Y,*

- *Probability measures P on X and Q on Y,*

- *A measurable function $g : X \times Y \to [0,1]$, and*

- *An accuracy parameter $\epsilon \in (0,1)$, a level parameter $\alpha \in (0,1)$, and a confidence parameter $\delta \in (0,1)$.*

Define

$$f(y) := E_P[g(x,y)]$$

Choose integers

$$n \geq \frac{\lg(2/\delta)}{\lg[1/(1-\alpha)]}, \text{ and } m \geq \frac{1}{2\epsilon^2} \ln \frac{4n}{\delta}. \tag{4.4}$$

Generate i.i.d. samples $y_1, \ldots, y_n \in Y$ according to Q and $x_1, \ldots, x_m \in X$ according to P. Define

$$\hat{f}_i := \frac{1}{m} \sum_{j=1}^{m} g(x_j, y_i), \ i = 1, \ldots, n, \text{ and}$$

$$\hat{f}_0 := \min_{1 \leq i \leq n} \hat{f}_i.$$

Then with confidence $1 - \delta$, it can be said that $\hat{f}_0$ is a probably approximate (Type 3) near minimum of $f(\cdot)$ to accuracy ϵ and level α.

The proof of the claim in Algorithm 2 is easy. Once the i.i.d. samples $y_1, \ldots, y_n$ are generated where n satisfies (4.4), one can define

$$\bar{f} := \min_{1 \leq i \leq n} f(y_i).$$

Then it follows from the reasoning used in Algorithm 1 that, with confidence $1 - \delta/2$ (*not* $1 - \delta$ – compare (4.4) with (4.3)), the number $\bar{f}$ is a probable near minimum of $f(\cdot)$ to level α. Now consider the *finite* family of functions $\mathcal{A} := \{g(\cdot, y_i), \ i = 1, \ldots, n\}$. It follows from (4.1) that the quantity $q(m, \epsilon; \mathcal{A})$ for this family is bounded by

$$q(m, \epsilon; \mathcal{A}) \leq 2n \, \exp(-2m\epsilon^2).$$

Hence, with confidence $1 - 2e^{-2m\epsilon^2}$, it can be asserted that the quantities $\hat{f}_i$ satisfy

$$|f(y_i) - \hat{f}_i| \leq \epsilon, \text{ for } i = 1, \ldots, n.$$

In particular, it follows that

$$|\hat{f}_0 - \bar{f}| \leq \epsilon. \tag{4.5}$$

Now the choice of m in (4.4) ensures that

$$2n \, \exp(-2m\epsilon^2) \leq \delta/2.$$

Thus (4.5) holds with confidence $1 - \delta/2$. Combining the two statements shows that, with confidence $1 - \delta$, $\hat{f}_0$ is a probably approximate (Type 3) near minimum of $f(\cdot)$ to accuracy ϵ and level α.

To state the next algorithm, recall the definitions of the symbols $g_y(x)$, $\mathcal{G}$, $\hat{E}(g_y; \mathbf{x})$ defined previously. Suppose the family $\mathcal{G}$ has the UCEM property, i.e., that $q(m, \epsilon; \mathcal{G}) \to 0$ as $m \to \infty$ for each $\epsilon > 0$.

Algorithm 3 *Given*

- *Sets X, Y,*

- *Probability measures P on X and Q on Y,*

- *A measurable function $g : X \times Y \to [0, 1]$, and*

- *An accuracy parameter $\epsilon \in (0, 1)$, a level parameter $\alpha \in (0, 1)$, and a confidence parameter $\delta \in (0, 1)$.*

Define

$$f(y) := E_P[g(x, y)]$$

Suppose the associated family of functions $\mathcal{G}$ has the UCEM property, and define $q(m, \epsilon; \mathcal{G})$ as in in (4.2). Select integers n, m such that

$$n \geq \frac{\lg(2/\delta)}{\lg[1/(1-\alpha)]}, \text{ and } q(m, \epsilon; \mathcal{G}) \leq \delta/2. \tag{4.6}$$

Generate i.i.d. samples $y_1, \ldots, y_n \in Y$ according to Q and $x_1, \ldots, x_m \in X$ according to P. Define

$$\hat{f}_i := \frac{1}{m} \sum_{j=1}^{m} f(x_j, y_i), \; i = 1, \ldots, n, \; \text{and}$$

$$\hat{f}_0 := \min_{1 \leq i \leq n} \hat{f}_i.$$

Then with confidence $1 - \delta$, it can be said that $\hat{f}_0$ is a probably approximate (Type 3) near minimum of $f(\cdot)$ to accuracy ϵ and level α.

It can be seen by comparing (4.4) and (4.6) that the only difference between Algorithms 2 and 3 is in the integer m of x-samples. The key point to note is that in Algorithm 3, the integer m is *independent* of the integer n, which in turn depends on the level parameter α.

5   Some Sufficient Conditions for the UCEM Property

In this section, brief review is given of some known sufficient conditions for a family of functions to have the UCEM property. These sufficient conditions are based on two powerful notions known as the Vapnik-Chervonenkis (VC-) dimension, and the Pollard (P-) dimension, respectively [20, 6]. For a thorough and contemporary treatment of the UCEM property and related issues, see [23].

5.1   Definitions of the VC-dimension and P-Dimension

Definition 4 *Suppose A is a family of measurable functions mapping X into $\{0,1\}$. A set $S = \{x_1, \ldots, x_n\}$ is said to be **shattered** by A if each of the 2^n functions mapping S into $\{0,1\}$ is the restriction to S of some function in A. Equivalently, S is shattered by A if, for every subset $A \subseteq S$, there exists a corresponding function $a_A(\cdot) \in A$ such that $a_A(x_i) = 1$ if $x_i \in A$ and $a_A(x_i) = 0$ if $x_i \notin A$. The **Vapnik-Chervonenkis (VC-) dimension** of A, denoted by VC-dim(A), is the largest integer n such that there exists a set of cardinality n that is shattered by A.*

See [23], Section 4.1 for several examples of the explicit computation of the VC-dimension.

The VC-dimension is defined for families of *binary-valued* functions. The corresponding notion for families of $[0,1]$-valued functions is referred to by various authors as the Pollard dimension or the pseudo-dimension; it is referred to here by the neutral symbol P-dimension.

Definition 5 *Suppose $\mathcal{A}$ is a family of measurable functions mapping X into $[0,1]$. A set $S = \{x_1, \ldots, x_n\}$ is said to be **P-shattered** by $\mathcal{A}$ if there exists a vector $\mathbf{c} \in [0,1]^n$ such that, for every binary vector $\mathbf{e} \in \{0,1\}^n$, there exists a corresponding function $a_{\mathbf{e}}(\cdot) \in \mathcal{A}$ such that*

$$a_{\mathbf{e}}(x_i) \geq c_i \ \text{if } e_i = 1, \text{ and } a_{\mathbf{e}}(x_i) < c_i \ \text{if } e_i = 0.$$

*The **P-dimension** of $\mathcal{A}$, denoted by P-dim($\mathcal{A}$), is the largest integer n such that there exists a set of cardinality n that is P-shattered by $\mathcal{A}$.*

It is easy to see that if every function in $\mathcal{A}$ is binary-valued, then VC-dim($\mathcal{A}$) = P-dim($\mathcal{A}$). More generally, the two dimensions are related as follows: Let $\eta : \Re \rightarrow \{0,1\}$ denote the **Heaviside** function (sometimes referred to also as the "step" function) defined as follows:

$$\eta(r) = 1, \ \text{if } r \geq 0, \text{and } \eta(r) = 0, \ \text{if } r = 0.$$

Given a family of functions $\mathcal{A}$ mapping X into $[0,1]$, define an associated family of functions $\bar{\mathcal{A}}$ mapping $X \times [0,1]$ into $\{0,1\}$ as follows: For each $a(\cdot) \in \mathcal{A}, x \in X, c \in [0,1]$, let

$$\bar{a}(x,c) = \eta[a(x) - c].$$

Now let a vary over $\mathcal{A}$; the corresponding family of functions $\bar{a}$ is the collection $\bar{\mathcal{A}}$.

Lemma 1 *With all symbols defined as above, we have*

$$P\text{-}dim(\mathcal{A}) = VC\text{-}dim(\bar{\mathcal{A}}).$$

The above observation is explicitly stated and proved in [12]; see also [23], Lemma 10.1, p. 300.

5.2 Finiteness of the VC- and P-Dimensions Implies the UCEM Property

The significance of the VC-dimension and the P-dimension arises from the fact that the finiteness of these dimensions is sufficient for a family of functions to have the UCEM property. Specifically, we have the following results:

Theorem 1 *[21, 22, 20, 23] Suppose $\mathcal{A}$ is a family of measurable functions X mapping into $\{0,1\}$, and that VC-dim($\mathcal{A}$) $\leq d < \infty$. Then $\mathcal{A}$ has the UCEM property, whatever be the probability measure P. Moreover,*

$$q(m, \epsilon; \mathcal{A}) \leq 4 \left(\frac{2em}{d} \right)^d \exp(-m\epsilon^2/8), \ \forall m, \epsilon. \tag{5.1}$$

The inequality $q(m, \epsilon; \mathcal{A}) \leq \delta$ is satisfied provided

$$m \geq \max \left\{ \frac{16}{\epsilon^2} \ln \frac{4}{\delta}, \frac{32d}{\epsilon^2} \ln \frac{32e}{\epsilon^2} \right\}. \tag{5.2}$$

Theorem 2 *[6, 23] Suppose $\mathcal{A}$ is a family of measurable functions X mapping into $[0,1]$, and that P-dim$(\mathcal{A}) \leq d < \infty$. Then $\mathcal{A}$ has the UCEM property, whatever be the probability measure P. Moreover, if $\epsilon < e/(2 \lg e) \approx 0.94$, we have*

$$q(m, \epsilon; \mathcal{A}) \leq 8 \left(\frac{16e}{\epsilon} \ln \frac{16e}{\epsilon} \right)^d \exp(-m\epsilon^2/32), \ \forall m, \ \forall \epsilon. \tag{5.3}$$

The inequality $q(m, \epsilon; \mathcal{A}) \leq \delta$ is satisfied provided

$$m \geq \frac{32}{\epsilon^2} \left[\ln \frac{8}{\delta} + d \left(\ln \frac{16e}{\epsilon} + \ln \ln \frac{16e}{\epsilon} \right) \right]. \tag{5.4}$$

The bounds given in Theorems 1 and 2 have the advantage that they are explicit and easy to apply. However, numerical experimentation shows that these bounds are also *very conservative*, perhaps by two or three orders of magnitude. Trying to improve these bounds for specific families of functions and/or for specific probability measures is an active area of research. A popular recent approach is to combine the methods of statistical mechanics with those of empirical process theory to obtain better bounds; see [11] as an illustration of this approach.

5.3   Upper Bounds for the VC-Dimension

Theorems 1 and 2 bring out the importance of the VC-dimension and the P-dimension in the context of the UCEM property. Moreover, Lemma 1 shows that the P-dimension of a family of $[0,1]$-valued functions is just the VC-dimension of an associated family of binary-valued functions. Because of these theorems, several researchers have explored the VC-dimension of various families of binary-valued functions. Over the years an impressive array of upper bounds have been derived for the VC-dimension of a wide variety of function families. Note that upper bounds are enough to apply Theorems 1 and 2. Many of these bounds are collected and rederived in [23], Chapter 10. However, there is one result that is particularly appropriate for the class of problems studied here; it is presented next.

A little terminology is first introduced towards this end. Suppose $X \subseteq \Re^k$, $Y \subseteq \Re^l$ for some integers k, l respectively, and suppose $\tau : \Re^k \times \Re^l \to \Re$ is a polynomial. A polynomial inequality is an expression of the form $\tau(x, y) > 0$. Note that, for each $x \in X, y \in Y$, the expression "$\tau(x, y) > 0$" evaluates to either "true" or "false." Now suppose $\tau_1(x, y), \ldots, \tau_t(x, y)$ are polynomials in x, y, and suppose that the degree *with respect to y* of each polynomial

$\tau_i(x, y)$ is no larger than r; (the degree with respect to x does not matter). Finally, suppose $\phi(x, y)$ is a **Boolean formula** obtained from the expressions "$\tau_i(x, y) > 0$" using the standard logical connectives $\neg$ (not), $\vee$ (or), $\wedge$ (and) and $\Rightarrow$ (implies). Thus, for each $x \in X, y \in Y$, each expression "$\tau_i(x, y) > 0$" evaluates to either "true" or "false," and then the overall Boolean formula $\phi(x, y)$ itself evaluates to either "true" or "false" according to the standard rules of Boolean logic. By associating the values 1 with "true" and 0 with "false," one can think of ϕ as a map from $X \times Y$ into $\{0, 1\}$. In this set-up, the t expressions "$\tau_i(x, y) > 0$" for $i = 1, \ldots, t$ are called "atomic formulas" of the overall Boolean formula $\phi(x, y)$. Note that the terminology is not very precise, since a given function $\phi : X \times Y \to \{0, 1\}$ can perhaps be written as a Boolean formula in more than one way. However, in what follows, we will always start with the τ_i and proceed towards ϕ, so that both the nature and the number of the atomic formulas is unambiguous. Now define, for each $y \in Y$,

$$A_y := \{x \in X : \phi(x, y) = 1\}, \text{ and}$$

$$\mathcal{A} := \{A_y : y \in Y\}.$$

Then $\mathcal{A}$ is a collection of subsets of X. The objective is to obtain an upper bound for the VC-dimension of $\mathcal{A}$.

The following theorem is a refinement of a result from [8], [9], and is proved in this precise form in [23], Corollary 10.2, p. 330.

Theorem 3 *With all symbols as above, we have*

$$VC\text{-}dim(\mathcal{A}) \leq 2l \lg(4ert). \tag{5.5}$$

By choosing the polynomials $\tau_1, \ldots, \tau_t$ in an appropriate manner, it is possible to capture a large number of controller synthesis problems within the scope of Theorem 3. This is illustrated in subsequent sections.

6   Robust Stabilization

In this section, the approach of the preceding section is applied to the problem of robustly stabilizing a given family of plants $\{G(x), x \in X\}$ using a single fixed controller selected from the family $\{K(y), y \in Y\}$. In conventional control theory, the problem "Does there exist a controller $K(y_0) \in \{K(y), y \in Y\}$ that stabilizes *every* plant in $\{G(x), x \in X\}$?" is thought of as having a "yes or no" answer. However, the present approach is geared towards optimization and not binary decisions. As in (2.1), define $\psi(G, K)$ to equal 1 if the pair (G, K) is unstable, and 0 if the pair (G, K) is stable. Then, for each $y \in Y$, the cost function

$$f(y) := E_P[\psi(G(x), K(y))] = E_P[g(x, y)]$$

equals the volume of the subset of $\{G(x),\ x \in X\}$ that *fails* to be stabilized by $K(y)$. Accordingly, the problem of minimizing $f(y)$ with respect to $y \in Y$ corresponds to choosing a controller that *destabilizes the smallest volume* of the plant family.

As before, define

$$g_y(x) := g(x,y) = \begin{cases} 1, & \text{if the pair } (G(x), K(y)) \text{ is unstable, and} \\ 0, & \text{if the pair } (G(x), K(y)) \text{ is stable.} \end{cases}$$

Now it is shown that, under some reasonable assumptions, the family of binary-valued functions $\mathcal{G} := \{g_y(\cdot), y \in Y\}$ has the UCEM property, so that a probably approximate near minimum of $f(\cdot)$ can be found to arbitrary accuracy, level and confidence using Algorithm 3. It is assumed that both the plant and the controller are single-input, single-output, and that each plant $G(x,s)$ is of the form

$$G(x,s) = \frac{n_G(x,s)}{d_G(x,s)}, \ \forall x \in X,$$

where n_G, d_G are *polynomials* in x, s. Next, it is assumed that all the plants are *strictly* proper and have McMillan degree α_s. In other words, it is assumed that the degree of d_G with respect to s is α_s for every $x \in X$, and that the degree of n_G with respect to s is no larger than $\alpha_s - 1$ for every $x \in X$. The assumptions about $K(y,s)$ are similar. It is assumed that

$$K(y,s) = \frac{n_K(y,s)}{d_K(y,s)}, \ \forall y \in Y$$

is a proper rational function of s, with McMillan degree β_s. Also, it is assumed that $n_K(y,s)$, $d_K(y,s)$ are polynomials in y of degree no larger than β_y. Finally, it is assumed that $X \subseteq \Re^k, Y \subseteq \Re^l$ for suitable integers k, l. The above assumptions are made solely in the interests of simplicity of exposition, and can be relaxed considerably. For instance, it is easy to handle the case where each coefficient is a *rational function* of x or y, rather than a polynomial. Such extensions are given in the full-length version of this paper, which can be obtained from the author.

Theorem 4 *Under the above assumptions, the family of binary-valued functions $\mathcal{G} := \{g_y : y \in Y\}$ has finite VC-dimension. In particular,*

$$\text{VC-dim}(\mathcal{G}) \leq 2l \lg[4e(\alpha_s + \beta_s)^2 \beta_y]. \tag{6.1}$$

Remarks It is important to note that the quantity $\alpha_s + \beta_s$, which is the order of the closed-loop system, appears *inside* the $\lg(\cdot)$ function. Thus the bounds given above will be rather small even for high-order control systems. However, the integer l, representing the number of "degrees of freedom" in the parameter y, appears linearly in the upper bound.

Proof The proof consists of writing down the conditions for the closed-loop system to be stable as a Boolean formula entailing several polynomial inequalities in x and y, and then appealing to Theorem 3. For a fixed $x \in X$, $y \in Y$, the pair $[G(x,s), K(y,s)]$ is stable if and only if two conditions hold: (i) the closed-loop transfer function is proper, and (ii) the characteristic polynomial of the closed-loop system is Hurwitz. If every plant $G(x)$ is *strictly* proper and every controller is proper, the first condition is automatically satisfied, and we are left with only the second condition. Now the closed-loop characteristic polynomial equals

$$\theta(x, y, s) := n_G(x, s)\, n_K(y, s) + d_G(x, s)\, d_K(y, s).$$

This is a polynomial of degree $\alpha_s + \beta_s$ in s, where each coefficient of θ is a polynomial in x, y; moreover, the degree of each coefficient with respect to y is no larger than β_y. The stability of the polynomial θ can be tested by forming its Hurwitz determinants as in [4], pp. 190 *ff.* Let $n_c := \alpha_s + \beta_s$ denote the degree of θ with respect to s, and write

$$\theta(x, y, s) = \sum_{i=0}^{n_c} a_{n_c - i}(x, y)\, s^i.$$

Let H_i denote the i-th Hurwitz determinant of θ; then H_i is of the form

$$H_i(x, y) = \begin{vmatrix} a_1 & a_3 & a_5 & \cdots & a_{2i-1} \\ a_0 & a_2 & a_4 & \cdots & a_{2i-2} \\ 0 & a_1 & a_3 & \cdots & a_{2i-3} \\ \vdots & \vdots & \vdots & \cdots & \vdots \\ 0 & 0 & 0 & \cdots & a_i \end{vmatrix},$$

where $a_i(x, y)$ is written as a_i in the interests of clarity. Note that a_i is taken as zero if $i > n_c$. Now θ is a Hurwitz polynomial if and only if $H_i(x, y) > 0$ for $i = 1, \ldots, n_c$. Hence the pair $[G(x,s), K(y,s)]$ is stable, and $\psi(x, y) = 0$, if and only if

$$[H_1(x, y) > 0] \ldots [H_{n_c}(x, y) > 0]. \tag{6.2}$$

This is a Boolean formula. Moreover, $\psi(x, y) = 1$ is just the *negation* of the above Boolean formula, and is thus another Boolean formula with just the same atomic polynomial inequalities, with the connectives changed according to De Morgan's formula. To apply Theorem 3, it is necessary to count the number of atomic polynomial inequalities, and their maximum degree with respect to y. First, the number of atomic inequalities is $n_c = \alpha_s + \beta_s$. Thus we take

$$t = \alpha_s + \beta_s.$$

Next, let us examine the degree of each of the atomic polynomial inequalities with respect to y. Each Hurwitz determinant $H_i(x, y)$ is the determinant of an $i \times i$ matrix, each of whose entries is a polynomial in y of degree β_y or less. Therefore the degree of H_i with respect to y is at most $\beta_y i$. Hence the maximum degree of the atomic polynomial inequalities with respect to y can be taken as

$$r = (\alpha_s + \beta_s)\beta_y.$$

Finally, applying Theorem 3 and in particular the bound (5.5) leads to the estimate

$$\text{VC-dim}(\mathcal{G}) \leq 2l \lg(4ert) \leq 2l \lg[4e(\alpha_s + \beta_s)^2 \beta_y].$$

This completes the proof. ∎

7   Weighted H_∞-Norm Minimization

By using arguments entirely analogous to those in Section 6, it is possible to estimate the P-dimension of the family $\mathcal{G}$ in the case where the objective is to minimize the weighted H_∞-norm of the closed-loop transfer function. Let $W(s)$ be a given weighting function, and define the performance measure to be

$$\psi(G,K) := \left\{ \begin{array}{ll} 1, & \text{if the pair } (G,K) \text{ is unstable,} \\ J(G,K)/[1+J(G,K)], & \text{if the pair } (G,K) \text{ is stable,} \end{array} \right.$$

where

$$J(G,K) = \parallel W(1+GK)^{-1} \parallel_\infty$$

and $\parallel \cdot \parallel_\infty$ denotes the H_∞-norm. Let $G = G(x,s)$, $K = K(y,s)$ be as in Section 6, wherein $G = n_G/d_G$, $K = n_K/d_K$, and the coefficients of n_G, d_G are *rational* functions of x without any poles in X, and the coefficients of n_K, d_K are *rational* functions of y without any poles in Y. As before, let β_y denote the maximum degree of any coefficient of n_K, d_K with respect to y. For convenience, let $n_c := \alpha_s + \beta_s$ denote the McMillan degree of the closed-loop transfer function, and let n_w denote the McMillan degree of the *weighted* closed-loop transfer function $W(1+GK)^{-1}$. Define the symbols $g_y(\cdot)$ and $\mathcal{G}$ as before.

Theorem 5 *With all symbols as above, we have*

$$P\text{-}dim(\mathcal{G}) \leq 2(l+1) \lg[8e\beta_y n_w(2n_c + 2n_w + 1)]. \tag{7.1}$$

Remarks As in the case of the bound (6.1) given in Theorem 4, the quantities n_c which is the order of the closed-loop system and n_w which is the order of the weighted closed-loop transfer function, both appear *inside* the $\lg(\cdot)$ function. Thus the bounds given above will be rather small even for high-order control systems. However, the integer l, representing the number of "degrees of freedom" in the parameter y, appears linearly in the upper bound.

Proof For each $y \in Y$, define the associated function $\bar{g}_y : X \times [0,1] \to \{0,1\}$ by

$$\bar{g}_y(x,c) = \eta[g_y(x) - c],$$

where $\eta(\cdot)$ is the Heaviside function. Let $\bar{\mathcal{G}} := \{\bar{g}_y : y \in Y\}$. Then it follows from Lemma 1 that

$$\text{P-dim}(\mathcal{G}) = \text{VC-dim}(\bar{\mathcal{G}}).$$

Hence we concentrate on estimating $\text{VC-dim}(\bar{\mathcal{G}})$.

For this purpose, we write

$$\eta[g_y(x) - c] = 0$$

as a Boolean formula involving several polynomial inequalities in x, y, c and then apply Theorem 3. Note that c is now an additional parameter in addition to y. Also,

$$\eta[g_y(x) - c] = 0 \Leftrightarrow g(x,y) - c < 0.$$

Let us write J as a shorthand for $J[G(x), K(y)]$. Then[2]

$$g(x,y) - c < 0 \Leftrightarrow [(c = 1)(J < \infty)] \vee [(c < 1)(J < c/(1-c))].$$

Now $J < \infty$ is just closed-loop stability. The results of Section 6 imply that this condition can be written in terms of at most n_c polynomial inequalities, each of which has degree at most $\beta_y n_c$ with respect to y. Next, the condition $J < c/(1-c)$ says that the weighted transfer function $F := W(1+GK)^{-1}$ has H_∞-norm less than $b := c/(1-c)$. This is the case if and only if

$$(G, K) \text{ is stable, } b > F(\infty), \text{ and } b^2 - F^*(jw)F(jw) > 0 \; \forall w.$$

Again, the first requirement involves up to n_c polynomial inequalities of degree at most $\beta_y n_c$ in y. Next, since $F(s)$ is a *stable* rational function, we can express the condition

$$b^2 - F^*(jw)F(jw) > 0 \; \forall w$$

equivalently as a *polynomial* inequality in ω of the form

$$p(\omega) > 0 \; \forall \omega,$$

where the coefficients of $p(\cdot)$ are rational functions of x, y, c. Now $p(\omega)$ is an *even* polynomial in ω of degree $2n_w$. Thus the condition $p(\omega) > 0 \; \forall \omega$ is equivalent (since $p(\infty) > 0$) to the requirement that $p(\cdot)$ does not have any real zeros. By replacing ω by $j\theta$ and defining $q(\theta) := p(j\theta) = p(\omega)$, the condition reduces to: $q(\cdot)$ does not have any imaginary axis zeros. Now let $q'(\theta)$ denote $dq(\theta)/d\theta$, and define $r(\theta) = q(\theta) + q'(\theta)$. Then the condition $p(\omega) > 0 \; \forall \omega$ is satisfied if and only if $r(\theta)$ has exactly n_w zeros in the left half-plane and exactly n_w zeros in the right half-plane. This is equivalent to the requirement that, out of the $2n_w$ Hurwitz determinants of $r(\cdot)$, there

---

[2]Strictly speaking, we should write $\neg(c < 1)$ instead of $c = 1$.

are exactly n_w negative determinants. This is a Boolean formula in these determinants. Moreover, each of these determinants is a rational function of x, y, c, and the maximum degree of any determinant with respect to y and c is $2\beta_y n_w$. Since none of these rational functions has any poles in $X \times Y \times [0, 1)$, it is possible to clear the denominators and to express the conditions in terms of $2n_w$ *polynomial* inequalities in x, y, c. Since $n_w \geq n_c$, we can now apply Theorem 3 with the number of parameters l replaced by $l + 1$,

$$t = \text{No. of inequalities} = 2n_c + 2n_w + 1,$$

$$r = \text{Max. degree w.r.t. } y = 2\beta_y n_w.$$

This completes the proof. ∎

The problem of weighted H_2-norm minimization can be handled similarly. The details are contained in the full-length version of the paper.

8   Conclusions

In this paper, a beginning has been made towards the application of recent results from statistical learning theory to the development of randomized algorithms for solving a wide variety of controller analysis and synthesis problems. What is given here is only a broad framework, and future researchers will undoubtedly be able to apply the methodology suggested here to many more problems.

References

[1] R. Braatz, P. Young, J. Doyle and M. Morari, "Computational complexity of the μ calculation," *IEEE Trans. Auto. Control*, 39, pp. 1000-1002, 1994.

[2] V. Blondel and J. N. Tsitsiklis, "NP-hardness of some linear control design problems," *SIAM J. Control and Opt.*, (to appear).

[3] J. Doyle, "Analysis of feedback systems with structured uncertainties," *Proc. IEEE*, 129, pp. 242-250, 1982.

[4] F. R. Gantmacher, *Matrix Theory*, Volume II, Chelsea, New York, 1959.

[5] M.R. Garey and D.S. Johnson, *Computers and Intractability: A Guide to the Theory of NP-Completeness*, W.H. Freeman, New York, 1979.

[6] D. Haussler, "Decision theoretic generalizations of the PAC model for neural net and other learning applications," *Information and Computation*, 100, pp. 78-150, 1992.

[7] W. Hoeffding, Probability inequalities for sums of bounded random variables. *J. Amer. Statist. Assoc.* 58, pp. 13-30, 1963.

[8] M. Karpinski and A.J. Macintyre, "Polynomial bounds for VC dimension of sigmoidal neural networks," *Proc. 27th ACM Symp. Thy. of Computing*, pp. 200-208, 1995.

[9] M. Karpinski and A.J. Macintyre, "Polynomial bounds for VC dimension of sigmoidal and general Pfaffian neural networks," *J. Comp. Sys. Sci.*, (to appear).

[10] P.P. Khargonekar and A. Tikku, "Randomized algorithms for robust control analysis have polynomial complexity," *Proc. Conf. on Decision and Control*, 1996 (to appear).

[11] A. Kowalczyk, H. Ferra and J. Szymanski, "Combining statistical physics with VC-bounds on generalisation in learning systems," *Proceeding of the Sixth Australian Conference on Neural Networks (ACNN'95*, pp. 41-44, Sydney, 1995.

[12] A.J. Macintyre and E.D. Sontag, "Finiteness results for sigmoidal neural networks," *Proc. 25th ACM Symp. Thy. of Computing*, pp. 325-334, 1993.

[13] C. Marrison and R. Stengel, "The use of random search and genetic algorithms to optimize stochastic robustness functions," *Proc. Amer. Control Conf.*, Baltimore, MD, pp. 1484-1489, 1994.

[14] R. Motwani and P. Raghavan, *Randomized Algorithms*, Cambridge U. Press, Cambridge, 1995.

[15] A. Nemirovskii, "Several NP-hard problems arising in robust stability analysis," *Math. of Control, Signals, and Systems*, 6(2), pp. 99-105, 1993.

[16] S. Poljak and J. Rohn, "Checking robust nonsingularity is NP-hard," *Math. Control, Signals, and Systems*, 6(1), pp. 1-9, 1993.

[17] L.R. Ray and R.F. Stengel, "Stochastic robustness of linear time-invariant control systems," *IEEE Trans. Auto. Control*, 36, pp. 82-87, 1991.

[18] J. M. Steele, "Empirical discrepancies and subadditive processes," *Ann. Prob.*, 6, pp. 118-127, 1978.

[19] R. Tempo, E.W. Bai and F. Dabbene, "Probabilistic robustness analysis: Explicit bounds for the minimum number of sampling points," *Proc. Conf. on Decision and Control*, 1996 (to appear).

[20] V.N. Vapnik, *Estimation of Dependences Based on Empirical Data*, Springer-Verlag, 1982.

[21] V.N. Vapnik and A.Ya. Chervonenkis, "On the uniform convergence of relative frequencies to their probabilities," *Theory of Prob. and its Appl.* 16(2), pp. 264-280, 1971.

[22] V.N. Vapnik and A.Ya. Chervonenkis, "Necessary and and sufficient conditions for the uniform convergence of means to their expectations," *Theory of Prob. and its Appl.*, 26(3), pp. 532-553, 1981.

[23] M. Vidyasagar, *A Theory of Learning and Generalization: With Applications to Neural Networks and Control Systems*, Springer-Verlag, London, 1997.

Probabilistic Search Algorithms for Robust Stability Analysis and Their Complexity Properties *

Pramod P. Khargonekar[†]   Ashok Tikku[‡]

Abstract

In this paper, we consider several robust control analysis and design problems. As has become well known over the last few years, most of these problems are NP hard. We show that if instead of worst-case guaranteed conclusions, one is willing to draw conclusions with a high degree of confidence, then the computational complexity decreases dramatically.

1   Introduction

Robust multivariable control analysis and design have been central problems in control theory over the last fifteen years. Most of the robust control research has been dominated by worst-case deterministic formulations of the control analysis and design problems. Despite the considerable successes of the worst-case deterministic approach such as $\mathcal{H}_\infty$ control theory, in robust control problems involving real parametric and/or structured uncertainty, most of the analysis and design problems remain open. Analytical results, such as the celebrated Kharitonov theorem, apply to rather specialized problems. For general and comprehensive formulations such as the structured singular value theory — μ theory — analysis and design problems have turned out to be very difficult, and the research focus has been on obtaining upper and lower bounds on the quantities of interest.

Recent research on computational complexity of robust control analysis and design problems indicates that these difficulties are, most likely, inherent

---

*This work was supported in part by Airforce Office of Scientific Research under contract no. F-49620-93-1-0246DEF and Army Research Office under grant no. DAAH04-93-G-0012

[†]Department of Electrical Engineering and Computer Science, University of Michigan, Ann Arbor, MI 48109-2122. Tel. (313) 764-4328, Fax (313) 763-8041, e-mail: *pramod@eecs.umich.edu*

[‡]Department of Electrical Engineering and Computer Science, University of Michigan, Ann Arbor, MI 48109-2122. Tel. (313) 763-1498, e-mail: *tikku@eecs.umich.edu*

to the problem formulations rather than a lack of ingenuity. For example, it has been shown that the problem of checking robust stability of a matrix polytope is NP hard [19, 21]. Similar results on computational complexity have been obtained for the μ computation problems [3].

In view of these results on the computational complexity of the worst-case deterministic formulations of robust control problems, it appears that a major change in the robust control paradigm is necessary. In particular, one should explore alternatives to the worst-case deterministic problem formulations. Recall that a typical result in robust control leads to a statement of the form that a "certain (stability, performance) property holds for all systems in a prescribed set (given by uncertain parameters in compact sets, unmodeled dynamics given by balls in function spaces, etc.") Instead, it may be more tractable to provide statements of the form that a "certain property holds for most of the systems with a high degree of confidence".

Probabilistic approaches to robust control, while not very common, are not new. Ray and Stengel [22] and Marrison and Stengel [18] have strongly advocated stochastic approaches. Hall *et. al.* [11] have analyzed the average steady state covariance matrix of linear systems depending on random parameters for certain probability distributions. Friedman *et. al.* [8] gave results on the average case $\mathcal{H}_2$ performance of linear control systems. Barmish and Lagoa [2] have given results on worst-case probability distributions for certain types of robustness analysis problems. This list is not exhaustive and the reader should consult references in these papers for related literature.

In theoretical computer science, e. g., computational learning theory, there has been a great deal of work on the use of randomized algorithms for NP hard problems. Motivated by this literature and the intractability of the current approaches to robust control, in this paper, we focus on computational complexity of probabilistic analysis of certain very general robust control problems. *The main results of this paper demonstrate that the computational complexity of a variety of robust control problems is very low, provided one is willing to settle for probabilistic statements.* (It should be noted that it is not necessary to believe that the uncertain parameters obey certain probabilistic distributions. As a matter of fact, the parameters need not be random variables at all. Rather, one can view the algorithms as being randomized and interpret the results as probabilistic statements on robustness properties.) For example, consider the question of deciding whether a given polytope of real matrices has eigenvalues in the open left half plane. This is a classic robust stability analysis problem which captures essential computational complexity difficulties of even more general robust control problems. Suppose one is willing to use a randomized algorithm, and based on the outcome of the algorithm wants to draw a conclusion of the form: **the volume of the set of uncertain parameters leading to unstable matrices is less than ϵ with probability greater than $1 - \delta$.** Thus, if ϵ, δ can be made as small as desired, then we can say with a high degree of confidence that the given set of matrices is stable. We show that there exists a (very simple) algorithm with precisely these properties whose complexity is polynomial in the size of matrices, $1/\delta$, and $1/\epsilon$. Thus, the complexity of obtaining high confidence robustness statements is very low. These are the first results which demonstrate low computational complexity for large classes of robust control problems analyzed using randomized algorithms.

We would like to note that Stengel and his coworkers [22, 18] have pointed out the fact that random sampling approaches to robust control do not suffer from complexity problems. However, they have not given concrete complexity results of the type obtained in this paper. Thus, the results in this paper rigorously justify the low complexity of random sampling approaches. Similar types of conclusions can be drawn from Tempo and Bai [25] who, independently and in parallel to us, have derived some results very similar to our results in Sections 3–5 on the number of samples needed to ensure that the probability of achieving a certain performance level is greater than a prescribed degree of confidence.

The results presented here are based on standard results from statistics and probability theory. We do not claim that the algorithms, based essentially on crude random sampling, suggested by the results are necessarily efficient or optimal in a practical sense. They certainly enjoy very low computational complexity but may still be inefficient. The reason is that computational complexity focuses only on growth rates rather than the actual number of computations. In this paper, we do propose some algorithms that go beyond crude random sampling. However, development of efficient algorithms is not the focus of this paper. We are currently working on algorithms which are more efficient for specific robust control problems.

Finally, we feel that this change in paradigm has advantages that go far beyond just computational complexity. In the traditional worst-case deterministic robust control theory, it is very often necessary to transform the real engineering analysis/design problem to fit a theoretical framework, such as, $\mathcal{H}_\infty$ control and μ synthesis. This results in a rather indirect attack on the real engineering problem of interest. Moreover, problems such as reduced order controller design, control of nonlinear plants, etc., are extremely difficult. We believe that creative combinations of the theoretical results and the probabilistic approach will result in tools that can address real engineering control problems.

This paper is organized as follows. In Section 2, we briefly set up the general robust stability analysis problem in the usual μ type setting. Section 3 gives the main results on robust stability analysis with mixed real and complex uncertainty. The problem in this section is to decide whether for a given level of uncertainty, the system is robustly stable. Then in Section 4, we give results on the problem of computing the robust stability margin which is the size of the smallest destabilizing perturbation. In Section 5 we treat a related but technically different problem. We give results on the problem of computing the largest real part of the eigenvalues of the uncertain system. This turns out to be a harder technical problem but results are similar to the previous results. Then, in Section 6, we give four different algorithms for analyzing the robust stability problems which are motivated by the theoretical results in the preceding sections. In Section 7, we extend the robust stability analysis results to the problem of controller design. We include three examples in Section 8 which have been purposely designed to explore the efficacy of the randomized algorithms. Examples include a 55 state, 20 parameter uncertain multivariable system and a reduced order controller design problem for a flexible system. It is hoped that these examples will give the reader a good feel for the potential of randomized algorithms. The paper concludes with a discussion of some key issues and future research directions.

2   General mixed real-complex robust analysis problem

Consider the feedback interconnection of the LTI system M and the *perturbation* Δ depicted in Figure 1. The perturbation Δ is unknown but has a

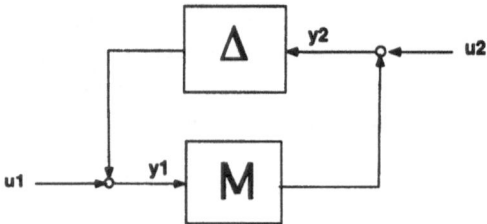

Figure 1: Nominal system in feedback with structured uncertainty.

certain block structure. Following the notation in [6, 26], let m_r, m_c, m_C be non-negative integers, and define the block structure as an m-tuple of positive integers ($m \equiv m_r + m_c + m_C$) given by

$$\mathcal{K} = (k_1, \ldots, k_{m_r}, k_{m_r+1}, \ldots, k_{m_r+m_c}, k_{m_r+m_c+1}, \ldots, k_m).$$

We will assume that Δ belongs to the class of allowable perturbations

$$\Delta = \left\{ \mathrm{diag} \left[\delta_1^r I_{k_1}, \ldots, \delta_{m_r}^r I_{k_{m_r}}, \delta_1^c I_{k_{m_r}+1}, \ldots, \delta_{m_c}^r I_{k_{m_r+m_c}}, \Delta_1^C, \ldots \Delta_{m_C}^C \right] \right.$$
$$\left. : \ \delta_i^r \in \mathbf{R}, \delta_i^c \in \mathbf{C}, \Delta_i^C \in \mathbf{C}^{k_{m_r+m_c+i} \times k_{m_r+m_c+i}} \right\}.$$

For a given real number γ, let Δ_γ denote the subset of perturbations in Δ with size at most γ:

$$\Delta_\gamma = \{\Delta \in \Delta \ : \ \bar{\sigma}(\Delta) \le \gamma\}.$$

Define the *robustness measure* γ_{opt} to be the size of the smallest allowable perturbation that destabilizes the feedback interconnection. Equivalently, γ_{opt}

may be defined as the largest non-negative real number γ for which the feedback interconnection is stable for all $\Delta \in \mathbf{\Delta}_\gamma$. And, finally, given a real number γ, define $\lambda_{\max}$ as the smallest real number λ_o for which the eigenvalues of the system matrix of the feedback interconnection have real parts less than λ_o for all $\Delta \in \mathbf{\Delta}_\gamma$.

Given the system M, we are interested in the following problems:

1) Given a real number γ, check if $\gamma_{opt} \leq \gamma$.

2) Compute γ_{opt}.

3) Compute $\lambda_{\max}$.

It has recently been shown that these problems are NP-hard [3, 19, 21].

In the sections that follow we will present and analyze randomized algorithms for these problems. Throughout the paper, we will conduct searches over a compact subset of parameters, say Q. A probability space is given by the triple $(Q, \mathcal{F}, \mathbf{P})$, where Q is the sample space, $\mathcal{F}$ is a collection of subsets of Q that is a σ-field, and $\mathbf{P}$ is a probability measure. In this paper we will restrict attention to probability measures induced by probability density functions which are strictly positive and absolutely continuous with respect to the Lebesgue measure on Q. For such a probability density function w and an event $E \in \mathcal{F}$, we will take

$$\mathbf{P}(E) = \mathrm{vol}_w(E) \equiv \int_E w \, dx$$

From here on we shall use the notation $\mathrm{vol}_w(E)$ to denote $\mathbf{P}(E)$.

3 Robust stability

We begin by addressing the robust stability problem. The problem is to check if $\gamma_{opt} \leq \gamma$, or, equivalently, if $\mathbf{U}$, the set of all destabilizing perturbations with norm less than γ,

$$\mathbf{U} = \{\Delta \in \mathbf{\Delta}_\gamma \; : \; \mathrm{feedback\,interconnection\,is\,unstable}\},$$

is non-empty.

For a given probability density function w on $\mathbf{\Delta}_\gamma$, let $\Delta_1^\circ, \Delta_2^\circ, \ldots, \Delta_n^\circ$ be the outcomes of n i.i.d. continuous random variables, $\Delta_1, \Delta_2, \ldots, \Delta_n$, that take values in $\mathbf{\Delta}_\gamma$ and have probability density function w. For the moment, let us assume we have an *oracle* that tells us whether or not $\Delta_i^\circ \in \mathbf{U}$. If any of the Δ_i° are in $\mathbf{U}$ then we may conclude that $\mathbf{U}$ is non-empty and hence $\gamma_{opt} \leq \gamma$. However, what if it is the case that none of the Δ_i° lie in $\mathbf{U}$? Clearly we cannot conclude that $\gamma_{opt} > \gamma$, but we can make the following probabilistic statement regarding the *volume* of the set $\mathbf{U}$.

Theorem 3.1 *Let $\epsilon, \delta \in (0,1)$, and $n \geq \frac{\ln\left(\frac{1}{\delta}\right)}{\ln\left(\frac{1}{1-\epsilon}\right)}$. Then:*

$$\text{vol}_w(\mathbf{U}) \geq \epsilon \text{ implies } Prob\left\{\Delta_1 \notin \mathbf{U} \& \Delta_2 \notin \mathbf{U} \& \ldots, \Delta_n \notin \mathbf{U}\right\} \leq \delta. \quad (1)$$

Proof: Define the random variables $Y_1, Y_2, \ldots, Y_n$ as

$$Y_i = \begin{cases} 1 & \text{if} \Delta_i \in \mathbf{U} \\ 0 & \text{otherwise} \end{cases}$$

The Y_i's are independent and have a Bernoulli distribution with parameter $\text{vol}_w(\mathbf{U})$. Thus

$$
\begin{aligned}
Prob\left\{\Delta_1 \notin \mathbf{U} \& \Delta_2 \notin \mathbf{U} \& \ldots \& \Delta_n \notin \mathbf{U}\right\} &= \Pi_{i=1}^n \, Prob\left\{Y_i = 0\right\} \\
&= (1 - \text{vol}_w(\mathbf{U}))^n \\
&\leq (1 - \epsilon)^n.
\end{aligned}
$$

The result follows on observing that $n \geq \frac{\ln\left(\frac{1}{\delta}\right)}{\ln\left(\frac{1}{1-\epsilon}\right)}$ implies $(1 - \epsilon)^n \leq \delta$. $\qquad\square$

Let us take ϵ and δ to be small. Then the probability of the event E

$$E = \text{event that } \Delta_1 \notin \mathbf{U} \& \Delta_2 \notin \mathbf{U} \& \ldots \& \Delta_n \notin \mathbf{U},$$

is smaller than δ provided $\text{vol}_w(\mathbf{U}) \geq \epsilon$. Therefore, if E does occur, it should lead us to believe that $\text{vol}_w(\mathbf{U}) \leq \epsilon$. And the main point is that n, the number of random samples needed to make this statement (1), is at most $\frac{\ln\left(\frac{1}{\delta}\right)}{\ln\left(\frac{1}{1-\epsilon}\right)}$, a function that is better than polynomial in $\frac{1}{\epsilon}$ and $\frac{1}{\delta}$.

As already indicated, we can conclude that $\gamma_{opt} \leq \gamma$ if it is the case that one of the Δ_i lies in $\mathbf{U}$. And if none of the Δ_i lie in $\mathbf{U}$, then we can conclude that $\gamma_{opt} > \gamma$, *i.e.* robust stability, with a high degree of confidence. But regardless of the outcomes of $\Delta_1, \Delta_2, \ldots \Delta_n$, we would still like to make some statement regarding the volume of $\mathbf{U}$ *i.e.* we would like to know what is the "percentage" of destabilizing perturbations with norm less than γ among all perturbations with norm less than γ. (Recall that the previous result makes no claims regarding $\text{vol}_w(\mathbf{U})$ if for some i, $\Delta_i \in \mathbf{U}$.)

Define the i.i.d. random variables $Y_1, Y_2, \ldots, Y_n$ as

$$Y_i = \begin{cases} 1 & \text{if } \Delta_i \in \mathbf{U} \\ 0 & \text{otherwise} \end{cases}$$

and let

$$\hat{V} := \frac{1}{n} \sum_{i=1}^{n} Y_i. \quad (2)$$

We will take the outcome of the random variable $\hat{V}$ to be an estimate of $\text{vol}_w(\mathbf{U})$.

Theorem 3.2 *Let* $\epsilon, \delta \in (0,1)$, *and*

$$n \geq \frac{1}{2\epsilon^2} \ln \left(\frac{2}{\delta} \right).$$

Define $\hat{V}$ *as in equation (2). Then*

$$Prob \left(|\hat{V} - \text{vol}_w(\mathbf{U})| \leq \epsilon \right) \geq 1 - \delta$$

This theorem can be established using the so-called *Hoeffding Bound*:

Theorem 3.3 *[13] Let* $Z_i, i = 1, 2, \ldots, n$, *be i.i.d. random variables with* $0 \leq Z_i \leq 1$. *Let* $S = \sum_{i=1}^{n} Z_i$. *Then*

$$\text{Prob } (S \geq \mathcal{E}(S) + nt) \leq e^{-2nt^2}$$

where $\mathcal{E}(\cdot)$ *denotes the expectation operator.*

It is an easy consequence of Theorem 3.3 that

$$\text{Prob } (|S - \mathcal{E}(S)| \geq nt) \leq 2e^{-2nt^2}. \tag{3}$$

Proof of Theorem 3.2: As in the proof of Theorem 3.1, each Y_i has a Bernoulli distribution with parameter $\text{vol}_w(\mathbf{U})$. Thus $\mathcal{E}(\hat{V}) = \text{vol}_w(\mathbf{U})$. It follows from equation (3) that

$$Prob \left(|\hat{V} - \text{vol}_w(\mathbf{U})| \leq \epsilon \right) \geq 1 - 2e^{-2n\epsilon^2}.$$

The result follows on observing that $n \geq \frac{1}{2\epsilon^2} \ln \left(\frac{2}{\delta} \right)$ implies $2e^{-2n\epsilon^2} \leq \delta$. □

A final point before we proceed to the next section: recall that we assumed the existence of an oracle that can determine whether or not Δ_i is in $\mathbf{U}$. A brute force way to check stability is by computing eigenvalues of the closed loop system matrix. There exist algorithms which calculate the eigenvalues of a matrix with $\mathcal{O}(N^3)$ computations where N is the dimension of the matrix. Thus, letting N denote the number of states of the system, it takes at most $\mathcal{O}(N^3)$ computations to check the stability of the feedback interconnection. Ignoring the complexity of generating samples, the total complexity for arriving at our conclusions, the computational complexity of the oracle times the number of samples at which it is invoked, is $\mathcal{O}(N^3) * \frac{\ln\left(\frac{1}{\delta}\right)}{\ln\left(\frac{1}{1-\epsilon}\right)}$ and $\mathcal{O}(N^3) * \frac{1}{2\epsilon^2} \ln \left(\frac{2}{\delta} \right)$. These expressions are polynomial in N, $\frac{1}{\epsilon}$, and $\frac{1}{\delta}$.

4　On computing stability margins

In this section we consider the problem of computing γ_{opt}. Recall that γ_{opt} is the largest non-negative real number γ for which the feedback interconnection in Figure 1 is stable for all $\Delta \in \mathbf{\Delta}_\gamma$. Our approach will be similar to that taken in the last section. Let γ_0 be an upper bound for γ_{opt}. Let $\Delta_1, \Delta_2, \ldots, \Delta_n$ be i.i.d. continuous random variables that take values in $\mathbf{\Delta}_{\gamma_0}$ and have probability density function w. Let

$$\hat{\gamma}(n) = \min\{\bar{\sigma}(\Delta_i) \ : \ \Delta_i \text{ makes feedback interconnection unstable, } 1 \le i \le n\}.$$

and let $\hat{\gamma}(n) = \gamma_0$ if there exists no destabilizing perturbation among $\{\Delta_i \ : \ 1 \le i \le n\}$. We will take the outcome of the random variable $\hat{\gamma}(n)$ as an estimate of γ_{opt}.

Theorem 4.1 *Let $\epsilon, \delta \in (0,1)$, and*

$$n \ge \frac{\ln\left(\frac{1}{\delta}\right)}{\ln\left(\frac{1}{1-\epsilon}\right)}.$$

Then

$$\text{Prob } (\text{vol}_w(\mathbf{B}) \le \epsilon) \ge 1 - \delta$$

where

$$\mathbf{B} = \{\Delta \in \mathbf{\Delta}_{\gamma_0} \ : \ \Delta \text{ makes feedback interconnection unstable}$$
$$\text{and } \bar{\sigma}(\Delta) < \hat{\gamma}(n)\}.$$

Proof: It is necessary to introduce further notation. For $\alpha \ge 0$ define

$$\mathbf{U}(\alpha) = \{\Delta \in \mathbf{\Delta}_{\gamma_0} \ : \ \Delta \text{ makes feedback interconnection unstable}$$
$$\text{and } \bar{\sigma}(\Delta) \le \gamma_{opt} + \alpha\},$$

and let $f(\alpha) = \text{vol}_w(\mathbf{U}(\alpha))$. Note that the sets $\mathbf{U}(\cdot)$ are nested: $\mathbf{U}(\alpha) \subseteq \mathbf{U}(\beta)$ for $\alpha \le \beta$, and so $f(\alpha)$ is a non-decreasing function of α and $f(0) = 0$.

Let us next prove that $f(\alpha)$ is a continuous function. Suppose $f(\alpha)$ is not left continuous at $\hat{\alpha}$. This implies that there exits an open set $\mathcal{N} \subseteq \mathbf{U}(\hat{\alpha})$ such that

$$\mathcal{N} \bigcap \left(\bigcup_{\zeta > 0} \mathbf{U}(\hat{\alpha} - \zeta) \right) = \phi. \tag{4}$$

Let Δ be some arbitrary member of $\mathcal{N}$. Equation (4) implies that $\bar{\sigma}(\Delta) > \gamma_{opt} + \hat{\alpha} - \zeta$ for all $\zeta > 0$, which implies $\bar{\sigma}(\Delta) \ge \gamma_{opt} + \hat{\alpha}$. But since $\Delta \in \mathcal{N} \subseteq \mathbf{U}(\hat{\alpha})$, this implies that $\bar{\sigma}(\Delta) = \gamma_{opt} + \hat{\alpha}$. It cannot be the case, however, that all matrices in an open set have the same norm. This is a contradiction proving that $f(\alpha)$ is left continuous at $\hat{\alpha}$. Next we will prove $f(\alpha)$ is right continuous at $\hat{\alpha}$ by showing that $\mathbf{U}(\hat{\alpha}) = \bigcup_{\zeta>0} \mathbf{U}(\hat{\alpha} + \zeta)$. It is clear that $\mathbf{U}(\hat{\alpha}) \subseteq \bigcup_{\zeta>0} \mathbf{U}(\hat{\alpha} + \zeta)$ since the sets $\mathbf{U}(\cdot)$ are nested. We have only to show that $\bigcup_{\zeta>0} \mathbf{U}(\hat{\alpha} + \zeta) \setminus \mathbf{U}(\hat{\alpha})$ is empty. Let $\Delta \in \bigcup_{\zeta>0} \mathbf{U}(\hat{\alpha} + \zeta)$. Then $\bar{\sigma}(\Delta) \le \gamma_{opt} + \hat{\alpha} + \zeta$ for all $\zeta > 0$. Thus $\bar{\sigma}(\Delta) \le \gamma_{opt} + \hat{\alpha}$ and so $\Delta \in \mathbf{U}(\hat{\alpha})$.

With this notation and these preliminary results in place, we can proceed with the proof of Theorem 4.1. Since $f(\alpha)$ is continuous and $f(0) = 0$, there exists $\hat{\alpha}$ such that $f(\hat{\alpha}) = \epsilon$. Suppose it is the case that $\hat{\gamma}(n) \leq \gamma_{opt} + \hat{\alpha}$. Then it is clear that $\mathbf{B} \subseteq \mathbf{U}(\hat{\alpha})$ and so $\text{vol}_w(\mathbf{B}) \leq \text{vol}_w(\mathbf{U}(\hat{\alpha})) = f(\hat{\alpha}) = \epsilon$. Therefore

$$
\begin{aligned}
Prob\,&\{\text{vol}_w(\mathbf{B}) > \epsilon\} \\
\leq\;\; & Prob\,\{\hat{\gamma}(n) > \gamma_{opt} + \hat{\alpha}\} \\
=\;\; & Prob\,\{\Delta_1 \notin \mathbf{U}(\hat{\alpha}) \;\&\; \Delta_2 \notin \mathbf{U}(\hat{\alpha}) \;\&\; \ldots \;\&\; \Delta_n \notin \mathbf{U}(\hat{\alpha})\} \\
=\;\; & (1 - f(\hat{\alpha}))^n = (1 - \epsilon)^n
\end{aligned}
$$

The result follows by observing that $n \geq \frac{\ln\left(\frac{1}{\delta}\right)}{\ln\left(\frac{1}{1-\epsilon}\right)}$ implies $(1 - \epsilon)^n \leq \delta$.   □

5 On computing the relative degree of stability

Lastly, we consider the problem of computing $\lambda_{\max}$ for a given γ. Recall that $\lambda_{\max}$ is the smallest real number λ_o for which the eigenvalues of the feedback interconnection have real parts less than λ_o for all $\Delta \in \mathbf{\Delta}_\gamma$.

Let $\Delta_1, \Delta_2, \ldots, \Delta_n$ be i.i.d. continuous random variables that take values in $\mathbf{\Delta}_\gamma$ and have probability density function w. Let

$$
\hat{\lambda}(n) = \max_{1 \leq i \leq n}\{\text{max real part of the eigenvalues of the system matrix} \\ \text{of the feedback interconnection with } \Delta = \Delta_i\}.
$$

We will take the outcome of the random variable $\hat{\lambda}(n)$ as an estimate of $\lambda_{\max}$.

Theorem 5.1 *Let $\epsilon, \delta \in (0, 1)$, and*

$$
n \geq \frac{\ln\left(\frac{1}{\delta}\right)}{\ln\left(\frac{1}{1-\epsilon}\right)}.
$$

Then

$$
Prob\,(\text{vol}_w(\mathbf{B}) \leq \epsilon) \geq 1 - \delta
$$

where

$$
\mathbf{B} = \{\Delta \in \mathbf{\Delta}_\gamma \;:\; \text{the system matrix of the feedback interconnection} \\ \text{has an eigenvalue with real part } > \hat{\lambda}(n)\}.
$$

The proof is similar to that of Theorem 4.1 and can be found in the appendix.

6 Simple adaptive random search algorithms

The algorithms suggested by the proofs of the preceding theorems are crude random search algorithms. While our results show that these algorithms have good complexity properties, they may be inefficient in practice. Thus, adaptive random search techniques should be employed whenever possible.

In this section we present several adaptive algorithms for the computation of the robustness measure γ_{opt}.

- First we present a crude random search algorithm for the sake of completeness.

  Algorithm 1: Crude random search

  The user must specify γ_0 (such that $\gamma_0 > \gamma_{opt}$), and **iterations**.

  Step 1. Initialize $\hat{\gamma} = \gamma_0$ and let $i = 1$.
  Step 2. Generate Δ_i uniformly from $\{\Delta \in \Delta \ : \ \bar{\sigma}(\Delta) \leq \gamma_0\}$.
  Step 3. If feedback system with $\Delta = \Delta_i$ is unstable then let $\hat{\gamma} = \min(\bar{\sigma}(\Delta_i), \hat{\gamma})$.
  Step 4. If $i < $ **iterations** then let $i = i + 1$ and go to Step 2.
  Step 5. End, and return the variable $\hat{\gamma}$.

  This algorithm uniformly selects a number of samples from $\{\Delta \in \Delta : \bar{\sigma}(\Delta) \leq \gamma_0\}$, and returns the norm of the smallest destabilizing perturbation found in the collection.

- Algorithm 2: Simple adaptive random search

  The user must specify γ_0 (such that $\gamma_0 > \gamma_{opt}$), and **iterations**.

  Step 1. Initialize $\hat{\gamma} = \gamma_0$ and let $i = 1$.
  Step 2. Generate Δ_i uniformly from $\{\Delta \in \Delta \ : \ \bar{\sigma}(\Delta) \leq \hat{\gamma}\}$.
  Step 3. If feedback system with $\Delta = \Delta_i$ is unstable then let $\hat{\gamma} = \bar{\sigma}(\Delta_i)$.
  Step 4. If $i < $ **iterations** then let $i = i + 1$ and go to Step 2.
  Step 5. End, and return the variable $\hat{\gamma}$.

  This adaptive search algorithm differs from the crude random search algorithm in that it shrinks the search space (the space from which the next sample will be taken) each time a destabilizing perturbation is found (as can be seen in Step 2.) In the examples, some of which are presented in the next section, we found it to be much more efficient than the crude random search algorithm.

- Algorithm 3: Non-uniformly weighted adaptive search

  The user must specify γ_0 (such that $\gamma_0 > \gamma_{opt}$), and **iterations**.

  Step 1. Initialize $\hat{\gamma} = \gamma_0$ and let $i = 1$.
  Step 2. Generate Δ_i from $\{\Delta \in \Delta \ : \ \bar{\sigma}(\Delta) \leq \hat{\gamma}\}$ so that $\bar{\sigma}(\Delta_i)$ is uniformly distributed on $[0, \hat{\gamma}]$.
  Step 3. If feedback system with $\Delta = \Delta_i$ is unstable then let $\hat{\gamma} = \bar{\sigma}(\Delta_i)$.

Step 4. If $i <$ iterations then let $i = i + 1$ and go to Step 2.
Step 5. End, and return the variable $\hat{\gamma}$.

This is another adaptive search algorithm which differs from Algorithm 2 in that it generates Δ_i from $\{\Delta \in \Delta : \bar{\sigma}(\Delta) \leq \hat{\gamma}\}$ so that $\bar{\sigma}(\Delta_i)$ is uniformly distributed on $[0, \hat{\gamma}]$. The reason for generating Δ with this distribution is to find smaller destabilizing perturbations more quickly. We found this algorithm to have faster initial convergence than Algorithm 2 when the initial search space is conservative. However, Algorithm 2 proved more efficient when the estimate neared γ_{opt}.

- Algorithm 4: Probabilistic bisection

The user must specify γ_0 (such that $\gamma_0 > \gamma_{opt}$), confcount, tol, and iterations.

Step 1. Initialize $\hat{\gamma} = \gamma_0$, $U = \gamma_0$, $L = 0$, and let $i = 1$, $k = 1$.
Step 2. Generate Δ_i uniformly from $\{\Delta \in \Delta : \bar{\sigma}(\Delta) \leq \hat{\gamma}\}$.
Step 3. If feedback system with $\Delta = \Delta_i$ is unstable then
  - Let $\hat{\gamma} = \min(\frac{\hat{\gamma}}{2}, \bar{\sigma}(\Delta_i))$ and $U = \bar{\sigma}(\Delta_i)$.
  - If $|U - L| \leq$ tol or $i =$ iterations then go to Step 7;
    else let $k = 1, i = i + 1$ and go to Step 2.
Step 4. If $i =$ iterations then let $i = i + 1$ and go to Step 7.
Step 5. If $K <$ confcount then let $k = k + 1$ and $i = i + 1$ and go Step 2.
Step 6. Let $L \leftarrow \hat{\gamma}$; $\hat{\gamma} \leftarrow \frac{U+L}{2}$;
If $|U - L| >$ tol then let $k = 1$ and $i = i + 1$ and go to Step 2.
Step 7. End, and return the variables $[L, U]$ and i.

In essence, this algorithm cuts the radius of the search space in half whenever a destabilizing perturbation is found. If it cannot find a destabilizing perturbation over this reduced search space, it increases the radius of the search space to a value in between the current value and the magnitude of the smallest destabilizing perturbation found so far. It iterates on the radius of the search space in this manner until some user specified tolerances are met. We do not have enough experience with this algorithm to comment on its efficiency, although initial experience with it is quite encouraging.

Finally, we note that in order to implement these random algorithms it is necessary to generate samples from Δ_γ. One way to do this is to embed Δ_γ in the unit ball in the Euclidean space $\mathbf{R}^M \times \mathbf{C}^N$ for some M, N. Since it is known how to generate samples from this until ball with uniform distribution, using acceptance/rejection sampling one can generate uniformly distributed samples from Δ_γ [24]. In general, sampling from a convex compact set is a standard problem in statistics and computer science. The interested reader is referred to [23, 16] and the references cited there.

7 Controller synthesis

So far we have considered only robust analysis questions. It is clear that we can apply these randomized algorithms to controller synthesis problems as well. Consider the standard feedback diagram depicted in Figure 2. Here

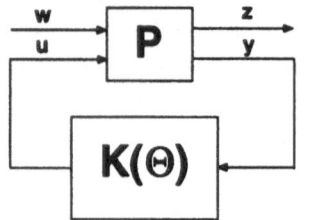

Figure 2: Standard feedback diagram.

P is a known plant, and $K(\theta)$ is a fixed structure controller parameterized by $\theta \in \Theta$. For example, $K(\theta)$ could be the usual PID controller with θ representing the PID gains. As usual, u and y denote the control inputs and measurable outputs, and w and z denote the performance inputs and outputs.

For the purposes of exposition we will consider the special case when the plant is linear time-invariant, and the goal is to design a reduced order controller $K(\theta)$ to achieve the performance objective $\|T_{zw}\|_\infty < 1$. It is possible to treat other design objectives with equal ease. Moreover, we will limit our search to a compact parameter space Θ.

Let $\mathbf{U}$ be the set of all controller parameters that meet the control objective:

$$\mathbf{U} = \{\theta \in \Theta \ : \ \|T_{zw}(\theta)\|_\infty < 1\}.$$

Our goal is to find a member of this set. Given a candidate control parameter vector θ, it is computationally easy (by checking the eigenvalues of a Hamiltonian matrix) to check if the control objective is attained.

Let $\Theta_1, \Theta_2, \ldots, \Theta_n$ be i.i.d. random variables in Θ with probability density function w.

Theorem 7.1 *Let $\epsilon, \delta \in (0,1)$, and $n \geq \dfrac{\ln(\frac{1}{\delta})}{\ln(\frac{1}{1-\epsilon})}$. Then:*

$$\mathrm{vol}_w(\mathbf{U}) \geq \epsilon \ \text{implies} \ Prob\,\{\Theta_1 \notin \mathbf{U} \ \& \ \Theta_2 \notin \mathbf{U} \ \& \ \ldots \ \& \ \Theta_n \notin \mathbf{U}\} \leq \delta.$$

This theorem leads to the following algorithm. Choose ϵ, δ as small as desired and choose n such that $n \geq \dfrac{\ln(\frac{1}{\delta})}{\ln(\frac{1}{1-\epsilon})}$. Select $\theta_1, \theta_2, \ldots, \theta_n$ as above and check if any one of them satisfies the desired performance requirement $\|T_{zw}(\theta)\|_\infty < 1$. If the answer is yes, then we are done. Otherwise, with a high degree of confidence, we can conclude that it is necessary to expand our search; *i.e.* search over larger parameter set Θ, use a higher order controller, or try a controller with a different structure.

Next, given a controller structure and a parameter set Θ we want to find the best controller in the class. Let

$$\hat{\Theta}(n) = \arg \min_{\Theta \in \{\Theta_1, \Theta_2, \ldots, \Theta_n\}} \|T_{zw}(\Theta)\|_\infty$$

We will take the outcome of the random variable $\hat{\Theta}$ as an estimate of the optimal control law, say θ_o. Once again, with an argument similar to that used to prove Theorem 5.1, we are able to derive the following result.

Theorem 7.2 *Let* $\epsilon, \delta \in (0, 1)$, *and*

$$n \geq \frac{\ln\left(\frac{1}{\delta}\right)}{\ln\left(\frac{1}{1-\epsilon}\right)}.$$

Then

$$\text{Prob } (\text{vol}_w(\mathbf{B}) \leq \epsilon) \geq 1 - \delta$$

where

$$\mathbf{B} = \left\{ \theta \in \Theta \ : \ \|T_{zw}(\theta)\|_\infty < \|T_{zw}(\hat{\Theta}(n))\|_\infty \right\}$$

We would like to remark that in principle this strategy will work for any control synthesis problem when there exists a computationally cheap oracle to verify if a candidate design meets the control objectives (or, additionally, evaluates the "loss function.") We could just as easily have addressed the $\mathcal{H}_2$ synthesis problem, or, for example, the problem of designing a PID controller to achieve a prescribed (or maximal) degree of closed-loop stability.

8 Examples and Practical Considerations

In this section we present our results from a study of three example problems. The first two are robust stability problems, and the third is a controller design problem.

8.1 Kharitonov example

Our first example is the well known robust stability result due to Kharitonov [14]. This example was chosen because it is very simple and extremely well understood analytically.

i	a_i	ϵ_i	δ_i
1	1.3482e+01	1.7901e+00	-1.4505e-01
2	8.6744e+01	2.8030e+00	1.4717e-02
3	3.5184e+02	1.0812e+00	-1.3025e-01
4	1.0026e+03	2.6989e-01	-1.8768e-02
5	2.1202e+03	5.9540e-02	-6.0206e-02
6	3.4284e+03	1.6867e+00	7.6873e-02
7	4.3064e+03	1.9491e+00	6.3835e-02
8	4.2239e+03	8.7473e-01	-1.8698e-02
9	3.2204e+03	1.5326e+00	-4.2824e-02
10	1.8794e+03	4.5754e-01	6.6580e-02
11	8.1512e+02	9.1157e-01	-8.7904e-03
12	2.5000e+02	2.6750e+00	-6.5514e-02
13	4.9919e+01	8.6000e-01	-6.0558e-02
14	5.6094e+00	6.4806e-03	-3.7869e-02
15	2.5964e-01	9.1899e-01	7.8414e-02

Table 1: Coefficients of the interval polynomial and smallest destabilizing perturbation found.

Let $p_0(s) = s^n + \sum_{i=1}^{n} a_i s^{n-i}$ be a stable nominal polynomial of degree 15, define the interval polynomial

$$p(s,r) = p_0(s) + r \sum_{i=1}^{15} [-\epsilon_i, \epsilon_i] s^{i-1},$$

and let $\mathcal{P}_r$ be the family of interval polynomials given by $\mathcal{P}_r = \{p(s, \hat{r}) : \hat{r} \leq r\}$. Let r_{max} denote the largest real number r such that all polynomials in $\mathcal{P}_r$ are stable. There exist standard analytical results to calculate r_{max} (see, for example, [9].) For our selection of a_i, ϵ_i (see Table 1) it turns out that $r_{max} = .14504$.

Let M be a rank one system with transfer function

$$M(s) = \frac{-1}{p_0(s)} (\epsilon_1 \ \epsilon_2 \ \cdots \ \epsilon_{15})'(1 \ s \ \cdots \ s^{15}),$$

and consider the class of allowable perturbations $\Delta = \{ \ \text{diag} \ [\delta_1, \delta_2, \ldots, \delta_{15}] : \delta_i \in \mathbf{R}\}$. It is not difficult to show that r_{max} is equal to γ_{opt} of Section 2.

Using the randomized algorithms discussed above, we analyzed the following questions.

- Let $r = .15$. We wanted to analyze the robust stability of the resulting class of interval polynomials. After only 69 random samples from Δ, we found that this collection of polynomials was not robustly stable. This is remarkable given how close $r = .15$ is to the stability margin $r_{max} = .14504$.

- Let $r = .14$. Then, of course, we would never find a perturbation leading to $p(s)$ having a root in the closed right half plane. If we set $\epsilon = .01$ and $\delta = .05$, then after 18,445 samples we could conclude from Theorem 3.2 that $Prob\,(\text{vol}_w(\mathbf{U}) \le .01) \ge .95$, *i.e.* with probability at least .95 we could conclude that the volume of destabilizing perturbations with norm less than .14 makes up less than 1% of the volume of all perturbations with norm less than .14.

- We computed r_{max} using a randomized algorithm. Using Algorithm 2 of Section 6, we generated 10,000 random perturbations from the allowable class Δ. The smallest destabilizing perturbation had magnitude .14505. It is given in Table 1. Thus, our estimate of r_{max} is .14505 which differs from the actual value by less than 0.02%!

Finally, we computed the μ upper and lower bounds for this system using MATLAB μtools software [17, 1]. These bounds indicate that γ_{opt} must lie in the interval $[0.2067, 1.5813]$. This is obviously wrong. Since this problem involves only real perturbations, we attribute this to possible numerical problems in frequency sweep grid size.

8.2   An example with 55 states and 20 real parameters

Next, we wanted to see how randomized algorithms perform on moderately sized problems. We generated a stable linear system M with 55 states and 20 inputs and outputs. The data for state space matrices (A, B, C) of M take up too much space and are thus omitted. The interested reader can obtain this data upon request from the authors. The class of allowable perturbations is:

$$\Delta = \{ \text{ diag } [\delta_1, \delta_2, \ldots, \delta_{20}] : \delta_i \in \mathbf{R} \}.$$

For this system, we computed the μ upper and lower bounds. They indicate that γ_{opt} must lie in the interval $[0.31591, \infty)$. Note that these bounds may be arbitrarily conservative. Next, we applied Algorithm 2 of Section 6 with a sample size of 10,000. The smallest destabilizing perturbation had magnitude 0.53555. This estimate of γ_{opt} (an upper bound) drastically reduces the conservativeness of the previous bounds.

8.3   A reduced order $\mathcal{H}_\infty$ controller design example

In this section we apply a simple randomized algorithm to a controller design example. The example is taken from the book by Doyle *et. al.* [5] (Chapter 12.4). Consider the feedback system in Figure 3. The plant P has transfer function

$$P(s) = \frac{-6.4750s^2 + 4.0302s + 175.7700}{s(5s^3 + 3.5682s^2 + 139.5021s + 0.0929)}.$$

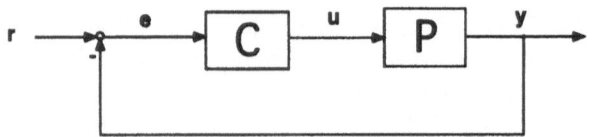

Figure 3: Feedback diagram.

Choosing appropriate weights W_1, W_2, the design is done using the following mixed sensitivity problem: design a controller C to make the *mixed sensitivity norm*

$$\left\| (\begin{array}{cc} W_1S & W_2CS \end{array})' \right\|_\infty$$

less than 1, where S is the sensitivity function: $S = (1 + PC)^{-1}$. This is an analytically solvable problem and the nearly optimal controller C has order 8 and achieved a mixed sensitivity norm of .9275 [5].

In this section our objective is to find an optimal third order controller C. Here optimality is in the sense of minimizing the $\mathcal{H}_\infty$ norm of the above mixed sensitivity function. As is well known, there is no tractable analytic solution to this problem. It is also known that this optimization problem is not convex in the controller parameters. We also do not compute the optimal controller but attempt to use random sampling to come up with a "good" third order controller.

We restricted our attention to strictly proper, stable, minimum-phase controllers. There are any number of ways for generating controllers randomly. We took the following approach: Randomly decide how many poles and zeros will be real. Generate the real poles and zeros by sampling uniformly from $(-\texttt{realbound}, 0)$ and the complex poles and zeros by sampling uniformly from $(-\texttt{realbound}, 0) \times (-\texttt{imagbound}, \texttt{imagbound})$ under a conjugate symmetry constraint. Finally, generate the gain by sampling uniformly from $(0, \texttt{gainbound})$.

For this problem we arbitrarily chose $\texttt{realbound} = \texttt{imagbound} = \texttt{gainbound} = 2$ and generated 10,000 candidate controllers. Out of these 10,000, there were 220 controllers that stabilized the closed loop system and achieved mixed sensitivity norm less than 2. The optimal controller was the 4869th candidate controller among the 10,000 samples. It achieved a mixed sensitivity norm of 1.0234 which compares very well with the optimal mixed sensitivity norm of .9275. The transfer function for our third order controller is

$$C(s) = \frac{1.6408s^2 + 2.4623s + 0.0838}{s^3 + 3.4880s^2 + 6.9434s + 3.9351}.$$

Finally, to compare this result against a traditional approach, we did a Hankel model reduction on the full 8th order controller to arrive at a 3rd order controller. This 3rd order Hankel–reduced controller achieved a mixed sensitivity norm of 5.1976. Clearly, in this example, the random search method produced a much superior reduced order controller.

9   Discussion and Conclusions

The main objectives of this paper were twofold. First, we wanted to show how adoption of randomized algorithms leads to significant reduction in computational complexity of robust control problems. Our results have certainly accomplished this goal. The second objective was to use these positive results to argue in favor of changing the robust control paradigm from worst-case deterministic analysis to average-case probabilistic formulations. For this point of view to be compelling, much more research is necessary. In particular, it is necessary to get a much better understanding of the following issues:

- As we have seen, the use of randomized algorithms leads to statements about the smallness of the size of "bad sets" with high confidence. A key issue here is the notion of size. *The theory naturally leads to the quantification of size in terms of volume of the parameter set using the same measure which is used to generate the samples.* It will be useful to know and understand the physical significance of smallness of this volume. This requires experience with real engineering problems as well as theoretical results on other measures of size.

  In the context of controller synthesis, obviously, it is critical to limit the initial search space such that the *relative volume* of the set of "good controllers" is not too small. Otherwise, the crude or adaptive random search methods may not be useful. This is a place where good heuristics are very useful, and one can bring to bear all the relevant analytical knowledge in choosing the most appropriate parameterization of controllers and limiting the search space. This is a topic for further research.

- We have used a very simple rejection sampling approach to establish the results. It is obvious that there are much smarter methods for sampling so that the bad parts of the parameter space for robust control analysis problems (or good parts of the parameter space for design problems) can be searched more efficiently. Algorithms such as genetic algorithms [10], ordinal optimization [15, 4, 12], etc. may have a large role to play here.

  The real potential of this change of paradigm is in dealing with real engineering problems. By using nonlinear simulations as oracles, we can allow the incorporation of realistic design objectives in the analysis and design process. One can then focus on the efficient searches through the space of controller parameters.

- We noted that the dimension of the parameter space plays no role in the computational complexity bounds. The dimension of the underlying objects, e. g., size of matrices, does play a role in the computational complexity. For problems considered here, the dependence on the matrix dimension is of polynomial type. Does it mean that we are no longer plagued by the curse of dimensionality? This is clearly tied up with the issue of how to measure size as discussed above. A full theoretical understanding of the true computational complexity with respect to problem dimensions will be useful. Numerical experience with large size problems will be critical in understanding this issue.

10   Appendix: Proof of Theorem 5.1

It is necessary to introduce further notation and some preliminary results. First, for $\Delta \in \mathbf{\Delta}_\gamma$, let $\lambda(M, \Delta)$ be the maximum real part of the eigenvalues of the system matrix of the feedback interconnection depicted in Figure 1. Next, for $\alpha \geq 0$ define

$$\mathbf{U}(\alpha) = \{\Delta \in \mathbf{\Delta}_\gamma \; : \; \lambda(M, \Delta) \geq \lambda_{\max} - \alpha\},$$

and let $f(\alpha) = \mathrm{vol}_w(\mathbf{U}(\alpha))$. Note that the sets $\mathbf{U}(\cdot)$ are nested: $\mathbf{U}(\alpha) \subseteq \mathbf{U}(\beta)$ for $\alpha \leq \beta$, and so $f(\alpha)$ is a non-decreasing function of α. We will restrict attention to the case when $\epsilon \geq f(0)$ since Theorem 5.1 follows trivially otherwise.

Finally, we will need the following results. We will delay their proofs until completion of the proof for Theorem 5.1.

(a) Let $\overline{\mathbf{U}}(\alpha) = \bigcap_{\zeta > 0} \mathbf{U}(\alpha + \zeta)$. Then $\overline{\mathbf{U}}(\alpha) = \mathbf{U}(\alpha)$.

(b) The function $f(\cdot)$ is right continuous. Moreover, there exists $\hat{\alpha} > 0$ such that $f(\hat{\alpha}) \geq \epsilon \geq f(\hat{\alpha} - \zeta)$ for all $\zeta > 0$.

(c) Define $\underline{\mathbf{U}}(\alpha) = \bigcup_{\zeta > 0} \mathbf{U}(\alpha - \zeta)$. Let $\hat{\alpha}$ be defined as in (b). Then $\mathrm{vol}_w(\underline{\mathbf{U}}(\hat{\alpha})) \leq \epsilon$.

With these results in place, we can proceed with the proof of Theorem 5.1. By (b), there exists $\hat{\alpha} > 0$ such that $f(\hat{\alpha}) \geq \epsilon \geq f(\hat{\alpha} - \zeta)$ for all $\zeta > 0$. Suppose it is the case that $\hat{\lambda}(n) \geq \lambda_{\max} - \hat{\alpha}$. Then the following argument shows that $\mathrm{vol}_w \mathbf{B} \leq \epsilon$.

$$
\begin{aligned}
\Delta \in \mathbf{B} \;\; &\Longrightarrow \;\; \lambda(M, \Delta) > \hat{\lambda}(n) \geq \lambda_{\max} - \hat{\alpha} \\
&\Longrightarrow \;\; \lambda(M, \Delta) > \lambda_{\max} - \hat{\alpha} + \zeta \text{ for some } \zeta > 0 \\
&\Longrightarrow \;\; \Delta \in \mathbf{U}(\hat{\alpha} - \zeta) \\
&\Longrightarrow \;\; \mathbf{B} \subseteq \bigcup_{\zeta > 0} \mathbf{U}(\hat{\alpha} - \zeta) = \underline{\mathbf{U}}(\hat{\alpha}) \\
&\Longrightarrow \;\; \mathrm{vol}_w \mathbf{B} \leq \mathrm{vol}_w(\underline{\mathbf{U}}(\hat{\alpha})) \leq \epsilon \quad \text{(this follows from (c))}
\end{aligned}
$$

Therefore

$$Prob\{\text{vol}_w(\mathbf{B}) > \epsilon\}$$
$$\leq \quad Prob\left\{\hat{\lambda}(n) < \lambda_{\max} - \hat{\alpha}\right\}$$
$$= \quad Prob\{\Delta_1 \notin \mathbf{U}(\hat{\alpha}) \ \& \ \Delta_2 \notin \mathbf{U}(\hat{\alpha}) \ \& \ \dots \ \& \ \Delta_n \notin \mathbf{U}(\hat{\alpha})\}$$
$$= \quad (1 - f(\hat{\alpha}))^n$$
$$\leq \quad (1 - \epsilon)^n$$

Theorem 5.1 follows by observing that $n \geq \ln\left(\frac{1}{\delta}\right) / \ln\left(\frac{1}{1-\epsilon}\right)$ implies $(1-\epsilon)^n \leq \delta$. $\qquad\square$

Proof of (a): It is clear that $\mathbf{U}(\alpha) \subseteq \overline{\mathbf{U}}(\alpha)$ since the sets $\mathbf{U}(\cdot)$ are nested. We have only to show that $\overline{\mathbf{U}}(\alpha) \setminus \mathbf{U}(\alpha)$ is empty. Let $\Delta \in \overline{\mathbf{U}}(\alpha)$. Then $\Delta \in \mathbf{U}(\alpha + \zeta)$ for all $\zeta > 0$, or, equivalently, $\lambda(M, \Delta) \geq \lambda_{\max} - \alpha - \zeta$ for all $\zeta > 0$. Thus $\lambda(M, \Delta) \geq \lambda_{\max} - \alpha$ and so $\Delta \in \mathbf{U}(\alpha)$. $\qquad\square$

Proof of (b): If there exists $\hat{\alpha}$ such that $f(\hat{\alpha}) = \epsilon$, then (b) trivially holds since $f(\cdot)$ is a non-decreasing function. Next, let us consider the case when there exists no such $\hat{\alpha}$. Fix two points a_1, b_1 such that $a_1 > b_1$ and $f(a_1) > \epsilon > f(b_1)$. (The points a_1, b_1 exist because of the restrictions placed on ϵ.) Then employ the following algorithm to generate two sequences: $\{a_i\}_1^\infty$, and $\{b_j\}_1^\infty$.

1) Set $i = 1$ and $j = 1$.
2) Let $c = \frac{a_i - b_j}{2}$.
   If $f(c) > \epsilon$ then
   - set $a_{i+1} = c$
   - $i \leftarrow i + 1$
   else
   - set $b_{j+1} = c$
   - $j \leftarrow j + 1$.
3) Go to step 2.

It is clear that $\{a_i\}_1^\infty$ is a decreasing sequence and $\{b_j\}_1^\infty$ is an increasing sequence. Moreover, they are convergent sequences that converge to the same point which we shall call $\hat{\alpha}$. All that remains is to verify that $\hat{\alpha}$ has the desired properties.

First we will show using proof by contradiction that $f(\hat{\alpha}) \geq \epsilon$. Suppose $f(\hat{\alpha}) < \epsilon$. Then there exists an open set $\mathcal{N} \subseteq \cap_{i=1}^\infty \mathbf{U}(a_i) \setminus \mathbf{U}(\hat{\alpha})$. But this implies $\overline{\mathbf{U}}(\hat{\alpha}) \neq \mathbf{U}(\hat{\alpha})$, a contradiction to (a). Finally, it is straightforward to verify that $f(\hat{\alpha} - \zeta) < \epsilon$ for all $\zeta > 0$. Fix $\zeta > 0$. There exists $j < \infty$ such that $b_j > \hat{\alpha} - \zeta$. Thus $f(\hat{\alpha} - \zeta) \leq f(b_j) < \epsilon$. $\qquad\square$

Proof of (c): Consider the sequence $\{b_j\}_1^\infty$ constructed in the proof of claim (b). Note that $\underline{\mathbf{U}}(\hat{\alpha}) = \bigcup_{j=1}^\infty \mathbf{U}(b_j)$. We will prove $\text{vol}_w(\underline{\mathbf{U}}(\hat{\alpha})) \leq \epsilon$ using proof by contradiction. Suppose that $\text{vol}_w(\underline{\mathbf{U}}(\hat{\alpha})) = \epsilon + \beta$ for some $\beta > 0$. Since $\bigcup_{j=1}^m \mathbf{U}(b_j) = \mathbf{U}(b_m)$, it is the case that $\bigcup_{j=1}^{m_1} \mathbf{U}(b_j) \subseteq \bigcup_{j=1}^{m_2} \mathbf{U}(b_j)$ for

$m_1 \leq m_2$. Therefore there exists sufficiently large m such that $\text{vol}_w \left(\underline{\mathbf{U}}(\hat{\alpha}) - \bigcup_{j=1}^{m} \mathbf{U}(b_j) \right) \leq \frac{\beta}{2}$ and so

$$\text{vol}_w \left(\underline{\mathbf{U}}(\hat{\alpha}) \right) \leq \text{vol}_w \left(\bigcup_{j=1}^{m} \mathbf{U}(b_j) \right) + \frac{\beta}{2} \leq \epsilon + \frac{\beta}{2}.$$

This is a contradiction thus proving the claim.   □

References

[1] G. Balas, J.C. Doyle, K. Glover, A. Packard, and R. Smith, μ-*Analysis and Synthesis Toolbox User's Guide*, MUSYN Inc., 1994.

[2] B. R. Barmish and C.M. Lagoa, "The uniform distribution: a rigorous justification for its use in robustness analysis," Technical Report ECE-95-12, Dept. of ECE, Univ. of Wisconsin-Madison, 1995.

[3] R. Braatz, P. Young, J. Doyle, and M. Morari, "Computational complexity of μ calculation," *Proc. of the Automatic Control Conference*, pp: 1682–1683, San Francisco, 1993.

[4] M. Deng and Y.-C. Ho, "Sampling-selection method for stochastic optimization problems," Preprint, Harvard University, 1995.

[5] J. C. Doyle, B. A. Francis, and A. R. Tannenbaum, *Feedback Control Theory*, Macmillan Publishing Company, New York, 1992.

[6] M. K. H. Fan, A. L. Tits, and J. C. Doyle, "Robustness in the presence of joint parametric uncertainty and unmodeled dynamics," *Proc. of the American Control Conference*, pp. 1195–1200, 1988.

[7] W. Feller, *An Introduction to Probability Theory and its Applications, Volume 2*, Wiley & Sons, New York, 1965.

[8] J. Friedman, P. Kabamba, and P. Khargonekar, "Worst-case and average $\mathcal{H}_2$ performance analysis against real constant parametric uncertainty," *Automatica*, vol. 31, pp 649–657, 1995.

[9] M. Fu and B. R. Barmish, "Maximal unidirectional perturbation bounds for stability of polynomials and matrices," *Systems & Control Letters*, 11 (1988) pp 173–179.

[10] D. E. Goldberg, *Genetic Algorithms in Search, Optimization, and Machine Learning*, Addison-Wesley Publishing Co., Reading, Mass., 1989.

[11] S. Hall, D. MacMartin, and D. Bernstein, "Covariance averaging in the analysis of uncertain systems," *IEEE Trans. Auto. Contr.*, vol. 38, pp 1858–1862, 1993.

[12] Y.-C. Ho, "Heuristics, rules of thumb, and the 80/20 proposition," *IEEE Trans. Auto. Contr.*, vol. 39, pp. 1025–1027, 1994.

[13] W. Hoeffding, "Probability inequalities for sums of bounded random variables," *Amer. Stat. Assoc. J.*, pp. 13–30, 1963.

[14] V. L. Kharitonov, "Asymptotic stability of an equilibrium position of a family of systems of linear differential equations," *Differencial'nye Uravnenija* 14 (11) (1978) pp. 2086–2088.

[15] T.W.E. Lau and Y.-C. Ho, "Universal alignment probabilities and subset selection for ordinal optimization," Preprint, Harvard University, 1996.

[16] L. Lovasz and M. Simonovits, "Random walks in a convex body and an improved volume algorithm," *Random Structures and Algorithms*, vol. 4, pp. 359-412, 1993.

[17] MATH WORKS. *MATLAB Reference Guide*. The MATH WORKS, Inc., 1992.

[18] C. Marrison and R. Stengel, "The use of random search and genetic algorithms to optimize stochastic robustness functions," *Proc. of the American Control Conference*, pp. 1484–1489, Baltimore, Maryland, 1994.

[19] A. Nemirovskii, "Several NP-hard problems arising in robust stability analysis," *Math. of Control, Signals, and Systems*, vol. 6, pp 99–105, 1993.

[20] A. Packard and J.C. Doyle, "The complex structured singular value," *Automatica*, vol. 29, pp. 71–110, 1993.

[21] S. Poljak and J. Rohn, "Checking robust nonsingularity is NP-hard," *Math. of Control, Signals, and Systems*, vol. 6, pp 1–9, 1993.

[22] L. R. Ray and R. F. Stengel, "Stochastic robustness of linear time-invariant control systems," *IEEE Trans. Auto. Contr.*, vol. 36, pp. 82-87, 1991.

[23] H. E. Romeijn, *Global Optimization by Random Walk Sampling Methods*, book nr. 32 of Tinbergen Institue Research Series, Thesis Publishers, Amsterdam, The Netherlands, 1992.

[24] R.Y. Rubinstein, *Monte Carlo Optimization, Simulation and Sensitivity of Queueing Networks*, Wiley and Sons, New York, 1986.

[25] R. Tempo and E. W. Bai, "Robustness analysis with nonlinear parametric uncertainty: A probabilistic approach," CENS-CNR Report 95/11, Poiltecnico di Torino, Torino, Italy, 1995.

[26] P. Young and J. Doyle, "Computation of μ with real and complex uncertainties," *Proc. of 29th Conf. on Decision and Control*, pp. 1230–1235, Honolulu, Hawaii, 1990.

Overview of complexity and decidability results for three classes of elementary nonlinear systems*

Vincent D. Blondel[†]   John N. Tsitsiklis[‡]

Abstract

It has become increasingly apparent this last decade that many problems in systems and control are NP-hard and, in some cases, undecidable. The inherent complexity of some of the most elementary problems in systems and control points to the necessity of using alternative approximate techniques to deal with problems that are unsolvable or intractable when exact solutions are sought.

We survey some of the decidability and complexity results available for three classes of discrete time nonlinear systems. In each case, we draw the line between the problems that are unsolvable, those that are NP-hard, and those for which polynomial time algorithms are known.

1 Introduction

We look at the decidability and the complexity of four particular control problem for three different classes of discrete time nonlinear systems. The first two problems that we consider are analysis problems, the other two are control design problems.

STATE GOES TO THE ORIGIN
Input: A system $x_{t+1} = f(x_t)$, a state ξ.
Question: Does the initial state $x_0 = \xi$ eventually reach the origin when driven by $x_{t+1} = f(x_t)$?

STABILITY: ALL STATES GO TO THE ORIGIN
Input: A system $x_{t+1} = f(x_t)$.

---

*This research was partly carried out while Blondel was visiting Tsitsiklis at MIT and was supported by the NATO under grant CRG-961115.

[†]Institute of Mathematics, University of Liège B37, B-4000 Liège, Belgium; Email: vblondel@ulg.ac.be

[‡]Laboratory for Information and Decision Systems, Massachusetts Institute of Technology, Cambridge, MA 02139, USA; Email: jnt@mit.edu

Question: Do all initial states $x_0 = \xi$ eventually reach the origin when driven by $x_{t+1} = f(x_t)$?

STATE CAN BE DRIVEN TO THE ORIGIN
Input: A system $x_{t+1} = f(x_t, u_t)$, a state ξ.
Question: Does there exists some $k \geq 0$ and controls u_t, $i = 0, \ldots, k-1$ such that the system $x_{t+1} = f(x_t, u_t)$ drives $x_0 = \xi$ to the origin?

NULL-CONTROLLABILITY. ALL STATES CAN BE DRIVEN TO THE ORIGIN
Input: A system $x_{t+1} = f(x_t, u_t)$.
Question: Does there exists, associated to every state ξ, some $k \geq 0$ and controls u_t, $i = 0, \ldots, k-1$ such that the system $x_{t+1} = f(x_t, u_t)$ drives $x_0 = \xi$ to the origin?

Asymptotic versions of these definitions are obtained by requiring the sequences to converge to the origin rather than reaching it exactly. The results surveyed in this paper are stated in their non-asymptotic version, most of them remain valid when stated in the asymptotic case.

For linear systems all four questions are decidable and can be decided efficiently (see, e.g., [Sontag, 1990]). On the other hand, no such algorithms exist for general nonlinear systems. Stated at the general level of nonlinear systems, these questions are not interesting because they are far too difficult to solve. For example, as pointed in [Sontag, 1995], the null-controllability question for general nonlinear systems encompasses the problem of solving an arbitrary nonlinear equation. Indeed, for a given function g, consider the system $x_{t+1} = g(u_t)$. Then the system is null-controllable if and only if g has a zero and so the null-controllability question for nonlinear systems is at least as hard as deciding the existence of a zeros for an arbitrary nonlinear functions, which is a far too general problem. For nonlinear control problems to lead to interesting questions we need to constraint the type of nonlinear systems considered.

In the next sections we consider nonlinear systems of the following type: systems with a single nonlinearity, systems of the neural network type, and piecewise-linear systems. In many of these cases control questions become intractable even for systems that are apparently weakly nonlinear. An overview of the results surveyed in this contribution is given in a summarising table.

Before proceeding to the results, let us say a few words on the notions of decidability and computational complexity. When we say that a certain problem is *decidable* we mean that there is an algorithm which, upon input of the data associated to the problem, provides an answer after finitely many steps. The precise definition of *algorithm* is not critical here, it may be, for instance, a

Turing machine, an unlimited register machine or any one of most of the other abstract computer models that are proposed in the literature. Most models proposed so far have been shown equivalent from the point of view of their computing capabilities.

When we say that a problem can be *decided in polynomial time*, or that it can be *decided efficiently*, we mean that there is a polynomial P and an algorithm which, upon input of any instance Σ of the problem, provides an answer after at most $P(size(\Sigma))$ computational steps. Again, the precise definition of the size of Σ, and the definition of what is meant by a computational step are not critical. The property of being decidable in polynomial time is robust across all reasonable definitions. The class P is the class of problems that can be decided in polynomial time. The class NP is a class of problems that includes all problems in P and includes a large number of problems of practical interest for which no polynomial time algorithms have yet been found. It is widely believed that $P \neq NP$. A problem is NP-hard if it is at least has hard as any problem in NP. A polynomial time algorithm for an NP-hard problem would immediately result in a polynomial time algorithm for all problems in NP. Finally, a problem is NP-complete if it is NP-hard and belongs to NP. For an introduction to computability, see [Davis, 1982] or [Hopcroft and Ullman, 1969]. For an introduction to computational complexity, see [Garey and Johnson, 1979] or the more recent reference [Papadimitriou, 1994].

This paper is partly based on a survey paper on computational complexity results for systems and control problems [Blondel and Tsitsiklis, 1997c]. A survey of complexity results for nonlinear systems is given in [Sontag, 1995]. See also [Tsitsiklis, 1994].

2   Systems with a single nonlinearity

Let us fix a scalar function $\nu : \mathbf{R} \mapsto \mathbf{R}$. We use the function ν to capture the nonlinearity in a system that has a single nonlinearity. Let $n \geq 1$, $A_0, A_1 \in \mathbf{R}^{n \times n}$, $c \in \mathbf{R}^n$, and consider the system

$$x_{t+1} = \left(A_0 + \nu(c^T x_t)A_1\right) x_t. \tag{1}$$

When ν is constant, the system (1) is linear and its stability can be decided easily. In Theorem 1 in [Blondel and Tsitsiklis, 1997a] the authors show that for most functions ν that are not constant, the stability of systems of the form (1) is NP-hard to decide.

Theorem 1: Let $\nu : \mathbf{R} \mapsto \mathbf{R}$ be a nonconstant scalar function such that

$$\lim_{x \to -\infty} \nu(x) \leq \nu(x) \leq \lim_{x \to +\infty} \nu(x)$$

for all $x \in \mathbf{R}$. Then, STABILITY of

$$x_{t+1} = \left(A_0 + \nu(c^T x_t) A_1 \right) x_t$$

is NP-hard to decide.

Each particular choice of a nonconstant function ν leads to a particular class of nonlinear systems for which stability is NP-hard to decide. In particular, one of the classes is the class of systems that are linear on each side of a hyperplane that divides the state space in two.

Corollary: The problem of deciding, for given matrices A_+, A_- and vector c, whether the system

$$x_{t+1} = \begin{cases} A_+ x_t & \text{when} \quad c^T x_t \geq 0, \\ A_- x_t & \text{when} \quad c^T x_t < 0, \end{cases}$$

is stable, is NP-hard.

A control implication of this result is obtained for linear systems controlled by bang-bang controllers. A linear system $x_{t+1} = A x_t + B u_t$ controlled by a bang-bang controller of the type

$$u_k = \begin{cases} K_0 x_t & \text{when} \quad y_t \geq 0, \\ K_1 x_t & \text{when} \quad y_t < 0, \end{cases}$$

leads to a closed-loop system

$$x_k = \begin{cases} (A + BK_0) x_t & \text{when} \quad y_t \geq 0, \\ (A + BK_1) x_t & \text{when} \quad y_t < 0. \end{cases}$$

¿From Theorem 1 we see that the stability of such systems is NP-hard to decide.

It is not clear when the stability of the systems (1) is actually *decidable*. Except for the trivial case where ν is constant, and the systems are then linear, the authors are not aware of any function ν for which stability of (1) is decidable. For the simple case where ν is piecewise constant, the problem is related to the difficult open problem of deciding the stability of all possible sequences of products of finitely many matrices, see [Blondel and Tsitsiklis, 1997b] for more details.

One can easily adapt the definition (1) to include the possibility of a control action. Let us consider systems of the type

$$x_{t+1} = \left(A_0 + \nu(c^T x_t) A_1 \right) x_t + B u_t. \tag{2}$$

When ν is a constant function, these systems are linear and control questions can be decided easily. When ν is a nonconstant function that satisfies the hypothesis of Theorem 1, it is clear that null-controllability of (2) for $B = 0$ is equivalent to the stability of (1), and so NULL-CONTROLLABILITY of (2) is NP-hard to decide. One can in fact say more than that. We will see in Section 4 that, when ν is a function that has a finite range of cardinality greater or equal to two, then the system (2) becomes piecewise linear and null-controllability of the system is undecidable. The decidability of the case where the range of ν is infinite is open.

3   Systems of the neural network type

Let us fix a scalar function $\sigma : \mathbf{R} \mapsto \mathbf{R}$. Let $n \geq 1$, $A \in \mathbf{R}^{n \times n}$, and consider the system

$$x_{t+1} = \sigma(Ax_t) \tag{3}$$

where σ is defined componentwise, i.e.,

$$\sigma \begin{pmatrix} q_1 \\ q_2 \\ \vdots \\ q_n \end{pmatrix} = \begin{pmatrix} \sigma(q_1) \\ \sigma(q_2) \\ \vdots \\ \sigma(q_n) \end{pmatrix}.$$

Systems of this type arise in a wide variety of situations. The dynamics of (3) depends heavily on the function σ. When σ is linear, the systems are linear and most dynamical properties are easy to check. When σ is the Heaviside function, the entries of the state vector take values in $\{0, 1\}$ and the system becomes finite state after the first iteration. The dynamics of such a system, and in fact of any system (3) with a function σ that has finite range, can be modeled by a directed graph whose nodes correspond to the finite states of the system and with directed edges constructed from the matrix A. Dynamical properties for such systems are easy to decide.

Recurrent artificial neural networks are modeled by equations (3) where the function σ is the activation function used in the network (see [Sontag, 1993]). Activation functions that are common in the neural network literature are the *saturated linear function*

$$\sigma(x) = \begin{cases} 0 & \text{when} \quad x \leq 0 \\ x & \text{when} \quad 0 < x < 1 \\ 1 & \text{when} \quad x \geq 1 \end{cases}$$

the standard sigmoid $\sigma(x) = 1/(1 + e^x)$, and the inverse trigonometric function $\sigma(x) = \arctan(x)$. All these functions are continuous and have a finite limit on both end of the real axis. These are features that are common in the context of artificial neural networks. Systems (3) with the saturated linear function also arises in the context of linear systems with saturation on the state, and in the analysis and design of fixed-point digital filters (see [Liu and Michel, 1994] for motivations and many references related to filter design). Finally, we also deal with the *cut function* $\sigma(x) = \max(0, x)$ which is probably the simplest piecewise linear function after the linear ones.

Although the difference between the systems (3) and linear systems looks minor when the function σ is weakly nonlinear (such as the cut function for example), the differences in the behavior is complete. In a work announced in [Siegelmann and Sontag, 1991] and completed in [Siegelmann and Sontag, 1995] it is shown that, when σ is the saturated linear function, systems of the type (3) are capable of simulating arbitrary Turing machines. In the simulation, the Turing machine is encoded in the matrix A and the tape content and machine configuration are encoded on some of the states of the system. The simulation of the machine is then obtained by simple iteration. Thus, as computational devices, linear saturated systems are as powerful as Turing machines. The problem of deciding if a given Turing machine halts on some particular tape configuration (the halting problem) is undecidable for Turing machines. Therefore, the problem of deciding if a given initial state of a saturated linear system eventually reaches a state that encodes a halting configuration, is also undecidable. One can show that this halting state can always be chosen to be the origin. And so one conclude (see [Sontag, 1995] for the sketch of a proof).

Theorem 2: STATE GOES TO THE ORIGIN is undecidable for saturated linear systems.

By using a universal Turing machine one can in fact prove the stronger result that STATE GOES TO THE ORIGIN is undecidable for some *particular* matrix A. There exists a particular matrix A (of size less than 1000×1000 and with integer entries) such that the problem of deciding if a given initial state $x_0 = \xi$ eventually hits the origin when driven by $x_{t+1} = \sigma(Ax_t)$, is undecidable.

The initial result by [Siegelmann and Sontag, 1995] has generated research activity in the direction of finding conditions on the function σ under which Turing machine simulation is possible by systems of the type (3). The fact that such simulations are possible is proved in a very elementary and simple way in [Hyotyniemi, 1997] in the case of the cut function. (Notice that the title of the reference [Hyotyniemi, 1997] involves the term "stability", but this term is actually used in a sense different than the usual notion of stability in systems theory).

In [Koiran, 1996], the author shows how to simulate Turing machines with systems of the type (3) and any function σ that eventually becomes constant on both ends of the real line and is twice differentiable with nonzero derivative on some open interval. The function $\sigma = \arctan$ and other classical function in the neural network literature do not satisfy these hypothesis. Conditions on σ under which systems (3) have Turing power are relaxed in [Kilian and Siegelmann, 1996] where the authors offer a sketch of a proof that Turing machines can be simulated by systems (3) with functions σ that belong to a class that encloses, among others, the functions just described and all the functions that are classically used in neural networks models. The function do not need to become ultimately constant but need to be monotone. Using an argument similar to that used for the case of the saturated linear function one then obtain:

Theorem 3: STATE GOES TO THE ORIGIN is undecidable for systems of the type $x_{t+1} = \sigma(Ax_t)$ when σ is the saturated linear function, the cut function, the sigmoid function, the zeroing function and any function that belongs to the classes defined in [Koiran, 1996] and [Kilian and Siegelmann, 1996].

At this point we feel safe to conjecture that, STATE GOES TO THE ORIGIN is undecidable for any function σ that is not linear and that contains an open set in its codomain (the case where σ has finite range is trivially decidable).

¿From Theorem 3, undecidability of STATE CAN BE DRIVEN TO THE ORIGIN for the controlled system

$$x_{t+1} = \sigma(Ax_t + Bu_t) \qquad (4)$$

is immediate to obtain. This result does however not have direct implications for the decidability of null-controllability (ALL STATES CAN BE DRIVEN TO THE ORIGIN) or for the decidability of stability (ALL STATES GO TO THE ORIGIN) of the systems (4) and (3). Despite various attempts and the fact that the undecidability of STABILITY for saturated linear systems was conjectured in [Sontag, 1995], it is yet unclear whether there exists functions σ for which STABILITY is undecidable. And if one exists, it is not clear if one exists that is continuous. The computational complexity of this problem is also an open question. Although the stability of (3) is strongly suspected to be NP-hard for most function σ, this result was never proved. Let us finally notice that, since undecidability of STABILITY would imply undecidability of NULL-CONTROLLABILITY, the later problem is probably easier to prove undecidable.

4 Piecewise linear systems

Let a finite partition of $\mathbf{R}^n$ be given by $\mathbf{R}^n = H_1 \cup H_2 \cup \cdots \cup H_m$, and suppose that different linear systems are associated to each partition, i.e., the overall nonlinear system is given by

$$x_{t+1} = A_i x_t \qquad \text{when} \qquad x_t \in H_i. \tag{5}$$

When the partitions H_i are definable in terms of a finite number of linear equalities and inequalities, the systems (5) are the *piecewise linear* systems introduced in [Sontag, 1981] as a unifying model for interconnection between automata and linear systems (see [Sontag, 1996] for an updated overview of results available for this model).

Particular classes of piecewise linear systems are obtained from (1) when ν is piecewise constant and from (3) when σ is piecewise linear. Hence, STATE CAN BE DRIVEN TO THE ORIGIN and STATE GOES TO THE ORIGIN are both undecidable and NP-hard for piecewise linear systems. Undecidability of these questions is obtained by using the fact that Turing machines can be simulated by systems of the type (3). These simulations are performed in [Siegelmann and Sontag, 1995] with linear saturated systems of state dimension approximately equal to 1000. In [Koiran *et al.*, 1994], the authors show that similar simulations of Turing machines are possible by iteration of piecewise *affine* systems of state dimension two, or by piecewise linear systems of dimension three. Hence, STATE GOES TO THE ORIGIN is undecidable for such systems.

As in the case of systems of the neural network type one can prove a stronger result by using a universal Turing machine. There exist a *particular* piecewise linear system with state dimension three (the system has approximately 800 partitions) such that the problem of deciding for this system if a given initial state $x_0 = \xi$ eventually hits the origin, is undecidable.

Theorem 4: There exist a particular piecewise linear system with state dimension three and with 800 partitions such that STATE GOES TO THE ORIGIN is undecidable.

The systems (5) are similar to the *piecewise constant derivative* systems analyzed in [Asarin *et al.*, 1995] and for which analogous undecidability results are available. A piecewise constant derivative system is given by a finite partition $\mathbf{R}^n = H_1 \cup H_2 \cup \cdots \cup H_m$, and by slope vectors b_i for every region H_i of the partition. On any given region of the partition, the state $x(t)$ of the system has a fixed constant derivative,

$$\frac{d\,x(t)}{dt} = b_i \qquad \text{when} \qquad x \in H_i.$$

The trajectories of such systems are continuous broken lines, with breaking points occurring on the boundaries of the regions. In [Asarin *et al.*, 1995] the authors show that, for given states x_b and x_e, the problem of deciding whether x_b is reached by a trajectory starting from x_b, is decidable for systems of dimension two, but is undecidable for systems of dimension three or more.

Suppose now that we add a control to the system and define

$$x_{t+1} = A_i x_t + B u_t \qquad \text{when} \qquad x_t \in H_i. \tag{6}$$

As already explained, it follows from Theorem 4 that STATE CAN BE DRIVEN TO THE ORIGIN is undecidable for such systems. This result is also obtained in

[Blondel and Tsitsiklis, 1997a] by using a different proof based on the undecidability of the Post correspondence problem.

POST'S CORRESPONDENCE PROBLEM.

Instance: A set of pairs of words $\{(U_i, V_i) : i = 1, \ldots, n\}$ over a finite alphabet.

Question: Does there exist a non-empty sequence of indices $i_1, i_2, \ldots, i_k$ where $1 \leq i_j \leq n$, such that $U_{i_1} U_{i_2} \cdots U_{i_k} = V_{i_1} V_{i_2} \cdots V_{i_k}$?

Post's correspondence problem is trivially decidable for one letter alphabets. Furthermore, it is easy to see that the solvability of the problem does not depend on the size of the alphabet, as long as the alphabet contains more than one letter. Post proved that the correspondence problem for an alphabet with more than one letter is undecidable (for a proof of this classical result see [Hopcroft and Ullman, 1969]). In a recent contribution ([Matiyasevich and Sénizergues, 1996]) this result has been improved by showing that the problem remains undecidable in the case where there are only seven pairs of words. On the other hand, the problem is known to be decidable for two pairs of words. The limit between decidability/undecidability is somewhere between three and seven pairs.

There is an obvious trade-off in piecewise linear systems between the state space dimension n and the number of partitions m. When there is only one partition, or when the state dimension is equal to one, STATE CAN BE DRIVEN TO THE ORIGIN and NULL-CONTROLLABILITY are easy to check. The proof technique used in [Blondel and Tsitsiklis, 1997a] is effective for obtaining bounds on n and m for which undecidability is attained. The next result is proved in [Blondel and Tsitsiklis, 1997d].

Theorem 5: Let n_p be any number of pairs of words for which POST'S COR-RESPONDENCE PROBLEM is undecidable. Let n be the state space dimension of a piecewise linear system defined on m partitions. If $n \geq 4$, $m \geq 2$ and $nm \geq 2 + 6n_p$, then, STATE CAN BE DRIVEN TO THE ORIGIN is undecidable.

As mentioned earlier we can take $n_p = 7$, and thus STATE CAN BE DRIVEN TO THE ORIGIN is undecidable when $nm \geq 44$. In particular, STATE CAN BE DRIVEN TO THE ORIGIN is undecidable for piecewise linear systems of state dimension 22 and with as few as 2 partitions.

Theorem 5 does not have direct implications for the problems NULL-CONTROL-LABILITY and STABILITY for which we require certain properties to be shared by *all* states. Piecewise linear systems on two partitions are obtained as special cases of systems with a single nonlinearity. It is therefore clear that STABILITY and NULL-CONTROLLABILITY are NP-hard for piecewise linear systems. But that doesn't settle the issue of the decidability of these problems. We now consider these two problems in turn. The first one is undecidable but decidability of the second problem is an unsolved question. The next result is proved in [Blondel and Tsitsiklis, 1997d].

Theorem 6: Let n_p be any number of pairs of words for which Post's correspondence problem is undecidable. Let n be the state space dimension of a piecewise linear system defined on m partitions. If $n \geq 4$, $m \geq 2$ and $nm \geq 26 + 6n_p$, then, NULL-CONTROLLABILITY, is undecidable.

We finally turn our attention to the decidability of STABILITY of piecewise linear systems. Consider the particular class of piecewise linear systems in which the partition consists of two regions separated by a hyperplane. The system is

$$x_{t+1} = \begin{cases} A_1 x_t & \text{when} & c^T x_t \geq 0 \\ A_2 x_t & \text{when} & c^T x_t < 0 \end{cases} \tag{7}$$

Deciding stability of nonlinear systems as simple as (7) is already a nontrivial task. We know that the problem is NP-hard but we do not know if it is decidable. The decidability of this problem is, as we now argue, intimately related to the problem of determining if all possible sequences of products of two given matrices are stable. Let us illustrate this with an example. We build a piecewise linear system with state vector (v_t, y_t, z_t), where v_t and y_t are scalars and z_t is a vector in $\mathbf{R}^n$. The system consists of two linear systems, each of which is enabled in one of two halfspaces, as determined by the sign of y_t

$$\begin{pmatrix} v_{t+1} \\ y_{t+1} \\ z_{t+1} \end{pmatrix} = \begin{pmatrix} 1/2 & 0 & 0 \\ -1/2 & 1 & 0 \\ 0 & 0 & A_+ \end{pmatrix} \begin{pmatrix} v_t \\ y_t \\ z_t \end{pmatrix} \quad \text{when } y_t \geq 0,$$

and

$$\begin{pmatrix} v_{t+1} \\ y_{t+1} \\ z_{t+1} \end{pmatrix} = \begin{pmatrix} 1/2 & 0 & 0 \\ 1/2 & 1 & 0 \\ 0 & 0 & A_- \end{pmatrix} \begin{pmatrix} v_t \\ y_t \\ z_t \end{pmatrix} \quad \text{when } y_t < 0.$$

Let us now look at the evolution of an initial state vector (v_0, y_0, z_0). Suppose that $v_0 = 1$ in which case we have $v_t = 2^{-t}$ for all t. Suppose in addition, that y_0 can take any value in $[-1, 1]$. Then, it is easily seen that y_1 can take any value in $[-1/2, 1/2]$, no matter what was the sign of y_0. Continuing inductively, we see that y_t can take any value in $[-2^{-t}, 2^{-t}]$, can have either sign, and this is independent of the signs of y_s for $s < t$. This shows that every possible sign sequence can be generated by suitable choice of y_0. Hence, the dynamics of the state subvector z_t are of the form $z_{t+1} = A_t z_t$, where A_t is an arbitrary matrix from $\{A_-, A_+\}$. We conclude that the state vector converges to zero, for all possible initial states, if and only if all sequences of products of the matrices A_- and A_+ (taken in an arbitrary order) converge to zero. Thus, a decision algorithm for STABILITY of piecewise linear systems would lead to a test for the stability of all possible sequences of products of two matrices.

5   Summary

AUTONOMOUS SYSTEMS		STABILITY	STATE GOES TO ORIGIN
$x_{t+1} = \left(A_0 + \nu(c^T x_t) A_1 \right) x_t$	Complexity	NP-hard for nonconstant ν	?
	Decidability	?	?
		for nonconstant ν	
$x_{t+1} = \sigma(A x_t)$	Complexity	?	?
	Decidability	Conjectured undecidable	Undecidable for most σ
$x_{t+1} = A_i x_t \ (x_t \in H_i)$	Complexity	NP-hard	?
	Decidability	?	Undecidable

CONTROLLED SYSTEMS		NULL-CONTROLLABILITY	STATE DRIVEN TO ORIGIN
$x_{t+1} = \left(A_0 + \nu(c^T x_t) A_1 \right) x_t$	Complexity	NP-hard	?
$\quad + B u_t$	Decidability	Undecidable for ν with finite range	Undecidable for ν with finite range
$x_{t+1} = \sigma(A x_t + B u_t)$	Complexity	?	?
	Decidability	?	Undecidable for most σ
$x_{t+1} = A_i x_t + B u_t \ (x_t \in H_i)$	Complexity	NP-hard	?
	Decidability	Undecidable	Undecidable

References

[Asarin *et al.*, 1995] Asarin, A., O. Maler and A. Pnueli (1995). Reachability analysis of dynamical systems having piecewise-constant derivatives, *Theoretical Computer Science*, **138**, 35–66.

[Blondel and Tsitsiklis, 1997a] Blondel, V. D. and J. N. Tsitsiklis (1997). Complexity of elementary hybrid systems, *Proc. of the 4th European Control Conference*, Brussels.

[Blondel and Tsitsiklis, 1997b] Blondel, V. D. and J. N. Tsitsiklis (1997). When is a pair of matrices mortal?, *Information Processing Letters*, **63**, 283-286.

[Blondel and Tsitsiklis, 1997c] Blondel, V. D. and J. N. Tsitsiklis (1997). Survey of complexity results for systems and control problems, (in preparation).

[Blondel and Tsitsiklis, 1997d] Blondel, V. D. and J. N. Tsitsiklis (1997). Decidability limits for low-dimensional piecewise linear systems, (submitted).

[Davis, 1982] Davis, M. (1982). *Computability and Unsolvability*, New York, Dover.

[Garey and Johnson, 1979] Garey, M. R. and D. S. Johnson (1979). *Computers and Intractability : A Guide to the Theory of NP-completeness*, Freeman and Co., New York.

[Hopcroft and Ullman, 1969] Hopcroft, J. E. and J. D. Ullman (1969). *Formal languages and their relation to automata*, Addison-Wesley.

[Hyotyniemu, 1997] Hyotyniemu, H. (1997). On unsolvability of nonlinear system stability, Proc. ECC conference, to appear.

[Kilian and Siegelmann, 1996] Kilian, J. and H. Siegelmann (1996). The dynamic universality of sigmoidal neural networks, *Information and Computation*, **128**, 48-56.

[Koiran, 1996] Koiran, P. (1996). A family of universal recurrent networks, *Theor. Comp. Sciences*, **168**, 473-480.

[Koiran *et al.*, 1994] Koiran, P., M. Cosnard and M. Garzon (1994). Computability properties of low-dimensional dynamical systems, *Theoretical Computer Science*, **132**, 113-128.

[Liu and Michel, 1994] Liu, D. and A. Michel (1994). Dynamical systems with saturation nonlinearities: analysis and design, Springer-Verlag, London, 1994.

[Matiyasevich and Sénizergues, 1996] Matiyasevich, Y. and G. Sénizergues (1996). Decision problem for semi-Thue systems with a few rules, preprint.

[Papadimitriou, 1994] Papadimitriou, C. H. (1994). *Computational complexity*, Addison-Wesley, Reading.

[Siegelmann and Sontag, 1991] Siegelmann, H. T. and E. D. Sontag (1991). Turing computability with neural nets, *Applied Mathematics Letters*, **4** , 77-80.

[Siegelmann and Sontag, 1995] Siegelmann, H. and E. Sontag (1995). On the computational power of neural nets, *J. Comp. Syst. Sci.*, 132–150.

[Sontag, 1981] Sontag, E. (1981). Nonlinear regulation: the piecewise linear approach, *IEEE Trans. Automat. Control*, **26**, 346–358.

[Sontag, 1990] Sontag, E. (1990). *Mathematical control theory*, Springer, New York.

[Sontag, 1993] Sontag, E. (1993). *Neural networks for control* in Essays on Control: Perspectives in the Theory and its Applications (H.L. Trentelman and J.C. Willems, eds.), Birkhauser, Boston, pp. 339-380.

[Sontag, 1995] Sontag, E. (1995). From linear to nonlinear: some complexity comparisons, *Proc. IEEE Conference Decision and Control*, New Orleans, 2916–2920.

[Sontag, 1996] Sontag, E. (1996). Interconnected automata and linear systems: A theoretical framework in discrete-time, in *Hybrid Systems III: Verification and Control* (R. Alur, T. Henzinger, and E.D. Sontag, eds.), Springer, 436–448.

[Tsitsiklis, 1994] Tsitsiklis, J. N. (1994). Complexity theoretic aspects of problems in control theory, *Transactions of the eleventh Army*, ARO Report.

Computational Complexity
in Robust Controller Synthesis

Shinji HARA   Yuji YAMADA*

Abstract

This paper is concerned with the computational complexity analysis in robust control problems. We first formulate a fairly general class of nonconvex optimization problem named "Matrix Product Eigenvalue Problem (MPEP)" and explain the connection to robust control problems. We next summarize the worst case computational complexity results and investigate the computational cost for the actual case. Finally, we make a comparison with an element-wise bounding for the BMI optimization problem and a matrix-based bounding for the MPEP.

1 Introduction

Motivated by increasing computer power and sophisticated algorithms, analysis and design of control systems via numerical optimization has been widespread [1, 4, 8, 15, 20]. In particular, Linear Matrix Inequalities (LMIs) have gained much attention in recent years [1, 4, 15], since computational problems involving LMIs can be solved very efficiently by recently developed convex optimization techniques for semidefinite programming [1, 20].

It is known that a number of controller synthesis problems with a single objective, and of order equal to the plant, or of state feedback, can be formulated in terms of LMIs and that most of these LMIs in fact provide necessary and sufficient conditions for the corresponding problems. However, these problems are not recast as LMIs in general provided that a multi-objective is required and/or that the order of controller is fixed and/or that some of the states are unavailable, (see e.g., [9, 16]). For example, let us consider the H_∞ control problem with scaling W. The problems marked $\bigcirc$ in Table 1 are known to be convex, while we have not been able to recast the problems marked $\times$ as convex optimization problems. The latter problems may be viewed as inherently nonconvex ones which are extremely difficult to solve in general.

---

*Department of Computational Intelligence and Systems Science, Tokyo Institute of Technology, 4259 Nagatsuta-cho, Midori-ku, Yokohama 226, Japan.   E-mail: hara@cs.dis.titech.ac.jp, yuji@cs.dis.titech.ac.jp

Table 1.1: Scaled H_∞ control

Scaling	Analysis	SF	AnyOrd	FixOrd
$W = I$	O	O	O	×
W: const.	O	O	×	×
$W(s)$	×	×	×	×

(O: convex, ×nonconvex)

Even though they can not be recast as LMIs, many control problems can be characterized by feasibility problems of finding a matrix satisfying certain matrix inequalities. Among these problems, there are some common properties when we consider specific control synthesis problems. Iwasaki and Skelton [9] formulated the matrix inequality problem for static output feedback controller synthesis as the dual LMI problem, and proposed a coordinate decent type algorithm. El Ghaoui et al. [5] formulated the similar problem as a trace minimization problem of the product of two positive definite symmetric matrices under LMI constraints, and proposed a cone complementarity linearization algorithm. Mesbahi and Papavassilopoulos [12] proposed an algorithm to solve the rank minimization problem under LMI constraints corresponding to the reduced order output feedback controller synthesis. Although these problems are shown to cover a large class of control problems, the algorithms stated above are not guaranteed to converge to the global solution, nor a locally optimal solution in general.

On the other hand, Safonov et. al. [16] defined the framework of Bilinear Matrix Inequality (BMI) and showed that a wide array of control synthesis problems are expressed as BMIs. Goh et al. [7] formulated a general class of BMIs as the BMI eigenvalue problem and discussed a global optimization approach for the BMI problem. The BMI eigenvalue problem covers the largest class of problems among existing formulations for control synthesis based on matrix inequality conditions. However, the internal structure of each control problem has been lost by posing the problem in general setting, and hence the BMI problem is the most difficult among these type of nonconvex optimization problems for control synthesis. Actually, the NP-hardness of the BMI problem has been proved [18]. Although global optimization algorithms based on the branch and bound technique have been proposed [2, 7, 17], the computational complexity analysis is very hard due to its generality. Therefore, the effectiveness of the approach has not been justified.

One of the reasons why we have not been able to discuss the computational complexity of the algorithm seems that the BMI framework is too general. The authors defined a subclass of the BMI problem as the Matrix Product Eigenvalue Problem (MPEP), and proposed a global algorithm whose iteration number is bounded by the iteration upper bound [24]. Note that the MPEP is related to the dual LMI problems in [9], and it has been shown [21, 24] that many fixed order controller design problems including performance and robustness specifications, e.g., the constantly scaled $\mathcal{H}_\infty$ problem and the gain scheduled controller synthesis problem [14] via fixed order output feedback, can be formulated as the MPEP. In other words, the MPEP is also general to formulate control synthesis problem, and the problem defines a reasonable subclass of BMI problems for robust control synthesis.

The purpose of this paper is to summarize our previous computational complexity analysis for the MPEP and discuss the computational complexity in robust control synthesis. Section 2 formulates the MPEP and makes the connection to robust control problems clear. Section 3 is devoted to the computational complexity analysis. We summarize the worst case computational complexity results and investigate the computational cost for the actual case. We make a comparison with an element-wise bounding for the BMI optimization problem and a matrix-based bounding for the MPEP in Section 4.

2 The Matrix Product Eigenvalue Problem

2.1 Problem Description

In this section, we formulate the Matrix Product Eigenvalue Problem (MPEP) introduced in [21, 24]. Let us first define a set of block-diagonal matrices $\mathcal{D}$ as[1],

$$\mathcal{D} := \{ \text{block-diag}(\Sigma_1, \ldots, \Sigma_l, \sigma_{l+1}I_{\nu_{l+1}}, \ldots, \sigma_{l+f}I_{\nu_{l+f}}) \mid \sigma_j > 0, \ \Sigma_i \in \mathcal{S}_{\nu_i} \} \quad (2.1)$$

where $\mathcal{S}_\nu$ is a set of $\nu \times \nu$ positive definite symmetric matrices given by

$$\mathcal{S}_\nu := \{ \Sigma \mid \Sigma \in \Re^{\nu \times \nu}, \ \Sigma = \Sigma^T > 0 \} \quad (2.2)$$

Then the MPEP is formulated as follows:

$$\textbf{MPEP} \quad \left| \quad \begin{array}{ll} \text{minimize} & \lambda_{max}^{1/2}(\Sigma\Lambda) \\ \text{subject to} & (\Sigma, \ \Lambda) \in \mathcal{Z}_c \subset \mathcal{D}^2 \end{array} \right. \quad (2.3)$$

where $\mathcal{Z}_c$ is a closed bounded convex subset on $\mathcal{D}^2$ and $\lambda_{max}^{1/2}(\cdot)$ denotes the square root of the maximum eigenvalue, or the square root of the spectral radius [2]. Then the optimal value γ_{opt} for the MPEP is defined as

$$\gamma_{opt} := \inf_{(\Sigma, \ \Lambda) \in \mathcal{Z}_c} \lambda_{max}^{1/2}(\Sigma\Lambda) \quad (2.4)$$

---

[1]We omit to specify the sizes of $\mathcal{D}$, $(l, f, \{\nu_i\})$ for the brevity.

[2]We use the square root of the maximum eigenvalue instead of the maximum eigenvalue itself, since the square root is directly related to the control performance as shown in Section 2.2.

or equivalently

$$\gamma_{\text{opt}} := \inf \left\{ \gamma > 0 \mid (\Sigma, \Lambda) \in \mathcal{Z}_c \subset \mathcal{D}^2 , \ \lambda_{max}(\Sigma\Lambda) \leq \gamma^2 \right\} \tag{2.5}$$

We here assume that the convex set $\mathcal{Z}_c$ satisfies the following monotonicity property, which is not restrictive when we consider control synthesis problems [21, 24]:

Property 2.1 (Monotonicity property) *For any given* $(\hat{\Sigma}, \hat{\Lambda}) \in \mathcal{D}^2$, *if*

$$(\hat{\Sigma}, \hat{\Lambda}) \in \mathcal{Z}_c$$

holds, then $(\Sigma, \Lambda) \in \mathcal{Z}_c$ *holds for all* $(\Sigma, \Lambda) \in \mathcal{D}^2$ *satisfying*

$$\Sigma \geq \hat{\Sigma}, \quad \Lambda \geq \hat{\Lambda}$$

It is noted that the MPEP is considered as a generalization to the problem for the positive definite matrix case from the Linear Multiplicative Programming (LMP) problem [10], which has been explored in the field of numerical optimization on the Euclidean space. The problem is known to be a concave minimization problem, and it is formulated as follows:

$$\textbf{LMP} \quad \left| \begin{array}{ll} \text{minimize} & f_1(x)f_2(x) \\ \text{subject to} & x \in \mathcal{C} \subset \Re^n \\ & f_1(x) > 0, \quad f_2(x) > 0 \end{array} \right.$$

where $\mathcal{C}$ is a convex subset on $\Re^n$, and $f_i(x)$, $i = 1, 2$ are linear functions on $x \in \Re^n$. Clearly, the MPEP addresses the LMP for the case where the set $\mathcal{D}$ is given by the class of scalar parameter. Although the LMP is nonconvex, it is known that there exist some practical algorithms to find the global solution [10, 11]. Note that, for the scalar case, i.e., $\mathcal{D} = \{ \sigma > 0 \}$, the relaxation problem proposed in [25, 26] for the MPEP corresponds to the one in [11] for the LMP. Also note that it is not difficult to extend the results of the scalar case to the diagonal case, i.e., $\mathcal{D} = \{ \text{diag}(\sigma_1 I, \ldots, \sigma_f I) \mid \sigma_i > 0 \}$, because each subproblem in this case is still defined on the vector space. For the diagonal case, global algorithms for the MPEP with diagonal set have been proposed [3, 19, 25, 26] to solve partial problems, i.e., the constantly scaled $\mathcal{H}_\infty$ control problems. On the other hand, the extension of the results of the LMP to the MPEP seems difficult, because the the nonconvexity in the MPEP is concerned with the product of positive definite matrices, and is not defined on the vector space.

We next discuss the relationship between the MPEP and the BMI problem. The MPEP can be reformulated in terms of BMI condition, since the spectral radius condition $\lambda_{max}(\Sigma\Lambda) \leq 1$ with $(\Sigma, \Lambda) \in \mathcal{D}^2$ can be rewritten as a BMI condition

$$\begin{bmatrix} \Sigma & \Sigma\Lambda \\ \Lambda\Sigma & \Lambda \end{bmatrix} \geq 0 \tag{2.6}$$

This implies that the MPEP defines a subclass of the so-called BMI problem [16], and hence we can expect to get a more specific algorithm than those of BMI approaches. Especially, we can construct an algorithm based on the space defined by block-diagonal positive definite matrices [24], while the relaxation problems proposed for the BMI problem [2, 7, 17] are element-wise. The detail of this point will be discussed in Section 4.

Moreover, it is possible to show that the MPEP with monotonicity property (the MPEP for short hereafter) addresses a lot of typical BMI problems for control synthesis, e.g., fixed order output controller synthesis [9], robust stabilization and performance synthesis using D-scaling [15, 6, 23, 25], multi-objective control synthesis [16], and so on (see Section 2.2 for details). In this sense, the MPEP can be considered as a fairly general and reasonable subclass of BMI problems for robust control synthesis.

2.2 Control Synthesis via MPEP

It is known that there are a lot of control synthesis problems leading to nonconvex optimization problems and that many of these problems are characterized by LMI conditions and additional nonconvex constraints such as the inverse of a positive definite symmetric matrix or the product of Lyapunov matrix and state feedback (or estimation) gain [9, 16]. These types of nonconvex optimization problems are mostly reduced to MPEPs by a simple manipulation. To illustrate this, we here provide two examples of control synthesis that can be recast as MPEPs. We use the discrete-time synthesis problems to explain the idea, the continues-time problems can also be treated in a similar manner.

We first show that the simplest static output feedback stabilization can be reduced to the MPEP with monotonicity property.

Consider a discrete-time plant given by

$$x_{k+1} = A_d x_k + B_d u_k, \quad y_k = C_d x_k \tag{2.7}$$

where $x \in \Re^n$, $u \in \Re^q$, and $y \in \Re^r$. Suppose that $\mathcal{D} = \mathcal{S}_n$, i.e., $\mathcal{D}$ has no structure. As is well known, the system (2.7) is stabilizable via static output feedback

$$u_k = K_d y_k \tag{2.8}$$

if and only if there exists a matrix pair $\Sigma \in \mathcal{D}$ and $K_d \in \Re^{q \times r}$ satisfying

$$(A_d + B_d K_d C_d)\Sigma(A_d + B_d K_d C_d)' - \Sigma < 0 \tag{2.9}$$

or equivalently, by the Schur complement,

$$\Phi(\Sigma, \, \Lambda, \, K_d) := \begin{bmatrix} \Sigma & A_d + B_d K_d C_d \\ (A_d + B_d K_d C_d)' & \Lambda \end{bmatrix} > 0 \tag{2.10}$$

$$\Lambda = \Sigma^{-1} \in \mathcal{D} \tag{2.11}$$

The first condition (2.10) is an LMI on $(\Sigma, \Lambda, K_d) \in \mathcal{D}^2 \times \Re^{q \times r}$. Let us define $\mathcal{Z}_c$ by

$$\mathcal{Z}_c := \left\{ (\Sigma, \Lambda) \mid {}^{\exists} K_d \in \Re^{q \times r} \text{ s.t. } \Phi(\Sigma, \Lambda, K_d) > 0 \right\}$$

Then, it is readily verified that the set $\mathcal{Z}_c$ satisfies the monotonicity property. On the other hand, we can replace the condition (2.11) by the inequality $\Lambda \leq \Sigma^{-1}$ without loss of generality, because if (2.11) holds for some $\Sigma \in \mathcal{D}$ and $K_d \in \Re^{q \times r}$, then the inequality in (2.10) always holds for all $\Lambda \in \mathcal{D}$ satisfying $\Lambda \leq \Sigma^{-1}$. Moreover, the condition $\Lambda \leq \Sigma^{-1}$ is equivalent to $\lambda_{max}(\Sigma\Lambda) \leq 1$. Consequently, we see that the stabilizability via static output feedback can be checked by solving the MPEP with monotonicity property, i.e, the system (2.7) is stabilizable via static output feedback if and only if the minimum of the MPEP is less than or equal to 1.

Note that more complex control specifications (e.g. $\mathcal{H}_2$, $\mathcal{H}_\infty$, etc) and the fixed order dynamic output feedback case can be recast as MPEPs in a similar manner. For example, we can show that the scaled LFT optimization introduced in [15] is formulated as as the MPEP with monotonicity property. A general statement of the scaled LFT optimization problem is as follows: Minimize γ subject to the scaled LFT condition

$$\lambda_{\max}\left[Z^{-1/2}\left(M_1 + M_2 K M_3^T\right) Z^{1/2}\right] < \gamma, \quad Z \in \mathcal{D}, \ K \in \Re^{p_2 \times p_3} \qquad (2.12)$$

where $M_1 \in \Re^{p_1 \times p_1}$, $M_2 \in \Re^{p_1 \times p_2}$, $M_3 \in \Re^{p_1 \times p_3}$ are given matrices defined by plant data, and $Z \in \mathcal{D}$ is the input/output similarity scaling that includes the Lyapunov matrix and the D-scaling. Note that the sizes of $\mathcal{D}$, $(l, m, \{\nu_i\})$, are given when the structure of the uncertainty and the order of the controller are specified. The problem can be formulated as the following MPEP by replacing $\gamma Z \in \mathcal{D}$ and γZ^{-1} with $\Sigma \in \mathcal{D}$ and Λ respectively:

$$\left|\begin{array}{ll} \text{minimize} & \lambda_{max}^{1/2}(\Sigma\Lambda) \\ \text{subject to} & \Phi(\Sigma, \Lambda, K) > 0 \\ & (\Sigma, \Lambda, K) \in \mathcal{D}^2 \times \Re^{q \times r} \end{array}\right. \qquad (2.13)$$

where

$$\Phi(\Sigma, \Lambda, K) := \begin{bmatrix} \Sigma & M_1 + M_2 K M_3^T \\ (M_1 + M_2 K M_3^T)^T & \Lambda \end{bmatrix} > 0 \qquad (2.14)$$

It is readily verified that the inequality (2.12) is equivalent to (2.14) and that the set

$$\mathcal{Z}_c := \left\{ (\Sigma, \Lambda) \mid {}^{\exists} K \in \Re^{q \times r} \text{ s.t. } \Phi(\Sigma, \Lambda, K) > 0 \right\}$$

satisfies the monotonicity property.

The plant data matrices, M_1, M_2 and M_3, in (2.13) for the robust stabilization synthesis problem via fixed order output feedback are given as follows: Consider the discrete-time generalized plant expressed as

$$x_{k+1} = Ax_k + B_1 w_k + B_2 u_k, \quad z = C_1 x_k + D_{11} w_k + D_{12} u_k, \quad y = C_2 x_k + D_{21} w_k \qquad (2.15)$$

where $x_k \in \Re^n$, $w_k \in \Re^p$, $z_k \in \Re^p$, $u_k \in \Re^q$, and $y_k \in \Re^r$. Let $\nu_1 = n$ and $\nu_2 + \cdots + \nu_{q+r} = p$, and define the following set of block-diagonally structured uncertainties:

$$\boldsymbol{\Delta}_s := \{ \Delta = \text{block-diag}(\delta_2 I_{\nu_1}, \ldots, \delta_l I_{\nu_l}, \Delta_1, \ldots, \Delta_m)$$
$$| \, \delta_i \in \mathbf{C}, \, \Delta \in \mathbf{C}^{\nu_i \times \nu_i}, \, \|\Delta\| \leq 1 \}$$

We also define the state space realization of a discrete time controller $C(z)$ of order n_c as follows:

$$C(z) := \left[\begin{array}{c|c} A_c & B_c \\ \hline C_c & D_c \end{array} \right]$$

Then, the system (2.15) with $w = \Delta z$ is robustly stabilizable via output feedback controller $u = C(z)y$ of order n_c against norm bounded structured uncertainty $\Delta \in \boldsymbol{\Delta}_s$ if there exist a positive definite symmetric matrix $\Sigma \in \mathcal{D}$ and a real matrix $K \in \Re^{(n_c+q) \times (n_c+r)}$ satisfying condition (2.14), where M_1, M_2 and M_3 are defined as

$$\left[\begin{array}{ccc} M_1 & M_2 & M_3 \end{array}\right] := \left[\begin{array}{ccc|ccc|ccc} A & 0 & B_1 & 0 & B_2 & 0 & C_2^T \\ 0 & 0 & 0 & I_{n_c} & 0 & I_{n_c} & 0 \\ C_1 & 0 & D_{11} & 0 & D_{12} & 0 & D_{21} \end{array} \right]$$

and K contains the controller parameters as follows:

$$K := \left[\begin{array}{cc} A_c & B_c \\ C_c & D_c \end{array} \right]$$

3   Computational Complexity Analysis

3.1 ϵ-global optimization

This section is concerned with the computational complexity of the MPEP with respect to the relative tolerance of the achievable performance. To this end, we introduce a notion of the ϵ-global optimization corresponding to finding a suboptimal solution within a specified small tolerance ϵ. Let $\epsilon \in (0, 1)$ denote a relative tolerance for the optimal value γ_{opt}. Then, the ϵ-global optimization is to find a sub-optimal value γ^* such that

$$(1 - \epsilon)\gamma^* \leq \gamma_{\text{opt}} \leq \gamma^* \tag{3.1}$$

where γ_{opt} is the exact optimal value defined in (2.4).

Our main concern here is to investigate the computational complexity with respect to the inverse of the relative tolerance, $1/\epsilon$, from the following reasons:

- In control synthesis problems, ϵ corresponds to the accuracy of the solution, and hence, we have to choose sufficiently small ϵ to obtain the the better accuracy of the global solution.

- The computational cost increases as ϵ gets smaller, or equivalently, ϵ inverse gets larger.

Hence, we can not say that an algorithm is efficient if the algorithm does not have a reasonable computational complexity with respect to $1/\epsilon$. In other words, the worst case iteration number required to get an ϵ-global solution should be at most a polynomial with respect to the inverse of the tolerance, $1/\epsilon$. We will provide two types of ϵ-global algorithms satisfying the above computational complex requirement.

3.2 Point-wise method

Before showing one of the algorithms, we define a level set for the MPEP. If we respectively replace Σ and Λ by $\gamma\Sigma$ and $\gamma\Lambda$ in the MPEP, then the optimal value γ_{opt} defined in (2.5) is given by

$$\gamma_{\mathrm{opt}} = \inf \left\{ \gamma > 0 \mid (\gamma\Sigma, \ \gamma\Lambda) \in \mathcal{Z}_c \subset \mathcal{D}^2 \ , \ \lambda_{max}(\Sigma\Lambda) \leq 1 \right\} \qquad (3.2)$$

and the level set on $\Sigma \in \mathcal{D}$ can be defined as

$$\mathcal{F}(\gamma) := \left\{ \Sigma \in \mathcal{D} \mid {}^{\exists}\Lambda \in \mathcal{D} \ \text{s.t.} \ (\gamma\Sigma, \ \gamma\Lambda) \in \mathcal{Z}_c \ \text{and} \ \lambda_{max}(\Sigma\Lambda) \leq 1 \right\} \quad (3.3)$$

Note that the set can be rewritten as

$$\mathcal{F}(\gamma) = \left\{ \Sigma \in \mathcal{D} \mid (\gamma\Sigma, \ \gamma\Sigma^{-1}) \in \mathcal{Z}_c \right\} \qquad (3.4)$$

since we can replace Λ by Σ^{-1} without loss of generality under the monotonicity property. We also define a function $\gamma_{\mathrm{s}} \colon \mathcal{D} \to \Re$ as

$$\gamma_{\mathrm{s}}(\Sigma) := \inf \left\{ \gamma \mid \Sigma \in \mathcal{F}(\gamma) \right\} \qquad (3.5)$$

Notice that, for any fixed $\Sigma \in \mathcal{D}$, $\gamma_{\mathrm{s}}(\Sigma)$ can be obtained by solving a convex optimization problem with $\Sigma \in \mathcal{D}$ fixed. Moreover, it is readily verified that

$$\gamma_{\mathrm{opt}} = \inf_{\Sigma \in \mathcal{D}} \gamma_{\mathrm{s}}(\Sigma)$$

holds.

The following lemma plays an important role to derive a point-wise global algorithm and analyze its computational complexity.

Lemma 3.1 *Suppose that $\hat{\Sigma} \in \mathcal{D}$ and $\kappa \in (0, \ 1)$ are arbitrarily given. Then we have*

$$\kappa\gamma_{\mathrm{s}}(\hat{\Sigma}) \leq \gamma_{\mathrm{s}}(\Sigma) \leq \kappa^{-1}\gamma_{\mathrm{s}}(\hat{\Sigma}), \ \ {}^{\forall}\Sigma \in \left\{ \Sigma \in \mathcal{D} \ \middle| \ \kappa\hat{\Sigma} \leq \Sigma \leq \kappa^{-1}\hat{\Sigma} \right\} \qquad (3.6)$$

Proof: See [21] for the proof. ∎

Let us define the following convex optimization problem with $\Sigma \in \mathcal{D}$ fixed:

$$\mathbf{P}_{pw}(\Sigma) \quad \left| \quad \begin{array}{ll} \text{minimize} & \gamma \\ \text{subject to} & (\gamma\Sigma, \; \gamma\Sigma^{-1}) \in \mathcal{Z}_c \end{array} \right. \tag{3.7}$$

Then, the following theorem shows the existence of a point-wise ϵ-global algorithm.

Theorem 3.1 *There exists a point-wise algorithm satisfying the following two properties for any given $\epsilon \in (0, 1)$:*

P1) *Solving convex optimization problems $\mathbf{P}_{pw}(\Sigma)$ for different values of fixed $\Sigma \in \mathcal{D}$ finite time (at most $\mathbf{N}_{pw}$ times) yields an ϵ-global solution γ^* satisfying (3.1)*

P2) *The order of Npw with respect to $1/\epsilon$ is given by*

$$\mathbf{N}_{pw} = O\left((1/\epsilon)^\xi\right) \tag{3.8}$$

where ξ is the number of scaler parameters in $\Sigma \in \mathcal{D}$, i.e.,

$$\xi = m + \sum_i^l \frac{\nu_i(\nu_i + 1)}{2} \tag{3.9}$$

We here show the proof and provide such an algorithm only for the simplest case, i.e., $\mathcal{D} = \{\, \sigma > 0 \,\}$. Those for the general case are found in [21].

Assume that $\mathcal{F}(\gamma)$ is bounded to satisfy

$$\mathcal{F}(\gamma) \subset \{\, \sigma \in [\underline{\sigma}, \; \bar{\sigma}] \mid 0 < \underline{\sigma} < \bar{\sigma} < \infty \,\}$$

for some $\gamma = \hat{\gamma}$ such that $\mathcal{F}(\hat{\gamma}) \neq \emptyset$, and let

$$\eta := \sqrt{\bar{\sigma}/\underline{\sigma}} \tag{3.10}$$

Then, we have the following theorem:

Theorem 3.2 *Algorithm 1 below provides an ϵ-global solution γ^* within given relative tolerance $\epsilon \in (0, 1)$ satisfying (3.1), and the iteration number required denoted by $\mathbf{N}$ satisfies*

$$\mathbf{N}_{min} \leq \mathbf{N} \leq \mathbf{N}_{max} \tag{3.11}$$

where

$$N_{min} \; := \; 2 \times \min \left\{ n: \text{natural number} \; \middle| \; \sum_{j=0}^{n-1} 3^j \geq \frac{\ln \eta}{2\ln\left(\frac{1}{1-\epsilon}\right)} \right\} - 1$$

$$N_{max} \; := \; 2 \times \min \left\{ n: \text{natural number} \; \middle| \; n \geq \frac{\ln \eta}{2\ln\left(\frac{1}{1-\epsilon}\right)} \right\} - 1$$

Algorithm 1

Initialize: Let $\sigma^{(0)} := \sqrt{\underline{\sigma} \cdot \bar{\sigma}}$. Set itr$\leftarrow 1$ and compute $\gamma_s(\sigma^{(0)})$.

$\gamma_{tmp} \leftarrow \gamma_s(\sigma^{(0)})$, $\quad \delta_0 := 1$, $\quad i \leftarrow 0$.

While $\sigma^{(i)} < \bar{\sigma}$ or $\sigma^{(-i)} > \underline{\sigma}$ do:

If $\sigma^{(i)} < \bar{\sigma}$: Let $\sigma^{(i+1)} := \delta_i^2(1-\epsilon)^{-2}\sigma^{(i)}$. Set itr$\leftarrow$itr$+1$ and compute $\gamma_s(\sigma^{(i+1)})$.

$\gamma_{tmp} \leftarrow \min\left(\gamma_s(\sigma^{(i+1)}), \; \gamma_{tmp}\right)$, $\quad \delta_{i+1} := \gamma_s(\sigma^{(i+1)})/\gamma_{tmp}$.

If $\sigma^{(-i)} > \underline{\sigma}$: Let $\sigma^{(-i-1)} := \delta_{-i}^{-2}(1-\epsilon)^2\sigma^{(-i)}$. Set itr$\leftarrow$itr$+1$ and compute $\gamma_s(\sigma^{(-i-1)})$.

$\gamma_{tmp} \leftarrow \min\left(\gamma_s(\sigma^{(-i-1)}), \; \gamma_{tmp}\right)$, $\quad \delta_{-i-1} := {}_{-}gss-i - 1/\gamma_{tmp}$.

Set $i \leftarrow i+1$.

end

Solution $\gamma^* \leftarrow \gamma_{tmp}$ and $N \leftarrow$ itr.

Proof: We will first show $N \leq N_{max}$. Let $k_u > 0$ and $k_l > 0$ satisfy $\sigma^{(k_u)} \geq \bar{\sigma}$ and $\sigma^{(-k_l)} \leq \underline{\sigma}$ in Algorithm 1, respectively, i.e., N is given by $N = k_u + k_l - 1$. From the definitions of $\sigma^{(i)}$ ($i \in [-k_l, \; k_u]$), the following conditions hold:

$$\sigma^{(k_u)} = \delta_{k_u-1}^2 \delta_{k_u-2}^2 \cdots \delta_2^2 \delta_1^2 (1-\epsilon)^{-2k_u} \sigma^{(0)} \tag{3.12}$$

$$\sigma^{(-k_l)} = \delta_{-(k_l-1)}^{-2} \delta_{-(k_l-2)}^{-2} \cdots \delta_{-2}^{-2} \delta_{-1}^{-2} (1-\epsilon)^{2k_l} \sigma^{(0)} \tag{3.13}$$

Since $\delta_i \geq 1$ holds for all $i \in [-k_l, \; k_u]$, $\sigma^{(k_u)} \geq \bar{\sigma}$ and $\sigma^{(-k_l)} \leq \underline{\sigma}$ always hold if

$$(1-\epsilon)^{-2k_u}\sigma^{(0)} \geq \bar{\sigma}, \quad (1-\epsilon)^{2k_l}\sigma^{(0)} \leq \underline{\sigma},$$

or equivalently,

$$k_u, \; k_l \geq \frac{\ln \eta}{2\ln\left(\frac{1}{1-\epsilon}\right)}$$

Therefore we conclude that $N \le N_{max}$ holds, where the equality holds when $\delta_i = 1$, $\forall i \in [-k_l, k_u]$.

We will next show $N_{min} \le N$. Let $k_u > 0$ and $k_l > 0$ satisfy $\sigma^{(k_u)} \ge \bar{\sigma}$ and $\sigma^{(-k_l)} \le \underline{\sigma}$ in Algorithm 1, respectively, i.e., N is given by $N = k_u + k_l - 1$. We first show that $\sigma^{(k_u)} \ge \bar{\sigma}$ and $\sigma^{(-k_l)} \le \underline{\sigma}$ hold only if k_u and k_l satisfy

$$\sum_{j=0}^{k_u-1} 3^j \ge \frac{\ln \eta}{2 \ln \left(\frac{1}{1-\epsilon} \right)} \tag{3.14}$$

$$\sum_{j=0}^{k_l-1} 3^j \ge \frac{\ln \eta}{2 \ln \left(\frac{1}{1-\epsilon} \right)} \tag{3.15}$$

respectively. Then we conclude that $N \ge N_{min}$ holds.

From Lemma 3.1, the following condition holds:

$$\begin{aligned} \gamma_s(\sigma^{(i)}) &= \gamma_s \left(\delta_{i-1}^2 (1-\epsilon)^{-2} \sigma^{(i-1)} \right) \\ &\le \delta_{i-1}^2 (1-\epsilon)^{-2} \gamma_s(\sigma^{(i-1)}), \quad i \in [1, k_u] \end{aligned} \tag{3.16}$$

Noting that $\gamma^* \le \gamma_s(\sigma^{(i)})$ holds for all $i \in [-k_l, k_u]$, the inequality in (3.16) implies

$$\gamma_s(\sigma^{(i)}) \le \left(\frac{\gamma_s(\sigma^{(i-1)})}{\gamma^*} \right)^2 (1-\epsilon)^{-2} \gamma_s(\sigma^{(i-1)}), \quad i \in [1, k_u] \tag{3.17}$$

From (3.17), we obtain

$$\gamma_s(\sigma^{(i)}) \le (1-\epsilon)^{-3^i+1} \gamma^*, \quad i \in [1, k_u]$$

Furthermore, since $\delta_i \le \gamma_s(\sigma^{(i)})/\gamma^*$ holds, we have

$$\delta_i \le (1-\epsilon)^{-3^i+1}, \quad i \in [1, k_u] \tag{3.18}$$

We can see from (3.12) that the following condition holds:

$$\bar{\sigma} \le \sigma^{(k_u)} \le (1-\epsilon)^{-2\left(1+3+3^2+\cdots+3^{k_u-1}\right)} \sigma^{(0)} \tag{3.19}$$

The Condition (3.19) implies that k_u satisfies (3.14) if $\bar{\sigma} \le \sigma^{(k_u)}$ holds. Similarly, we can obtain the other condition concerning k_l, i.e., k_l satisfies (3.15) if $\bar{\sigma} \le \sigma^{(k_u)}$ holds. Clearly, these two conditions imply that $N \ge N_{min}$ holds, where the equality holds when $\gamma^* = \gamma_s(\sigma^{(0)})$, $\delta_i = (1-\epsilon)^{-3^i+1}$ ($\forall i \in [1, k_u]$), and $\delta_{-i} = (1-\epsilon)^{-3^i+1}$ ($\forall i \in [1, k_l]$). ∎

Remark 3.1 *A similar algorithm has been proposed in our previous work of [27] for the $\mathcal{H}_\infty$ control problem with constant diagonal scaling, showing that its worst case iteration number is given by $\mathbf{N}_{max}$. Note that $\mathbf{N}_{max}$ can be achieved when $\delta_i = 1$ at each step in Algorithm 1. What is new in this theorem is that we have an iteration lower bound of the algorithm, i.e., $\mathbf{N}_{min}$.*

Remark 3.2 *The orders of $\mathbf{N}_{min}$ and $\mathbf{N}_{max}$ with respect to the inverse of the tolerance, $1/\epsilon$, are given by*

$$\mathbf{N}_{min} = O\left(\ln\left(1/\epsilon\right)\right), \quad \mathbf{N}_{max} = O\left(1/\epsilon\right)$$

The order of $\mathbf{N}_{max}$ is coincident with that in Theorem 3.1, i.e., the proof for $\mathbf{N} \leq \mathbf{N}_{max}$ is a proof of Theorem 3.1 for the simplest case. Those orders also imply that the average computational complexity of Algorithm 1 lies between the log and the linear orders with respect to $1/\epsilon$.

3.3   LMI-based method

This section affords an ϵ-global algorithm whose computational complexity is less than that of the point-wise algorithm in the previous subsection. Recall that the worst case order of the algorithm in [25] named "triangle covering method" is half of that for the rectangle method (a point-wise method) for solving H_∞ control problems with constant diagonal scaling [25]. We will try to generalize the triangular area in [25] to adopt the MPEP.

A convex set on $\mathcal{D}$ corresponding to the triangular area in [25] can be defined as follows: For given $\Theta \in \mathcal{D}$ and $\gamma > 0$, let us define the following set of $(\Sigma, \Lambda) \in \mathcal{D}^2$:

$$\mathcal{Q}_c(\Theta, \gamma) := \left\{ (\Sigma, \Lambda) \in \mathcal{D}^2 \ \middle| \ \Sigma + \Theta\Lambda\Theta \leq 2\gamma\Theta \right\} \qquad (3.20)$$

The set $\mathcal{Q}_c(\Theta, \gamma)$ is convex and corresponds to the set $\mathcal{C}(\theta, \gamma)$ in the triangle covering method in [25].

Let $\mathcal{U}(\Theta, \gamma)$ be the intersection between the sets $\mathcal{Z}_c$ and $\mathcal{Q}_c(\Theta, \gamma)$, i.e.,

$$\mathcal{U}(\Theta, \gamma) := \mathcal{Z}_c \cap \mathcal{Q}_c(\Theta, \gamma) \qquad (3.21)$$

Then, we have the following lemma with respect to the set $\mathcal{U}$ by observing an equivalent relation

$$\lambda_{max}(\Sigma\Lambda) \leq \gamma^2 \ \Leftrightarrow \ \exists \Theta \in \mathcal{D} \ \text{s.t.} \ \Theta\Sigma\Theta + \Lambda \leq 2\gamma\Theta \qquad (3.22)$$

Lemma 3.2 *For any given γ, the level set $\mathcal{F}(\gamma)$ defined by (3.3) is non-empty, i.e.,*

$$\mathcal{F}(\gamma) \neq \emptyset$$

holds if and only if there exists $\Theta \in \mathcal{D}$ satisfying

$$\mathcal{U}(\Theta, \gamma) = \left\{ (\Sigma, \Lambda) \in \mathcal{Z}_c \mid \Sigma + \Theta\Lambda\Theta \leq 2\gamma\Theta \right\} \neq \emptyset \qquad (3.23)$$

Note that we can decide whether the set $\mathcal{U}(\Theta, \gamma)$ is empty or not for given $\Theta \in \mathcal{D}$ by solving a convex optimization problem, since all the constraints are convex. It is possible from Lemma 3.2 to conclude that γ is feasible, i.e., $\mathcal{F}(\gamma) \neq \emptyset$, if $\mathcal{U}(\Theta, \gamma) \neq \emptyset$ for given $\Theta \in \mathcal{D}$, similarly to the triangle covering method in [25]. Otherwise, i.e., $\mathcal{U}(\Theta, \gamma) = \emptyset$, we can determine an infeasibility of the MPEP on the space defined by two positive definite matrices for a level $\gamma_\kappa \in (0, \gamma)$, as shown in the following lemma:

Lemma 3.3 $\mathcal{U}(\Theta, \gamma) = \emptyset$ *holds for given* $\Theta \in \mathcal{D}$ *and* $\gamma > 0$ *if and only if*

$$\{ X \in \mathcal{D} \mid X_l \leq X \leq X_u \} \cap \mathcal{F}(\gamma_\kappa) = \emptyset, \quad {}^\forall \kappa \in (0, 1) \qquad (3.24)$$

where

$$\gamma_\kappa := \kappa\gamma$$

and $X_l \in \mathcal{D}$ *and* $X_u \in \mathcal{D}$ *are defined as*

$$X_l := \left(\frac{1 - \sqrt{1 - \kappa^2}}{\kappa} \right) \Theta, \quad X_u := \left(\frac{1 + \sqrt{1 - \kappa^2}}{\kappa} \right) \Theta$$

Let us define the following convex optimization problem with $\Theta \in \mathcal{D}$ fixed:

$$\mathbf{P}_{lmi}(\Theta) \quad \left| \quad \begin{array}{ll} \text{minimize} & \gamma \\ \text{subject to} & (\Sigma, \Lambda) \in \mathcal{Z}_c \\ & \Sigma + \Theta\Lambda\Theta \leq 2\gamma\Theta \end{array} \right. \qquad (3.25)$$

The following theorem gives a computational complexity result based on $\mathbf{P}_{lmi}(\Theta)$, which is generalized from that of the triangle covering method in [25] for the diagonal case:

Theorem 3.3 *There exists an LMI-based algorithm satisfying the following two properties for any given* $\epsilon \in (0, 1)$:

P1) *Solving convex optimization problems* $\mathbf{P}_{lmi}(\Theta)$ *for different values of fixed* $\Theta \in \mathcal{D}$ *finite time (at most* $\mathbf{N}_{lmi}$ *times) yields an* ϵ-*global solution* γ^* *satisfying (3.1).*

P2) *The order of* $\mathbf{N}_{lmi}$ *with respect to* $1/\epsilon$ *is given by*

$$\mathbf{N}_{lmi} = O\left((1/\epsilon)^{\xi/2} \right)$$

where ξ *is the number of scaler parameters in* $\Sigma \in \mathcal{D}$ *defined by (3.9).*

Proof: See [21] for the proof and a corresponding LMI-based ϵ-global algorithm.   ∎

The above theorem shows that the iteration number of an LMI-based algorithm grows no faster than a polynomial of the inverse of the tolerance with fixed number of scaler parameters in the block-diagonal matrices and that its order is twice as small as a point-wise algorithm in the previous subsection.

3.4   A Numerical Example

Consider a mechanical system consisting of two masses connected by a spring [15]. A state space realization of the system is given by the following descriptor form:

$$
\begin{bmatrix} 1 & 0 & 0 & 0 \\ 0 & 1 & 0 & 0 \\ 0 & 0 & m_1 & 0 \\ 0 & 0 & 0 & m_2 \end{bmatrix} \begin{bmatrix} \dot{x}_1 \\ \ddot{x}_1 \\ \dot{x}_2 \\ \ddot{x}_2 \end{bmatrix} = \begin{bmatrix} 0 & 0 & 1 & 0 \\ 0 & 0 & 0 & 1 \\ -k & k & 0 & 0 \\ k & -k & 0 & 0 \end{bmatrix} \begin{bmatrix} x_1 \\ \dot{x}_1 \\ x_2 \\ \dot{x}_2 \end{bmatrix} + \begin{bmatrix} 0 \\ 0 \\ 1 \\ 0 \end{bmatrix} (u + \beta_d d)
$$

$$
y = x_2 + \beta_v v
$$

where m_1 and m_2 are masses and k is the spring constant, and the process and measurement noises are denoted by d and v with the noise to signal ratios β_d and β_v.

We here suppose that two masses m_1 and m_2 take the same value m, i.e., $m_1 = m_2 = m$. The control objective is to regulate the output of the plant, without excessive actuator power, in the presence of process and measurement noises, against uncertainty in the parameter m. Hence, we respectively set the disturbance input and the controlled output as

$$
r = \begin{bmatrix} d & v \end{bmatrix}^T, \quad e = \begin{bmatrix} \beta_x x_2 & \beta_u u \end{bmatrix}^T
$$

with appropriate weights β_x and β_u. We also suppose that there exists an uncertainty in m, which is given by

$$
m = m_0 + 0.2 \cdot \delta_m, \quad |\delta_m| \le 1
$$

We will solve the problem as the constantly scaled $\mathcal{H}_\infty$ problem, where the generalized plant $G(s)$ is expressed as

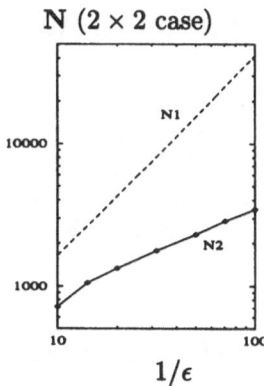

Figure 3.1: The iteration numbers

$$
G(s) =
\left[
\begin{array}{cccc|cccc|c}
0 & 0 & 1 & 0 & 0 & 0 & 0 & 0 & 0 \\
0 & 0 & 0 & 1 & 0 & 0 & 0 & 0 & 0 \\
-k/m_0 & k/m_0 & 0 & 0 & -0.2/m_0 & 0 & \beta_d/m_0 & 0 & 1/m_0 \\
k/m_0 & -k/m_0 & 0 & 0 & 0 & 0.2/m_0 & 0 & 0 & 0 \\
-k/m_0 & k/m_0 & 0 & 0 & -0.2/m_0 & 0 & \beta_d/m_0 & 0 & 1/m_0 \\
k/m_0 & -k/m_0 & 0 & 0 & 0 & 0.2/m_0 & 0 & 0 & 0 \\
0 & \beta_x & 0 & 0 & 0 & 0 & 0 & 0 & 0 \\
0 & 0 & 0 & 0 & 0 & 0 & 0 & 0 & \beta_u \\
\hline
0 & 1 & 0 & 0 & 0 & 0 & 0 & \beta_v & 0
\end{array}
\right]
$$

and the set of scaling matrices is given by

$$
\mathcal{D} = \left\{ \text{block diag}(\Sigma,\ I_2) \ \middle|\ \Sigma = \begin{bmatrix} \sigma_{11} & \sigma_{12} \\ \sigma_{12} & \sigma_{22} \end{bmatrix} \in \mathcal{S}_2 \right\}
$$

We have chosen the following nominal values for the plant and the disturbance model:

$$m_0 = 1, \quad k = 5,$$
$$\beta_d = 1, \quad \beta_v = 0.1, \quad \beta_x = 0.082, \quad \beta_u = 0.408$$

Fig. 3.1 illustrates the the number of iterations required. $N1$ and $N2$ respectively denote the theoretical (worst case) and numerical results. Since the set $\mathcal{D}$ consists of 3 scaler parameters, the order of $N1$ is given by $N1 = O\left((1/\epsilon)^{\frac{3}{2}}\right)$. We see from Fig. 3.1 that there is a large difference between the theoretical worst case result and the numerical result and that the slope of $N2$ is half of $N1$, i.e., the increase in $N2$ with $1/\epsilon$ is twice as slow when compared to $N1$ on a log scale. Such a situation happens most of the practical cases, since the complexity of the actual case is expected to be much less than that for the worst case as shown in Theorem 3.2.

4   Matrix-Based vs Element-Wise Bounding

In this section, we compare our approach explained in previous sections with
the BMI approaches [2, 7, 17]. We will show that an essential difference of the
two approaches lies in their relaxation problems in the branch and bounding
methods. In the sequel, we will prove that the matrix-based bounding in our
approach has some advantages over the element-wise bounding in the BMI
approach for solving the MPEP.

4.1   An Equivalent BMI Formulation

Let us define the following optimization problem of a linear objective function
$\gamma > 0$ under a BMI constraint and quasi-convex constraints:

$$\textbf{BMI}_{op} \quad \left| \quad \begin{array}{ll} \text{minimize} & \gamma \\ \text{subject to} & (\gamma X,\ \gamma Y) \in \mathcal{Z}_c \subset \mathcal{D}^2 \\ & \begin{bmatrix} X & XY \\ YX & Y \end{bmatrix} \geq 0 \end{array} \right. \qquad (4.1)$$

With the assumption of the monotonicity property on the set $\mathcal{Z}_c$, we can
readily see from (2.6) and (3.2) that the $\textbf{BMI}_{op}$ is equivalent to the MPEP in
the sense that

$$\inf\left\{\ \gamma \mid (4.1) \text{ holds }\right\} = \inf\left\{\ \lambda_{max}^{1/2}\left(\Sigma\Lambda\right) \mid (\Sigma,\ \Lambda) \in \mathcal{Z}_c \subset \mathcal{D}^2\ \right\} = \gamma_{\text{opt}}$$
$$(4.2)$$

holds [21, 24]. Therefore, seeking a solution on $(\Sigma,\ \Lambda) \in \mathcal{D}^2$ for the MPEP
corresponds to seeking a solution on $(X,\ Y) \in \mathcal{D}^2$ for $\textbf{BMI}_{op}$.

As both the BMI problem and the MPEP are nonconvex optimization problems which may have plural local minima, we have to search over a solution
on all the entire domain directly or secondhand in general; this is the difference between convex and nonconvex optimization problems, which makes the
nonconvex optimization problem extremely difficult. A general and perhaps
an efficient way to do this process is the so-called branch and bound method
(see e.g., [13]). Roughly speaking, the branch and bound method consists of
the following two operations [13]:

Branching: Subdivide the bounded domain to be searched into a finite number of smaller subspaces, e.g. hyper rectangles.

Bounding: For each subspace, find upper and lower bounds on the objective
function by using upper bound and lower bound functions. If the lower
bound on certain subspace is larger than the current optimal value, we
can prune the subspace from the domain.

If we intend to apply the branch and bound process to a nonconvex optimization problem, we have to construct relaxation problems which give upper and

lower bounds respectively on a given restricted subspace, and these relaxation problems should be defined by convex problems which approximates the nonconvex problem on the subspace. The performance of the global algorithm based on the branch and bound technique clearly depends on the relaxation problem. In this section, we discuss relaxation problems which enable us to apply a branch and bound process proposed for the BMI eigenvalue problem and the MPEP, which are respectively characterized by the "element-wise bounding" and the "matrix-based bounding."

4.2 Element-Wise Bounding

To state relaxation problems for the BMI eigenvalue problem, let us define a biaffine function $\Phi \colon \Re^{n_x} \times \Re^{n_y} \to \Re^{m \times m}$ for given symmetric matrices $\Phi_{ij} = \Phi_{ij}^T \in \Re^{m \times m}$:

$$\Phi(x, y) := \Phi_{00} + \sum_{i=1}^{n_x} x_i \Phi_{i0} + \sum_{j=1}^{n_y} y_j \Phi_{0j} + \sum_{i=1}^{n_x} \sum_{j=1}^{n_y} x_i y_j \Phi_{ij} \qquad (4.3)$$

Then the BMI eigenvalue problem is formulated as follows [7]: Given Φ_{ij}, minimize $\lambda_{max}^{1/2} (\Phi(x, y))$ over $(x, y) \in \mathcal{Q}_D \subset \Re^{n_x} \times \Re^{n_y}$, where $\mathcal{Q}_D$ is a given closed bounded hyper rectangle. For the BMI eigenvalue problem, Goh et al. [7] proposed a relaxation problem by "element-wisely" replacing biaffine terms $x_i y_j$ by w_{ij} with additional bound constraints on x_i, y_j and w_{ij}. They showed that the relaxation problem provides an lower bound of the objective function in a given subspace on the domain. The relaxation problem has been improved by Fujioka and Hoshijima [2] and Takano et al. [17]. They proposed methods to construct the convex hull on the nonconvex constraint $w_{ij} = x_i y_j$ in a given hyper rectangle to obtain a tighter lower bound.

The common property between these relaxation problems is that these methods partially approximate the nonconvex constraint $w_{ij} = x_i y_j$ in terms of a convex region including the nonconvex area on a given hyper rectangle and that the relaxation problems are based on the element-wise bounding. However, the precision of the approximation becomes worse as the number of nonconvex constraints gets increase. Moreover, we have to join two [7] or four [2, 17] additional linear constraints per one bilinear term to solve the relaxation problem using the element-wise bounding. This fact makes the relaxation problem based on the element-wise bounding conservative. If we formulate $\mathbf{BMI}_{op}$ in terms of a biaffine function, the number of nonconvex constraints increases rapidly compared with the number of parameters. For instance, consider the simplest matrix case, i.e., XY, $(X, Y) \in \mathcal{S}_2^2$. The condition includes seven biaffine terms, since XY is written as

$$XY = \begin{bmatrix} x_{11} & x_{12} \\ x_{12} & x_{22} \end{bmatrix} \begin{bmatrix} y_{11} & y_{12} \\ y_{12} & y_{22} \end{bmatrix} = \begin{bmatrix} x_{11}y_{11} + x_{12}y_{12} & x_{11}y_{12} + x_{12}y_{22} \\ x_{12}y_{11} + x_{22}y_{12} & x_{12}y_{12} + x_{22}y_2 \end{bmatrix}$$
$$(4.4)$$

This implies that we have to search over 7 nonconvex areas in spite of the simplest case. In general, the bilinear terms appeared in XY, $(X, Y) \in \mathcal{S}_n^2$ is $n^3 - n(n-1)/2$, which is fairly large even if n is small. This is one of the drawbacks of the element-wise bounding.

4.3   Matrix-Based Bounding

On the other hand, the relaxation problem for the MPEP is characterized by the matrix-based bounding, i.e., the parameter bounds of the subspace are determined by block-diagonal positive definite matrices based on Lemma 3.3.

Recall that Lemma 3.3 implies that γ_κ gives a lower bound of the objective function in the interval between X_l and X_u which are defined by $\kappa \in (0,\ 1)$ and $\Theta \in \mathcal{D}$. In other words, we can find a lower bound on the objective function for a given interval defined by positive definite matrices X_l and X_u by solving a convex subproblem; minimize γ subject to $\mathcal{U}(\Theta,\ \gamma) \neq \emptyset$ with $\Theta \in \mathcal{D}$ fixed. Also, we see from Lemma 3.2 that the optimal value for this convex subproblem also provides an upper bounds on the objective function for the MPEP. It should be noted that the convex subproblem is based on the area defined by block-diagonal positive definite matrices, i.e., the relaxation problems is characterized by the matrix-based bounding. If the space defined by κ and $\Theta \in \mathcal{D}$, i.e., the space defined by $X_l \in \mathcal{D}$ and $X_u \in \mathcal{D}$, covers the entire domain, then γ_κ guarantees a lower bound of the optimal objective function on the entire domain. Otherwise, we have to subdivide the domain into smaller subspace to obtain a better lower bound. A subdivision rule is proposed in [21, 24, 25] which shows that we need to solve at most $O\left((1/\epsilon)^{\frac{\xi}{2}}\right)$ convex subproblems, where $\epsilon \in (0,\ 1)$ is a given relative tolerance and ξ is the number of scalar parameters in $X \in \mathcal{D}$ (See Theorem 3.3. Note that this is the worst case order, and the actual order may be much less as discussed in Section 3.4.

In comparison with the worst case analysis [2] for the BMI eigenvalue problem, the worst case order for the MPEP with respect to the inverse of the accuracy, $1/\epsilon$, is better than that for the BMI branch and bound algorithm based on the element-wise bounding in [2] which requires $O\left((1/\epsilon)^{\xi}\right)$ operations in the worst case. This implies that the worst case order of the algorithm using the matrix-based bounding is twice as small as that for the BMI branch and bound algorithm based on the element-wise bounding and is at least superior in the worst case. We will compare the actual case computational complexities between the two by numerical experiments in the next subsection.

4.4   A Numerical Example

We consider the same numerical example treated in Section 3.4. The order of the generalized plant is four, and hence the matrix size of LMIs to be solved in the standard $\mathcal{H}_\infty$ problem is 32, containing two Lyapunov matrices of size 4×4 as variables.

We here compare the two algorithms, the FH algorithm in [2] and the YH algorithms in [24]. The reason why we only focus on the FH algorithm among the BMI global optimization algorithms [2, 7, 17] using the element-wise bounding method is a follows: Using convex hull seems the tightest to approximate the nonconvex region, and the FH algorithm is at least better than the one in [7].

Fig. 4.1 compares the number of convex subproblems N vs. the inverse of the relative tolerance $1/\epsilon$ for the optimal objective function. We only counted the number of branches corresponding to the number of lower bound problems for the FH algorithm. Moreover, the given size of domain to be sought for the FH algorithm is smaller than that for the YH algorithm in this numerical example in order to make a fair comparison. Nevertheless, the iteration number of the YH algorithm is fewer than that for the FH algorithm for each tolerance ϵ.

Although there is only a little difference of the iteration number between the two algorithms, there is a significant difference of the CPU-time as shown in Fig. 4.2, which compares the total CPU-time required to find a solution within tolerance ϵ. We have to join four additional linear constraints per one nonconvex constraint to solve one lower bound problem for the FH algorithm as already mentioned. Since the number of nonconvex constraints is seven for this case (see equation (4.4)), the matrix size of LMI to be solved becomes 60 in this case, although that in the YH algorithm does not increase and it is given by 32. Hence, the CPU-time of the FH algorithm per one iteration (11.7 sec.) is large in comparison with the one for the YH algorithm (1.77 sec.).

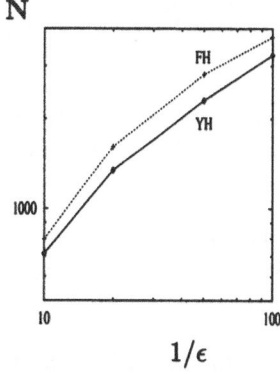

Figure 4.1: Iteration ♯ N vs. $1/\epsilon$ for the FH and YH Algorithms

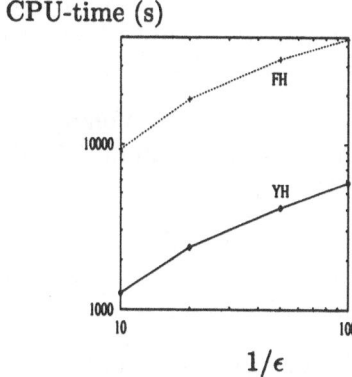

Figure 4.2: Total CPU-time (s) vs. $1/\epsilon$ for the FH and YH Algorithms

5   Conclusion

We have formulated the Matrix Product Eigenvalue Problem (MPEP) as a
fairly general class of nonconvex optimization problem for robust control synthesis. We have summarized the worst case computational complexity results
and investigated the computational cost for the actual case. We have also
made a comparison with an element-wise bounding for the BMI optimization
problem and a matrix-based bounding for the MPEP. One of the interesting
future topics is the computational complexity analysis for the average case.

References

[1] S.P. Boyd et al., *Linear Matrix Inequalities in System and Control Theory*,
     SIAM, 1994.

[2] H. Fujioka and K. Hoshijima, "Bounds for the BMI Eigenvalue Problem,"
     Trans. SICE, **33**(7):616–621, 1997.

[3] Hisaya Fujioka and Kohji Yamashita, "Computational Aspects of Con-
     stantly Scaled Sampled-Data $\mathcal{H}_\infty$ Control Synthesis," Proc. IEEE CDC,
     440–445, 1996.

[4] P. Gahinet and P. Apkarian, "An LMI-based Parameterization of all $\mathcal{H}_\infty$
     Controllers with Applications," Proc. IEEE CDC, 656–661, 1993.

[5] L. El Ghaoui, F. Oustry, and M. AitRami, "A Cone Complementarity
     Linearization Algorithm for Static Output-Feedback and Related Prob-
     lems," IEEE Trans. Automat. Contr., **AC-42**(8):1171–1176, 1997.

[6] L. El Ghaoui, V. Balakrishnan, E. Feron and S.P. Boyd, "On maximizing
     a robustness measure for structured nonlinear perturbations," Proc. ACC,
     2923–2924, 1992.

[7] K.G. Goh, M.G. Safonov, and G.P. Papavassilopoulos, "A Global Opti-
     mization Approach for the BMI Problem," Proc. IEEE CDC, 2009–2014,
     1994.

[8] T. Iwasaki and R.E. Skelton, "All Controllers for the General $\mathcal{H}_\infty$ Control
     Problem: LMI Existence Conditions and State Space Formulas," Auto-
     matica, **30**(8):1307–1318, 1994.

[9] T. Iwasaki and R.E. Skelton, "The XY-centering algorithm for the dual
     LMI problem: A new approach to fixed order control design," Int. J.
     Control, **62**(6): 1257–1272, 1995.

[10] H. Konno and T. Kuno, "Linear multiplicative programming," Mathe-
     matical Programming, **56**: 51–64, 1992.

[11] T. Kuno and H. Konno, "A Parametric Successive Underestimation Method for Convex Multiplicative Programming Problems," Journal of Global Optimization, 1: 267–286, 1991.

[12] M. Mesbahi and G.P. Papavassilopoulos, "On the Rank Minimization Problem Over a Positive Semidefinite Linear Matrix Inequality," IEEE Trans. Automat. Contr., AC-42(2):239–243, 1997.

[13] G.L. Nemhauser, A.H.G. Rinnooy Kan and M.J. Todd (Eds.), *Optimization: Handbooks in Operations Research and Management Science*, 1, Elsevier, 1989.

[14] A. Packard, "Gain scheduling via linear fractional transformations," Syst. Contr. Lett., 22:79–92, 1994.

[15] A. Packard, K. Zhou, P. Pandey and G. Becker, "A Collection of robust control problem leading to LMIs," Proc. IEEE CDC, 1245–1250, 1991.

[16] M.G. Safonov, K.G. Goh, and J.H. Ly, "Control System Synthesis via Bilinear Matrix Inequalities," Proc. ACC, 45/49, 1994.

[17] Y. Takano, T. Watanabe and K. Yasuda, "Branch and Bound Technique for Global Solution of BMI," *Trans. SICE*, 33(7):701–708, 1997 (in Japanese).

[18] O. Toker and H. Özbay, "On the NP-Hardness of Solving Bilinear Matrix Inequalities and Simultaneous Stabilization with Static Output Feedback," Proc. ACC, 2525–2526, 1995.

[19] H.D. Tuan and S. Hosoe, "DC optimization approach to robust controls: The optimal scaling value problem," *Proc. ACC*, 1997.

[20] L. Vandenberghe and S. P. Boyd, "Semidefinite Programming," SIAM Review, 38(1):49–95, 1996.

[21] Y. Yamada, "Global Optimization for Robust Control Synthesis based on the Matrix Product Eigenvalue Problem," Ph. D. dissertation, Tokyo Inst. of Tech., 1998.

[22] Y. Yamada and S. Hara, "A Global Algorithm for Scaled Spectral Norm Optimization," Proc. The 25th SICE Symp. on Control Theory, 177/182, 1996.

[23] Y. Yamada and S. Hara, "Global Optimization for $\mathcal{H}_\infty$ Control with Block-diagonal Constant Scaling," Proc. IEEE CDC, 1325–1330, 1996.

[24] Y. Yamada and S. Hara, "The Matrix Product Eigenvalue Problem – Global optimization for the spectral radius of a matrix product under convex constraints –," Proc. IEEE CDC, 4926–4931, 1997.

[25] Y. Yamada and S. Hara, "Global Optimization for $\mathcal{H}_\infty$ Control with Constant Diagonal Scaling," *IEEE Trans. AC*, **AC-43**(2): 191–203, 1998.

[26] Y. Yamada, S. Hara, and H. Fujioka, "Global Optimization for Constantly Scaled $\mathcal{H}_\infty$ Control Problems," Proc. ACC, 427–430, 1995.

[27] Y. Yamada, S. Hara, and H. Fujioka, "ϵ-Feasibility for $\mathcal{H}_\infty$ Control Problem with Constant Diagonal Scaling," *Trans. SICE*, **33**(3): 155–162, 1997.

Switching Surfaces and Groebner Bases[1]

Tryphon Georgiou   Allen Tannenbaum[†]

Abstract

A number of problems in control can be reduced to finding suitable real solutions of algebraic equations. In particular, such a problem arises in the context of switching surfaces in optimal control. Recently, a powerful new methodology for doing symbolic manipulations with polynomial data has been developed and tested, namely the use of Groebner bases. In this note, we apply the Groebner basis technique to find effective solutions to the classical problem of time-optimal control.

This paper is dedicated with warm friendship to Bruce Francis and M. Vidyasagar on the occasion of their 50th birthday.

1   Introduction

Optimal control is one of the most widely used and studied methodologies in modern systems theory. As is well-known, time-optimal problems lead to switching surfaces which typically are defined or may be approximated by polynomial equations [1]. The problem of determining on which side a given trajectory is in relation to the switching surface is of course key in developing the control strategy. Since the complexity of the switching surfaces can grow to be quite large, this may become quickly a formidable task. Here is where new techniques in computational algebraic geometry may become vital in effectively solving this problem.

In the paper, we would like to introduce Groebner bases in this context which will reduce the problem to a combinatorial one. Groebner bases have already been employed in a number of applications in robotics and motion planning [4, 10]. Recently, thanks largely to the efforts of Bill Helton, they have also been used in systems problems. Here we would like to propose them as a tool in optimal control.

---

[1]This work was supported in part by grants from the National Science Foundation ECS-99700588, ECS-9505995, NSF-LIS, by the Air Force Office of Scientific Research AF/F49620-94-1-0461, AF/F49620-98-1-0168, by the Army Research Office DAAH04-94-G-0054, DAAG55-98-1-0169, and MURI Grant.

[†]Department of Electrical and Computer Engineering, University of Minnesota, Minneapolis, MN 55455

This paper is intended to be of a tutorial nature. Applications studies will be published elsewhere. Our main purpose here is to indicate a novel technique in geometry for an important problem in control. The contents of this paper is as follows. In Section 2, we give the relevant control background. Section 3 introduces the basic notions of algebraic geometry, elimination theory, and Groebner bases. In Section 4, these notions are applied to indicate a general solution to the time optimal control problem. In Section 5, we make some conclusions and indicate the future course of this work.

2   Time-Optimal Control and Switching Surfaces

We focus on the classical problem of time-optimal control for a system consisting of a chain of integrators. It is standard that for such a system, minimum-time optimal control with a bounded input, leads to "bang-bang" control with at most n switchings – n being the order of the system. The control algorithm usually requires explicit determination of the switching surfaces where the sign of the control input changes. Explicit expressions for switching strategy are in all but the simplest cases prohibitively complicated (e.g., see [7], [8]).

Consider the linear system with saturated control input

$$
\begin{aligned}
\dot{x}_1(t) &= x_2(t) \\
\dot{x}_2(t) &= x_3(t) \\
\dot{x}_3(t) &= u(t), \text{ where } |u(t)| \leq 1,
\end{aligned}
$$

and as objective to drive the system from an initial condition $x(0)$ to a target $x(t_f)$, in minimum time t_f. In this case the Hamiltonian is $\mathcal{H} = 1 + \lambda_1 x_2 + \lambda_2 x_3 + \lambda_3 u$. The co-state equations become

$$
\begin{aligned}
\dot{\lambda}_1(t) &= 0 \\
\dot{\lambda}_2(t) &= -\lambda_1(t) \\
\dot{\lambda}_3(t) &= -\lambda_2(t),
\end{aligned} \tag{1}
$$

while the optimal $u(t)$ is given by $u(t) = -\text{sign}\,(\lambda_3(t))$.

A closed form expression for the optimal $u(t)$ as a function of $x(t)$ can be worked out (e.g., [7], see also [8]). Such an expression in fact tests the location of the state vector with regard to a switching surface. Bang-bang switching in practice is not desirable because of the incapacitating effect of noise and chattering. This issue has been addressed by a number of authors (see [8] and the references therein) and will not be discussed herein. While various remedies have been proposed and applied, the basic issue of knowing the switching surfaces is still instrumental in most methodologies.

The approach we take herein is algebraic in nature. The idea is to test directly whether a particular switching strategy is feasible. There are only two possible strategies where the input alternates between $+1$ and -1, taking the values $+1, -1, +1, \ldots$, or $-1, +1, -1, \ldots$, respectively. In each case, taking into account the maximal number of switchings, one can easily derive an expression for the final value of the state as a function of the switching times. This expression is then analyzed against the requirement of a given $x(t_f)$.

For this standard time-optimal control problem, it is well-known and easy to see by analyzing (1) that, in general, there are no singular intervals, and that the control input switches at most 3 times. Designate by t_1, t_2 and t_3, the length of the successive intervals where $u(t)$ stays constant. Any set of initial and final conditions can be translated to having $x(0) = 0$ and a given value for $x(t_f)$ and this is the setting from here on. The particular choice (among the only two possible ones),

$$u(t) = \begin{cases} +1 \text{ for } 0 \leq t < t_1 \\ -1 \text{ for } t_1 \leq t < t_1 + t_2 \\ +1 \text{ for } t_1 + t_2 \leq t < t_1 + t_2 + t_3 =: t_f \end{cases}$$

drives the chain of integrators for the origin to the final point $x(t_f)$ given by

$$
\begin{aligned}
x_3(t_f) &= t_1 - t_2 + t_3 \\
x_2(t_f) &= \frac{t_1^2}{2} + t_1 t_2 - \frac{t_2^2}{2} - t_2 t_3 + \frac{t_3^2}{2} + t_3 t_1 \\
x_1(t_f) &= \frac{t_1^3}{6} - \frac{t_2^3}{6} + \frac{t_3^3}{6} \\
&\quad + \frac{t_1^2}{2} t_2 + \frac{t_1^2}{2} t_3 + \frac{t_2^2}{2} t_1 - \frac{t_2^2}{2} t_3 + \frac{t_3^2}{2} t_1 - \frac{t_3^2}{2} t_2 + t_1 t_2 t_3. \quad (2)
\end{aligned}
$$

It turns out that the selection between alternating values $+1, -1, +1, \ldots$ or, $-1, +1, -1, \ldots$ for the optimal input $u(t)$ depends on whether (2) have a solution for a specified final condition $x(t_f) = (x_1, x_2, x_3)'$.

3   Groebner Bases and Elimination Theory

Algebraic geometry is concerned with the properties of geometric objects (*varieties*) defined as the common zeros of systems of polynomials. As such it is intimately related to the study of rings of polynomials and the associated ideal theory [5].

More precisely, let k denote an algebraically closed field (e.g., the field of complex numbers $\mathbf{C}$). Then one may show that affine geometry (the study of subvarieties of affine space k^n) is equivalent to the ideal theory of the polynomial ring $k[x_1, \ldots, x_n]$ (see [5] especially the discussion of the Hilbert Nullenstellensatz).

Clearly, the ability to manipulate polynomials and to understand the geometry of the underlying varieties can be very important in a number of applied fields (e.g., the kinematic map in robotics is typically polynomial). See also [9, 10] and the references therein for a variety of applications of geometry to systems theory. In this paper, we will show how a key problem in optimal control may be reduced to a problem in affine geometry.

The problem with algebraic geometry (until recently) was that despite its vast number of deep results, very little could actually be effectively computed. Because of this, it has not lived up to its potential to have a major impact on more applied fields. The advent of *Groebner bases* with powerful fast computers has largely remedied this situation. Groebner bases were introduced in the 1960's by B. Buchberger and named in honor of his doctoral advisor W. Groebner. They were also essentially discovered by H. Hironaka at around the same time in connection with his work on resolution of singularities. We follow the treatments in [3, 4, 2].

The method of Groebner bases allows one to treat a number of key problems for reasonably sized systems of polynomial equations among these being:

1. Find all common solutions in k^n of a system of polynomial equations

$$f_1(x_1, \ldots, x_n) = \cdots = f_m(x_1, \ldots, x_n) = 0.$$

2. Determine the (finite set of) generators of a given polynomial idea.

3. For a given polynomial f and an ideal I, determine if $f \in I$.

4. Let $g_i(t_1, \ldots, t_m)$, $i = 1, \ldots, n$ be a finite set of rational functions. Suppose $V \subset k^n$ is defined parametrically as $x_i = g_i(t_1, \ldots, t_m)$, $i = 1, \ldots, n$. Find the system of polynomial equations which define the variety V.

3.1   Groebner Bases

Motivated by the long division in the polynomial ring of one variable, one needs to order monomials in polynomial rings of several variables $k[x_1, \ldots, x_n]$.

Let $\mathbf{Z}_+^n$ denote the set of n-tuples of non-negative integers. Let $\alpha, \beta \in \mathbf{Z}_+^n$. For $\alpha = (\alpha_1, \ldots, \alpha_n)$, and set $x^\alpha = x_1^{\alpha_1} \cdots x_n^{\alpha_n}$. Let $>$ denote a total (linear) ordering on $\mathbf{Z}_+^n$ (this means that exactly one of the following statements is true: $\alpha > \beta$, $\alpha < \beta$, or $\alpha = \beta$. Moreover we say that $x^\alpha > x^\beta$ if $\alpha > \beta$. Then a **monomial ordering** on $\mathbf{Z}_+^n$ is a total ordering such that if $\alpha > \beta$ and $\gamma \in \mathbf{Z}_+^n$, then $\alpha + \gamma > \beta + \gamma$, and $>$ is a well-ordering, i.e., every nonempty subset of $\mathbf{Z}_+^n$ has a smallest element. One of the most commonly used monomial ordering is that defined by the ordinary lexicographical order

$>_{lex}$ on $\mathbf{Z_+}^n$. Recall that this means $\alpha >_{lex} \beta$ if the left most non-zero element of $\alpha - \beta$ is positive.

We now fix a monomial order on $\mathbf{Z_+}^n$. Then the **multidegree** of an element $f = \sum_\alpha a_\alpha x^\alpha \in k[x_1,\ldots,x_n]$ (denoted by multideg(f)) is defined to be the maximum α such that $a_\alpha \neq 0$. The **leading term** of f (denoted by LT(f)) is the monomial $a_{\text{multideg}(f)} \cdot x^{\text{multideg}(f)}$. We now come to the following crucial definition:

Definition. A finite set of polynomials $f_1,\ldots,f_m$ of an ideal $I \subset k[x_1,\ldots,x_n]$ is called a *Groebner basis* if the ideal generated by LT(f_i) for $i = 1,\ldots,m$ is equal to the ideal generated by the leading terms of all the elements of I.

The crucial result is that:

Theorem 1 *Every non-trivial ideal has a Groebner basis. Moreover, any Groebner basis of I is a basis of I.*

Notice that the use of Groebner bases reduces the study of generators of polynomial ideals (and so affine algebraic geometry) to that of the combinatorial properties of monomial ideals. Therein lies the power of this method assuming that one can easily compute a Groebner basis (see [3, 4]).

In what follows, we will indicate how Groebner basis techniques may be used to solve polynomial equations.

3.2 Elimination Theory

Elimination theory is a classical method in algebraic geometry for eliminating variables from systems of polynomial equations and as such is is a key method in finding their solutions. Groebner bases give a powerful method for carrying out this procedure systematically.

More precisely, let $I \subset k[x_1,\ldots,x_n]$ be an ideal. The jth **elimination ideal** of I is defined to be

$$I_j = I \cap k[x_{j+1},\ldots,x_n].$$

Suppose that I is generated by $f_1,\ldots,f_m$. Then I_j is the set of all consequences of the solutions of $f_1 = \ldots = f_m = 0$ in which the variables $x_1,\ldots,x_j$ are eliminated. Thus in order to eliminate $x_1,\ldots,x_j$, we need to find nonzero polynomials in I_j. This is where the Groebner basis methodology plays the key role:

Theorem 2 *For $I \subset k[x_1,\ldots,x_n]$ an ideal, and G a Groebner basis with respect the the lexicographical order with $x_1 > \ldots > x_n$, for every $j = 0,\ldots,n$*

$$G_j := G \cap k[x_1,\ldots,x_n]$$

is a Groebner basis of I_j. (Note we take $I_0 = I$.)

Thus using this Elimination theory, we may eliminate the variables one at a time until we are left with a polynomial in x_n, which we may solve. We must of course then extend the solution to the original system. For an ideal I we set
$$V(I) := \{(z_1, \ldots, z_n) \in k^n : f(z_1, \ldots, z_n) \forall f \in I\}.$$
Again this can be done in a systematic matter via the following result.

Theorem 3 *Let $I \subset k[x_1, \ldots, x_n]$ be generated by $f_1, \ldots, f_m$. Let I_1 be the first elimination ideal of I as defined above. For each $i = 1, \ldots, m$ write f_i as*

$$f_i = g_i(x_2, \ldots, x_n)x_1^{n_1} + \text{lower order terms in } x_1.$$

Suppose that $(z_2, \ldots, z_n) \in V(I_1)$. Then if there exists some i such that $g_i(z_2, \ldots, z_n) \neq 0$, then we may extend $(z_2, \ldots, z_n)$ to a solution of $(z_1, \ldots, z_n) \in V(I)$.

This ends our brief discussion of Groebner bases and elimination theory. We should note that there are symbolic implementations of this methodology on such standard packages as Mathematica or Maple.

4   Application to Time Optimal Control

In this section, we indicate the solution to the time optimal control problem formulated in Section 2. Even though we work out the case of 3rd order system, the method we propose is completely general, and should extend in a straightforward manner to any number of switchings.

The idea as outlined in Section 3, is to use the Groebner basis technique to eliminate the variables t_1, t_2, t_3 lexicographically, and then use Theorem 3 to check whether we may extend the solutions. We used the Mathematica symbolic computation program to eliminate the variables.

Accordingly, using the notation of Section 2, consider the ideal of the ring $C[t_1, t_2, t_3]$ generated by the polynomials

$$
\begin{aligned}
f_1 &= t_1 - t_2 + t_3 - x_3 \\
f_2 &= \frac{t_1^2}{2} + t_1 t_2 - \frac{t_2^2}{2} - t_2 t_3 + \frac{t_3^2}{2} + t_3 t_1 - x_2 \\
f_3 &= \frac{t_1^3}{6} - \frac{t_2^3}{6} + \frac{t_3^3}{6} \\
&\quad + \frac{t_1^2}{2}t_2 + \frac{t_1^2}{2}t_3 + \frac{t_2^2}{2}t_1 - \frac{t_2^2}{2}t_3 + \frac{t_3^2}{2}t_1 - \frac{t_3^2}{2}t_2 + t_1 t_2 t_3 - x_1,
\end{aligned}
$$

where $x_1 = x_1(t_f), x_2 = x_2(t_f), x_3 = x_3(t_f)$. Since the leading terms of f_1, f_2, f_3 all have constant leading coefficients (note we are using the lexicographical order $t_1 > t_2 > t_3$), from Theorem 3, we can extend any solution of the variety $V(I_1)$ of the first elimination ideal gotten by eliminating t_1 to a solution of $V(I)$.

Now, one can compute that the first elimination ideal $I_1 \subset \mathbf{C}[t_2, t_3]$ is generated by

$$
\begin{aligned}
h_1 &= 2t_2^2 - 4t_2t_3 + 4t_2x_3 + x_3^2 - 2x_2, & (3) \\
h_2 &= 6t_2^3 - 6t_2^2t_3 - 6t_2t_3^2 + 12t_2^2x_3 + 6t_2x_3^2 + x_3^3 - 6x_1.
\end{aligned}
$$

Notice that the leading coefficients of the leading terms of the generators of I_1 are constants, and so once again relying upon Theorem 3 we see that we may extend any solution in the variety $V(I_2)$ defined by the second elimination ideal $I_2 \subset \mathbf{C}[t_3]$ to a solution in $V(I_1)$. Thus any solution $V(I_2)$ extends to a solution of the full system $f_1 = f_2 = f_3 = 0$.

Finally, one can compute that I_2 is generated by the polynomial (in t_3)

$$
\begin{aligned}
0 = \ & 72x_2^3 + x_2^2(-72t_3^2 - 144t_3x_3 + 36x_3^2) \\
& + x_2(-72t_3^4 + 288t_3x_1 + 144t_3^3x_3 - 144x_1x_3 + 72t_3^2x_3^2 \\
& - 48t_3x_3^3 + 6x_3^4) + 144t_3^3x_1 - 72x_1^2 - 432t_3^2x_1x_3 + 36t_3^4x_3^2 \\
& + 288t_3x_1x_3^2 - 96t_3^3x_3^3 - 48x_1x_3^3 + 54t_3^2x_3^4 - 12t_3x_3^5 + x_3^6. \quad (4)
\end{aligned}
$$

Thus the Groebner basis technique provides a complete algorithmic procedure to solve the algebraic aspects of the switching surfaces problem. There is still one key aspect that must be resolved, namely we must check that we can find positive solutions t_1, t_2, t_3 to the algebraic equations. This problem may in fact be treated using the work of Kannai [6] who provides an implementable multidimensional Routh-type criterion for polynomials. However, in this context direct computation of all solutions of the full system $f_1 = f_2 = f_3 = 0$ is also feasible. In particular, if (4) has a positive root t_3, we substitute t_3 into (3) and determine whether h_1 has a positive root t_2. Finally, we substitute t_2, t_3 into f_1 to determine whether f_1 has a positive root as well. Thus, if there is a positive solution t_1, t_2, t_3, then the value of the optimal control u assumes the values $+1, -1, +1$ successively (over time intervals of length t_1, t_2, t_3 respectively), and in particular, the present value for the optimal control is $u(0) = +1$. If no positive solution exists the present value of the optimal control is $u(0) = -1$.

5   Conclusions

This note provided a general approach to the switching control strategy in time-optimal control. The key idea is to use the Groebner basis technique which allows one to algorithmically work with systems of polynomials in several variables. These results are still preliminary, and we will provide full details elsewhere. We expect that this approach will lead to a complete solution of the problem of identifying switching surfaces, in the sense that we will be able to provide a symbolic computer program which which will allow one to solve the problem for a reasonable number of variables (with "reasonable" a function of the computing power of the machine doing the computation!).

Very importantly, the employment of Groebner bases is the basis of computational algebraic geometry which certainly has a variety of practical applications for problems where polynomial manipulations play an essential role. They will clearly play an ever increasing role in the systems and control area.

References

[1] M. Athans and P. Falb, *Optimal Control: An Introduction to the Theory and its Applications*, New York: McGraw-Hill, 1966.

[2] D. Bayer and D. Mumford, "What can be computed in algebraic geometry?," in *Computational Algebraic Geometry and Commutative Algebra*, edited by D. Eisenbud and L. Robbiano, Cambridge University Press, Cambrdige, 1993, pages 1-48.

[3] T. Becker and V. Weispfenning, *Groebner Bases,* Springer-Verlag, New York, 1993.

[4] D. Cox, J. LIttle, and D. O'Shea, *Ideals, Varieties, and Algorithms, Second Edition* Springer-Verlag, New York, 1997.

[5] R. Hartshorne, *Algebraic Geometry,* Springer-Verlag, New York, 1976.

[6] Y. Kannai, "Causality and stability of linear systems described by partial differential operators," *SIAM J. Control Opt.* **20**: 669-674, 1982.

[7] E.B. Lee and L. Markus, *Foundations of Optimal Control Theory*, Malabar, FL:Krieger, 1967.

[8] L.Y Pao and G.F. Franklin, "Proximate time-optimal control of third-order servomechanisms," *IEEE Transactions on Automatic Control*, **38**(4): 560-280, 1993.

[9] A. Tannenbaum, *Invariance and Systems Theory: Algebraic and Geometric Aspects,* Lecture Notes in Mathematics **845**, Springer-Verlag, New York, 1981.

[10] A. Tannenbaum and Y. Yomdin, "Robotic manipulators and the geometry of real semi-algebraic sets," *IEEE Journal of Robotics and Automation* **RA-3** (1987), 301-308.

Part B

Hybrid Systems

From Digital Control to Digital Signal Processing*

Bruce A. Francis†

Abstract

The problem of designing multirate filter banks for subband coding is reviewed and a new approach is suggested based on recent techniques from optimal sampled-data control design.

1   Introduction

The theme of this paper is that some recent work on optimal sampled-data control systems carries over to a family of digital signal processing problems. A number of DSP problems could be considered; multirate filter banks for subband coding is chosen. Sampled-data control systems are periodic (not time-invariant) continuous-time systems, while multirate filter banks are periodic (not time-invariant) discrete-time systems. The control-theoretic concept of model matching carries over very conveniently.

The paper has three parts: 1) a quick review of optimal sampled-data control, 2) some background on the signal processing problem of compression, and subband coding in particular, and 3) an account of some recent work on optimal subband coders.

2   Optimal Sampled-Data Control Systems

An interesting place to start is the experimental facility called *Daisy* (Figure 1). Daisy is an experimental testbed facility at the University of Toronto's Institute for Aerospace Studies (UTIAS) whose dynamics are meant to emulate those of a real large flexible space structure (LFSS). The purpose of the facility is to test advanced identification and multivariable control design methods. Modeled roughly to resemble the flower of the same name, Daisy consists of a

---

*This work was supported by the Natural Sciences and Engineering Research Council (Canada).

†Electrical and Computer Engineering, University of Toronto, Toronto, Canada M5S 1A4. francis@control.utoronto.ca, www.control.utoronto.ca/~francis

Figure 1: Daisy (and Benoit Boulet).

rigid hub (the "stem") mounted on a spherical joint and on top of which are ten ribs (the "petals") attached through passive two-degree-of-freedom rotary joints and low-stiffness springs. Each rib is coupled to its two neighbors via low-stiffness springs. The hub would represent the rigid part of a LFSS, while the ribs would model its flexibilities.

Concerning Daisy's actuators, each rib is equipped with four unidirectional air jet thrusters that are essentially on-off devices, each capable of delivering a torque of 0.8 Nm at the rib joint. Pulse-width modulation (PWM) of the thrust is used to apply desired torques on the ribs. The four thrusters are aligned by pairs to implement two orthogonal bidirectional actuators. The hub actuators consist of three torque wheels driven by DC motors whose axes are orthogonal. Concerning the sensors, mounted at the tip of each rib is an infra-red emitting diode. Two hub-mounted infra-red CCD cameras measure the positions of these diodes via ten lenses. The cameras are linked to a computer that from the kinematics of Daisy computes the 20 rib angles relative to the hub in real-time from the sampled infra-red video frames. The hub orientation and angular velocity can be measured with position and velocity encoders.

Like all control systems, Daisy can be viewed as in Figure 2. Here $\mathbf{G}$ denotes the generalized plant and $\mathbf{K}$ the controller (boldface caps stand for linear transformations in the time domain); $z(t)$ is the vector of signals to be controlled, such as deviations from setpoint; $w(t)$ the vector of exogenous inputs; $y(t)$ the vector of measured signals; and $u(t)$ the vector of actuating signals. System $\mathbf{G}$ is linear time-invariant. If $\mathbf{K}$ is LTI, then so is the input-output system from w to z.

Let us recall how controllers are designed based on input-output optimiza-

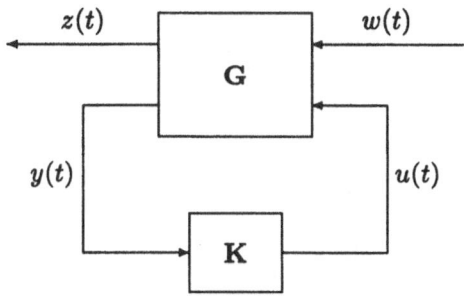

Figure 2: General control system configuration.

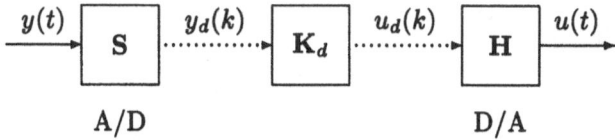

Figure 3: A digital controller.

tion. The possibilities include these three:

1. $\mathcal{H}_2$ optimization: Minimize $\|z\|_2$ for $w(t)$ an impulse; or equivalently minimize rms (z) for $w(t)$ white noise.

2. $\mathcal{H}_\infty$ optimization: Minimize the operator norm $\mathcal{L}_2 \longrightarrow \mathcal{L}_2$ from w to z.

3. $\mathcal{L}_1$ optimization: Minimize the operator norm $\mathcal{L}_\infty \longrightarrow \mathcal{L}_\infty$ from w to z.

To use these performance measures as the basis of controller design, one needs of course lots of engineering judgement in the choice of weights.

These optimization techniques have reached a certain level of maturity for analog controller design. Let us turn now to digital controllers. Figure 3 shows the prototype: $\mathbf{S}$ denotes the periodic sampler, of period T, defined by $y_d(k) := y(kT)$; $\mathbf{H}$ is the synchronized zero-order hold, $u(t) := u_d(k)$, $kT \leq t < (k+1)T$; and $\mathbf{K}_d$ is a discrete-time controller. The controller seen in continuous time is $\mathbf{HK}_d\mathbf{S}$, which is T-periodic, that is, it commutes with T-delay. Inserting this in Figure 2 gives Figure 4. If $\mathbf{K}_d$ is LTI in discrete time, then the input-output system from w to z is T-periodic. By extension from analog $\mathcal{H}_\infty$ optimization, we are led to attempt to design $\mathbf{K}_d$ to minimize the operator norm $\mathcal{L}_2 \longrightarrow \mathcal{L}_2$ from w to z. Note that this handles intersample behavior directly.

In studying this T-periodic system, the key technical tool is *blocking* (also called *lifting*). The idea is to break a signal $u(t)$ up into blocks of time duration

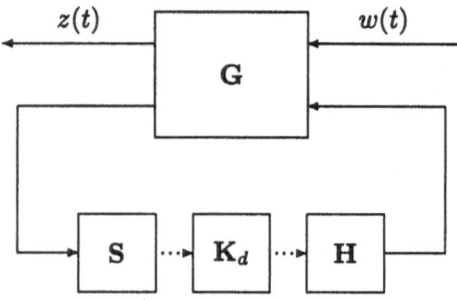

Figure 4: Sampled-data control system.

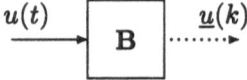

Figure 5: Blocking.

Figure 6: Example use of blocking.

T. Define $\underline{u}(0)$ to be that piece of $u(t)$ on $[0, T)$, $\underline{u}(1)$ the piece of $u(t)$ on $[T, 2T)$, and so on. Each $\underline{u}(k)$ is infinite-dimensional and forms a discrete-time signal $\underline{u}$. This process defines a linear transformation $\mathbf{B}$ mapping continuous time to discrete time as in Figure 5.

Figure 6 shows an example use of $\mathbf{B}$. The sampled-data system with an-tialiasing filter $\mathbf{F}$ and sampled-data controller is converted via $\mathbf{B}$ as shown into a purely discrete-time input-output system. If $\mathbf{F}$ and $\mathbf{K}_d$ are LTI, so is the input-output system in this figure. In particular, it has a transfer function, al-beit operator-valued. Blocking allows a complete solution to the control design problem in Figure 4.

The idea of blocking has an interesting history. I believe it starts with Kranc in 1957 for a multirate sampled-data control system.

3 Multirate DSP

Multirate DSP is not a recent subject. An early influential book is that by Crochiere and Rabiner [2]. The subject has received new stimulus in the 1990s because of its connection, first seen by Mallat, to the popular topic of wavelets. This section reviews some basics leading to the subject of subband coding for signal compression.

First a few general remarks about compression. Compression (or source coding) means reducing the number of bits in the digital representation of a signal. We all use *compress* or *zip* to compress our text files. Other examples of compression algorithms are JPEG for still images and MPEG for video and audio. Compression techniques are either lossless or lossy. For lossy methods, a common measure of signal quality is the SNR (perhaps a better term would be input signal to compression error ratio). Let $x(k)$ be a signal and $\hat{x}(k)$ its compressed version by some algorithm. Then

$$
\begin{aligned}
SNR \quad &:= \quad 10\log_{10} \frac{\text{encoder input signal energy}}{\text{error signal energy}} \\
&= \quad 10\log_{10} \frac{\sum x(k)^2}{\sum (x(k) - \hat{x}(k))^2} \\
&= \quad 20\log_{10} \frac{\|x\|_2}{\|x - \hat{x}\|_2} \\
&= \quad \frac{\|x\|_2}{\|x - \hat{x}\|_2} \text{in dB.}
\end{aligned}
$$

Thus the worst-case SNR equals $\min_x \frac{\|x\|_2}{\|x-\hat{x}\|_2}$, so the worst-case $1/SNR$ equals $\max_x \frac{\|x-\hat{x}\|_2}{\|x\|_2}$, namely, the ℓ_2-induced norm from input to error. This suggests an obvious analogy to $\mathcal{H}_\infty$ optimal control.

Now we turn to some basics of discrete-time signal processing. A discrete-time signal, say real-valued, can be represented either as

$$
x = \{\ldots, x(-1)|x(0), x(1), x(2), \ldots\},
$$

where the vertical line separates negative time from non-negative time, or as an infinite vector

$$
x = \begin{bmatrix} \vdots \\ x(-1) \\ \hline x(0) \\ x(1) \\ \vdots \end{bmatrix}.
$$

Denote the vector space of all such signals by $\ell(Z)$. The subspace of finite-energy signals is $\ell_2(Z)$. A linear system, then, is a linear transformation **G**: $\ell(Z) \longrightarrow \ell(Z)$. For example, the unit time delay system **U**. A linear system **G** has a matrix representation:

$$[\mathbf{G}] = \left[\begin{array}{cc|cc} & \vdots & \vdots & \\ \cdots & g(-1,-1) & g(-1,0) & \cdots \\ \hline \cdots & g(0,-1) & g(0,0) & \cdots \\ & \vdots & \vdots & \end{array} \right].$$

The vertical line separates negative time from non-negative time in the input, and the horizontal line does the same in the output. Thus, column j equals the output when the input equals the impulse at time j. Clearly, **G** is causal iff its matrix is lower triangular, and LTI iff it is Toeplitz. If **G** is LTI, its matrix looks like

$$[\mathbf{G}] = \left[\begin{array}{cc|cc} & \vdots & \vdots & \\ \cdots & g(0) & g(-1) & \cdots \\ \hline \cdots & g(1) & g(0) & \cdots \\ & \vdots & \vdots & \end{array} \right]$$

where $g(k)$ is the impulse-response function. A *filter bank* is either a single-input, multi-output or multi-input, single-output linear system.

What makes a system multirate is the presence of downsamplers and/or upsamplers. The downsampler **D** (sometimes denoted $\downarrow 2$ in a block diagram) is the discrete sampler mapping $x(k)$ to $y(k) = x(2k)$. Its matrix is

$$[\mathbf{D}] = \left[\begin{array}{cc|ccc} \vdots & \vdots & \vdots & \vdots & \vdots \\ \cdots & 1 & 0 & 0 & 0 & 0 & \cdots \\ \hline \cdots & 0 & 0 & 1 & 0 & 0 & \cdots \\ \cdots & 0 & 0 & 0 & 0 & 1 & \cdots \\ \vdots & \vdots & \vdots & \vdots & \vdots \end{array} \right],$$

so it is stable, non-causal, and time-varying. In the frequency domain, the input and output are related by

$$Y(e^{j\omega}) = \frac{1}{2}[X(e^{j\omega/2}) + X(e^{j(\omega-2\pi)/2})],$$

and thus aliasing is present in general.

The upsampler is the mathematical adjoint **D*** (sometimes denoted $\uparrow 2$), that is,

$$\mathbf{D}^* : \{\ldots, x(-1)|x(0), x(1), x(2), \ldots\}$$

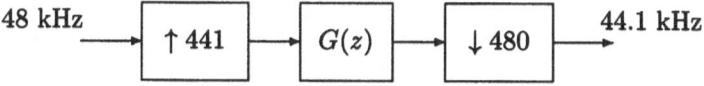

Figure 7: Sample rate changer.

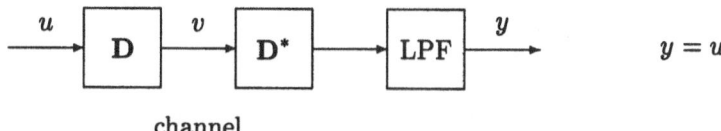

channel

Figure 8: Communication at half the rate.

$$\mapsto \{\ldots, 0, x(-1), 0|x(0), 0, x(1), 0, x(2), \ldots\}.$$

Again, it is stable, non-causal, time-varying. The input-output relationship in the frequency domain is

$$Y(e^{j\omega}) = X(e^{j2\omega}).$$

As a result, $\mathbf{DD}^* = I$, whereas

$$[\mathbf{D}^*\mathbf{D}] = \begin{bmatrix} & \vdots & \vdots & \vdots & \vdots & \vdots & \\ \cdots & 1 & 0 & 0 & 0 & 0 & \cdots \\ \cdots & 0 & 0 & 0 & 0 & 0 & \cdots \\ \cdots & 0 & 0 & 1 & 0 & 0 & \cdots \\ \cdots & 0 & 0 & 0 & 0 & 0 & \cdots \\ \cdots & 0 & 0 & 0 & 0 & 1 & \cdots \\ & \vdots & \vdots & \vdots & \vdots & \vdots & \end{bmatrix},$$

which is time-varying, in fact 2-periodic, i.e., commutes with $\mathbf{U}^2$.

Let us turn to some example uses of down- and upsampling. The first is *sample rate conversion*. A common problem in audio is to transform a signal sampled at, say, 48 kHz into the same signal sampled at 44.1 kHz. The common way to do this (there's a chip to do it) is as shown in Figure 7, where $G(z)$ is a suitable lowpass filter.

Another application is the system shown in Figure 8. Suppose the input $u(k)$ is bandlimited to $|f| < 0.25$, half the total discrete frequency range. Then it can be downsampled to $v(k)$ without loss of information, transmitted at half the rate (samples per second), then reconstructed at the receiver as shown using a lowpass filter (LPF). A similar process applies to a signal bandlimited to the high frequency range $|f - 0.5| < 0.25$. Combining these two processes gives the famous *quadrature mirror filter* of Figure 9. This has the perfect

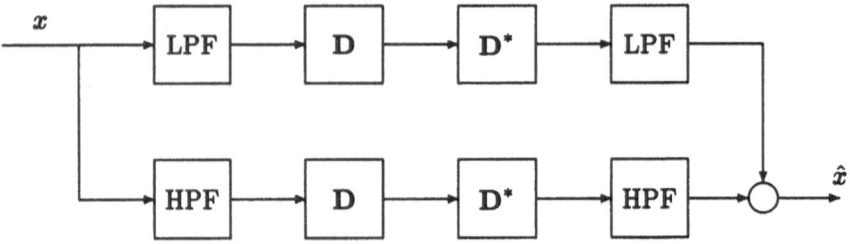

Figure 9: Quadrature mirror filter.

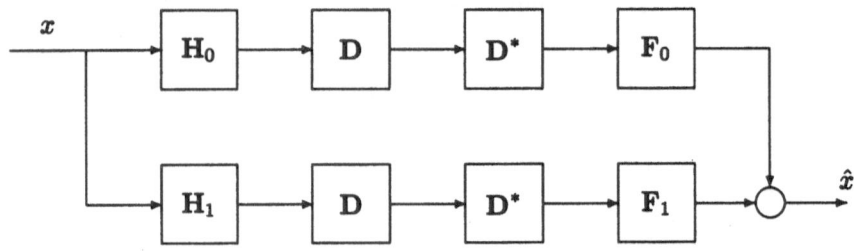

Figure 10: Multirate filter bank.

reconstruction property: $\hat{x} = x$. The two filters on the left perform a subband decomposition of the input. The two downsampled signals would be quantized appropriately, depending on the dynamic range of the two signals.

The quadrature mirror filter with ideal lowpass and highpass filters can be generalized to the multirate filter bank in Figure 10. The four filters, $\mathbf{H_0}, \mathbf{H_1}, \mathbf{F_0}, \mathbf{F_1}$, are LTI. The system from x to $\hat{x}$: is

$$\mathbf{T} = \mathbf{F_0 D^* D H_0} + \mathbf{F_1 D^* D H_1},$$

which is 2-periodic. Of course, this setup can be extended to M channels (MPEG audio has 32 channels, for example).

Two theoretical problems solved in the 1980s are 1) the problem of alias cancellation, that is, characterizing the filters such that $\mathbf{T}$ is LTI, and 2) the problem of perfect reconstruction, that is, designing filters so that, for some integer m, $\hat{x}(k) = x(k - m)$. In the latter, the ideal system from x to $\hat{x}$ is

$$\mathbf{T_d} = \mathbf{U}^m, \quad T_d(z) = z^{-m}.$$

Conventional design treats separately the different forms of distortion: aliasing, magnitude distortion, and phase distortion.

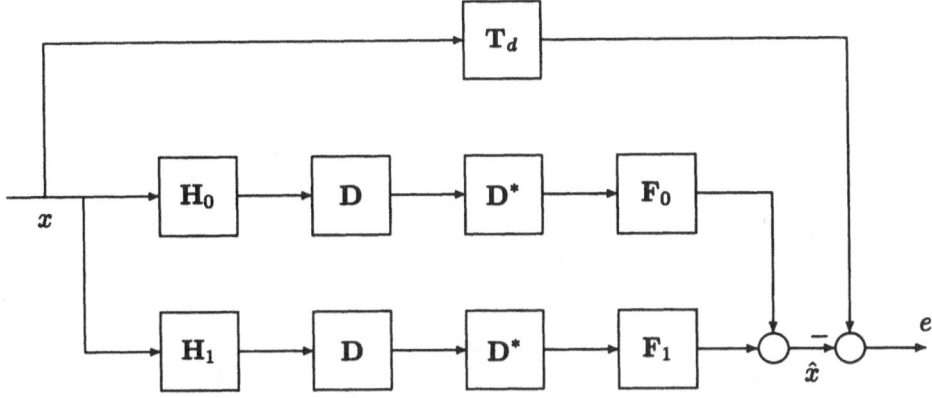

Figure 11: Error system.

4 Recent Work

Now we turn to some recent work on the design of multirate filter banks, borrowing some ideas from control. Comparing the system in Figure 10 to the ideal system $\mathbf{T}_d$ gives the error system of Figure 11. Let J denote the $\ell_2(\mathbf{Z})$-induced norm from x to e (equivalently, the rms-induced norm). We take a two-step design approach for the four filters. First, design the analysis filters $\mathbf{H}_0$, $\mathbf{H}_1$ for good coding of the input, for example, to maximize the coding gain; then design the synthesis filters $\mathbf{F}_0$, $\mathbf{F}_1$ to achieve as close to perfect reconstruction as possible. More precisely:

- Given (causal, stable) analysis filters $H_0(z)$ and $H_1(z)$ and given a tolerable time delay m;

- Design (causal, stable) synthesis filters $F_0(z)$ and $F_1(z)$ to minimize J.

Define

$$J_{opt} = \min_{F_0(z),\ F_1(z)} J = \min_{F_0(z),\ F_1(z)} \max_{\|x\|_2=1} \|e\|_2.$$

Since J_{opt} depends on the time delay m, we may write $J_{opt}(m)$. It's easy to show that $J_{opt}(m) \leq 1$ and $J_{opt}(m)$ is a non-increasing function of m. So one interesting question is, when is $\lim_{m\to\infty} J_{opt}(m) = 0$?

Since the system in Figure 11 is 2-periodic, blocking is immediately suggested as a means to convert to an LTI system. Introduce the blocking operator

$$\begin{bmatrix} \mathbf{D} \\ \mathbf{DU} \end{bmatrix} : \{\cdots, x(-1)|x(0), x(1), \cdots\}$$

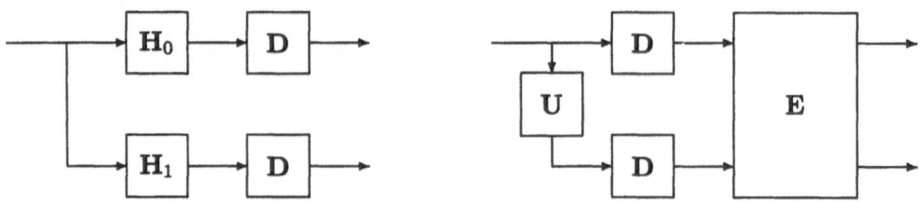

Figure 12: Analysis side.

$$\mapsto \left\{ \cdots, \begin{bmatrix} x(-2) \\ x(-3) \end{bmatrix} \middle| \begin{bmatrix} x(0) \\ x(-1) \end{bmatrix}, \cdots \right\}.$$

On the analysis side, introduce $\mathbf{E}$ so that the two systems in Figure 12 are input-output equivalent. That is, define $\mathbf{E}$ via

$$\begin{bmatrix} \mathbf{DH_0} \\ \mathbf{DH_1} \end{bmatrix} = \mathbf{E} \begin{bmatrix} \mathbf{D} \\ \mathbf{DU} \end{bmatrix}.$$

This has the unique solution

$$\mathbf{E} = \begin{bmatrix} \mathbf{DH_0} \\ \mathbf{DH_1} \end{bmatrix} \begin{bmatrix} \mathbf{D^*} & \mathbf{U^*D^*} \end{bmatrix} = \begin{bmatrix} \mathbf{DH_0D^*} & \mathbf{DH_0U^*D^*} \\ \mathbf{DH_1D^*} & \mathbf{DH_1U^*D^*} \end{bmatrix}.$$

Similarly, on the synthesis side there is a unique system $\mathbf{R}$ such that,

$$\begin{bmatrix} \mathbf{F_0D^*} & \mathbf{F_1D^*} \end{bmatrix} = \begin{bmatrix} \mathbf{UD^*} & \mathbf{D^*} \end{bmatrix} \mathbf{R} = \mathbf{U} \begin{bmatrix} \mathbf{D^*} & \mathbf{U^*D^*} \end{bmatrix} \mathbf{R}$$

namely,

$$\mathbf{R} = \begin{bmatrix} \mathbf{D} \\ \mathbf{DU} \end{bmatrix} \mathbf{U^*} \begin{bmatrix} \mathbf{F_0D^*} & \mathbf{F_1D^*} \end{bmatrix} = \begin{bmatrix} \mathbf{DU^*F_0D^*} & \mathbf{DU^*F_1D^*} \\ \mathbf{DF_0D^*} & \mathbf{DF_1D^*} \end{bmatrix}.$$

Both $\mathbf{E}$ and $\mathbf{R}$ can be shown to be LTI. Using these constructions in the error system leads to the system in Figure 13.

Now the blocking operator

$$\begin{bmatrix} \mathbf{D} \\ \mathbf{DU} \end{bmatrix} : \ell_2 \to \ell_2^2$$

is an isometry with inverse

$$\begin{bmatrix} \mathbf{D^*} & \mathbf{U^*D^*} \end{bmatrix} : \ell_2^2 \to \ell_2.$$

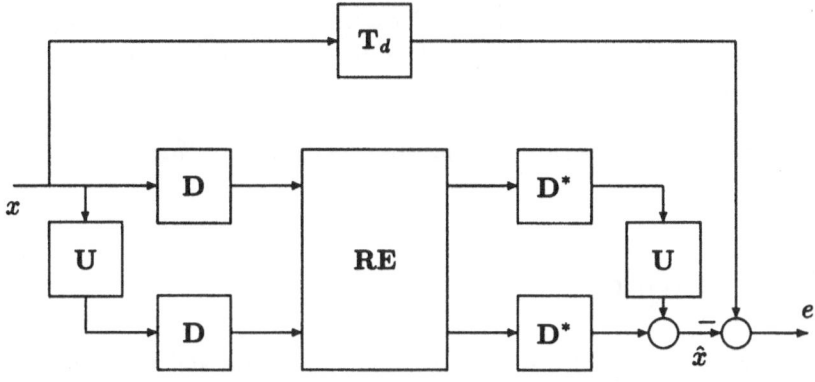

Figure 13: Equivalent error system.

So the induced norm of the error system is unaltered if it is premultiplied by

$$\begin{bmatrix} \mathbf{D} \\ \mathbf{DU} \end{bmatrix} \mathbf{U}^*$$

and postmultiplied by

$$[\ \mathbf{D}^* \quad \mathbf{U}^*\mathbf{D}^*\ ].$$

The resulting system is simply $\mathbf{W} - \mathbf{RE}$, where

$$\mathbf{W} := \begin{bmatrix} \mathbf{D} \\ \mathbf{DU} \end{bmatrix} \mathbf{U}^*\mathbf{T}_d [\ \mathbf{D}^* \quad \mathbf{U}^*\mathbf{D}^*\ ].$$

This system has a very simple form, as displayed in the theorem below. Each of these three systems—**E, R, W**—is LTI and 2×2. So the equivalent design problem is to compute $R(z)$ to minimize the $\mathcal{H}_\infty$-norm of $W(z) - R(z)E(z)$, a well-known control problem.

In summary:

Theorem 1 *Let*

$$W(z) = \begin{cases} z^{-d} \begin{bmatrix} 1 & 0 \\ 0 & 1 \end{bmatrix}, & \text{if } m = 2d + 1 \\ z^{-d} \begin{bmatrix} 0 & z \\ 1 & 0 \end{bmatrix}, & \text{if } m = 2d \end{cases}.$$

Then $J_{opt} = \min_{R(z)} \|W - RE\|_\infty.$

The next theorem says that under a mild condition on the analysis filters, arbitrarily close to perfect reconstruction can be achieved if the latency is sufficiently large.

Theorem 2 *Assume $E(e^{j\omega})$ is nonsingular for all ω. Then $\lim_{m \to \infty} J_{opt}(m) = 0$.*

5 Example

Multirate filter banks are used in subband coding and decoding of speech, images, and video for transmission and storage. Subband coders have been adopted in many new international coding standards [6]. The problem of designing multirate filter banks that provide good frequency resolution while allowing for exact or near perfect reconstruction of the input is quite challenging. Several techniques have been developed for the design of linear phase FIR filter banks with the perfect reconstruction (PR) property. Among these, those designed by Johnston [3] are widely used in many practical applications, for example in CCITT Rec. G.722, a subband coding standard for digital transmission of wideband audio [4]. In these filter banks aliasing is canceled and phase distortion is eliminated, while magnitude distortion and stopband leakage in the analysis filters are minimized. Compared to FIR PR systems with a linear phase analysis bank and with the same length analysis filters [7], Johnston filters have much greater stopband attenuation. Furthermore, in many practical applications such as audio and image coding, one does not require PR as long as the reconstruction errors are imperceptible.

Once the analysis filters have been chosen, it makes sense to design the synthesis filters to achieve as close to PR as possible, that is, to maximize the worst-case SNR. The question studied in this section is this: If Johnston filters are adopted at the analysis side to provide good coding, how close to optimal would it then be to choose Johnston filters at the synthesis end? We answer the question by studying an example in detail.

All filters are causal and frequency f is normalized ($0 \le f < 1$). The analysis filter $H_0(z)$ is chosen to be Johnston's FIR filter named 32D, which has 32 coefficients, passband edge 0.25, and stopband edge 0.293. The minimum stopband attenuation is 38 dB. Then as in [3] we take $H_1(z) = H_0(-z)$, $F_0(z) = 2H_0(z)$, $F_1(z) = -2H_0(z)$. (The factor 2 in F_0, F_1 makes the overall system gain equal to 1.) Since the Johnston filters have 32 coefficients, their order is 31 and the inherent overall delay is $m = 31$.

We now fix H_0 and H_1, and design F_0 and F_1 as in the preceding section. We take $m = 31$ for a fair comparison, for which $optSNR = 58.0$ dB. (As we shall see, this is an improvement of 8 dB over the Johnston filters.) The optimal F_0 and F_1 (as computed using μ-Tools) are IIR of order 185, which were reduced to order 41 with negligible error by truncation of a balanced state realization. IIR filters of order 41 are probably not practical; it is therefore desirable to approximate them by low-order FIR filters. For the filters at hand, simply truncating the impulse responses at 32 terms works well. A better way would be to constrain F_0 and F_1 to be FIR of a certain order from the start, and formulate the problem as one of convex optimization. Nevertheless, we continue with the order 41 IIR filters.

Let us introduce the following convenient notation:

O - the system in Figure 10 when F_0 and F_1 are the order-41 optimal IIR filters,

J - when F_0 and F_1 are the order-31 Johnston filters.

We shall compare **O** and **J** with respect to distortion and SNR.

Distortion Here we apply results derived in [5]. The frequency-domain relation between $x(k)$ and $\hat{x}(k)$ in Figure 10 is

$$\hat{X}(f) = A_0(f)X(f) + A_1(f)X(f - 0.5),$$

where

$$A_0(f) := \frac{1}{2}[H_0(f)F_0(f) + H_1(f)F_1(f)],$$

$$A_1(f) := \frac{1}{2}[H_0(f - 0.5)F_0(f) + H_1(f - 0.5)F_1(f)].$$

The desired output is $x(k - m)$, or $e^{-j2\pi fm}X(f)$ in the frequency domain. Thus the error is

$$
\begin{aligned}
E(f) &= e^{-j2\pi fm}X(f) - \hat{X}(f) \\
&= \underbrace{[e^{-j2\pi fm} - A_0(f)]}_{\text{error TF}} X(f) - \underbrace{A_1(f)X(f - 0.5)}_{\text{aliasing term}}.
\end{aligned}
$$

The reconstructed signal $\hat{X}(f)$ therefore suffers from aliasing, magnitude, and phase distortion in general.

Aliasing Distortion The natural definition of *aliasing distortion* at frequency f is $AD(f) := |A_1(f)|$ and of *maximum aliasing distortion* is $maxAD := \max_f AD(f)$. It is proved in [5] that for any design $maxAD \leq minSNR^{-1}$. For **J**, $AD(f) = 0$ by design.

	O	**J**
$maxAD$	-64.4 dB	$-\infty$ dB

Magnitude and Phase Distortions If aliasing is cancelled, the overall system is LTI with frequency-response function $A_0(f)$. If $|A_0(f)| \neq 1$, the filter bank is said to have magnitude distortion, and if "$A_0(f) \neq -2\pi fm$, it is said to have phase distortion. This suggests the following indices for *magnitude* and *phase distortions* at frequency f: $MD(f) := |1 - |A_0(f)||$, $PD(f) := |-2\pi fm - {}^{"}A_0(f)|$. Actually a more natural index capturing both magnitude and phase distortion together is the *distortion* index $D(f) := |e^{-j2\pi fm} - A_0(f)|$. Note that a design would be worthless if $D(f)$ is not less than 1; indeed, by taking the synthesis filters to be zero we can achieve $D(f) = 1$ because $A_0(f) = 0$. So it is natural to assume that $D(f) < 1$. Also define $maxMD$, $maxPD$, and $maxD$ by maximizing over f. It is proved in [5] that for any design $maxMD \leq minSNR^{-1}$, $maxPD \leq \arcsin(minSNR^{-1})$, and $maxD \leq minSNR^{-1}$.

	O	**J**
$maxMD$	-58.2 dB	-49.8 dB
$maxPD$	0.071°	0
$maxD$	-58.1 dB	-49.8 dB

In summary, to achieve zero aliasing and phase distortions, Johnston filters pay the penalty of 8 dB in magnitude distortion.

SNR Now we shall compare SNR for the three systems. First the minimum SNR over all inputs:

	O	**J**
$minSNR$	58.0 dB	49.8 dB

Thus Johnston filters are inferior to the optimal ones by 8 dB.

The optimal filters are worst-case optimal for wideband inputs, that is, there is no constraint on $x(k)$ other than it have finite energy. If one knows the dominant frequency range of the inputs, one can modify the optimization design procedure by introducing a shaping filter $W(z)$ and allowing the pre-filtered input x_{pf} ($x = \mathbf{W}x_{pf}$) to be an arbitrary finite-energy signal. The worst-case SNR is then $minSNR := \min_{x_{pf} \in \ell_2} \|x_{pf}\|_2 / \|e\|_2$.

In conclusion, a two-step procedure is advocated for the design of maximally decimated filter banks: First, select suitable filters for the analysis side to get good coding gain; then design the synthesis filters to optimize the SNR.

6   Concluding Remark

Multirate filter banks have interesting generalizations. One is a pyramid structure obtained by iterating on one output of a filter bank. This pyramid structure is in turn central in wavelet theory: A wavelet transform can be computed as the output of a pyramid. Also, the pyramid leads to a multiresolution decomposition of signal space. This line of ideas is leading to a host of interesting applications, including texture image synthesis [1].

References

[1] J.S. De Bonet. Multiresolution sampling procedure for analysis and synthesis of texture images. In *SIGGRAPH Proceedings*, pages 361–368, 1997.

[2] R.E. Crochiere and L.R. Rabiner. *Multirate Digital Signal Processing*. Prentice Hall, Englewood Cliffs, New Jersey, 1983.

[3] J.D. Johnston. A filter family designed for use in quadrature mirror filter banks. In *Proceedings of IEEE International Conference on Acoustics, Speech and Signal Processing*, pages 291–294, April 1980.

[4] Paul Mermelstein. G.722, a new CCITT coding standard for digital transmission of wideband audio signals. *IEEE Communications Magazine*, 26:8–15, January 1988.

[5] S. Mirabbasi, T. Chen, and B. Francis. Controlling distortions in maximally decimated filter banks. *IEEE Trans. on Circuits and Systems, Part II*, pages 597–600, 1997.

[6] K. Nayebi, T.P. Barnwell, and M.J.T. Smith. Low delay fir filter banks: Design and evaluation. *IEEE Trans. on Signal Processing*, 42(1):24–31, January 1994.

[7] T.Q. Nguyen and P.P. Vaidyanathan. Two-channel perfect reconstruction FIR QMF structures which yield linear-phase FIR analysis and synthesis filters. *IEEE Trans. on Acoustics, Speech, and Signal Processing*, 37:676–690, May 1989.

From Sampled-Data Control to Signal Processing

Yutaka Yamamoto*   Pramod P. Khargonekar†

1   Introduction

Sampled-data control theory has recently gone through a major breakthrough. While the classical treatment of digital control systems mostly deals with sample-point behavior only, the modern approach enables us to consider inter-sample behavior directly. For example, the classical sampled-data control design mostly employs either one of the following procedures:

- Execute a continuous-time design to obtain a continuous-time controller. Adopt a discretization method (e.g., Tustin transformation) to obtain a discrete-time controller. The sampling rate must be fast enough to allow for a sufficient approximation by the discretized controller of the original continuous-time controller. When the sampling rate becomes slower, there is a danger of getting even an unstable closed loop system. Tools for exact quantitative analysis/synthesis of the desired sampling period vs performance are quite limited.

- Discretize the plant by some means, e.g., via impulse invariant approximation, and then execute a discrete-time controller design. Inter-sample behavior is usually ignored. When the sampling rate gets faster, it has the potential of inducing undesirable oscillatory behavior.

In contrast to these, the modern sampled-data control theory directly deals with the inherent inter-sample behavior. The difficulty encountered here is that a sampled-data control system possesses two different time-sets: continuous and discrete, inherited from the plant and controller, respectively. This in turn induces a *time-varying* (albeit periodic) nature to the overall closed-loop system. The classical analysis/synthesis tools such as the transfer

*Department of Applied Analysis and Complex Dynamical Systems, Graduate School of Informatics, Kyoto University, Kyoto 606-8501, JAPAN. E-mail: yy@kuamp.kyoto-u.ac.jp This author was supported in part by the Sound Technology Promotion Foundation.

†Department of Electrical Engineering and Computer Science, The University of Michigan, Ann Arbor, MI 48109-2122, USA. E-mail: pramod@eecs.umich.edu This author was supported in part by grant by the Army Research Office no. DAAH04-93-G-0012.

function, frequency response concepts are not readily available. This drawback has been successfully circumvented by *lifting* (or *blocking*) which enables us to describe sampled-data systems as a *time-invariant* discrete-time system concept. The crux here is that lifting enables us to view continuous-time signals as function-space valued *sequences*. This introduces infinite-dimensionality to the system description; however, the advantage obtained through time-invariance usually outweighs the disadvantages or inconveniences in handling infinite-dimensionality. Furthermore, in many design problems, this infinite-dimensionality can be dealt with in the solution, and essentially finite-dimensional solution procedures are obtained.

Motivated by this success we attempt to apply this theory to digital signal processing. Digital signal processing and sampled-data control theory possess much in common. Both deal with continuous-time signals, and design a controller or a filter to shape them according to some prespecified criteria. Frequency domain consideration is important in this regard. As in traditional digital control, modern digital signal processing does not usually deal with continuous-time behavior directly. Rather, it is mostly dealt with using the notion of *aliasing* based on the assumption that the reference signals are sufficiently band-limited. Such a treatment tends to be indirect and it is desirable to build a link between modern sampled-data control theory and digital signal processing.

This paper gives an overview of such a connection between the two branches. We start with a brief review of the sampled-data control theory, giving some basic accounts and an example that illustrates the advantage of the modern theory. We then turn to describe some new developments in signal reconstruction problem.

2 Brief Review of Sampled-Data Control Theory

Consider the general sampled-data control system as shown in Fig. 1. Here $P(s)$ is a continuous-time plant, and its measured output y is sampled and then processed by the discrete-time controller $K(z)$, converted by the hold element $\mathcal{H}$, and then fed back to the plant P.

Let $f(t)$ be a locally square integrable signal on $[0, \infty)$ and h be a fixed sampling period. The *lifting operator* $\mathcal{L}$ decomposes f into a sequence of functions in $L^2[0, h)$:

$$\mathcal{L} : f \mapsto \{f(kh + \theta)\}_{k=0}^{\infty}, \quad 0 \leq \theta < h \tag{1}$$

Regarding them as infinite-dimensional vectors applied to a system at discrete-time instants kh, $k = 0, 1, 2, \ldots$, we can describe sampled-data control system

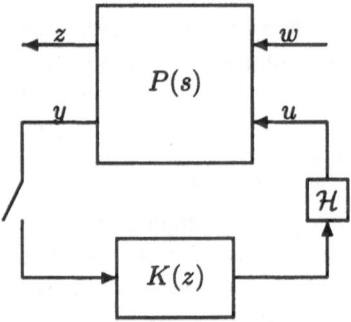

Figure 1: Sampled Feedback System

by a discrete-time state equation. For example, a linear continuous-time system

$$\dot{x}(t) = A_c x(t) + B_c u(t), \quad y(t) = C_c x(t).$$

can be described by time-invariant *discrete-time* equations as

$$x_{k+1} = e^{A_c h} x_k + \int_0^h e^{A_c(h-\tau)} B_c u_k(\tau) d\tau$$

$$y_k(\theta) = C_c e^{A_c \theta} x_k + \int_0^\theta C_c e^{A_c(\theta-\tau)} B_c u_k(\tau) d\tau$$

where $x_k = x_k(0)$. See, e.g., [2, 13]. Then all the components in Fig. 1 can be described as linear time-invariant *discrete-time* systems, so that the overall closed-loop system can also be described by a time-invariant model:

$$x_{k+1} = A x_k + B u_k, \quad y_k(\cdot) = C x_k + D u_k. \tag{2}$$

The difference here is that A, B, C, D are operators rather than matrices. We can then introduce its *transfer function* as

$$G(z) := D + C(zI - A)^{-1} B. \tag{3}$$

Actually A happens to be a matrix, and the system is stable when A has eigenvalues inside the unit circle only. When this is the case, one can substitute $e^{j\omega h}$ for z in $G(z)$, and obtain a *frequency response operator* [14]:

$$G(e^{j\omega h}) : L^2[0, h] \to L^2[0, h] \tag{4}$$

Its *gain* at each frequency ω is the induced norm:

$$\|G(e^{j\omega h})\| = \sup_{v \in L^2[0,h]} \frac{\|G(e^{j\omega h})v\|}{\|v\|}. \tag{5}$$

It is known that this gain gives, at each frequency, the largest magnification among all alias components [14]. The H^∞ norm of G is the supremum of $\|G(e^{j\omega h})\|$:

$$\|G\|_\infty := \sup_{0 \leq \omega \leq 2\pi/h} \|G(e^{j\omega h})\|. \tag{6}$$

Given a configuration as in Fig. 1, the H^∞ control objective is to find a controller (filter) $K(z)$ such that

$$\|\mathcal{F}_l(P, K)\|_\infty < \gamma \tag{7}$$

where $\mathcal{F}_l(P, K)$ denotes the closed-loop transfer function operator from w to z in Fig. 1. The H^2 design problem can be defined similarly.

The infinite-dimensional H^∞ design problem, as well as the computation of the frequency response, is generally reducible to a *finite-dimensional*, discrete-time problem. The procedure is as follows: When $\|\mathcal{F}_l(P, K))\|_\infty > \|D\|$, $\|\mathcal{F}_l(P, K))\|_\infty$ is indeed given by its maximal singular value. This is obtained by composing the expression (2) with its dual system expression

$$p_k = A^* p_{k+1} + C^* u_k, \quad y_k(\cdot) = B^* p_k + D^* u_k. \tag{8}$$

Here operators A, B, C and their adjoints have either finite-dimensional range or domain. Using the fact that the nonzero spectra of T^*T and TT^* coincide, operators appearing in the singular value equation

$$(\gamma^2 I - G^* G)v = 0$$

are reducible to matrices, except the term arising from D. It becomes necessary to compute

$$R_\gamma^{-1} = (\gamma^2 I - D^* D)^{-1}$$

using a result of an H^∞ problem for delay systems [16]. While this can be done by employing an differential equation expression, it complicates the formula. This infinite-dimensionality becomes an issue in what follows.

3 A Design Example

To illustrate the difference from the classical design methods, we give an H^2 type design example for the plant $P(s) = 1/(s^2 + 2s + 1)$ both in classical and modern methods. We execute

- Sampled-data (continuous-time based) H^2 design, and

- discrete-time H^2 design.

for $h = 0.2$.

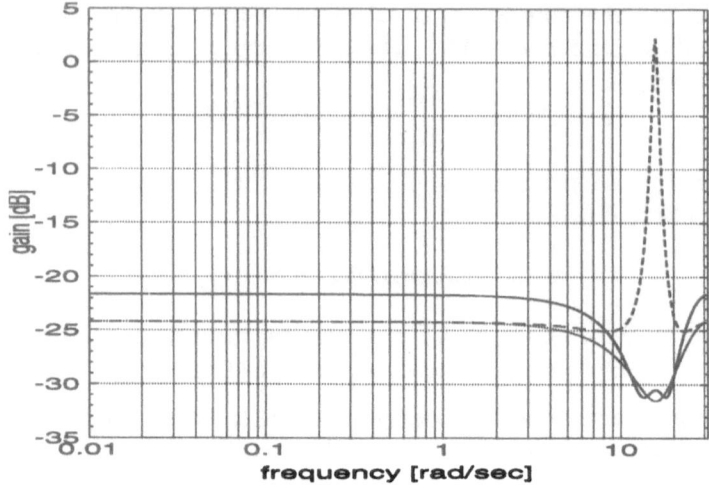

Figure 2: Frequency Responses $h = 0.2$

Fig. 2 and 3 show the frequency and time responses of the closed-loop systems, respectively. In Fig. 2, the solid curve shows the response of the sampled-design, whereas the thin dash curve shows the discrete-time frequency response when the designed controller K is connected with the discretized plant G_d (i.e., purely discrete-time frequency response). At a first glance, it appears that the discretized design performs better, but it is actually seen to be worse when we compute the real (continuous-time) frequency response of G connected with K_d. The dash curve shows this frequency response; it is similar to the discrete-time frequency response in the low frequency range, but exhibits a very sharp peak around the Nyquist frequency ($\pi/h \sim 15.7$ rad/sec, i.e., $1/2h = 2.5$Hz).

In fact, the impulse responses Fig. 3 exhibit a clear difference between them. The solid curve shows the sampled-data design, and the dashed curve the discrete-time one. The latter shows a very oscillatory behavior. Also, both responses decay to zero very rapidly *at sampling instants*. The difference is that the latter exhibits very large ripples, with period approximately 0.4 sec. This corresponds to $1/0.4$ Hz, which is the same as $(2\pi)/0.4 = \pi/h$ rad/sec, i.e., Nyquist frequency. This is precisely captured in the lifted frequency response in Fig. 2.

4   Application to Digital Signal Processing

Let us now proceed to the application of the sampled-data control theory to digital signal processing. In particular, we consider a signal reconstruction

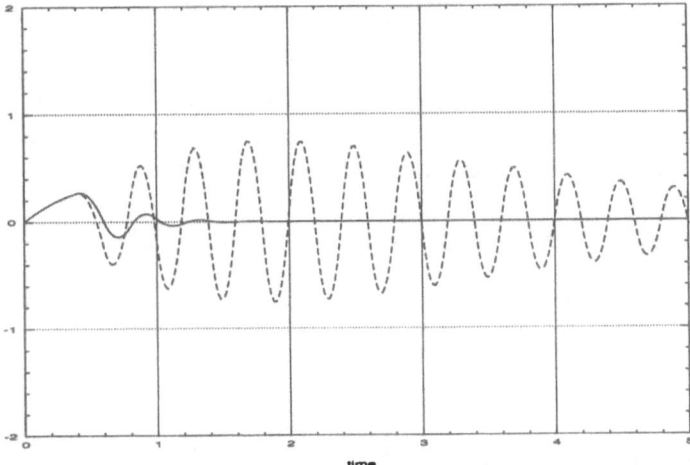

Figure 3: Impulse Responses $h = 0.2$

problem:

We are given a class of analog signals that are to be encoded, stored or transmitted digitally. The objective is then to reconstruct the original analog signal as much as possible with respect to some performance criteria. In this process analog to digital conversion is accompanied by sampling. The problem is that there is information loss associated with sampling. If the sampling period is h, then resolution in time is clearly limited by this value. This is described by the Shannon sampling theorem (see [15] for some historical accounts). On the other hand, it is also known that if the analog signal is bandlimited within the Nyquist frequency range, then it can be perfectly reconstructed. The formula is given as the convolution of the train of impulses consisting of sampled values with a special convolution kernel called the sinc function.

However, Shannon's formula works only in an ideal situation, and there are various practical limitations that do not allow an exact realization of the Shannon filter. They are roughly as follows:

- The Shannon filter is not causal since the sinc function has nonzero values on $(-\infty, 0]$.

- To deal with this non-causality, we can shift sinc function by finite time period. But the resulting (shifted and truncated at the origin) impulse response is still an approximation. Furthermore, it is not realizable via finite-dimensional system. Approximation in this regard is inevitable.

Summarizing, we need to approximate an ideal reconstruction filter in the discrete-time setting. A first trial would be to conduct the least square approximation in the frequency domain. It turns out [6] that such an approximation will yield a finite-length truncation of the impulse response, and results in the well known Gibbs phenomenon. To remedy this, one introduces various window functions so that the slope and shoulder characteristics become satisfactory. To execute this design method, one often takes the following route:

1. Choose a sufficiently wide bandwidth to avoid aliasing.

2. Design an FIR filter that allows for sufficiently close approximation for ideal reconstruction.

3. Combine this with 0-order hold.

The reason for preference for FIR filters is that they can realize a complete linear phase-shift property. We can then concentrate our attention mainly upon the gain characteristic. It should be however noted that this linear phase property is realizable only within the discrete-time setting, and when signals pass through a D/A converter and analog amplifiers, there is usually a nontrivial amount of phase distortion. Such a distortion is difficult to account for in the discrete-time domain.

The second problem is that, while the least square performance criterion in the frequency domain is often employed for the performance evaluation, it is quite difficult for this measure to control the filter behavior in a specific frequency range. H^∞ performance criterion aims at capturing worst-case behavior of systems, and is often advantageous in this regard. This criterion was introduced in signal processing only recently [11]. An application of discrete-time H^∞ design method to multirate filter banks was studied by [3], and this direction appears to be promising.

The third point to be noted is that by the problem arising from the combination of analog and digital components, it is desirable to give a direct design method that optimizes continuous-time behavior without involving discrete-time approximation. This has been made possible in the sampled-data control theory only recently, and our approach is to place the present problem in this framework.

5   Single-Rate Signal Reconstruction

Let us start with the single-rate signal reconstruction problem. Consider the block diagram Fig. 4.

The original analog signal $w(t)$ goes through the anti-aliasing prefilter $F_a(s)$ and gets sampled to become a discrete-time signal y_d, and then processed by a digital filter $K(z)$. It is then converted back to an analog signal via the

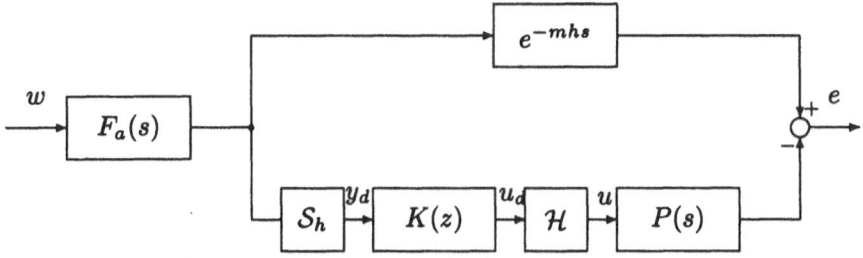

Figure 4: Signal Reconstruction Problem

hold device $\mathcal{H}$ and an amplifier $P(s)$. Comparing this with the original signal $F_a(s)w$ is usually too stringent a requirement. We thus allow a finite amount of delay. The obtained signal is thus compared with delayed analog signal $e^{-mhs}F(s)w$. Here we assumed, without loosing much generality, that the delay is an integer multiple of the sampling period h.

An H^∞ signal reconstruction problem is the following: Let $\mathcal{T}_{ew}(K)$ denote the (lifted) transfer function from w to e with controller $K(z)$. Given $\gamma > 0$, find a controller (filter) $K(z)$ such that

$$\|\mathcal{T}_{ew}(K)\|_\infty < \gamma.$$

We now proceed to reduce the above problem to a finite-dimensional discrete-time H^∞ problem. The block diagram in Fig. 4 is clearly equivalent to that in Fig. 5.

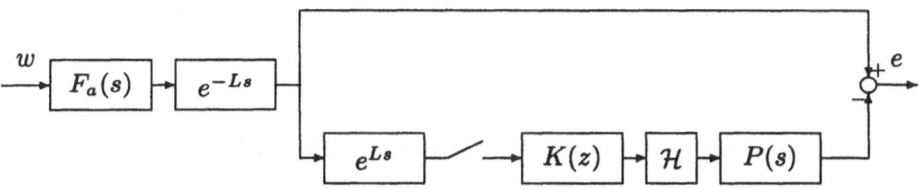

Figure 5: Modified Problem

Since e^{-mhs} is an inner function, we can interchange this with $F_a(s)$, and move e^{Ls} after the sampler as z^m. This leads to the diagram in Fig. 6. Therefore, our problem has been reduced to the design of a *non-causal controller* $z^m K(z)$ such that

$$\|\mathcal{T}_{ew'}(z^m K)\|_\infty < \gamma.$$

in Fig. 6.

We now give a state-space solution to this problem. Let (A_1, B_1, C_1) and (A_2, B_2, C_2) be minimal realizations of $F_a(s)$ and $P(s)$, respectively. The

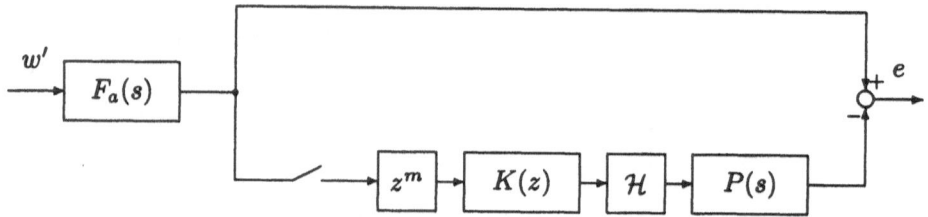

Figure 6: Modified Finite-Dimensional Problem

generalized plant in Fig. 6 is then given by

$$
\begin{bmatrix} e \\ y \end{bmatrix} = \left[\begin{array}{cc|cc} A_1 & 0 & B_1 & 0 \\ 0 & A_2 & 0 & B_2 \\ \hline C_1 & -C_2 & 0 & 0 \\ C_1 & 0 & 0 & 0 \end{array} \right] \begin{bmatrix} w' \\ u \end{bmatrix} \tag{9}
$$

Define an operator $\mathcal{D} : L^2[0,h] \to L^2[0,h]$ by

$$
\mathcal{D} : w \mapsto \int_0^\theta C_1 e^{A_1(\theta-\tau)} B_1 w(\tau) d\tau \tag{10}
$$

This operator $\mathcal{D}$ is clearly compact.

The following theorem is obtained in [10]:

Theorem 5.1 *Suppose $\gamma > \|\mathcal{D}\|$. Then there exist discrete-time rational and causal transfer functions $G_{d,ij}$, $i,j = 1,2$ such that $\|T_{ew}\|_\infty < \gamma$ is equivalent to*

$$
\|z^{-m} G_{d,11} - G_{d,12} K G_{d,21}\|_\infty < \gamma. \tag{11}
$$

The latter problem is clearly solvable via standard discrete-time H^∞ solutions.

Proof   Write $K'(z)$ for $z^m K(z)$. Then by the equivalence between Fig. 5 and Fig. 6, condition $\|T_{ew}(K)\|_\infty < \gamma$ is equivalent to $\|T_{ew'}(K')\|_\infty < \gamma$. Since we assumed $\gamma > \|\mathcal{D}\|$, standard techniques [1, 7, 8] of converting the problem $\|T_{ew}(z^m K)\|_\infty < \gamma$ to an equivalent discrete-time problem work. Let

$$
G_d = \left[\begin{array}{c|cc} A_d & B_{1d} & B_{2d} \\ \hline C_{1d} & D_{11d} & D_{12d} \\ C_{2d} & D_{21d} & D_{22d} \end{array} \right] =: \begin{bmatrix} G_{d,11} & G_{d,12} \\ G_{d,21} & G_{d,22} \end{bmatrix}
$$

denote a discrete-time generalized plant derived via one of the methods in [7, 8]. An important point here is that the methods given in [7, 8] preserve the one-block structure. To be precise, they possess the property that

$$
A_d = \begin{bmatrix} e^{A_1 h} & 0 \\ 0 & e^{A_2 h} \end{bmatrix}
$$

$$
\begin{aligned}
B_{2d} &= \begin{bmatrix} 0 \\ \int_0^h e^{A_2(h-\tau)} B_2 d\tau \end{bmatrix} \\
C_{2d} &= \begin{bmatrix} C_1 & 0 \end{bmatrix} \\
D_{d21} &= 0, \quad D_{d22} = 0.
\end{aligned}
$$

This readily implies that $G_{d,22}$, i.e., $(2,2)$-block of G_d, is identically zero. Therefore, the closed-loop transfer function $T_{ew'}(K')$ becomes

$$
\begin{aligned}
T_{ew'}(K') &= G_{d,11} - G_{d,12} K' G_{d,21} \\
&= G_{d,11} - G_{d,12} z^m K G_{d,21}.
\end{aligned}
$$

Hence multiplying an inner function z^{-m} on both sides, we see that $\|T_{ew'}(z^m K)\|_\infty < \gamma$ reduces to condition (11). $\quad\square$

6 Numerical Example for Single-Rate Reconstruction

A numerical example has been computed for

$$
F_a(s) = \frac{1}{1+10s}, \quad P(s) = \frac{1}{1+s}
$$

with sampling period $h = 0.1$ and $m = 2$. The frequency response, in the sense of [14], from input w to error e is shown on Fig. 7(a). It exhibits a very high attenuation of error.

Steady-state output of $P(s)$ against input $w = \sin 0.1t$ is shown in Fig. 7(b). The solid curve is the output while the dashed curve is the input. This exhibits a very high precision reconstruction.

7 Extension to the Multirate Filter Bank

The single-rate solution described above can be extended to the multirate filter bank problem. Consider the block diagram Fig. 8.

The diagram is similar to the single-rate case Fig. 4, and the objective is the same: *For a given $\gamma > 0$, find a filter $K(z)$ such that*

$$
\|\mathcal{T}_{ew}(K)\|_\infty < \gamma.
$$

The difference is that the system involves the up/down sampling components denoted by $\uparrow 2$ and $\downarrow 2$. The *downsampler* $\downarrow 2$ acts on a sequence $x = \{x_n\}$ as

$$
(\downarrow 2x)(n) := x_{2n}, \tag{12}
$$

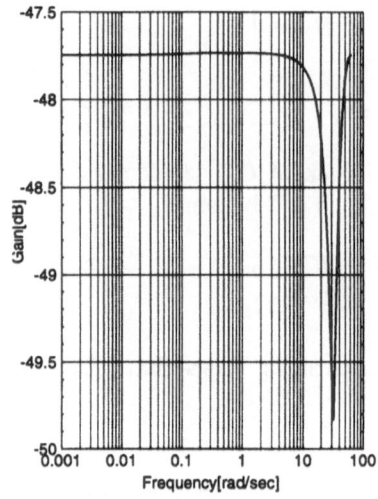

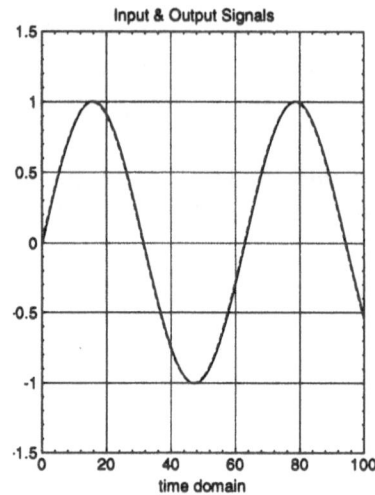

(a) Frequency Response from w to e

(b) Response against $\sin 0.1t$

Figure 7: Responses of the Designed Filter

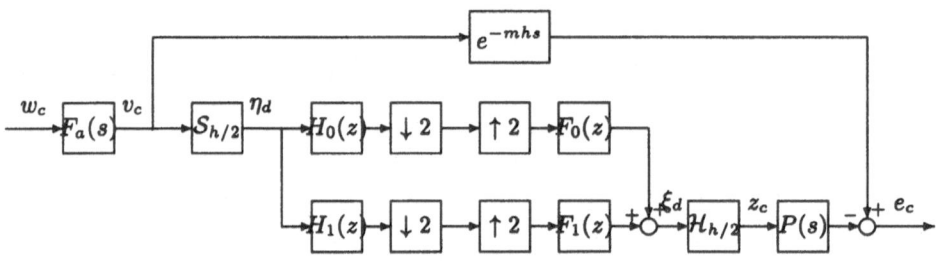

Figure 8: Multirate Filter Bank

that is, it extracts every odd numbered point. The *upsampler* acts on x by augmenting 0 in between every two elements:

$$(\uparrow 2x)(n) := \begin{cases} x_{n/2}, & n : \text{even} \\ 0, & n : \text{odd} \end{cases} . \tag{13}$$

The underlying idea is the following: The original continuous-time signal w_c first goes through an analog anti-aliasing filter $F_a(s)$. It is then sampled and then processed by digital filters $H_0(z)$ and $H_1(z)$. Usually, the former is low-pass and the latter is high-pass. Making use of this frequency selective nature, the filtered signal is then downsampled, without much information loss. This part is called the *analysis bank*. It then goes through a transmission channel, or stored in some memory device. The role of the *synthesis bank* is to reconstruct the incoming signal η_d or w_c as much as possible. The signal is first upsampled to go back to the original sampling rate, goes through the synthesis filters $F_0(z)$

and $F_1(z)$, and then becomes continuous-time signal via the hold device. It finally goes through a buffer amplifier or a low-pass filter to reduce remaining alias components. As in the single-rate case, we compare this signal with delayed continuous-time signal $e^{-mhs}F_a(s)w_c$, and minimize the H^∞ norm of T_{ew}.

The H^∞ design problem in the discrete-time setting was first studied by Chen and Francis [3]. In contrast to the single-rate case, the system in Fig. 8 is time-varying due to the up/down sampling components. They showed that this can be reduced to a single-rate problem by invoking the so-called polyphase representation. An extra complication arises here due to the continuous-time components $F_a(s), P(s)$ and $\mathcal{H}_{h/2}$, and the delay operator e^{-mhs}. We show that these defects can be circumvented. We will make use of both continuous-time and discrete-time lifting and give a characterization of polyphase representation in terms of lifting.

7.1 Reduction to a Single-Rate Problem

We start by giving some definitions.

Let M be a fixed positive integer. The *discrete-time lifting* [9] $\mathbf{L}_M$ by factor M is the correspondence that associates to $\sum_{n=0}^{\infty} a_n z^{-n}$ a vector valued z-transform as follows:

$$\mathbf{L}_M \left[\sum_{n=0}^{\infty} a_n z^{-n} \right] := \sum_{n=0}^{\infty} \begin{bmatrix} a_{nM} \\ \vdots \\ a_{nM+M-1} \end{bmatrix} \zeta^{-n}. \tag{14}$$

It stacks and combines M values $\{a_{nM}, \ldots, a_{nM+M-1}\}$ together as a vector-valued z-transform. To avoid confusion, we employ a different symbol ζ. Later it will be seen that it is natural to identify ζ with z^M. The lifted sequence $\mathbf{L}_M[\sum_{n=0}^{\infty} a_n z^{-n}]$ belongs to the space $\mathbf{R}^M[[\zeta^{-1}]]$ of formal power series with coefficients in $\mathbf{R}^M$.

There is a natural isomorphism $\mathbf{T} : \mathbf{R}^M[[\zeta^{-1}]] \cong (\mathbf{R}[[\zeta^{-1}]])^M$. The *polyphase components* of $\sum_{n=0}^{\infty} a_n z^{-n}$ are the M entries of $\mathbf{TL}_M[\sum_{n=0}^{\infty} a_n z^{-n}]$, i.e.,

$$\mathbf{TL}_M[\sum_{n=0}^{\infty} a_n z^{-n}] = \begin{bmatrix} \sum_{n=0}^{\infty} a_{nM} \zeta^{-n} \\ \sum_{n=0}^{\infty} a_{nM+1} \zeta^{-n} \\ \vdots \\ \sum_{n=0}^{\infty} a_{nM+M-1} \zeta^{-n} \end{bmatrix}$$

We denote this composed mapping $\mathbf{TL}_M$ by $\mathcal{P}_M$, and call it the *polyphase operator*. The following lemma is straightforward:

Lemma 7.1

$$[1, z^{-1}, \ldots, z^{-M+1}] \mathcal{P}_M \left[\sum_{n=0}^{\infty} a_n z^{-n} \right] \Bigg|_{\zeta = z^M}$$

$$= [1, z^{-1}, \ldots, z^{-M+1}] \begin{bmatrix} \sum_{n=0}^{\infty} a_{nM} z^{-nM} \\ \sum_{n=0}^{\infty} a_{nM+1} z^{-nM} \\ \vdots \\ \sum_{n=0}^{\infty} a_{nM+M-1} z^{-nM} \end{bmatrix} = \sum_{n=0}^{\infty} a_n z^{-n}.$$

Proof   Direct calculation.   □

Let P_1 be the projection to the first component: $\mathrm{R}^M[[\zeta^{-1}]] \to \mathrm{R}[[\zeta^{-1}]]$:

$$[x_1, \ldots, x_M]^T \mapsto x_1.$$

The *downsampler (decimator)* by factor M is the mapping $P_1 \mathcal{P}_M = P_1 T \mathbf{L}_M$ and is denoted by $\downarrow M$. It extracts every M value of the original sequence $\sum_{n=0}^{\infty} a_n z^{-n}$, i.e.,

$$\downarrow M : \sum_{n=0}^{\infty} a_n z^{-n} \mapsto \sum_{n=0}^{\infty} a_{nM} \zeta^{-n} \tag{15}$$

The z-transform symbol is changed to ζ, so it actually "skips" intermediate $M - 1$ signal values. The *upsampler (extender, interpolator)* $\uparrow M$ is the dual concept, and is defined by

$$(\uparrow M) : \sum_{n=0}^{\infty} a_n \zeta^{-n} \mapsto \mathbf{L}_M^{-1} \left(\sum_{n=0}^{\infty} \begin{bmatrix} a_n \\ 0 \\ \vdots \\ 0 \end{bmatrix} \zeta^{-n} \right)$$

$$= \sum_{n=0}^{\infty} a_n z^{-nM} \tag{16}$$

In the end result, the z-transform variable is changed to z by $\mathbf{L}_M^{-1}$, and the timing is increased by factor M. In terms of sequence, the result looks

$$a_0, 0, \ldots, 0, a_1, 0, \ldots, 0, a_2, 0, \ldots$$

We now state our main result:

Theorem 7.2 *The H^∞ norm of the input/output operator in Fig. 8 is the same as that given by Fig. 9, where $E(\zeta)$ and $R(\zeta)$ satisfy*

$$\begin{bmatrix} H_0(z) \\ H_1(z) \end{bmatrix} = E(z^2) \begin{bmatrix} 1 \\ z^{-1} \end{bmatrix} \tag{17}$$

$$\begin{bmatrix} F_0(z) & F_1(z) \end{bmatrix} = \begin{bmatrix} 1 & z^{-1} \end{bmatrix} R(z^2) \tag{18}$$

and

$$H_h(\theta) := \begin{cases} 1 & 0 \le \theta < h/2 \\ 0 & h/2 \le \theta < h \end{cases}. \tag{19}$$

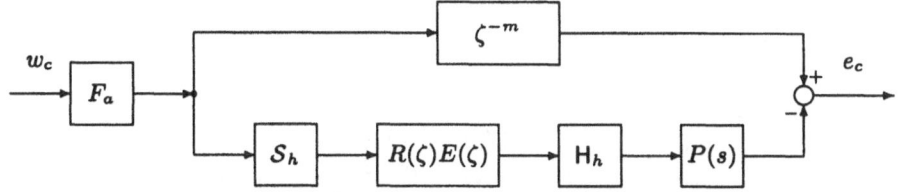

Figure 9: Reduced Single-Rate Problem

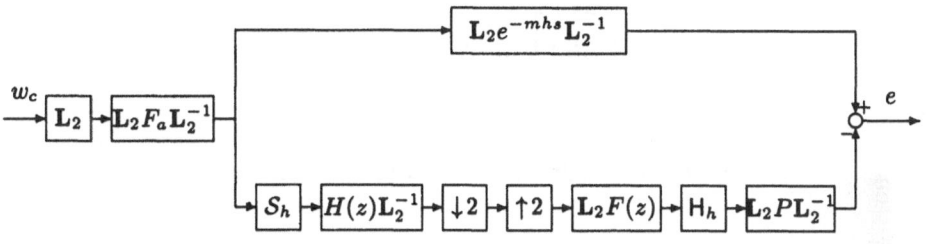

Figure 10: Lifted Multirate Problem

Proof We first unify the sampling rates to h. To this end, lift the signals by $\mathbf{L}_2$. This yields the diagram Fig. 10.

Here all the continuous-time signals are lifted, but the discrete-time signals before and after the up/down samplers are not lifted. Let us first consider the synthesis part. This is depicted in Fig. 11.

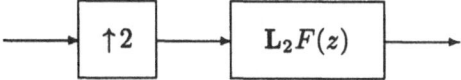

Figure 11: Synthesis Part

Now let (A, B, C, D) be a minimal realization of $F(z)$, and $\{u_k\}$ the input before the upsampler. Then it is easy to see that the state space realization of Fig. 11 is given by

$$\left[\begin{array}{c|c} A^2 & AB \\ \hline C & D \\ CA & CB \end{array}\right]. \tag{20}$$

Observe the identity

$$D + C(zI - A)^{-1}B = D + C(z^2I - A^2)AB + z^{-1}(CB + CA(z^2I - A^2)^{-1}AB).$$

This means that (20) simply gives the polyphase components of F, i.e., $\mathcal{P}_2(F) = R(\zeta)$. Therefore, the lifted output after the hold element is given by $H_h R(\zeta)\eta$ where η is the input to the upsampler.

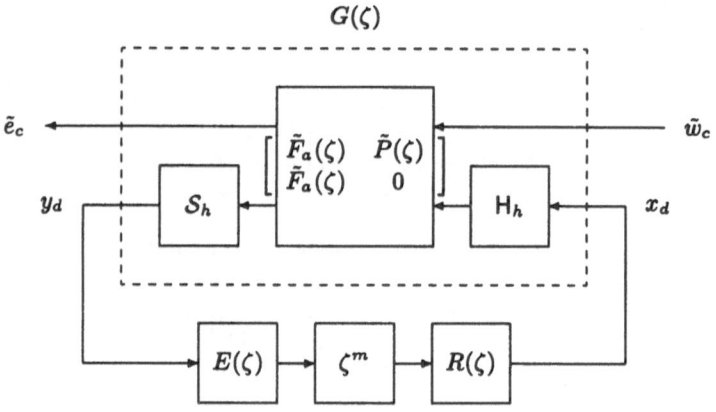

Figure 12: Modified Finite Dimensional Problem

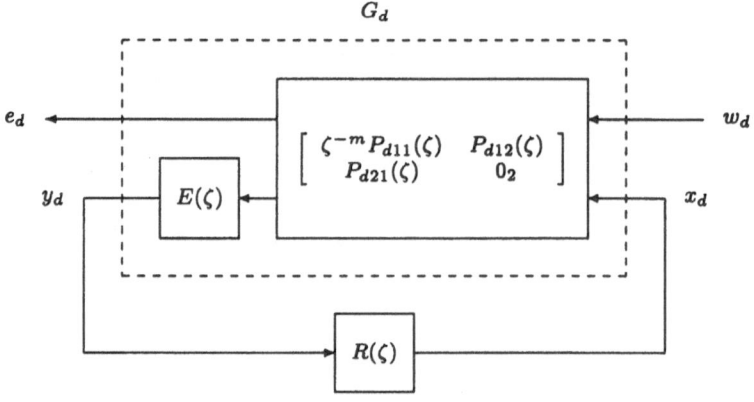

Figure 13: Equivalent Discrete-Time Problem

The analysis part is the adjoint of the synthesis part. Therefore, its input/output relation is given precisely by $\mathcal{P}_2(H) = E(\zeta)$. This yields the block diagram Fig. 9.   □

Observe now that the problem depicted in Fig. 9 is precisely in the situation we dealt with in the preceding section. Hence we can convert this problem to a finite-dimensional problem as given in Fig. 12

The state space realizations of $E(\zeta)$ and $R(\zeta)$ are as given in [3]. Then by the same argument as in the previous section, we can convert our problem to the block diagram Fig. 13.

Hence we obtain the following theorem.

Theorem 7.3 Let F_a, H_0, H_1, m, and h be as above, and

$$\mathbf{D}_{11} \quad : \quad L^2[0, h] \to L^2[0, h] :$$

$$w \mapsto C_c \int_0^\theta e^{A_c(\theta-\xi)} B_c w(\xi)\, d\xi$$

where (A_c, B_c, C_c) is a minimal realization of $F_a(s)$. Suppose $\|\mathbf{D}_{11}\| < 1$. Then there exists a discrete-time rational causal transfer function $P_d(\zeta)$ of the form

$$P_d(\zeta) = \begin{bmatrix} P_{d11}(\zeta) & P_{d12}(\zeta) \\ P_{d21}(\zeta) & 0 \end{bmatrix} \tag{21}$$

such that the following statements are equivalent:

(i) T_{ew} is internally stable and $\|T_{ew}\| < 1$.

(ii) $\mathcal{F}(G_d, R_d)$ is internally stable and $\|\mathcal{F}(G_d, R_d)\|_\infty < 1$, where G_d is given by

$$G_d(\zeta) := \begin{bmatrix} \zeta^{-m} P_{d11}(\zeta) & P_{d12}(\zeta) \\ E(\zeta) P_{d21}(\zeta) & 0 \end{bmatrix}$$

8 A Numerical Example

We have executed a design for the third-order Butterworth filter $H_0(z)$ with cut-off frequency 100π rad/sec, $m = 8$ and $h = 0.01$, and $H_1(z) = H_0(-z)$. The frequency response is shown in Fig. 14 $P(s)$ and $F_a(s)$ are as follows:

$$P(s) = \frac{1}{(\frac{s}{200\pi} + 1)^2}, \quad F_a(s) = \frac{1}{(\frac{s}{10\pi} + 1)},$$

Fig. 15(a) shows the frequency responses of F_0, F_1, and the time response against the input

$$w(t) = \sin(20\pi t) + \sin(10\pi t)$$

is shown in Fig. 15(b). It exhibits a fairly high fidelity in reconstruction. Observe also that the DC gain of F_0 is about 6 dB, compensating the gain-loss induced by the down/up sampling.

References

[1] B. Bamieh and J. B. Pearson, "A general framework for linear periodic systems with applications to H_∞ sampled-data control," *IEEE Trans. Autom. Control*, **AC-37**: 418–435, 1992.

[2] T. Chen and B. A. Francis, *Optimal Sampled-Data Control Systems*, Springer, 1995.

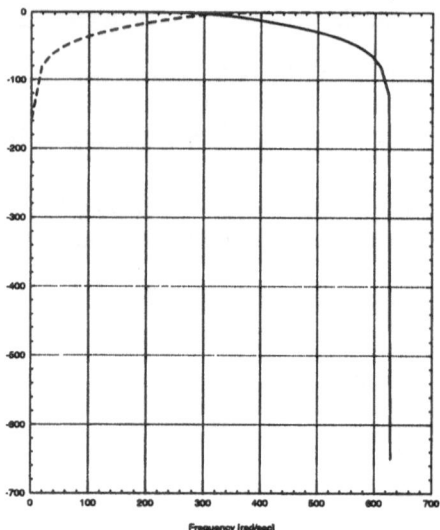

Figure 14: Butterworth 3rd Order H_0, H_1

[3] T. Chen and B. A. Francis, "Design of multirate filter banks by $\mathcal{H}_\infty$ opimization," *IEEE Trans. Signal Processing*, **SP-43**: 2822–2830, 1995.

[4] J. H. Davis, "Stability conditions derived from spectral theory: discrete systems with periodic feedback," *SIAM J. Control*, **10**: 1–13, 1972.

[5] S. Dasgupta, "A glimpse of multirate signal processing," *Semi-plenary Lecture at 4th ECC*: 329–358, 1997.

[6] N. J. Fliege, *Multirate Digital Signal Processing*, Wiley, 1994.

[7] Y. Hayakawa, S. Hara and Y. Yamamoto, "H_∞ type problem for sampled-data control systems—a solution via minimum energy characterization," *IEEE Trans. Autom. Control*, **39**: pp. 2278–2284, 1994.

[8] P. T. Kabamba and S. Hara, "Worst case analysis and design of sampled data control systems," *IEEE Trans. Autom. Control*, **AC-38**: 1337–1357, 1993.

[9] P. P. Khargonekar, K. Poolla and A. Tannenbaum, "Robust control of linear time-invariant plants using periodic compensation," *IEEE Trans. Autom. Control*, **AC-30**: 1088-1096, 1985.

[10] P. P. Khargonekar and Y. Yamamoto, "Delayed signal reconstruction using sampled-data control," *Proc. 35th IEEE CDC*, pp. 1259–1263 (1996)

[11] R. G. Shenoy, D. Burnside and T. W. Parks, "Linear periodic systems and multirate filter design," *IEEE Trans. Signal Processing*, **SP-42**: 2242–2256, 1994.

[12] P. P. Vaidyanathan *Multirate Systems and Filter Banks*, Prentice-Hall, 1993.

[13] Y. Yamamoto, "A function space approach to sampled-data control systems and tracking problems," *IEEE Trans. Autom. Control*, **AC-39**: 703–712, 1994.

[14] Y. Yamamoto and P. P. Khargonekar, "Frequency response of sampled-data systems," *IEEE Trans. Autom. Control*, **AC-41**: 166–176, 1996.

[15] A. I. Zayed, *Advances in Shannon's Sampling Theory*, CRC Press, 1993.

[16] K. Zhou and P. P. Khargonekar, "On the weighted sensitivity minimization problem for delay systems," *Syst. Control Lett.*, **8**: 307-312, 1987.

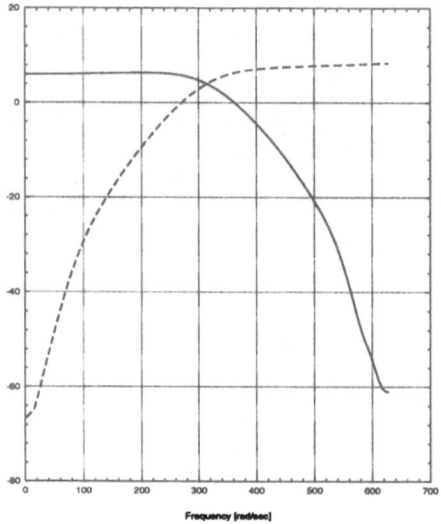

(a) Frequency Response of F_0, F_1

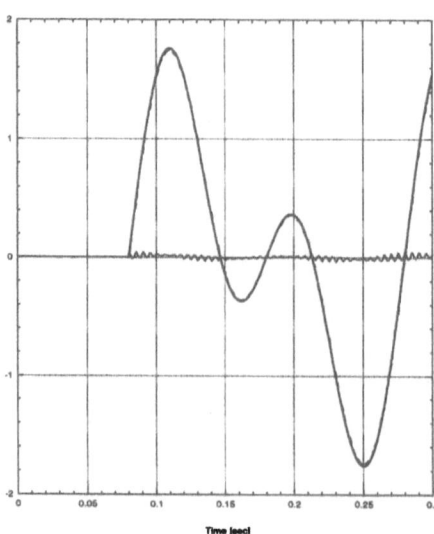

(b) Time Response

Figure 15: Responses of the Designed Multirate Filter

Optimum Biorthogonal Subband Coders

Soura Dasgupta*

Abstract

This paper considers biorthogonal maximally decimated uniform filter banks as subband coders. Subject to assumptions of optimum bit allocation and white decorrelated quantizer distortion, it derives the filter bank that maximizes the coding gain. Ingredients to the solution include design techniques developed by Vaidyanathan for optimal orthonormal subband coders and a generalization of the half whitening process that is known to maximize the coding gain of 1-channel biorthogonal filter banks.

1 Introduction

This paper provides a solution to the design of biorthogonal, maximally decimated uniform filter banks that achieve maximum coding gain. Consider figure 1 depicting a M channel uniform filter bank, with analysis filters $H_i(z^{-1})$ and synthesis filters $F_i(z^{-1})$. The Q_i depict quantizers allocated b_i bits and with the average bit rate fixed at

$$b = \frac{1}{M} \sum_{i=0}^{M-1} b_i. \tag{1}$$

The goal is to minimize the distortion generated by the quantizer in the mean square sense, i.e. to minimize the variance of $\hat{x}(k) - x(k)$. The coding gain refers to the ratio of the distortion variance had $x(k)$ been simply transmitted after a b bit quantization, and the distortion variance induced by the filter bank.

It is well known, [1] that the setting of fig. 1 is equivalent to that of fig 2 where $E(z^{-1})$ and $R_i(z^{-1})$ are $M \times M$ transfer function matrices representing the type 1 and type 2 polyphase matrix of the analysis and synthesis filters respectively. We say that the filter bank is biorthogonal if $R_i(z^{-1})E(z^{-1}) = I$. Subject to this requirement we will select $E(z^{-1})$ to maximize the coding gain.

The coding gain maximization problem has been considered by several authors, e.g. [3]-[7]. Some of these e.g. [5]-[7] involve tree structured filter

---

*Department of Electrical and Computer Engineering, The University of Iowa, Iowa City, IA-52242, USA. He was visiting the Department of Systems Engineering, Australian National University, ACT 0200, Australia, when part of this work was completed. Supported in part by NSF grants ECS-9211593 and ECS-9350346.

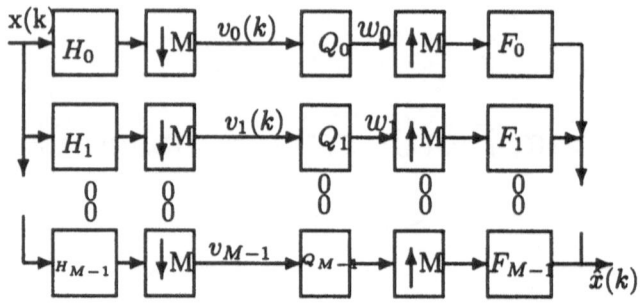

Figure 1: A Maximally Decimated Uniform Filter Bank

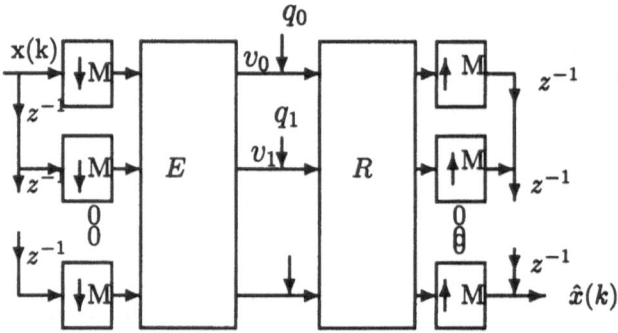

Figure 2: Polyphase Description

banks as opposed to the uniform filter banks considered here. The paper that has most influenced our thinking is by Vaidyanathan, [2], who considers coding gain maximization under the added constraint of orthonormality, i.e. when $E(e^{-j\omega})$ is unitary at all ω. Having given necessary and sufficient conditions on orthonormal $E(z^{-1})$ that maximize the coding gain, [2] gives an elegant sequence of compaction filter based design that synthesize the optimizing H_i. This reference thus generalizes the known work on transform coders, which is a special case of orthonormal filter banks but with constant E.

It is well known that relaxing the orthonormality condition to mere biorthogonality improves the coding gain. For example, while in the orthonormal case of M=1, no coding gain is forthcoming, even in this one-channel case coding gain can be obtained under the biorthogonality constraint. The filter that achieves this is the classical half whitening filter, [8]. Accordingly, in this paper, we address the coding gain maximization problem under the sole condition that the filter bank be biorthogonal. Specifically we show that some of the necessary conditions provided by [2] remain necessary in the biorthogonal case and provide a *generalized half whitening* process that we show achieves coding gain maximization. We show that our procedure benefits from the procedure in [2] in that a part of the optimal design uses the filters derived in [2]. As with all such design, we assume an optimum bit allocation strategy, [1]. In addition we assume that the quantizer noise processes are mutually decorrelated and white, an assumption that applies to high bit rates.

Section 2 gives preliminaries. Section 3 shows the necessity of some of the conditions given in [2] and provides a reformulation. Section 4 gives the solution. Section 5 is the conclusion. Proofs are omitted.

2 Preliminaries

Our analysis assumes the white uncorrelated quantizer model, generally valid at high bit rates, [8]. Thus in figure 1, if the quantizer Q_i has been allocated b_i bits and that its output obeys

$$w_i(k) = v_i(k) + q_i(k)$$

where

$$E[q_i(k)q_j(l)] = 2^{-2b_i}\sigma_{v_i}^2 \delta(k-l)\delta(i-j),$$

δ being the Kronecker delta and $\sigma_{v_i}^2$ the variance of the subband signal v_i. With the average bit rate fixed at

$$b = \frac{1}{M} \sum_{i=0}^{M-1} b_i, \tag{2}$$

we will assume that the optimum bit allocation strategy is enforced, [1], i.e.

$$2^{-2b_i}\sigma_{v_i}^2 \|F_i\|^2 = 2^{-2b_l}\sigma_{v_l}^2 \|F_l\|^2 \quad \forall i, l. \tag{3}$$

Here $\|F_i\|$ is the l_2-norm of the impulse response of F_i. Under these conditions, the coding gain for a general choice of F_i and H_i is given by, [2],

$$G_{SBC} = \frac{\sigma_x^2}{\left(\prod_{i=0}^{M-1} \sigma_{v_i}^2 \|F_i\|^2\right)^{1/M}} \tag{4}$$

where σ_x^2 is the variance of the filter bank input x.

Turn now to the polyphase representation of the filter bank given in figure 2. Define for each i in $\{0, \cdots, M-1\}$ e_i to be the M-vector with $(i+1)$-th element 1 and the rest zero. Then it is readily seen that

$$\|F_i\|^2 = \frac{1}{2\pi} \int_0^{2\pi} e_i' R'(e^{j\omega}) R(e^{-j\omega}) e_i d\omega. \tag{5}$$

Similarly, with $S_x(\omega)$ the power spectral density (psd) matrix of the vector signal $[x_0(k), \cdots x_{M-1}(k)]'$,

$$\sigma_{v_i}^2 = \frac{1}{2\pi} \int_0^{2\pi} e_i' E(e^{-j\omega}) S_x(\omega) E'(e^{j\omega}) e_i d\omega. \tag{6}$$

Then the optimal biorthogonal subband coder design problem reduces to the selection of $E(e^{-j\omega})$ and $R(e^{-j\omega})$, with

$$R(e^{-j\omega}) = E^{-1}(e^{-j\omega}) \qquad (7)$$

such that under (5,6)

$$J = \prod_{i=0}^{M-1} \sigma_{v_i}^2 \|F_i\|^2 \qquad (8)$$

is minimized. In orthonormal subband coding one imposes the additional restriction that $E(e^{-j\omega})$ be orthonormal, i.e. obey for all ω

$$E(e^{-j\omega})E'(e^{j\omega}) = I.$$

At this point it is instructive to consider the solution to the optimum orthonormal subband coding problem given in [2]. Since $S_x(\omega)$ is positive definite, Hermitian symmetric, (we will exclude singular cases), there exists an orthonormal $\Omega(e^{j\omega})$ and real positive $\lambda_i(\omega)$ obeying for all ω

$$\lambda_0(\omega) \geq \lambda_1(\omega) \geq \cdots \geq \lambda_{M-1}(\omega) \qquad (9)$$

such that under

$$\Lambda(\omega) = diag\ \{\lambda_0(\omega), \lambda_1(\omega), \cdots, \lambda_{M-1}(\omega)\} \qquad (10)$$

$$S_x(\omega) = \Omega'(e^{j\omega})\Lambda(\omega)\Omega(e^{-j\omega}). \qquad (11)$$

In the sequel $\Lambda(\omega)$ will be referred to as the *principal spectrum* of $[x_0(k), \cdots x_{M-1}(k)]'$. Then the solution given in [2] entails that

$$E(e^{-j\omega}) = \Omega(e^{-j\omega}). \qquad (12)$$

Consequently the psd matrix, $S_v(\omega)$ of $[v_0(k), \cdots v_{M-1}(k)]'$ obeys

$$S_v(\omega) = \Lambda(\omega). \qquad (13)$$

[2] also provides the direct design of the $H_i(z)$ leading to the $E(e^{-j\omega})$ as in (12) through a sequence of compaction filter design.

The diagonal nature of $\Lambda(\omega)$ assures that the subband signals $v_i(k)$ obey,

$$E[v_i(k)v_j(l)] = 0, \quad \forall i \neq j, k, l.$$

In fact it is shown in [2] that this *subband decorrelation* property is necessary for the optimal orthonormal subband coding gain to be attained. In Section 3 we show that subband decorrelation remains a necessary condition even in the biorthogonal case considered here, though of course the optimum subband signals will both obey (13).

It turns out however, that the principal spectrum continues to play an all important role in optimum biorthogonal design. We will assume that its elements are piecewise continuous with the lack of continuity occurring only at a finite number of frequencies.

3 Necessity of Subband Decorrelation

In this Section, we first establish the necessity of subband decorrelation even in the biorthogonal case. This leads to a reformulation that facilitates the eventual solution. For every invertible $E(e^{-j\omega})$, there exists real positive definite diagonal $\Gamma(\omega)$ with diagonal elements $\{\gamma_0(\omega), \gamma_1(\omega), \cdots, \gamma_{M-1}(\omega)\}$ and orthonormal $U(e^{-j\omega})$, $V(e^{-j\omega})$ such that

$$E(e^{-j\omega}) = U(e^{-j\omega})\Gamma(\omega)V(e^{-j\omega}). Thus \tag{14}$$

$$R(e^{-j\omega}) = V'(e^{j\omega})\Gamma(\omega)^{-1}U'(e^{j\omega}). \tag{15}$$

For a fixed $\Gamma(\omega)$ consider the selection of $U(e^{-j\omega})$ and $V(e^{-j\omega})$ to optimize, J. The following holds.

Theorem 3.1 *For every $\Gamma(\omega)$ as above, under (14, 15) there exist $P_i(\omega), i = 1, 2$ such that*

- *$P_i(\omega)$ is a permutation matrix at each ω.*

- *With $U(e^{-j\omega}) = P_1(\omega)$ and $V(e^{-j\omega})\Omega'(e^{j\omega}) = P_2(\omega)$, (5) and (6) are simultaneously minimized.*

Further for J to be minimum both $U(e^{-j\omega})$ and $V(e^{-j\omega})\Omega'(e^{j\omega})$ must take permutation matrix values at all but isolated values of ω.

Should $U(e^{-j\omega})$ and $V(e^{-j\omega})\Omega'(e^{j\omega})$ be permutation matrices at each ω, then

$$S_v(\omega) = E(e^{-j\omega})S_x(\omega)E'(e^{j\omega}) \tag{16}$$

is diagonal at all frequencies. Thus subband decorrelation is indeed necessary, if attention is restricted to piecewise continuous $U(e^{-j\omega})$ and $V(e^{-j\omega})$.

This permits a reformulation of the optimization problem. Without loss of generality one can confine attention to the class of $E(e^{-j\omega})$whose members force

$$S_v(\omega) = D(\omega)\Lambda(\omega)D(\omega) \tag{17}$$

for some real positive definite diagonal $D(\omega)$ with diagonal elements $\{d_0(\omega), d_1(\omega), \cdots, d_{M-1}(\omega)\}$. Then we have the following Lemma.

Lemma 3.1 *$E(e^{-j\omega})$ obeys (16, 17) iff there exists an orthonormal $U(e^{-j\omega})$ such that for all ω*

$$E(e^{-j\omega}) = D(\omega)\Lambda^{1/2}(\omega)U'(e^{j\omega})\Lambda^{-1/2}(\omega)\Omega(\omega). \tag{18}$$

Thus our task reduces to selecting $D(\omega)$ and $U(e^{-j\omega})$. Under (18)

$$\sigma_{v_i}^2 = \frac{1}{2\pi} \int_0^{2\pi} d_i^2(\omega)\lambda_i(\omega)d\omega \tag{19}$$

$$\|F_i\|^2 = \frac{1}{2\pi} \int_0^{2\pi} \frac{\lambda_i(\omega) e_i' U'(e^{j\omega}) \Lambda^{-1}(\omega) U(e^{-j\omega}) e_i}{d_i^2(\omega)} d\omega. \tag{20}$$

For any $U(e^{-j\omega})$, from the Cauchy-Schwarz inequality one has then that with

$$g_i = \int_0^{2\pi} \lambda_i(\omega) \sqrt{e_i' U'(e^{j\omega}) \Lambda^{-1}(\omega) U(e^{-j\omega}) e_i} \, d\omega \tag{21}$$

$$\sigma_{v_i}^2 \|F_i\|^2 \geq \frac{1}{4\pi^2} g_i^2. \tag{22}$$

Equality is achieved by choosing

$$d_i^4(\omega) = e_i' U'(e^{j\omega}) \Lambda^{-1}(\omega) U(e^{-j\omega}) e_i. \tag{23}$$

Thus the problem reformulates to: Find orthonormal $U(e^{-j\omega})$ to minimize

$$\prod_{i=0}^{M-1} g_i. \tag{24}$$

Then (23) gives $d_i(\omega)$ and (18) the analysis banks type 1 polyphase representation.

We will call (23) and the minimization of (24) *generalized half whitening*. Indeed in the $M = 1$ case this specializes to the traditional half whitening process, [8].

4   Generalized Half Whitening

In this Section we solve the optimization of (24). We need the following Lemma.

Lemma 4.1 *Suppose $U(e^{-j\omega})$ is orthonormal and $\Lambda(\omega)$ is as above. Then*

$$\sqrt{e_i' U'(e^{j\omega}) \Lambda^{-1}(\omega) U(e^{-j\omega}) e_i} \geq e_i' U'(e^{j\omega}) \Lambda^{-1/2}(\omega) U(e^{-j\omega}) e_i$$

with equality holding if $U(e^{-j\omega})$ is a permutation at each ω.

Define

$$f_i = \int_0^{2\pi} \lambda_i(\omega) e_i' U'(e^{j\omega}) \Lambda^{-1/2}(\omega) U(e^{-j\omega}) e_i \, d\omega \tag{25}$$

Then as long as the minimum of (26) below occurs for $U(e^{-j\omega})$ that has only permutation values, it suffices to choose $U(e^{-j\omega})$ to minimize (26).

$$\prod_{i=0}^{M-1} f_i. \tag{26}$$

Indeed the next Theorem shows that the optimizing $U(e^{-j\omega})$ does indeed take only permutation values.

Theorem 4.1 *The $U(e^{-j\omega})$ minimizing (26) is a permutation matrix at each ω. Further suppose f_i^* are the values of f_i corresponding to the minimum of (26). Then, for each ω, $U(e^{-j\omega})$ is any permutation matrix that permits the ordering*

$$\frac{f_i^*}{e_i'U'(e^{j\omega})\Lambda(\omega)U(e^{-j\omega})e_i} \leq \frac{f_{i+1}^*}{e_{i+1}'U'(e^{j\omega})\Lambda(\omega)U(e^{-j\omega})e_{i+1}} \qquad (27)$$

for all i in $\{0, \cdots, M-1\}$

How to select these permutation values for $U(e^{-j\omega})$. Below we give a solution, for $M = 2$. To avoid notational complexity only a sketch is given for $M > 2$. In the sequel $P(i,j)$ will be the permutation matrix that exchanges indices i and j. For $M = 2$, generate a sequence of functions $U_k(e^{-j\omega})$ in the following way.

$$U_0(e^{-j\omega}) = I$$

$$f_i^*(k) = \int_0^{2\pi} \lambda_i(\omega)e_i'U_k'(e^{j\omega})\Lambda^{-1/2}(\omega)U_k(e^{-j\omega})e_i\,d\omega$$

and

$$U_{k+1}(e^{-j\omega}) = \begin{cases} I; & \frac{f_0^*(k)}{f_1^*(k)} \leq \frac{\lambda_0(\omega)}{\lambda_1(\omega)} \\ P(0,1); & else \end{cases} \qquad (28)$$

One can show this sequence of functions $U_k(\omega)$ converges to the optimizing $U(e^{-j\omega})$, for $M = 2$. One can view this as a process of selecting at each frequency whether to swap rows 0 and 1 of the analysis bank. For $M > 2$ this must be continued in an obvious way to determining whether for each $0 \leq i \leq M-2$ and $i < j \leq M-1$ whether rows i and j must be swapped at the various frequencies. Again one can show that for $M > 2$ also, this process converges to the optimizing $U(e^{-j\omega})$.

Thus the procedure for selecting optimum biorthogonal subband coders is complete. To summarize:

- Choose the matrix $U(e^{-j\omega})$ using the procedure of this Section.

- Choose

$$d_i(\omega) = \left[e_i'U'(e^{j\omega})\Lambda^{-1}(\omega)U(e^{-j\omega})e_i\right]^{\frac{1}{4}}. \qquad (29)$$

- Choose $E(e^{-j\omega})$ as in (18). Observe $\Omega(e^{-j\omega})$ can be obtained from the compaction filter based design of [2].

Notice should

$$U(e^{-j\omega})(\omega) = I, \quad \forall\omega$$

then,

$$d_i(\omega) = [\lambda_i(\omega)]^{-\frac{1}{4}}$$

i.e. the half whitener for $\lambda_i(\omega)$. Such a selection will be called *simple half whitening*. Below we give a necessary and sufficient condition for simple half whitening to be optimizing.

Theorem 4.2 *The biorthogonal coding gain is optimized by simple half whitening iff for all ω and i*

$$\frac{\int_0^{2\pi} \sqrt{\lambda_i(\omega)}d\omega}{\lambda_i(\omega)} \leq \frac{\int_0^{2\pi} \sqrt{\lambda_{i+1}(\omega)}d\omega}{\lambda_{i+1}(\omega)}$$

5 Conclusion

We have provided the solution to the maximization of the coding gain of subband coders represented by maximally decimated uniform filter banks. Assumptions underlying our work include optimum bit allocation and white decorrelated quantizer distortions. The paper thus generalizes the work of Vaidyanathan, [2] for optimal orthonormal subband coders. The solution involves augmenting the design generated by [2] with a generalization of the traditional half whitening process. Extensions to tree structured and time varying filter banks are areas that bear further investigation.

References

[1] P.P. Vaidyanathan, *Multirate Systems and Filter Banks*, Prentice Hall, 1992.

[2] P.P. Vaidyanathan, "Optimal orthonormal filter banks", *Proceedings of ICASSP*, 1996.

[3] H. S. Malavar and D. H. Staelin, "The LOT: Transform coding without blocking effects", *IEEE Transactions on Accoustics Speech and Signal Processing*, ASSP-38, pp 553-559, 1990.

[4] R. L. de Queiroz and H. S. Malavar , " On the asymptotic performance of hierarchical transforms", *IEEE Transactions on Signal Processing*, pp 2620-2622, 1992.

[5] K. Ramachandran and M. Vetterli, "Best wavelet packet bases in a rate distortion sense", *IEEE Transactions on Signal Processing*, pp 160-174, 1993.

[6] A. Tewfik, D. Sinha and P.E. Jorgensen, "On the optimal choice of wavelet for signal representations", *IEEE Transactions on Information Theory*, pp 747-765, 1992.

[7] R. A. Gopinath, J. Odegard and C. S. Burrus, "Optimal wavelet representation of signals and the wavelet sampling theorem", *IEEE Transactions on Circuits and Systems*, pp 262-277, 1994.

[8] N.S. Jayant and P. Noll, *Digital Coding of Waveforms*, Prentice Hall, 1984.

Controller Switching Based on Output Predictions *

Judith Hocherman-Frommer   Sanjeev R. Kulkarni

Peter J. Ramadge [†]

Abstract

We analyse a switching control system for controlling a plant with unknown parameters so that the output asymptotically tracks a reference signal. The controller is selected on-line from a given set of controllers according to a switching rule based on output prediction errors. We provide sufficient conditions under which the switched closed loop control system is exponentially stable and asymptotically achieves good tracking control even if the switching does not stop.

1   Introduction

In this paper we consider the problem of controlling a fixed linear continuous-time plant with unknown parameters so that the plant output asymptotically tracks, with some desired accuracy, a bounded reference input. The control strategy that we analyze is based on switching, at certain decision times, among a family of fixed controllers based on a switching logic that attempts to select a good predictor for the plant.

We have available a family of model and controller pairs (Σ_p, Γ_p), $p \in \mathcal{P}$. The index set $\mathcal{P}$ may be finite, countable, or a compact subset of some metric space. Controller Γ_p stabilizes model Σ_p and yields desired asymptotic tracking performance for a class of admissible reference signals. For each model we run a corresponding predictor O_p driven by the inputs and outputs of the plant and the resultant prediction errors are used to form a real-valued performance measure for predictor p. Then at certain decision times a supervisor uses the performance measures to select a controller from the family $\{\Gamma_p, p \in \mathcal{P}\}$ to be connected in feedback with the plant. The resultant switched control system must ensure boundedness of the process states and satisfy an asymptotic tracking performance criterion.

---

*This work was supported in part by the National Science Foundation under grants IRI-9457645 and ECS-9216450, and by EPRI under grant RP8030-18.

†Department of Electrical Engineering, Princeton University, Princeton NJ 08544

We consider the simplest case in which the transfer function of the unknown plant exactly matches that of one of the known deterministic models $\{\Sigma_p,\ p \in \mathcal{P}\}$. This may be regarded as a case of purely parametric uncertainty. Although unrealistic, this situation is of theoretical interest since it provides a lower bound on what can be expected in practice.

The above architecture for on-line controller switching has been proposed and examined in several special cases in [13], [14], [15], [8], [12], [19], [2]. In [13], [14] the author studies the problem of tracking a constant set point for siso lti systems. At a sequence of decision times, the performance of the predictors is compared and the controller corresponding to the best predictor at that time is selected. The sequence of selections is not required to converge and in general will not do so. Nevertheless, the system variables remain bounded and the output of the siso system converges to the constant set-point. In [12] and [19] switching is used to select a controller structure matched to the similarity invariants of the plant, and in [2], [16], [17] it is used to improve the transient performance of stable adaptive control schemes. In [8] convergent decision rules are studied. It is shown that for mimo lti systems there exists a convergent selection rule under which the supervised control system is stable and satisfies a performance criterion with respect to a class of admissible inputs. Several other controller switching strategies have been examined in the literature. Generally these involve strategies that use a predefined search sequence, e.g., [5], [10], [11].

This paper is a summary of results originally obtained in [6]. We examine an asymptotic tracking problem in a general setting that extends previous work on set point control problems. In particular, we show that the output of the prediction based supervised control system is the sum of the output of a time-varying system in which at every moment Σ_p is controlled by the concurrent controller Γ_p, and a prediction error term. The stability of the above mentioned time-varying system can be assured using standard results from the literature on slowly time-varying systems. We give two main results. Theorem 4.2 provides sufficient conditions on the models and controllers such that even if switching between the candidate controllers does not stop, selecting a good predictor will imply good tracking control. As a special case, this provides a more direct proof of a result given in [14]. Theorem 5.1 shows that under a mild additional assumption, the key condition in our set of sufficient conditions is always satisfied if the switched control system is required to asymptotically exactly track the reference signal. The proofs of all results as well as additional details can be found in [6].

2   Formulation

We can select the input $u(t)$ and observe the output $y(t)$ of an unknown siso system Σ, hereafter called the "plant", with McMillan Degree at most n and a stabilizable and detectable $\bar{n}$-dimensional state space realization:

$$
\begin{aligned}
\Sigma \; : \quad \dot{x}(t) &= Ax(t) + Bu(t) \; ; \quad x(t_0) = x_0 \\
y(t) &= Cx(t)
\end{aligned}
\tag{1}
$$

The objective is to select the input $u(t)$ so that the output $y(t)$ asymptotically adequately tracks a reference signal $r(t)$ generated as the output of a finite dimensional autonomous linear system of the form:

$$
\begin{aligned}
\Xi \; : \quad \dot{\phi}(t) &= f\phi(t), \quad \phi(0) = \phi_0, \\
r(t) &= g\phi(t),
\end{aligned}
\tag{2}
$$

where for each initial condition ϕ_0 the state trajectory of Ξ is bounded.

We are given a family of linear time-invariant systems $O_p, p \in \mathcal{P}$, with a common state realization of the form:

$$
\begin{aligned}
O_p \; : \quad \dot{w}(t) &= Mw(t) + Nu(t) + Ky(t) \; ; \quad w(t_0) = w_0 \\
\hat{y}_p(t) &= C_p w(t) \\
e_p(t) &= C_p w(t) - y(t),
\end{aligned}
\tag{3}
$$

where $w(t) \in \mathbf{R}^{k_1}$, $u(t) \in \mathbf{R}$, $y(t) \in \mathbf{R}$; and the dimensions of the matrices M, N, K and C_p are appropriate. This system will be used as a predictor of the plant output. The predictor outputs are: $\hat{y}_p$, the prediction of $y(t)$, and e_p, the corresponding prediction error, $p \in \mathcal{P}$.

If we set $y(t) = \hat{y}_p(t)$ in O_p, then we obtain a realization of a linear time-invariant system $\Sigma_p, p \in \mathcal{P}$:

$$
\begin{aligned}
\Sigma_p \; : \quad \dot{x}_p(t) &= A_p x_p(t) + Nu(t) \; ; \quad x_p(t_0) = x_0 \\
y_p(t) &= C_p x_p(t)
\end{aligned}
\tag{4}
$$

where $A_p \triangleq M + KC_p$. We call Σ_p the pth model.

Similarly, we are given a family of linear time-invariant systems $\Delta_p, p \in \mathcal{P}$, with a common state realization of the form:

$$
\begin{aligned}
\Delta_p \; : \quad \dot{z}(t) &= Fz(t) + Gy(t) + Lu(t) + Rr(t) \; ; \quad z(t_0) = z_0 \\
u_p(t) &= H_p z(t) + S_p y(t) + T_p r(t)
\end{aligned}
\tag{5}
$$

where $z(t) \in \mathbf{R}^{k_2}$ and the dimensions of F, G, L, R, H_p, S_p and T_p are appropriate. Under the feedback connection $u(t) = u_p(t)$ in Δ_p we obtain a realization of a linear time-invariant system Γ_p which will call the pth controller.

The assumption of a common state realization for the predictors and controllers is standard in adaptive control. See for example [14], [15]. Further details are also given in [6].

The equations of the system (Σ_p, Γ_p) consisting of Σ_p in feedback connection with Γ_p are:

$$\dot{x}_{pp}(t) = A_{pp}x_{pp}(t) + \begin{pmatrix} R + LT_p \\ NT_p \end{pmatrix} r(t) \tag{6}$$

$$y_{pp}(t) = (0, \ C_p)\, x_{pp}(t)$$

where

$$A_{pp} \triangleq \begin{pmatrix} F + LH_p & GC_p + LS_pC_p \\ NH_p & M + KC_p + NS_pC_p \end{pmatrix}. \tag{7}$$

The models and controllers, parameterized as indicated above, will be assumed to satisfy the following constraints:

Assumption A1 The matrices M and F are Hurwitz.

Assumption A2 The matrix A_{pp} is Hurwitz with stability margin γ.

Assumption A3 The controlled system (Σ_p, Γ_p) yields acceptable asymptotic tracking performance over the admissible class of reference signals.

Assumption A4 $\mathcal{P}$ is a compact metric space with metric $d(\cdot, \cdot)$.

Assumption A5 The functions $f_1: p \mapsto C_p$, $f_2: p \mapsto H_p$, $f_3: p \mapsto S_p$, and $f_4: p \mapsto T_p$ are continuous with respect to the metric on $\mathcal{P}$ and any matrix norm.

Together, **A3** and **A2** impose the constraint that the pth controller should stabilize the pth model with a stability margin γ that is independent of p and satisfy the asymptotic tracking criterion. Examples of suitable criteria include: $\lim_{t\to\infty} |y(t) - r(t)| = 0$ (asymptotic exact tracking); and $\limsup_{t\to\infty} |y(t) - r(t)| \leq \epsilon$ (asymptotic ϵ-tracking). Other criteria are also clearly possible. Assumption **A4** can be satisfied by the discrete metric if $\mathcal{P}$ is finite. If $\mathcal{P}$ is a closed and bounded subset of $\mathbf{R}^k$ for some positive integer k, then we might take $d(\cdot, \cdot)$ to be the metric induced by a suitable norm on $\mathbf{R}^k$. If $\mathcal{P}$ is finite and $d(\cdot, \cdot)$ is the discrete metric, then Assumption **A5** is trivially satisfied. However, when $\mathcal{P}$ is a compact subset of $\mathbf{R}^k$ it is a nontrivial assumption – it requires that we have designed the controllers so that they satisfy **A2** and **A3** *and* vary continuously with respect to $p \in \mathcal{P}$. To show that such parameterizations exist it is sufficient to assume that there exists a design procedure for determining a controller transfer function $g(s)$ from a stabilizable and detectable plant realization such that **A2** and **A3** are satisfied and such that the parameters in $g(s)$ vary continuously as the parameters in the plant realization vary over some open set containing their nominal values. In this case, we can take p to be the vector of entries in C_p and $f_1: p \mapsto C_p$ is then obviously continuous. Suppose that p_o is the parameter of a nominal plant;

then Σ_{p_o} is a stabilizable and detectable realization of the corresponding plant transfer function. Let $\mathcal{P}_o$ be the closure of any open ball containing p_o such that $\{\Sigma_p : p \in \mathcal{P}_o\}$ is contained in an open set about Σ_{p_o} on which the controller design procedure is continuous. Then let the maps $f_2 : p \mapsto H_p$, $f_3 : p \mapsto S_p$, and $f_4 : p \mapsto T_p$ select the appropriate values of H_p, S_p, and T_p from the designed controller transfer function using standard methods. By choice of $\mathcal{P}_o$ these functions are continuous. More generally, we can take $\mathcal{P} = \cup_{k=1}^N \mathcal{P}_k$ where the $\mathcal{P}_k$ are constructed like $\mathcal{P}_o$.

At each term τ_k of a strictly monotone increasing sequence of switching times (that may depend on the initial condition of the plant, predictor, and controller), the controller connected to the plant will be "switched" among the family parameterized by $p \in \mathcal{P}$. Given $\tau_D > 0$ we require that the sequence $\{\tau_k, k \geq 0\}$ satisfies $\tau_{k+1} - \tau_k \geq \tau_D$, for each $k \geq 0$. Such a sequence is said to be τ_D-*admissible*. It will not be important exactly how the switching times are selected, only that they satisfy this form of constraint.

The selection of which controller to be connected into feedback with the plant at switching time τ_k is based on performance indices for each of the predictors. The performance index $J(t, p)$ of predictor p at time t is a function of the prediction error signal $e_p(s), s \in [0, t]$. For example, for fixed $\lambda > 0$ we might set

$$J(t,p) = \int_0^t e^{-\lambda(t-s)} |e_a(s)|^2 ds \qquad (8)$$

Then at each switching time τ_k, an index $q_k \in \mathcal{P}$ is selected based on the values $\{J(t,p), t \in [0, \tau_k], p \in \mathcal{P}\}$ according to a specified decision rule. The controller Γ_{q_k}, driven by the reference input r, is then connected in feedback with the plant over the time interval $[\tau_k, \tau_{k+1})$. The simplest instance of a decision rule is a fixed function $g : \mathbf{R}^{\mathcal{P}} \to \mathcal{P}$ with $q_k = g(J(\tau_k, p), p \in \mathcal{P})$. For example, the rule used in [14] is (roughly)

$$q_k = \operatorname{argmin}_{p \in \mathcal{P}} \{J(\tau_k, p)\} \qquad (9)$$

More complex rules can easily be envisioned, see e.g., [8], [12]. However, all that will be important for our investigation is that the rule has certain basic properties.

Let $\sigma(t)$ denote the piecewise-constant signal taking values in $\mathcal{P}$ that specifies the controller in use at time t and let $\bar{\sigma}$ denote its set of limit points in $\mathcal{P}$. By assumption A4 , $\bar{\sigma}$ is nonempty. Let $\mathcal{P}^* \subseteq \mathcal{P}$ denote the set of predictors for which the prediction error decays to zero along the system trajectory, i.e., $\mathcal{P}^* = \{p \in \mathcal{P} : e_p \to 0\}$. Clearly $\bar{\sigma}$ and $\mathcal{P}^*$ depend on the initial conditions.

We restrict attention to switching rules that satisfy one or more of the following:

Assumption R0 For all initial states of the plant and the predictor, $e_{\sigma(t)} \to 0$ as $t \to \infty$.

Assumption R1 There exist constants $C, \alpha > 0$ such that for all plant initial states $x(0)$ and predictor initial states $w(0)$, $|e_{\sigma(t)}(t)| < Ce^{-\alpha t}(\|x(0)\| + \|w(0)\|)$.

Assumption R2 $\bar{\sigma} \subseteq \mathcal{P}^*$.

Assumptions **R0**, **R1** and **R2** require that the performance measures and decision rule result in the selection of a "good" predictor. Clearly, assumption **R1** implies **R0**. Assumption **R2** requires that if $q \in \bar{\sigma}$, then $e_q(t) \to 0$ as $t \to \infty$, i.e., the only predictors that are selected "in the limit" are those that do "good prediction" along the state trajectory. Since $\bar{\sigma}$ is nonempty, under assumption **R2** so is $\mathcal{P}^*$ and under assumptions **A4** and **A5**, **R2** implies **R0**.

These assumptions implicitly define classes of switching rules and performance measures. The question of whether these classes are nontrivial (e.g. nonempty) is closely connected to the stability of the closed loop controlled system. This is discussed in [6].

In summary, the closed loop switched system is described by the following set of equations:

$$\tilde{x}(t) = \tilde{A}_{\sigma(t)}\tilde{x}(t) + \tilde{R}_{\sigma(t)}r(t) \quad \tilde{x}(0) = \tilde{x}_0 \tag{10}$$

$$y(t) = \tilde{C}\tilde{x}(t) \tag{11}$$

$$e_{\sigma(t)}(t) = \tilde{E}_{\sigma(t)}\tilde{x}(t) \tag{12}$$

$$\dot{\phi}(t) = f\phi(t), \quad \phi(0) = \phi_0 \tag{13}$$

$$r(t) = g\phi(t) \tag{14}$$

$$J(t,p) = h_t(e_p(s)|_{s=0}^t) \tag{15}$$

$$\sigma(t) = g_k(J(s,p)|_{s=0}^{\tau_k}, p \in \mathcal{P}), \quad \text{if } t \in [\tau_k, \tau_{k+1}) \tag{16}$$

In the above:

$$\tilde{x}(t) = \begin{pmatrix} x(t) \\ z(t) \\ w(t) \end{pmatrix} \tag{17}$$

$$\tilde{A}_{\sigma(t)} = \begin{pmatrix} A + BS_{\sigma(t)}C & BH_{\sigma(t)} & 0 \\ GC + LS_{\sigma(t)}C & F + LH_{\sigma(t)} & 0 \\ KC + NS_{\sigma(t)}C & NH_{\sigma(t)} & M \end{pmatrix}$$

$$\tilde{R}_{\sigma(t)} = \begin{pmatrix} BT_{\sigma(t)} \\ R + LT_{\sigma(t)} \\ NT_{\sigma(t)} \end{pmatrix}$$

$$\tilde{C} = (C, 0, 0)$$

$$\tilde{E}_{\sigma(t)} = (-C, 0, C_{\sigma(t)})$$

and $h_t, t \geq 0$ are functions mapping continuous real valued functions on the interval $[0, t]$ into the real line, and g_k is a sequence of functions mapping $\mathbf{R}^{\mathcal{P} \times [0, \tau_k]}$ into $\mathcal{P}$.

In the sequel it will be necessary to consider the joint trajectories of several state space systems. If $x_1(t) \in \mathbf{R}^{k_1}$ and $x_2(t) \in \mathbf{R}^{k_2}$ are vector valued signals, then the notation $(x_1(t), x_2(t))$ will denote the vector in $\mathbf{R}^{k_1 + k_2}$ formed by concatenating the vectors x_1 and x_2.

3 Stability Analysis

To show that the nonlinear system (10–16) is exponentially stable we first show that along any trajectory of the closed loop system the plant output is the sum of: the response of a switched linear system to the reference input r and the zero state response of the same system to a prediction error related disturbance. This is the content of the following result.

Proposition 3.1 *Let* $\tilde{x}(t) = (x(t), z(t), w(t))$ *be the state trajectory of the closed loop system (10–16) with initial condition* $\tilde{x}_0 = (x_0, w_0, z_0)$, *and let* σ *be the associated switching signal. Then along this trajectory*

$$\begin{pmatrix} z(t) \\ w(t) \end{pmatrix} = x_s^\sigma(t) + x_s^{\sigma, e}(t) \tag{18}$$

$$y(t) = y_s^\sigma(t) + \varepsilon(t), \quad t \geq 0$$

where x_s^σ *and* y_s^σ *are the state and output respectively of the switched linear system* $(\Sigma_{\sigma(t)}, \Gamma_{\sigma(t)})$ *with input* r *and initial condition* (z_0, x_0), *and* $x_s^{\sigma, e}$ *and* ε *are the state and output of this system to a disturbance* $e_{\sigma(t)}(t)$.

Proof: See [6].   □

In light of Proposition 3.1, we now examine the stability of the time-varying linear system $(\Sigma_{\sigma(t)}, \Gamma_{\sigma(t)})$ for each admissible switching signal $\sigma(t)$.

Recall that a time-varying linear system $\dot{x}(t) = A(t)x(t)$ is exponentially stable if there exist constants $k_1, k_2 > 0$ such that for all $t \geq \mu \geq 0$, $\|\Phi(t, \mu)\| \leq k_1 e^{-k_2(t-\mu)}$, where $\|\Phi(\cdot, \cdot)\|$ denotes the state transition matrix of the system. In the case at hand, $A(t) = A_{\sigma(t)\sigma(t)}$ is piecewise constant and by assumption **A2** for each $p \in \mathcal{P}$ there exist constants $a_p \geq 0$ and $\lambda_p \geq \gamma/2$ such that for all $t \geq 0$, $\|e^{A_{pp}t}\| \leq e^{(a_p - \lambda_p t)}$. Sufficient (conservative) conditions under which a linear time-varying system of this form is exponentially stable are given in the following lemma.

Lemma 3.2 *Assume that condition* **A2** *holds. If one of the following conditions is satisfied:*

1. *The finite dwell time satisfies $\tau_D > \sup_{p \in \mathcal{P}} \left\{ \frac{a_p}{\lambda_p} \right\}$;*

2. $\sup_{p \in \mathcal{P}} \|A_{pp}\| < \left(2 \sup_{p \in \mathcal{P}} \left\{ \frac{a_p}{\lambda_p} \right\} \exp \left[\sup_{p \in \mathcal{P}} \{a_p\} \right] \right)^{-1},$

then the time-varying linear system $(\Sigma_{\sigma(t)}, \Gamma_{\sigma(t)})$ is exponentially stable for any admissible switching signal $\sigma(t)$. Moreover, the constants in the exponential bound do not depend on the switching signal σ.

Proof: Standard result, see e.g., [3], [4], [7], [14], [18].   □

Fix an initial condition ϕ_0 of Ξ. Then each initial condition of the form (w_1, z_1, ϕ_0) for the asymptotically stable linear time-invariant system $(\Sigma_p, \Gamma_p, \Xi)$ gives raise to a trajectory with a nonempty positive limit set $\Omega(w_1, z_1, \phi_0)$. Moreover, by the asymptotic stability of the above system this limit set depends only on ϕ_0. The next lemma gives a trivial result on the convergence of the state trajectory of $(\Sigma_{\sigma(t)}, \Gamma_{\sigma(t)}, \Xi)$ that will be useful in a later example.

Lemma 3.3 *Assume that **A2** and one of the conditions of Lemma 3.2 holds. Suppose that for a fixed initial condition ϕ_0 of Ξ, the LTI systems $(\Sigma_q, \Gamma_q, \Xi)$, $q \in Q \subseteq \mathcal{P}$, have a common ω-limit set $\bar{\Omega}$. Then if $\sigma(t)$ takes values in Q, and the initial condition of Ξ is fixed to be ϕ_0, the set $\bar{\Omega}$ is the unique ω-limit set for the time-varying linear system $(\Sigma_{\sigma(t)}, \Gamma_{\sigma(t)}, \Xi)$.*

Proof: See [6].   □

We now use Proposition 3.1 and Lemma 3.2 to show the stability of the switched nonlinear system.

Proposition 3.4 *If **A2** and one of the conditions of Lemma 3.2 hold, then for every initial state of the nonlinear switched system (10–16):*

1. *If $e_{\sigma(t)}(t)$ is bounded, then $\tilde{x}$ is bounded.*

2. *If **R0** holds, then $\lim_{t \to \infty} \|(z(t), w(t)) - x_s^s(t)\| = 0$ and $\lim_{t \to \infty} \|y(t) - y_s^s(t)\| = 0$; and*

3. *If **R1** holds, then that the nonlinear closed loop system is globally exponentially stable in the sense that there exist constants $C, \beta > 0$ such that with $r \equiv 0$ and for all initial conditions, the state trajectory $\tilde{x}(t)$ of (10) satisfies $\|\tilde{x}(t)\| \leq C e^{-\beta t} \|\tilde{x}(0)\|$;*

Proof: See [6]   □

4 Sufficient Conditions for Tracking Performance

The reduction and stability result of the previous section indicates that under the conditions of Lemma 3.2, if $e_{\sigma(t)} \to 0$, then asymptotically y behaves like y_s^σ. Hence we need only show that y_s^σ adequately tracks the reference signal r. Now according to assumption **A3** the controllers Γ_p are designed so that the time-invariant closed loop systems (Γ_p, Σ_p) adequately track r. Thus it will be sufficient to determine conditions under which the system $(\Sigma_{\sigma(t)}, \Gamma_{\sigma(t)})$ inherits this property.

To this end, fix an admissible switching signal $\sigma(t)$. Then for a fixed initial state $(z(0), w(0), \phi(0))$ of the LTV system $(\Sigma_{\sigma(t)}, \Gamma_{\sigma(t)}, \Xi)$ let Ω_s^σ denote the ω-limit set, i.e., the set of positive limit points, of the resultant state trajectory $(z_s^\sigma(t), w_s^\sigma(t), \phi(t))$. In terms of Ω_s^σ we now indicate sufficient conditions under which the output y_s^σ of the system $(\Sigma_{\sigma(t)}, \Gamma_{\sigma(t)}, \Xi)$ will adequately asymptotically track r.

Proposition 4.1 *Assume that* **A2**, **A4**, **A5**, *and one of the conditions of Lemma 3.2 hold. Let $\sigma(t)$ be any admissible switching signal satisfying the following two conditions:*

Condition C1 *For each $p, q \in \bar{\sigma}$ and $(z, w, \phi) \in \Omega_s^\sigma$,*

$$(C_p - C_q)w = 0 \tag{19}$$

Condition C2 *For each $p, q \in \bar{\sigma}$ and $(z, w, \phi) \in \Omega_s^\sigma$,*

$$(H_p - H_q)z + (S_p - S_q)C_p w + (T_p - T_q)g\phi = 0 \tag{20}$$

Then for each $p \in \bar{\sigma}$, $\lim_{t\to\infty} |y_s^\sigma(t) - y_{pp}(t)| = 0$.

Proof: See [6]. □

Condition **C1** requires, roughly, that if two predictors are chosen "in the limit" then they agree on the ω-limit set. This is a natural condition that is easily seen to be implied, for example, by **R2**. Condition **C2** is similar but more restrictive. It requires that if two controllers are chosen "in the limit" then they agree on the ω-limit set. Unlike **C1**, there is no a priori reason to suggest that this will be a natural consequence of the switching rule. Hence it is imposing an additional constraint on the model/controller pairs. Of course, the problem is that one may not know the ω-limit set in advance, in which case it may be necessary to check that (20) is satisfied on a larger set that is known to contain the possible ω-limit sets. Note that when the assumptions of

Proposition 3.4 are satisfied, part (1) of the proposition implies that **C2** can be verified by examining the possible ω-limit sets of the nonlinear switched system. In fact, for some simple cases **C2** can be verified quite easily in this way as shown in the example after the proof of Theorem 4.2 below.

Our first result is that the standard assumptions together with **R2** and **C2** are sufficient to ensure that the output of the nonlinear switch system $y(t)$ asymptotically tracks the reference signal r. Indeed, the following theorem indicates that the asymptotic tracking performance of the closed loop switched system is as good as one of the time-invariant linear systems (Σ_p, Γ_p) and by assumption **A3** this is adequate.

Theorem 4.2 *Assume that* **A1-A5** *and either of the conditions of Lemma 3.2 are satisfied. Furthermore, assume that for each possible initial condition of the nonlinear switched system the resulting switching sequence is such that* **R2** *and* **C2** *are satisfied. Then the closed loop switched control system (10–16) satisfies* $\lim_{t\to\infty} |y(t) - y_{pp}(t)| = 0$ *for some* $p \in \mathcal{P}$ *with* $e_p(t) \to 0$. *Hence asymptotically* y *adequately tracks* r.

Proof: See [6].   □

Example 1
The set point control problem illustrated in Figure 1 is studied in [14]. We are given a family of siso model transfer functions $b_p(s)/a_p(s)$ and a family of siso controller transfer functions $\gamma_p(s)/\rho_p(s)$ such that for each $p \in \mathcal{P}$, the control system illustrated in Figure 1 with $\sigma = p = p^*$ is stable and for any constant reference signal r, $\lim_{t\to\infty} |y(t) - r| = 0$.

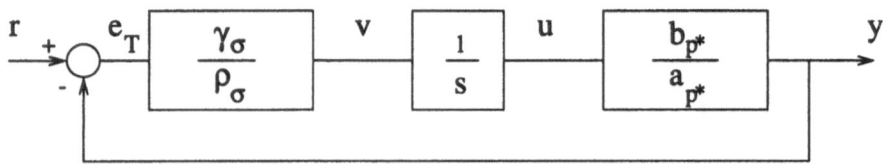

Figure 1: Set point control system.

As shown in the figure the actual plant input u is generated by an integrating subsystem, and the control signal v is generated by a switched controller Γ_σ driven by the tracking error $e_T(t) = r - y(t)$. In our notation, Σ is the cascade of the integrator and the actual plant. Thus the controller output and (augmented) plant input is denoted by v.

A common state realization of the family of predictors and controllers takes the form:

$$\dot{w} = Mw + Nv + Ky \qquad (21)$$

$$\dot{z} = Fz + Lv + Ge_T \tag{22}$$

$$v_p = H_p z + T_p e_T \tag{23}$$

$$e_p = C_p w - y \tag{24}$$

By the problem set-up and the construction of these realizations, assumptions **A1-A3** hold. We assume that the parameterization of the models and controllers satisfies assumptions **A4** and **A5**. That condition **C2** holds in this case can be verified by direct computation.

First, it is easy to see that the ω-limit set for the system (Σ_p, Γ_p) is just the single equilibrium state

$$x^\infty \triangleq \begin{pmatrix} w_p^\infty \\ z_p^\infty \end{pmatrix} = \begin{pmatrix} -M^{-1}K \\ 0 \end{pmatrix} r \tag{25}$$

and this is independent of p. Moreover, in steady state $y = r$, $v_p = 0$, and $e_p = 0$. So $C_p w_p^\infty = r$.

Let σ be a τ_D-admissible switching signal. Then by Lemma 3.3, under either of the conditions of Lemma 3.2, every state trajectory of the switched system $(\Sigma_{\sigma(t)}, \Gamma_{\sigma(t)})$ converges to the ω-limit set $\Omega(r) = x^\infty$. By Proposition 3.4 we have $\Omega^{z,w} = \Omega(r)$. Thus to check **C2**, we just need to check that (20) holds at the point x^∞. This is trivially true: for each $p, q \in \sigma$, $H_p z_p^* + T_p(r - C_p w_p^\infty) = 0 = H_q z_p^* + T_q(r - C_q w_p^\infty)$.

It follows from the above and Theorem 4.2 that any parameterization of the controllers for which **A4** and **A5** are satisfied and any switching rule satisfying **R2** and having a sufficiently large dwell-time τ_D will ensure that the state of the closed loop switched system is bounded and $\lim_{t\to\infty} |y(t) - r| = 0$.

5 Asymptotic Exact Tracking

An interesting additional result can be obtained when we restrict attention to asymptotic exact tracking, i.e., we require $\lim_{t\to\infty} |y(t) - r(t)| = 0$. In this case we will assume that the following additional mild restriction holds:

Assumption A6 The models Σ_p have no zeros in common with the eigenvalues of the reference signal generator.

Our second result is the following. In the case of asymptotic exact tracking in order to conclude that the nonlinear switched system will asymptotically exactly track the reference signal $r(t)$ it is only necessary to verify the structural conditions **A1-A6** and the switching rule constraints that **R2** and one of the conditions of Lemma 3.2 hold. In particular, it is not necessary to verify that condition **C2** holds.

Theorem 5.1 *Consider an asymptotic exact tracking problem for a reference signal given by (2). Assume that conditions* **A1-A6, R2** *and one of the conditions of Lemma 3.2 hold. Then the closed loop switched control system satisfies* $\lim_{t\to\infty}|y(t)-r(t)|=0$.

Proof: See [6].   □

We illustrate Theorem 5.1 with the following example.

Example 2

Consider the set point control problem illustrated in Figure 2. Assume, as in Example 1, that we are given a set of siso models and a set of siso controllers, such that for each $p^* \in \mathcal{P}$ the controlled system with $\sigma = p^*$ is stable, and for any constant reference input r, $\lim_{t\to\infty}|y(t)-r|=0$. The true plant Σ_{p^*} is one of the models and p^* is unknown. We assume that none of the models has a zero at 0. This example, although similar to Example 1, requires a different (and more difficult) analysis.

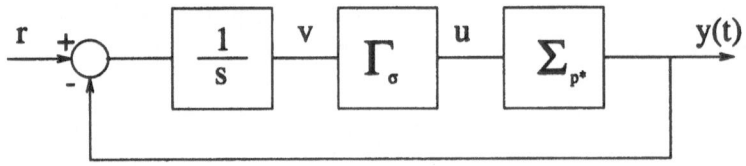

Figure 2: Set point control system, Example 2.

The plant input $u_{\sigma(t)}$ is generated by a switched controller $\Gamma_{\sigma(t)}$, whose input signal, $v(t)$, is generated by an integrating subsystem driven by the tracking error $r - y(t)$.

In our notation Γ is the cascade of the integrator and the controller. A common state realization of the system above takes the form

$$\begin{aligned} \dot{z}(t) &= Fz(t) + Lu(t) + G(r - y(t)) \quad\quad (26)\\ u_p(t) &= H_p z(t) \end{aligned}$$

$$\begin{aligned} \dot{w}(t) &= Mw(t) + Ky(t) + Nu(t)\\ e_p &= C_p w(t) - r \end{aligned}$$

Assume that **A1-A5** are satisfied by the problem set up, that one of the conditions of Lemma 3.2 holds and that the switching rule ensures that condition **R2** holds. Stability and boundedness of the switched nonlinear system then follows by Proposition 3.4. Since we have assumed that the models do not have a zero at the origin, **A6** is satisfied. Hence by Theorem 5.1, $\lim_{t \to \infty} |y(t) - r| = 0$. This latter result can also be shown by direct computation of the fixed points of the systems (Σ_p, Γ_p). However, in this case these fixed points depend on $p \in \mathcal{P}$ and the analysis method used in Example 1 does not apply.

6 Conclusion

We have presented a simple setting in which to analyze the stability and tracking performance of predictor based controller switching rules. Our approach is to relate control performance to prediction performance by separating the output into an exogenous input response and an error term.

Our first main result, Theorem 4.2, gives a set of sufficient conditions under which good asymptotic tracking performance of the switched system is achieved. As a special case, this results leads to a simpler proof of a recent result of Morse [14], concerning a set point control problem.

Our second main result, Theorem 5.1, shows that in the special case where asymptotic exact tracking is required and the plant has no zeros in common with the poles of the reference signal generator, the main assumption of Theorem 4.2 is in fact always satisfied so that the switched system achieves asymptotic exact tracking.

References

[1] Shankar Sastry and Mark Bodson, *Adaptive Control: Stability, Convergence, and Robustness*, Prentice Hall, Englewood Cliffs, New Jersey, 1989.

[2] K. Ciliz and K. Narendra, Multipole model based adaptive control of robotic manipulators, *Proc. of the 33rd IEEE conference on Decision and Control*, pp. 1305-1310, 1994.

[3] W. A. Coppel, Dichotomies in stability theory, Lecture Notes in Mathematics No. 629, Springer,1978.

[4] C. A. Desoer, Slowly varying system $\dot{x} = A(t)x$, *IEEE Trans. on Automatic Control*, pp. 780-781, Dec. 1969.

[5] M. Fu and B. R. Barmish, Adaptive stabilization of linear systems via switching controls, *IEEE Trans. on Automatic Control*, pp. 1079-1103, Dec. 1986.

[6] J. Hocherman-Frommer, S. R. Kulkarni, P. J. Ramadge, "Controller switching basd on output prdiction errors," To appear *IEEE Transactions on Automatic Control*, May 1998.

[7] A. Ilchmann, D. H. Owens and D. Pratzel-Walters, "Sufficient conditions for stability of linear time varying systems," *Systems & Control Letters*, No. 9, pp. 157-163, 1987.

[8] S. Kulkarni and P. J. Ramadge, "Model and controller selection policies based on output prediction errors," *IEEE Transactions on Automatic Control*, November 1996.

[9] S. Kulkarni and P. J. Ramadge, "Prediction error based controller selection policies," *IEEE Conference on Decision and Control*, New Orleans, Dec. 1995.

[10] B. Mårtensson, "The order of any stabilizing regulator is sufficient a priori information for adaptive stabilization," *Systems and Control Letters*, Vol. 6, No. 2, pp. 87-91, 1985.

[11] D. E. Miller, "Adaptive stabilization using a nonlinear time-varying controller," *IEEE Trans. on Automatic Control*, 39 (7), pp. 1347-1359, July 1994.

[12] A. S. Morse, D. Q. Mayne, and G. C. Goodwin, "Applications of hysteresis switching in parameter adaptive control," *IEEE Trans. Automatic Control*, 37 (9), pp. 1343-1354, Sept. 1992.

[13] A. S. Morse, Supervisory control of families of linear set-point controllers, *Proc. of the 32nd IEEE conference on Decision and Control*, pp.1055-1060, 1993.

[14] A. S. Morse, "Supervisory control of families of linear set-point controllers – part 1: exact matching," preprint, March 1993.

[15] A. S. Morse, "Control using logic-based switching," *Proc. of 1995 European Control Conference*, Rome, Italy, Sept. 1995.

[16] K. Narendra and J. Balakrishnan, "Improving transient response of adaptive control systems using multiple models and switching," *Proc. of the 32rd IEEE Conf. on Decision and Control*, San Antonio, Texas, Dec. 1993.

[17] K. Narendra and J. Balakrishnan, "Intelligent control using fixed and adaptive models," *Proc. of the 33rd IEEE Conf. on Decision and Control*, pp. 1680-1685, Lake Buena Vista, Florida, Dec. 1994.

[18] V. Solo, "On the stability of slowly time-varying linear systems," *Mathematics of Control, Signals and Systems*, No. 7, pp. 331-350, 1994.

[19] S. R. Weller and G. C. Goodwin, "Hysteresis switching adaptive control of linear multivariable systems," *IEEE Trans. Automatic Control*, 39 (7), pp. 1360-1375, July 1994.

[20] W. M. Wonham, *Linear Multivariable Control: a Geometric Approach*, Second Ed., Springer-Verlag, New York, 1979.

Hybrid Control of Inverted Pendulums

K. J. Åström*

1 Introduction

Much effort in automatic control has focused on regulation and servo problems. Even if these problems represent a large class of control problems there are many situations which require much more general and complex formulations. This has been referred to as task oriented control, because there are many different tasks that have to be accomplished. Typical examples of task level control are found in automated highways, robotics, unmanned autonomous vehicles. Task oriented control typically leads to a hierarchical structure where the low level consists of a collection of conventional controllers. Selection of the particular low level controller is done at higher levels which is often of discrete nature. There are often many layers in the hierarchy. Systems of this type can be represented as hybrid systems. There is currently much interest in hybrid control systems from a wide variety of groups. In this paper we emphasize that experimental work on hybrid control can give useful insights and that pendulums which are standard tools of most control laboratories can be very useful for this purpose.

It is a nontrivial problem to decompose a high level task into low level control problems. Sometimes this can be accomplished by a control principle which gives a broad characterization of the global control problem. There are many nice examples of control principles although the concept is not yet formalized. Two examples are given below.

In missile guidance the task is to make a missile home in on a target. There are several ways to do this. One principle is to always steer towards the target. This is a simple principle, but it leads to trajectories, called dog-curves, where the cross acceleration of the missile becomes very large close to the target. A much better principle is to steer so that the angle between the centerline of the missile and the line of sight to the target is constant. This leads to trajectories which require much smaller cross accelerations in the critical phase close to the target. The discovery of this principle led to a drastic improvement in missile guidances.

Control of Raibert's one-legged robot [5] was greatly simplified through the

---

*Department of Automatic Control, Lund Institute of Technology

notion of foot print and the control principle base on that. The footprint is a line on the ground whose endpoint are the position of the center of mass of the machine when the foot touches the ground and when it leaves the ground. Constant velocity is achieved by controlling the vehicle so that the foot touches the ground in the center of the footprint. The vehicle accelerates if the foot touches the ground behind the center of the footprint.

In spite of their usefulness control principles are not discussed much in literature. To arrive at a suitable control principle for a particular problem it is necessary to consider the fundamental properties of a task. Investigation of control principles is thus also a nice way to merge system theory and physics.

Inverted pendulums are good laboratory tools to illustrate task oriented control. They are sufficiently complicated to be interesting and yet so simple that many tasks can be accomplished with a reasonable effort. Since pendulums are readily available in laboratories the results can also be demonstrated experimentally. There are several types of pendulums conventional pendulum on a cart, the Furuta rotating pendulum, the Pendubot. Additional complications are obtained by using two pendulums on the bases, doubly or triply inverted pendulums.

There are several control problems that are of interest, for example: stabilization, rotation with varying velocity, stabilization and manual control of pivot, energy control, catching a swinging pendulum in the in upright position, swinging up an inverted pendulum. The first three tasks are conventional regulation or servo problem. Since energy is such a fundamental property of an electro-mechanical system it is naturally to use energy control as a fundamental idea when controlling pendulums.

2 Energy Control

Energy is a basic quantity of electro-mechanical systems. It is therefore of interest to control the energy of a system. This is also relatively easy to do. Many control tasks can be accomplished by controlling the energy of the instead of controlling its position and velocity directly, see [7].

Simple Pendulum on a Cart

Consider for example the problem of swinging up an inverted pendulum.

$$J\ddot{\theta} - mgl\sin\theta + mul\cos\theta = 0 \tag{1}$$

where θ is the angle the pendulum forms with the vertical, m is the mass, J is the moment of inertia with respect to the pivot, l the distance from the center of mass to the pivot and u is the acceleration of the pivot.

The energy of the pendulum is

$$E = \frac{1}{2}J\dot\theta^2 + mgl(\cos\theta - 1) \tag{2}$$

The time derivative of the energy along the equations of motion is given by

$$\frac{dE}{dt} = J\dot\theta\ddot\theta - mgl\dot\theta\sin\theta = -mul\dot\theta\cos\theta \tag{3}$$

It follows from Equation (3) that it is easy to control the energy. The system is simply an integrator with varying gain. Controllability is lost when the coefficient of u in the right hand side of (3) vanishes. This occurs for $\dot\theta = 0$ or $\theta = \pm\pi/2$, i.e., when the pendulum is horizontal or when it reverses its velocity. Control action is most effective when the angle θ is 0 or π and the velocity is large.

There are many control strategies that can be used to drive the energy towards a specified value. One possibility is the linear strategy

$$u = k(E - E_0)\dot\theta\cos\theta \tag{4}$$

which is associated with the Lyapunov function $V = \frac{1}{2}(E - E_0)^2$. A more aggressive strategy is

$$u = a\mathrm{sign}\left((E - E_0)\dot\theta\cos\theta\right) \tag{5}$$

which is associated with the Lyapunov-like function $V = |E - E_0|$ The control law (5) gives chattering for small errors. A control law which avoids this is

$$u = \mathrm{sat}_a\left(k(E - E_0)\mathrm{sign}(\dot\theta\cos\theta)\right) \tag{6}$$

where sat_a denotes a linear function which saturates at a.

General Lagrangian System

The energy control for a single pendulum is very simple. It leads to a first order system described by an integrator whose gain depends on the angle and its rate of change. The only difficulty is that the gain may vanish. This will only happen at isolated time instants because the time variation of the gain is generated by the motion of the pendulum. The ideas can be extended to control of more complicated configurations with rotating and multiple pendulums. In this section we will briefly discuss two generalizations.

To illustrate the ideas we consider a general mechanical system described by the equation

$$M(q,\dot q)\ddot q + C(q,\dot q)\dot q + \frac{\partial U(q)}{\partial q} = T \tag{7}$$

where q is a vector of generalized coordinates, $M(q, \dot{q})$ is the inertia matrix, $C(q, \dot{q})$ the damping matrix, $U(q)$ the potential energy and T the external control torques, see [4]. The potential energy is

$$E = \frac{1}{2}\dot{q}'M(q, \dot{q})\dot{q} + U(q) \tag{8}$$

Taking derivatives with respect to time we get

$$\begin{aligned}
\frac{dE}{dt} &= \dot{q}'M(q, \dot{q})\ddot{q} + \frac{1}{2}\dot{q}'\dot{M}(q, \dot{q})\dot{q} + \dot{q}'\frac{\partial U(q)}{\partial q} \\
&= \frac{1}{2}\dot{q}'(T - C(q, \dot{q})\dot{q}) + \frac{1}{2}\dot{q}'\dot{M}(q, \dot{q})\dot{q} \\
&= \frac{1}{2}\dot{q}'\left(\dot{M}(q, \dot{q}) - 2C(q, \dot{q})\right)\dot{q} + \dot{q}'T
\end{aligned} \tag{9}$$

In [6] it is shown that the matrix $\dot{M}(q, \dot{q}) - 2C(q, \dot{q})$ is skew symmetric. It thus follows that

$$\frac{dE}{dt} = \dot{q}'T$$

The control torques depend on the control signal u and we thus have a problem of the type we have discussed previously. It happens very frequently that T is linear or affine in the control variable u which simplifies the problems further.

3 Hybrid Strategies for Swing Up

Now consider the task of swinging up inverted pendulums. We will discuss a number of different strategies for different pendulum configurations.

A Simple Pendulum on a Cart

A simple strategy is to apply energy control and to switch to a linear stabilization strategy when the pendulum is close to the upright position. Energy control drives the motion towards the manifold

$$E = \frac{1}{2}J(\dot{\theta})^2 + mgl(\cos\theta - 1) = 0$$

Trajectory on this manifold pass through the upright position. The velocity is also zero at this position. Small errors in the strategy gives trajectories which are close to the desired position with zero velocity at the upright position.

The swing-up strategy is investigated in detail in [1]. It is shown that the crucial parameter is the acceleration of the pivot. It is shown that if the acceleration is larger than 2g swing-up can be accomplished with two switches of the control signal and that the pendulum will make one swing. For accelerations between $4g/3$ and $2g$ the control signal will make three switches and the pendulum makes one swing. Several swings are required for lower accelerations.

Algorithms for switching between several different strategies are investigated in [3]. In this paper it is shown that a systematic scheme can be obtained by associating each control strategy C_i with a Lyapunov like function W_i, where the functions W_i may have discontinuous derivatives. It is shown in [3] that stability can be guaranteed by switching to a Lyapunov function with a smaller value. Scalings of the Lyapunov functions can be used as design parameters. There are some subtleties that can lead to chattering that must be dealt with properly. In [3] it is shown that the simple swing-up strategy discussed above will give the desired result.

It is straight forward to devise more sophisticated swing up strategies that also control the position of the pivot. One possibility is a mixture of the following strategies.

S' Stabilization and control of the pivot

S   Stabilization

E   Energy Control

P   Controlling the pivot

The strategies S and E have already been discussed. The S' is similar to S but in addition it provides control of the pivot. The strategy S can be designed to have a larger catching region than S'. The strategy P simply attempts to bring the pivot to the desired position. This strategy can be applied during swing-up when the pendulum is close to horizontal and controllability of the pendulum is lost. The results of [3] can be used to safely switch between the different strategies. Scaling can be used as a design parameter. It turns out that there is a lot of freedom to mix the strategies.

Two Pendulums on a Cart

The ideas can be extended to many other configurations, rotating pendulums, and different configurations of multiple pendulums. Here we will consider two pendulums on a cart. Let m_1 and m_2 be the masses of the pendulums and l_1 and l_2 the distances from the pivots to the center of masses for the pendulums we find that the derivatives of the energies of the pendulums are given by

$$\frac{dE_1}{dt} = -u m_1 l_1 \dot{\theta}_1 \cos\theta_1$$
$$\frac{dE_2}{dt} = -u m_2 l_2 \dot{\theta}_2 \cos\theta_2 \tag{10}$$

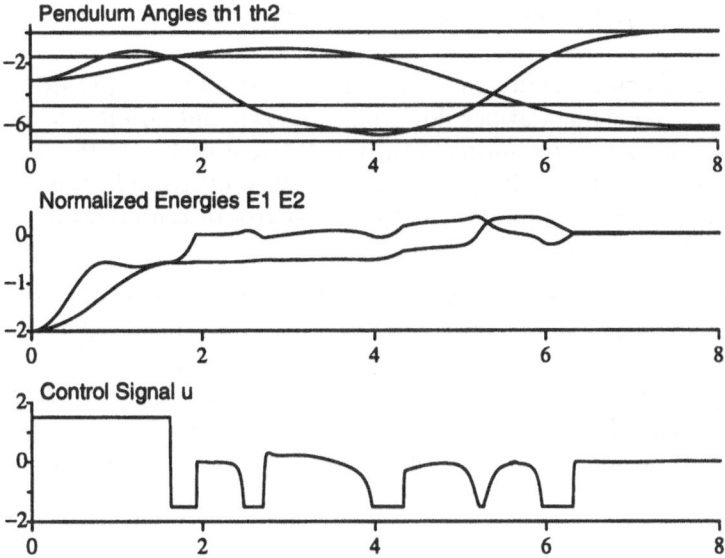

Figure 1: Simulation of the strategy for swinging up two pendulums on the same cart.

The energies are defined in such a way that they are zero when the pendulums are upright. A control strategy that drives E_1 and E_2 to zero can easily be obtained from the Lyapunov function

$$V = \frac{1}{2}(E_1^2 + E_2^2)$$

The derivative of this function is given by

$$\frac{dV}{dt} = -\left(m_1 l_1 E_1 \dot{\theta}_1 \cos\theta_1 + m_2 l_2 E_2 \dot{\theta}_2 \cos\theta_2\right)u$$

The control law

$$u = \text{sat}_a k\left(m_1 l_1 E_1 \dot{\theta}_1 \cos\theta_1 + m_2 l_2 E_2 \dot{\theta}_2 \cos\theta_2\right) \tag{11}$$

drives the Lyapunov function to zero. This implies that both pendulums will obtain their appropriate energies.

The energy control strategy can be combined with different switching strategies. There are many different alternatives, a pendulum can be caught and stabilized as soon as it is in the upright position. Stabilization of one pendulum can then be combined with energy control for the remaining pendulum. Figure 1 illustrates swing-up of two pendulums with $m_1 = m_2 = 1$ and $l_1 = 1$ and $l_2 = 4$. The control strategy is given by Equation (11) with parameters $a = 1.5g$ and $k = 20$. Notice that the control strategy brings the energies of both pendulums to their desired values. Also notice that the pendulums approach the upright position from different directions.

Rotating Double Inverted Pendulum

Swinging up a double pendulum is much richer and more complicated problem. There are many configurations down-up, up-down and up-up, where the first preposition denotes the position of the pendulum attached to the base. There are also many control principles such as energy control of each pendulum, stabilization of one pendulum and energy control of the other, alignment of the pendulums, catching one pendulum etc.

In joint work with Furuta we have used several of these ideas to swing up double inverted pendulums on a rotating base. One successful strategy is to combine alignment of the pendulum with total energy control, catching the inner pendulum and swing up of the outer pendulum. Both pendulums then swing together like a single pendulum. The inner pendulum is caught in the upright position and the outer pendulum is caught in the upright position after one extra turn.

4   Analysis of Systems with Switching

By applying Theorem 1 we are assured that the hybrid strategy will accomplish a given task for example singing up a pendulum. There is however many design choices and there are some subtleties that may be caused by the switching. Therefore it is desirable to analyze the properties of the system in some detail. As an illustration we will analyze the system obtained when the energy control strategy given by Equation (6) is applied to a single pendulum.

Introducing $v = dx/dt$ and scaling the equations we find that the closed loop system can be describe by the equations

$$\frac{d^2x}{dt^2} = \sin x - u \cos x$$
$$u = -\mathrm{sat}_a\left(kE\ \mathrm{sgn}(v \cos x)\right)$$
$$v = \frac{dx}{dt} \tag{12}$$
$$E = \frac{1}{2}v^2 - 1 + \cos x$$

Furthermore introduce

$$h(x, v) = k(\frac{1}{2}v^2 - 1 + \cos x)\ \mathrm{sgn}(v \cos x)$$
$$g(x, v) = \sin x - \cos x\ \mathrm{sat}_a h(x, v)$$

The equations of the closed loop system can be written as

$$\frac{d}{dt}\begin{pmatrix} x \\ v \end{pmatrix} = \begin{pmatrix} v \\ g(x, v) \end{pmatrix} \tag{13}$$

This equation is nontrivial because the function f is discontinuous when $v = 0$ or $x = \pm\pi/2$. Since the system is of second order we can, however, obtain a complete understanding by phase plane analysis. To do this we have to generalize the Poincare classification of the equilibria. These generalizations are of interest in a much broader context. Having obtained suitable concepts we will then proceed to investigate the particular problem. It follows from Equation (12) that

$$\frac{dE}{dt} = v(\frac{dv}{dt} - \sin x) = -kEv \cos x \; \operatorname{sat}_a \operatorname{sgn}(v \cos x) = -kEl(v, x) \qquad (14)$$

The manifold $E = 0$ is thus an invariant. Since

$$l(x, v) = \min\left(|v \cos x|, a\right) \geq 0$$

it follows that it is attracting. The trajectories approaches the manifold monotonously, except possibly for $v = 0$ and $\cos x = 0$. The energy could possibly stop decreasing for $v = 0$ or $x = \pm\pi/2$. The point $v = 0$, $x = 0$ is an equilibrium but $x = \pm\pi/2$ are not equilibria.

Characterization of Singularities

Consider the ordinary differential equation

$$\frac{dx}{dt} = f(x) \qquad (15)$$

Existence and uniqueness for discontinuous right hand side are discussed in the classic book by in Caratheodory. To characterize the solution we typically start by investigating the local behavior around the equilibria, i.e. the points where F is zero. When f is discontinuous considerable insight can be obtained by investigating the local properties around the discontinuities. For this purpose we introduce the notion of **singularities** as the set of points where the function f is discontinuous. Let S be this set and assume that the function f is such that the boundary ∂S is smooth. Let $n(s)$ be the normal of S at x. Define

$$f_+(x) = \begin{cases} f(x) & \text{if } x \in S, \\ \lim_{\epsilon \downarrow 0} f(x + \epsilon n), & \text{if } x \in \partial S. \end{cases} \qquad (16)$$

Similarly we introduce

$$f_+(x) = \begin{cases} f(x) & \text{if } x \in S, \\ \lim_{\epsilon \uparrow 0} f(x + \epsilon n), & \text{if } x \in \partial S. \end{cases} \qquad (17)$$

Furthermore let $\langle x, y \rangle$ denote the scalar product. A singular point is called **permeable** if $\langle f_+, n \rangle$ and $\langle f_+, n \rangle$ have the same signs. Furthermore a permeable point x is called smooth if f_+ and f_- are parallel. The velocity may however be discontinuous at a smooth point. A singularity is called **attract-ing** if $\langle f_+, n \rangle < 0$ and $\langle f_+, n \rangle > 0$ and it is called **repelling** if $\langle f_+, n \rangle > 0$ and $\langle f_+, n \rangle < 0$.

The classification is useful to describe the local behavior of the solutions in the neighborhood of the singular set. The orbits pass through the singular set S at a permeable point, they are attracted to S at an attracting point and they move away from S at a repelling point. The case of attracting points have been explored substantially in the literature on chattering solutions, see [2].

When one of the functions f_+ or f_- vanishes on in the singular set the local behavior on either side of the set can be obtained in the conventional way. In this way the ordinary notions of node, saddle, focus and center can be extended. We are thus able to talk about **half node**, **half focus** etc. We will use positive and negative to refer to the side where the singularity occurs. More complicated fractional singularities may also occur.

Applications to the Pendulum

Having obtained appropriate concepts we will now return to the particular case of Equation (12) which describes energy control of the pendulum. In this case the singular points are the line $v = 0$ and we choose $S = v$ which means that the normal points in the direction of the positive v-axis. Notice that the sgn function is also discontinuous for $x = \pi/2$. This is however not a singularity because the term is multiplied by $\cos x$.

The function f is given by

$$f(x,v) = \begin{pmatrix} v \\ g(x,v) \end{pmatrix}$$

It is only the second component of the vector field f that is discontinuous. To explore the singularities we introduce the functions g_+ and g_- defined by.

$$g_+(x) = \lim_{v \downarrow 0} g(x,v) = \sin x + |\cos x|\, \mathrm{sat}_a\big(k(1 - \cos x)\big) \qquad (18)$$

$$g_-(x) = \lim_{v \uparrow 0} g(x,v) = \sin x - |\cos x|\, \mathrm{sat}_a\big(k(1 - \cos x)\big) \qquad (19)$$

There are singularities for $v = 0$. The nature of the singularity depends on the parameters a and k. Tedious but straight forward calculations give.

The Equilibrium at the Origin

The system has an equilibrium at $v = 0$ and $x = 0$. In the neighborhood of the origin we have

$$E \approx \frac{1}{2}(v^2 + x^2)$$

Both the function f and its Jacobian are continuous. The equilibrium is thus a proper equilibrium and we have

$$A = \frac{\partial f}{\partial x} = \begin{pmatrix} 0 & 1 \\ 1 & 0 \end{pmatrix}$$

The equilibrium is thus an unstable saddle with a stable eigenvector $e_s = \begin{pmatrix} 1 & -1 \end{pmatrix}^T$ and and unstable eigenvector $e_{us} = \begin{pmatrix} 1 & 1 \end{pmatrix}^T$. Next we will consider the line segments with reflecting singularities. Because of the symmetry of the problem it is sufficient to consider the case $x > 0$. The function g_+ is positive for $0 < x \leq \pi$, but the function g_- may have sign changes. We can separate two cases.

One Reflecting Segment

We have g_- The function g_- has always a zero in the interval $\pi/2 \leq x \leq \pi$. For small values of k this zero is close to π for small k and it decreases with increasing k. If $k < k_0 \approx 3.33$ this is the only singular segment. If k is then the zero is $x = \pi - \arctan a$. For $k < k_0 \approx 3.33$ we find that there is one singular segment $|x| > x_0$ on the circle $v = 0$. The segment is short for small k and it increases with increasing k. The function g_- vanishes at the edges of the segment and there are singularities in $v < 0$ which are half centers. The part of the circle which is not a reflecting segment are transparent.

$$A_- = \lim_{v \uparrow 0} \begin{pmatrix} 0 & 1 \\ \frac{\partial g(x,v)}{\partial x} & \frac{\partial g(x,v)}{\partial v} \end{pmatrix} = \begin{pmatrix} 0 & 1 \\ g'_-(x) & 0 \end{pmatrix} \tag{20}$$

The character of the generalized equilibria are thus determined by the sign of g_-. There is a half saddle if $g_- > 0$ and a half focus if $g_- < 0$.

Three Reflecting Segment

When $k > k_0 = 3.33$ and $a > k_0(1 - \cos x_0) = 1.272$ the function g_- has two zeros of g_- in the interval $0 < x \leq \pi/2$. One zero is approximately at $x = 2/k$ the other is approximately $\pi/2 - 1/k$. If $k > (n^2 + 1 - \sqrt{a^2 + 1})/n$ this zero is at $\pi/2 - \arctan a$. In this case there is one reflecting segment in the interval $0 < x \leq \pi/2$, by symmetry there is also a reflecting segment in $-\pi/2 < x < 0$. We thus have three reflecting segments in this case. The function g_- vanishes at the ends of the segment and we thus have half singularities at these points.

Degenerate Cases

There are generalized equilibria at the ends of the reflecting line segment in the interval $0 < x < \pi/2$. The equilibrium to the left is a half center and the one to the right is a half saddle. There is a degenerate case when the generalized equilibria in coincide. If the controller does not saturate it happens when the function g_- has a double root. This occurs at $x = \arccos(\sqrt{5} - 1)/2 = 0.9046$ for $k = k_0 = 3.3302$ and $a \geq 1.2720$. The Jacobian is

$$A = \begin{pmatrix} 0 & 1 \\ 0 & 0 \end{pmatrix}$$

which corresponds to a degenerate half node with a horizontal tangent.

If $k > k_0$ and a is sufficiently small it may happen that the derivative of g_- is discontinuous at the point x_0 where g_- vanishes. We have $g(x_0) = 0$, $g'_-(x_0-) < 0$ and $g'_-(x_0+) > 0$. Therefore we have a quarter center for $v < 0$ and $x < x_0$ and a quarter saddle for $v < 0$ and $x > x_0$ with the stable eigenvector is $(1 \quad -\sqrt{b})$ where $b = \lim_{x \downarrow x_0} g'_-(x)$.

Local Behavior

Having obtained the singularities and the equilibria we can now describe the local behavior of the solutions in the neighborhood of the singularity $v = 0$. We will separate the cases when there are three or five equilibria apart from the attracting orbit $E = 0$. In both cases there is a proper saddle for $x = 0$.

The local properties for the different cases are illustrated in Figure 2. In this figure we have used the symmetry and shown the behavior for the whole segment $-\pi \leq x \leq \pi$. Case a) corresponds to the situation when there is only one reflecting segment and case b) corresponds to the case when there are three reflecting segments. There are also degenerate cases when the reflecting segments around the upright position goes to zero. The local behaviors for these cases are shown in c) and d).

Global Behavior

We will now illustrate the global behavior of the system for some particular values of the parameters.

Consider first the case $a = 0.25$. The acceleration of the pivot is so small that it takes many swings to bring the pendulum to the upright position. Interesting cases are $k = 0.5279$, $k = 3.3302$ and $k = 8.3731$. With $k = 1$ there is only one reflecting segment $2.8966 < x < \pi$. There is a saddle at the origin, and a half at $x = \pi - \arctan 0.25 = 2.8966$. The line segment $2.8966 < x < \pi$ is repelling. The phase plane of the system is shown in Figure 3 for $a = 0.25$ and $k = 1$.

Figure 2: Local behavior around the singularity $v = 0$ Case a) corresponds to $k < k_0 = 3.3302$ and all a or $k > k_0$ and $a <$, case b) to $k = k_0 = 3.3302$ and $a = 1.2720$, and case c) to $k > k_0$ and $a >$.

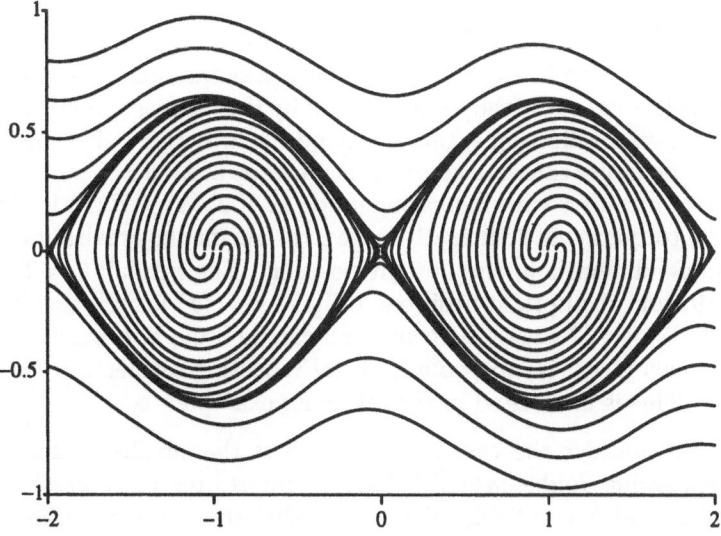

Figure 3: Phase plane of the system for $k = 1$ and $a = 0.25$. In this case there is only one reflecting segment

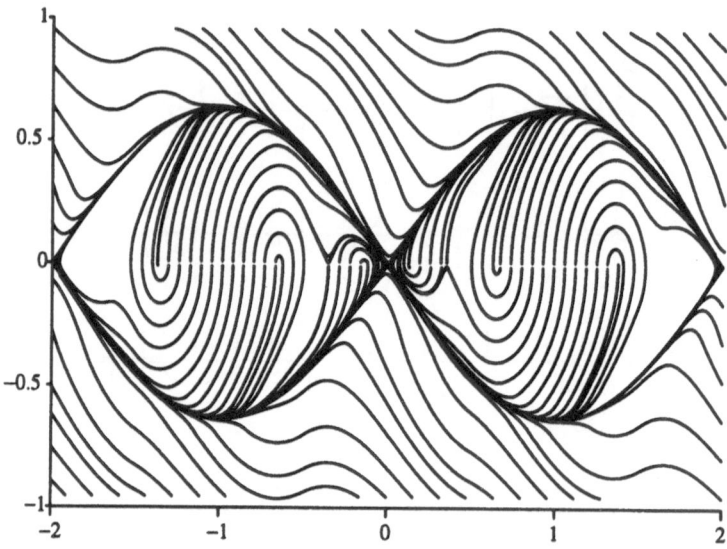

Figure 4: Phase plane of the system for $k = 5$ and $a = 2$. In this case there are three reflecting segments. For $v = 0$ and positive x they are given by $0.2033 < x < 1.1072$ and $2.0344 < x < \pi$.

It is useful to give a physical interpretation of the behavior of the system. The torque $T_g = \sin x$ is the torque due to the gravitational forces and the $T_c = \cos x \operatorname{sat}_a\big(k(1 - \cos x)\big)$ is the torque caused by the control actions when $v = 0$. There is always a reflecting segment around $x = \pi$ because the control torque will always be larger than the gravitational torque around $x = \pi$. The width of the segment increases with increasing a and k. If $T_g(x) > T_c(x)$ for $|x| < \pi/2$, the control torque is not sufficiently large to overcome the gravitational torques when the pendulum is in the upright position. The pendulum will therefore always fall from the upright position. If $T_g(x) < T_c(x)$ for $|x| < \pi/2$ there is a segment in $|x| < \pi/2$ where the pendulum can be brought towards the upright position. This gives rise to two additional reflecting segments in the interval $|x| < \pi/2$. This is illustrated in the following example.

Next we consider the case $a = 2$ which is the smallest value of a that gives single-swing-double-switch behavior. Interesting cases are $k = 0.5528$, $k = 1.4472$, and $k = 3.3302$. For $k > 3.3302$ there are reflecting segments for $v = 0$, $0.2033 < x < 1.1072$ and $2.0344 < x < \pi$. The segments $v = 0$, $0 < x < 0.2033$ and $1.1072 < x < 2.0344$ are permeable. Furthermore there is is a half center at 0.2033, a half saddle at $\arctan 2 = 1.1072$ and a half center at $\pi - \arctan 2 = 2.0344$. Figure 4 shows the phase plane for $a = 2$ and $k = 5$.

5 Conclusions

Several control tasks for inverted pendulums have been discussed. It has been demonstrated that they are natural tasks for applications of hybrid control and that energy control is a control principle that can be used to structure the system.

References

[1] Karl Johan Åström and K. Furuta. Swinging up a pendulum by energy control. In *IFAC'96, Preprints 13th World Congress of IFAC*, volume E, pages 37–42, San Francisco, California, July 1996.

[2] A. F. Filippov. *Differential Equations with Discontinuous Righthand Sides*. Kluwer Academic Publishers, 1988.

[3] Jörgen Malmborg, Bo Bernhardsson, and Karl Johan Åström. A stabilizing switching scheme for multi controller systems. In *IFAC'96, Preprints 13th World Congress of IFAC*, pages 229–234, San Francisco, California, July 1996.

[4] Jerrold E Marsden. *Lectures on Mechanics*. Cambridge University Press, 1992, Cambridge, U.K., 1992.

[5] Marc H. Raibert. *Legged Robots that balance*. MIT Press, Cambridge, MA, 1986.

[6] Mark W. Spong and M. Vidyasagar. *Robot Dynamics and Control*. John Wiley & Sons, 1989.

[7] Magnus Wiklund, Anders Kristenson, and Karl Johan Åström. A new strategy for swinging up an inverted pendulum. In *Preprints IFAC 12th World Congress*, Sydney, Australia, July 1993.

Variable Structure Control for Sampled-data Systems

Katsuhisa Furuta   Yaodong Pan*

Abstract

For a continuous-time system, a sliding sector is designed as a subset of the system state space, where some norm of the state decreases. The continuous-time VS control law is designed to move the system state from the outside to the inside of the sliding sector. The sector is defined as the PR-Sliding Sector where the norm is defined as the quadratic form of the state with the symmetric matrix P and its derivative is less than negative of a quadratic form of the state with the matrix R. In the paper, the discrete-time VS controller for the sampled-data system is designed as an extension of the continuous-time VS controller. The discrete-time sliding sector is to be defined as a subset of the continuous-time sliding sector. The discrete-time VS control law is equal to the continuous-time VS control law at every sampling instant. It is proved that such discrete-time sliding sector for a sampled-data system exists and the proposed discrete-time VS controller quadratically stabilizes the sampled-data system if the sampling interval and the feedback coefficient are chosen suitably. Simulation result is given to show the effectiveness of the proposed VS controller for sampled-data systems.

1   Introduction

The continuous-time variable structure (VS) control system is robust to parameter uncertainty and external disturbance because of the sliding motion on the predefined hyperplane (Utkin, 1964) (DeCarlo *et al.*, 1988). Such sliding motion does not exist (Kotta, 1989) (Sarpurk *et al.*, 1987) when the VS control is implemented in a discrete-time system. To design a VS controller for discrete-time system, a sliding sector was proposed in (Furuta, 1990), the discrete-time VS controller was designed to move the system state from the outside to the inside of a sliding sector.

The PR-sliding sector was first proposed in (Furuta and Pan, 1995), where the norm is defined as the quadratic form of the state with the symmetric

---

*Department of Mechanical and Environmental Informatics, Tokyo Institute of Technology, 2-12-1, Oh-okayama, Meguro-ku, Tokyo 152, JAPAN

matrix P and its derivative is less than negative of a quadratic form of the state with the matrix R without any control action. Both of the continuous-time PR-sliding sector and the discrete-time PR-sliding sector were given. A continuous-time and a discrete-time VS controller were designed by using the PR-sliding sectors in (Furuta and Pan, 1995).

In this paper, a discrete-time VS controller for sampled-data system will be designed as an extension of the continuous-time VS controller with PR-sliding sector. Firstly, a continuous-time sliding sector is shown to be designed by using Riccati equation, and the continuous-time VS control law is designed so as to move the system state from the outside to the inside of the PR-sliding sector while the norm keeps decreasing and thus to quadratically stabilize the system. Then the discrete-time $\bar{P}\bar{R}$-sliding sector is defined as a subset of the continuous-time PR-sliding sector, where the continuous-time VS control law is implemented at every sampling instant. It will be shown that such discrete-time $\bar{P}\bar{R}$-sliding sector exists and the discrete-time VS controller for the sampled-data system quadratically stabilizes the sampled-data system if the sampling interval and some feedback coefficient in the control law are determined to satisfy some inequalities.

The organization of the paper is as follows: Section 2 presents the continuous-time PR-sliding sector, designs a PR-sliding sector by using Riccati equation, and proposes a continuous-time VS control law based on the PR-sliding sector. Section 3 extends the continuous-time VS controller to the sampled-data system and shows the stability of the discrete-time VS controller with discrete-time $\bar{P}\bar{R}$-sliding sector for sampled-data systems. In Section 4, an inverted pendulum system is taken as a simulation model, and simulation result is given.

2 VS Controller for Continuous-time Systems

For the clarity of the presentation, a linear time-invariant continuous-time single input system:

$$\dot{x}(t) = Ax(t) + Bu(t) \tag{1}$$

will be considered, where $x(t) \in R^n$ and $u(t) \in R^1$ are state and input vectors, respectively. A and B are constant matrices of appropriate dimensions, and the pair (A, B) is controllable.

2.1 PR-Sliding Sector

A norm defined on R^n will be used throughout the paper. The following is its definition for continuous-time systems.

Definition 1 *The P-Norm* $||\cdot||_p$ *of the system state is defined as*

$$||x||_p = (x^T P x)^{\frac{1}{2}}, \quad x \in R^n \tag{2}$$

where $P \in R^{n \times n}$ *is a positive definite symmetric matrix.*

The square of the P-norm is denoted as

$$L = ||x||_p^2 = x^T P x > 0, \quad \forall x \in R^n, x \neq 0$$

which will be considered as a Lyapunov function candidate. If the autonomous system of (1) is quadratically stable, then there exist a positive definite symmetric matrix P and a positive semi-definite symmetric matrix $R = C^T C$ such that

$$\dot{L} = x^T (A^T P + PA) x \leq -x^T R x, \quad \forall x \in R^n$$

where $P \in R^{n \times n}$, $R \in R^{n \times n}$, and $C \in R^{l \times n}$, $l \geq 1$, (C, A) is observable pair. But for an unstable one, this inequality does not hold. It is possible to decompose all state space into two parts such that one satisfies $\dot{L} > -x^T R x$ for some element $x \in R^n$, and the other one satisfies $\dot{L} \leq -x^T R x$ for some other element $x \in R^n$. The latter ones form a special subset in which the P-norm $||x||_p$ decreases. Accordingly the special subset is defined as a PR-sliding sector.

Definition 2 *The PR-Sliding Sector is a subset of* R^n *defined as*

$$S = \{x|\ x^T (A^T P + PA) x \leq -x^T R x, x \in R^n \} \tag{3}$$

where $P \in R^{n \times n}$ *is a positive definite symmetric matrix,* $R \in R^{n \times n}$ *is a positive semi-definite symmetric matrix, and* $R = C^T C$, $C \in R^{l \times n}$, $l \geq 1$, (C, A) *is observable pair.*

Inside the PR-sliding sector, the P-norm $||x||_p$ of the plant (1) without any control action decreases because

$$\dot{L} \leq -x^T R x \leq 0, \quad \forall x \in S, x \neq 0. \tag{4}$$

The existence of such PR-sliding sector is described in the following theorem.

Theorem 1 *For any controllable plant (1), the PR-sliding sector defined in (3) exists for any positive definite symmetric matrix P and any positive semi-definite symmetric matrix R described in Definition 2, and can be rewritten as*

$$S = \{x|\ s^2(x) \leq \delta^2(x) \} \tag{5}$$

where

$$s^2(x) = x^T P_1 x \geq 0 \tag{6}$$
$$\delta^2(x) = x^T P_2 x \geq 0, \tag{7}$$

P_1 *and* P_2 *are* $n \times n$ *positive semi-definite symmetric matrices.*

Proof: Denote

$$\Omega = A^T P + PA + R. \tag{8}$$

Then the PR-sliding sector defined in (3) is determined by

$$x^T \Omega x \le 0.$$

For the matrix Ω, there exists a real orthogonal matrix $U \in R^{n \times n}$ such that

$$U^T \Omega U = diag(r_1, r_2, ..., r_n)$$

holds where $r_i (i = 1, 2, ...n)$ are the characteristic roots of Ω, which are all real because Ω is a symmetric matrix.

Assume

$$\begin{aligned}
\bar{P}_1 &= diag(\frac{|r_1| + r_1}{2}, \frac{|r_2| + r_2}{2}, ..., \frac{|r_n| + r_n}{2}) \\
\bar{P}_2 &= diag(\frac{|r_1| - r_1}{2}, \frac{|r_2| - r_2}{2}, ..., \frac{|r_n| - r_n}{2})
\end{aligned}$$

i.e, $\bar{P}_1$ and $\bar{P}_2$ are composed of the positive eigenvalues and the negative eigenvalues of Ω, respectively. Then the following holds.

$$U^T \Omega U = \bar{P}_1 - \bar{P}_2$$

where $\bar{P}_i \ge 0$ $(i = 1, 2)$. Denoting $P_i = U\bar{P}_i U^T$ $(n = 1, 2)$ gives

$$\begin{aligned}
\Omega &= P_1 - P_2, \\
P_i &\ge 0, \quad (i = 1, 2).
\end{aligned}$$

The PR-sliding sector defined in (3) thus can be rewritten as (5) where $s^2(t)$ and $\delta^2(t)$ are determined by (6) and (7), respectively. Which implies that such PR-sliding sector exists for any controllable plant. ■

Corollary 1 *Let $n_i = rank(P_i)$ $(i = 1, 2)$ and $n_3 = n - n_1 - n_2$. Then n_1 and n_2 are the numbers of the positive eigenvalues and the negative eigenvalues of Ω, respectively, n_3 is the number of the eigenvalues of Ω in the origin.*

Remark 1 *There are some special PR-sliding sectors:*

- *The PR-sliding sector S is equal to R^n if $n_1 = 0$.*

- *The PR-sliding sector S is reduced to a PR-sliding surface determined by $x^T P_1 x = 0$ if $n_2 = 0$ and $n - n_1 > 0$.*

- *The PR-sliding sector S is reduced to the equilibrium point $x = 0$ if $n_1 = n$.*

If $n_1 = 1$, the PR-sliding sector for the continuous-time system is defined as

$$S = \{x : |s(x)| \leq |\delta(x)| \} \tag{9}$$

where $s(x)$ is a linear function on x, $\delta^2(x)$ is a quadratic function on x:

$$
\begin{aligned}
s(x) &= Sx, \quad S = [\ s_1 \quad s_2 \quad \cdots \quad s_n\ ], \tag{10} \\
\delta(x) &= \sqrt{x^T P_2 x}. \tag{11}
\end{aligned}
$$

The PR-sliding sector shown in (9) with $n_1 = 1$ is called the simplified PR-sliding sector. If $n_2 = n_1 = 1$, then the simplified PR-sliding sector is determined by two linear functions $s(x)$ and $\delta(x)$ on x, i.e.

$$
\left\{
\begin{aligned}
s(x) &= Sx = [\ s_1 \quad s_2 \quad \cdots \quad s_n\ ] x \\
\delta(x) &= Dx = [\ d_1 \quad d_2 \quad \cdots \quad d_n\ ] x.
\end{aligned}
\right. \tag{12}
$$

S and D satisfy $S^T S = P_1$ and $D^T D = P_2$, respectively. The following theorem shows the simplified PR-sliding sector with $n_1 = 1$ does exist for any controllable plant.

Theorem 2 *For any controllable plant with single input described by (1), there exist a linear function $s(x)$ and a quadratic function $\delta^2(x)$ such that the PR-sliding sector defined by (3) or (5) can be rewritten as the simplified form (9) for some $n \times n$ positive definite symmetric matrix P and some $n \times n$ positive semi-definite symmetric matrix R where $R = C^T C$, $C \in R^{l \times n}$, $l \geq 1$, (C, A) is an observable pair.*

Proof: (A, B) is assumed to be controllable thus we can find a feedback control law $u = -Kx$ such that

$$\bar{A} = A - BK$$

is stable and all of its eigenvalues are real and distinct. Then there exists an invertible transformation $T \in R^{n \times n}$ to transform $\bar{A}$ to a dialog matrix $\tilde{A}$, i.e.

$$T \bar{A} T^{-1} = diag(-a_1^2, -a_2^2, \cdots, -a_n^2) =^d \tilde{A} \tag{13}$$

where $a_i (i = 1, 2, \ldots, n)$ are positive.

Choosing $\tilde{P}$ and $\tilde{R}$ as

$$
\begin{aligned}
\tilde{P} &= \frac{1}{2} I_n > 0, \\
\tilde{R} &= diag(-a_1^2, -a_2^2, \cdots, -a_n^2) > 0,
\end{aligned}
$$

then we have

$$\tilde{A}^T \tilde{P} + \tilde{P} \tilde{A} = -\tilde{R}.$$

Therefore if we choose P and R as

$$P = \frac{1}{2}T^T T > 0, \tag{14}$$

$$R = T^T diag(-a_1^2, -a_2^2, \cdots, -a_n^2)T \geq 0, \tag{15}$$

then

$$A^T P + PA = K^T B^T P + PBK - R.$$

Thus the sliding sector defined in (3) may be rewritten as

$$S = \{x \mid |s(x)| \leq |\delta(x)|\} \tag{16}$$

where $s(x)$ and $\delta(x)$ are respectively defined as

$$s(x) = Sx, \tag{17}$$

$$\delta(x) = \sqrt{x^t \Delta x}, \tag{18}$$

$1 \times n$ matrix S and $n \times n$ matrix Δ are determined by

$$S = K + B^T P$$

$$\Delta = R + K^T K + PBB^T P$$

where $R = C^T C$, $C = diag(a_1, a_2, \cdots, a_n) > 0$ $(a_i > 0, i = 1, 2, ..., n)$ which ensures that (C, A) is observable. ∎

2.2 Design of PR-Sliding Sector by Riccati Equation

In the proof of Theorem 2, a simplified PR-sliding sector has been designed for any controllable plant described by (1). In this subsection, a more effective algorithm to design a simplified PR-sliding sector will be proposed by using Riccati equation for any controllable plant.

Consider a quadratic performance index as

$$J = \int_0^{+\infty} (x(t)Qx^T(t) + u^2(t))dt \tag{19}$$

where $Q \in R^{n \times n}$ is a positive definite symmetric matrix.

As pair (A, B) in plant (1) is controllable, the optimal solution to minimize the performance index (19) for the plant (1) exists and has the form

$$u(t) = -Kx(t), \tag{20}$$

the gain matrix K is defined as

$$K = B^T P \tag{21}$$

where $P \in R^{n \times n}$ is a symmetric positive definite matrix , which is the solution of the following Riccati equation:

$$PA + A^T P - PBB^T P = -Q. \tag{22}$$

If the solution P of the Riccati equation (22) is chosen to define the P-norm as in Definition 1, then for the zero-input , (22) gives

$$
\begin{aligned}
\dot{L}(t) &= x^T(t)(A^T P + PA)x(t) \\
&= x^T(t)K^T Kx(t) - x^T(t)Qx(t).
\end{aligned} \tag{23}
$$

Theorem 3 *If the solution P of the Riccati equation (22) is used to define the P-norm and the matrix R is chosen to be $(1-r)Q$ $(0 < r < 1)$, then the PR-sliding sector defined in (5) can be rewritten as*

$$S = \{x| \; |s(x)| \le |\delta(x)| \; \} \tag{24}$$

where

$$
\begin{aligned}
s(x) &= Sx(t) \quad , \quad S = K \\
\delta(x) &= \sqrt{x^T(t)rQx(t)}
\end{aligned}
$$

Q is the symmetric positive definite matrix in the performance index (19), K is the optimal feedback gain matrix determined by (21).

Proof: According to the equation (23), the following holds when the input $u(t) = 0$.

$$
\begin{aligned}
\dot{L}(t) &= x^T(t)(A^T P + PA)x(t) \\
&= x^T(t)K^T Kx(t) - x^T(t)Qx(t) \\
&= s^2(x) - x^T(t)rQx(t) - x^T(t)(1-r)Qx(t) \\
&= s^2(x) - \delta^2(x) - x^T(t)Rx(t)
\end{aligned}
$$

Therefore, (24) defines a simplified PR-sliding sector. ∎

Corollary 2 *For controllable plant (1), find a matrix $C \in R^{n \times 1}$ such that pair (C, A) is observable, then the simplified PR-sliding sector can be rewritten as*

$$S = \{x| \; |s(x)| \le |\delta(x)| \; \} \tag{25}$$

which is determined by two linear functions on x, i.e.

$$
\begin{aligned}
s(x) &= Sx(t) \\
\delta(x) &= Cx(t)
\end{aligned} \tag{26}
$$

where $S = B^T P$, P is the solution of the following Riccati function:

$$PA + A^T P - PBB^T P + 2C^T C = 0,$$

which is also used to define the P-norm. R is chosen to be $C^T C$.

2.3   VS Controller with PR-Sliding Sector

Based on the PR-sliding sector defined in Subsection 2.1 and designed in Subsection 2.2, a VS controller may be designed:

1. to move the system state from the outside to the inside of the PR-sliding sector,

2. to ensure the P-norm be decreasing at any state.

Theorem 4 *Corresponding to the PR-sliding sector $\mathcal{S}$ given by (24), the VS control law*

$$u(t) = \begin{cases} 0 & x \in \mathcal{S}, \\ -(SB)^{-1}(SAx + ks(x)) & x \bar{\in} \mathcal{S}, \end{cases} \tag{27}$$

ensures the moving from the outside to the inside of the PR-sliding sector and the decreasing of the P-norm at any state if (A, B) is a controllable pair and the constant $k > 0$ satisfies

$$k > \max\{(SB)^{-1}/2, k_0\} \tag{28}$$

where positive constant k_0 satisfies the following quadratic inequality.

$$2k_0 r Q + S^T S A + A^T S^T S > 0.$$

Therefore the control law (27) results in a quadratic stable VS controller, where the PR-sliding sector $\mathcal{S}$ and its parameters: Q, P, R, r, $s(x)$, and $\delta(x)$ are defined in Theorem 3. SB is equal to $B^T PB$ which is nonsingular according to the theorem.

Proof: Consider the square of the P-norm as a Lyapunov function, i.e.:

$$L(t) = x^T(t)Px(t) > 0, \quad \forall x \in R^n, x \neq 0. \tag{29}$$

Inside the PR-sliding sector, i.e. $|s(x)| \leq |\delta(x)|$, the input $u(t)$ is zero. In this case, the derivate of the Lyapunov function $L(t)$ is

$$\dot{L}(t) = s^2(x) - \delta^2(x) - x^T(t)Rx(t) \leq -x^T(t)Rx(t) \leq 0.$$

Outside the PR-sliding sector, i.e. $|s(x)| > |\delta(x)|$, the input $u(t)$ is determined by

$$u(t) = -(SB)^{-1}(SAx + ks(x)).$$

The derivate of the linear function $s(x)$ satisfies that

$$\dot{s}(x) = S\dot{x}(t) = SAx(t) + SBu(t) = -ks(x)$$

Thus following inequality holds.

$$\frac{d}{dt}s^2(x) = 2s(x)\dot{s}(x) = -2ks^2(x) < 0$$

which means that the absolute value of the linear function $s(x)$ will decrease outside the PR-sliding sector with the VS control law (27) and the system state will move toward the inside of the PR-sliding sector, and can move inside it in a finite time if the decreasing rate of $\delta(x)$ is slower than the one of $s(x)$ for some large enough k.

While the system state is being moved from the outside to the inside of the PR-sliding sector by the VS control law (27), the P-norm is decreasing because according to the definition of the PR-sliding sector (24),

$$
\begin{aligned}
\dot{L}(t) &= (Ax(t) + Bu(t))^T Px(t) + x^T(t)P(Ax(t) + Bu(t)) \\
&= x^T(t)(A^T P + PA)x(t) + 2x^T(t)PBu(t) \\
&= s^2(x) - \delta^2(x) - x^T(t)Rx(t) + 2s(x)u(t) \\
&= -(2(SB)^{-1}k - 1)s^2(x) - 2s(x)(SB)^{-1}SAx(t) - \delta^2(x) \\
&\qquad\qquad\qquad\qquad\qquad\qquad\qquad\qquad\qquad -x^T(t)Rx(t) \\
&\leq -(2(SB)^{-1}k - 1)\delta^2(x) - 2x^T(t)S^T(SB)^{-1}SAx(t) - \delta^2(x) \\
&\qquad\qquad\qquad\qquad\qquad\qquad\qquad\qquad\qquad -x^T(t)Rx(t) \\
&= -2(SB)^{-1}k\delta^2(x) - 2x^T(t)S^T(SB)^{-1}SAx(t) - x^T(t)Rx(t) \\
&= -(SB)^{-1}x^T(t)(2krQ + S^T SA + A^T S^T S)x(t) - x^T(t)Rx(t) \\
&< -x^T(t)Rx(t) \\
&< 0
\end{aligned}
$$

where $SB = B^T PB > 0$.

Thus the system state will be moved from outside to the inside of the PR-sliding sector if the system state is outside the PR-sliding sector and at any state, the Lyapunov function (29) keeps decreasing with the VS control law (27), i.e.

$$\dot{L}(t) \leq -x^T(t)Rx(t), \quad \forall x \in R^n$$

Therefore the VS controller designed in the theorem quadratically stabilize the continuous-time plant (1).

$\blacksquare$

Remark 2 *Typically, a VS controller based on a sliding mode should guarantee that the system state converges to the sliding hyperplane in a finite time and stays in there forever, in which the system is designed to be stable. For the VS controller with PR-sliding sector, it is easy to modify the VS control law (27) as*

$$u(t) = \begin{cases} 0 & x \in \mathcal{S}, \\ -(SB)^{-1}(SAx + k\text{sign}(s(x))) & x\bar{\in}\mathcal{S}, \end{cases}$$

with which, the system state can move from the outside to the inside of the PR-sliding sector in a finite time and stay inside it forever. But the PR-sliding sector is designed in which the P-norm decreases and the VS controller proposed in Theorem 4 is designed such that the P-norm decreases at any state. Therefore it does not matter whether the system state moves to the inside from the outside of the PR-sliding sector in a finite time or in infinite time. Moreover, using the VS control law shown in above equation may result in chattering nearby the origin. This is the reason why the VS control law is designed in the form of (27).

3   VS Controller for Sampled-data Systems

The sampled-data system of the continuous-time system (1) can be described by following state equation:

$$x_{k+1} = \Phi x_k + \Gamma u_k \tag{30}$$

where zero-order holder is used, $x_k = x(k\tau)$, $u_k = u(k\tau)$, $(k = 0, 1, \ldots)$, $\Phi = e^{A\tau}$, $\Gamma = \int_0^\tau e^{At} B dt$, and τ is the sampling interval.

In general, implementing a continuous-time VS controller with sliding mode in a sampled-data system may not only result in chattering, but also may produce an unstable mode (Furuta and Pan, 1994). But the VS controller with the PR-sliding sector proposed in the last section can be extended to the sampled-data system with just a little modification. To design the discrete-time $\bar{P}\bar{R}$-sliding sector and the discrete-time VS controller for the sampled-data system, we write the matrices Φ and Γ in (30) as:

$$\Phi = e^{A\tau} = A + A\tau + \sum_{n=2}^{\infty} \frac{A^n}{n!}\tau^n$$

$$\Gamma = \int_0^\tau e^{At} B dt = B\tau + \sum_{n=1}^{\infty} \frac{A^n}{(n+1)!}\tau^{n+1}$$

3.1   Discrete-time $\bar{P}\bar{R}$-Sliding Sector

Similar with the definitions of P-norm and PR-sliding sector for the continuous-time system, two definitions are given for the discrete-time system.

Definition 3 *The $\bar{P}$-Norm $\|\cdot\|_{\bar{p}}$ of the discrete-time system state is defined as*

$$\|x_k\|_{\bar{p}} = (x_k^T \bar{P} x_k)^{\frac{1}{2}}, \quad x_k \in R^n \tag{31}$$

where $\bar{P}$ is a $n \times n$ positive definite symmetric matrix.

Definition 4 *A* Discrete-time $\bar{P}\bar{R}$-Sliding Sector *is defined as*

$$\bar{S} = \{x \mid x_k^T(\Phi^T\bar{P}\Phi - \bar{P})x_k \leq -x_k^T\bar{R}x_k, x_k \in R^n \} \tag{32}$$

where $\bar{P}$ is a $n \times n$ positive definite symmetric matrix and $\bar{R}$ is a $n \times n$ positive semi-definite symmetric matrix, and $\bar{R} = \bar{C}^T\bar{C}$, $(\bar{C}, \Phi)$ is observable.

Inside the $\bar{P}\bar{R}$-sliding sector, the $\bar{P}$-norm decreases without any control action, i.e.,

$$\begin{aligned} L_{k+1} - L_k &= x_k^T(\Phi^T\bar{P}\Phi - \bar{P})x_k \\ &\leq -x_k^T\bar{R}x_k \end{aligned}$$

where the function L_k is determined by

$$L_k = \|x_k\|_{\bar{P}}^2 = x_k^T\bar{P}x_k > 0, \quad \forall x_k \in R^n, x \neq 0 \tag{33}$$

which is a discrete-time Lyapunov function candidate.

For discrete-time system (30), the $\bar{P}\bar{R}$-sliding sector and the VS control law can be designed in a way being similar with the one for the continuous-time system, see (Furuta and Pan, 1995). In this paper, the discrete-time system (30) is a sampled-data one, the discrete-time $\bar{P}\bar{R}$-sliding sector and the discrete-time VS controller for the sampled-data system will be designed as the extension of the continuous-time ones proposed in the last section.

It is assumed that the P-norm is equal to the $\bar{P}$-norm at every sampling instant, and the decrease rate in the PR-sliding sector and the one in the $\bar{P}\bar{R}$-sliding sector are the same, i.e.

$$\begin{aligned} P &= \bar{P}, \\ R &= \tau\bar{R}, \end{aligned}$$

then for the zero control input, the discrete-time Lyapunov function candidate L_k satisfies:

$$\begin{aligned} \Delta L_{k+1} &= L_{k+1} - L_k = x_k^T(\Phi^T P\Phi - P)x_k \\ &= x_k^T(\tau A^T P + \tau PA + \tau^2 F(\tau))x_k \\ &= \tau(s^2(x_k) - \delta^2(x_k) - x_k^T R x_k) + \tau^2 x_k^T F(\tau)x_k \\ &= \tau(s^2(x_k) - x_k^T rQ x_k + \tau x_k^T F(\tau)x_k) - x_k^T\bar{R}x_k \end{aligned}$$

where $F(\tau)$ is defined as

$$\begin{aligned} F(\tau) &= (I + A^T\tau)P\sum_{n=2}^{\infty}\frac{A^n}{n!}\tau^{n-2} + \sum_{n=2}^{\infty}\frac{(A^T)^n}{n!}\tau^{n-2}P(I + A\tau) \\ &\quad + \tau^2\sum_{n=2}^{\infty}\frac{(A^T)^n}{n!}\tau^{n-2}P\sum_{n=2}^{\infty}\frac{A^n}{n!}\tau^{n-2}. \end{aligned}$$

As Q is positive definite, there exists a large enough positive constant f_0 such that

$$\min_{0 < \tau \leq \tau_0} \{f_0 r Q - F(\tau)\} \geq 0 \tag{34}$$

where T_0 is maximum value of the possible sampling interval, r is the positive constant defined in Theorem 3. Thus we have

$$
\begin{aligned}
\Delta L_{k+1} &\leq \tau(s^2(x_k) - (1 - \tau f_0)x_k^T r Q x_k) - x_k^T \bar{R} x_k \\
&= \tau(s^2(x_k) - (1 - \tau f_0)\delta^2(x_k)) - x_k^T \bar{R} x_k
\end{aligned}
$$

Inside the discrete-time $\bar{P}\bar{R}$-sliding sector, the right side of the above inequality should be less than and equal to $-x_k^T \bar{R} x_k$ such that

$$\Delta L_{k+1} \leq -x_k^T \bar{R} x_k$$

It is clear that the above inequality holds if

$$s^2(x_k) \leq \beta \delta^2(x_k)$$

and the sampling interval τ satisfies

$$\tau \leq \min\{\tau_o, \frac{1 - \beta}{f_0}\}. \tag{35}$$

where β is a positive constant satisfying $0 < \beta < 1$.

Thus the discrete-time $\bar{P}\bar{R}$-sliding sector for sampled-data systems is assumed to be a subset of the continuous-time PR-sliding sector (24), that is

$$S_d = \{x_k | \ |s(x_k)| \leq \beta |\delta(x_k)| \ \} \tag{36}$$

where $\beta(0 < \beta < 1)$ is a real constant, the functions $s(x_k)$ and $\delta(x_k)$ on x_k are respectively the same as the functions $s(x)$ and $\delta(x)$ on x defined in the PR-sliding sector S.

The result of this subsection is concluded in the following theorem.

Theorem 5 *For plant (30), the subset of R^n defined in (36) is a discrete-time $\bar{P}\bar{R}$-sliding sector for the sampled-data system (30) with $\bar{P} = P$ and $\bar{R} = \tau R$, inside which the discrete-time $\bar{P}$-norm is decreasing if the sampling interval τ satisfies (35). Where P and R are matrices defining a continuous-time PR-sliding sector in (5) for the continuous-time plant (1).*

3.2   Discrete-time VS Controller with $\bar{P}\bar{R}$-Sliding Sector

Based on the discrete-time $\bar{P}\bar{R}$-sliding sector $\mathcal{S}_d$ given by (36), the discrete-time VS control law is designed to be equal to the continuous-time VS control law (27) at every sampling instant, i.e.

$$u_k = \begin{cases} 0 & x_k \in \mathcal{S}_d, \\ -(SB)^{-1}(SAx_k + ks(x_k)) & x_k \bar{\in} \mathcal{S}_d, \end{cases} \tag{37}$$

Theorem 6 *The discrete-time VS control law (37) ensures the moving from the outside to the inside of the discrete-time $\bar{P}\bar{R}$-sliding sector and the decreasing of the P-norm at any state if (A,B) is controllable pair, the sampling interval τ satisfies that*

$$\tau < min\{\tau_0, \frac{1-\beta}{f_0}, \frac{1}{k}, \frac{k}{g_0}, \frac{2k\beta^2}{h_0}\}$$

and $1 - k\tau + \frac{\tau^2 g_0}{\beta}$ is chosen to satisfy

$$0 < 1 - k\tau + \frac{\tau^2 g_0}{\beta} < 1.$$

Where the real constants g_0 and h_0 are chosen to ensure the followings hold.

$$\max_{0 < \tau < \tau_0} \{g_0^2 rQ - G^T(\tau)G(\tau)\} > 0$$

$$\max_{0 < \tau < \tau_0} \{h_0 rQ - H(\tau)\} > 0$$

$G(\tau)$ and $H(\tau)$ are determined by

$$G(\tau) = \tau^2(\sum_{n=2}^{\infty} \frac{A^n}{n!}\tau^{n-2} + S(\sum_{n=2}^{\infty} \frac{A^{n-1}}{(n+1)!}\tau^{n-2}B(SB)^{-1}(SA+kS)$$

$$H(\tau) = F(\tau) - 2P\sum_{n=2}^{\infty} \frac{A^{n-2}}{n!}\tau^{n-2}B(SB)^{-1}(SA+kS)$$

$$- \frac{2}{SB}\sum_{n=1}^{\infty} \frac{(A^T)^n}{n!}\tau^{n-1}P\sum_{n=1}^{\infty} \frac{A^{n-1}}{n!}\tau^{n-1}B(SA+kS)$$

$$+ (SB)^{-2}(A^TS^T+kS^T)B^T\sum_{n=1}^{\infty} \frac{(A^T)^{n-1}}{n!}\tau^{n-1}P$$

$$\sum_{n=1}^{\infty} \frac{A^{n-1}}{n!}\tau^{n-1}B(SA+kS)$$

β, f_0, τ_0, and $F(\tau)$ are defined in the last subsection. Therefore the discrete-time VS control law (37) with the discrete-time $\bar{P}\bar{R}$-sliding sector (36) quadratically stabilizes the sampled-data system (30).

Proof: Consider the discrete-time Lyapunov function L_k defined in (33).

Inside the discrete-time $\bar{P}\bar{R}$-sliding sector, i.e. $|s(x_k)| \leq \beta|\delta(x_k)|$, the input u_k is zero. In this case, it has been proven in the last subsection that

$$\Delta L_{k+1} \leq -x_k^T \bar{R} x_k \tag{38}$$

Outside the discrete-time $\bar{P}\bar{R}$-sliding sector, i.e. $|s(x_k)| > \beta|\delta(x_k)|$, the input u_k is determined by

$$u_k = -(SB)^{-1}(SAx_k + ks(x_k)).$$

In this case, the linear function $s(x_{k+1})$ is

$$
\begin{aligned}
|s(x_{k+1})| &= |Sx_{k+1}| = |S\Phi x(t) + S\Gamma u(t)| \\
&= |(1-k\tau)s(x_k) + \tau^2 G x_k| \\
&< |(1-k\tau)s(x_k) + \tau^2 g_0 \sqrt{x_k^T Q x_k}| \\
&= |(1-k\tau)s(x_k) + \tau^2 g_0 |\delta(x_k)||| \\
&\leq (1-k\tau + \frac{\tau^2 g_0}{\beta})|s(x_k)| \\
&< |s(x_k)|
\end{aligned}
\tag{39}
$$

which means that the absolute value of the linear function s_k will decrease outside the discrete-time $\bar{P}\bar{R}$-sliding sector with the VS control law (37).

Outside the $\bar{P}\bar{R}$-sliding sector, the discrete-time Lyapunov function L_k will decrease with the discrete-time VS control law (37) because

$$
\begin{aligned}
\Delta L_{k+1} &= L_{k+1} - L_k \\
&= x_{k+1}^T P x_{k+1} - x_k^T P x_k \\
&= x_k^T (\Phi^T P \Phi - P) x_k + 2x_k^T \Phi^T P \Gamma u_k + u_k^T \Gamma^T P \Gamma u_k \\
&= \tau(s^2(x_k) - \delta^2(x_k) - x_k^T R x_k) - 2\tau k s^2(x_k) + \tau^2 x_k^T H x_k \\
&< -\tau x_k^T R x_k - 2\tau k s^2(x_k) + \tau^2 x_k^T H x_k \\
&< -x_k^T \bar{R} x_k - 2\tau k \beta^2 \delta^2(x_k) + \tau^2 h_0 \delta^2(x_k) \\
&< -x_k^T \bar{R} x_k.
\end{aligned}
\tag{40}
$$

According to (38), (39), and (40), we have

$$
\begin{aligned}
|s_{k+1}| &< |s_k|, \quad if x_k \bar{\in} S_d \\
\Delta L_{k+1} &\leq -x_k^T \bar{R} x_k, \quad \forall x_k \in R^n
\end{aligned}
$$

Thus it can concluded that the system state will be moved from outside to the inside of the discrete-time $\bar{P}\bar{R}$-sliding sector if the system state is outside the discrete-time $\bar{P}\bar{R}$-sliding sector, the discrete-time Lyapunov function (33) keeps decreasing at any state with the discrete-time VS control law (37). Therefore the discrete-time VS controller with $\bar{P}\bar{R}$-sliding sector quadratically stabilize the sampled-data system (30). ∎

4   Example

A rotational inverted pendulum system shown in Figure 1 is taken as the simulation model. An additional dynamics shown in Figure 2 is included in the inverted pendulum system as the uncertainty. For detail about the inverted pendulum system, see (Hara *et al.*, 1998).

The state equation of the pendulum system with the additional dynamics is

$$\dot{x} = Ax + bu$$

where

$$A = \begin{pmatrix} 0 & 0 & 0 & 1 & 0 & 0 \\ 0 & 0 & 0 & 0 & 1 & 0 \\ 0 & 0 & 0 & 0 & 0 & 1 \\ 0 & 8742.36 & 0 & -1.00645 & 917.565 & 0 \\ 0 & -4518.6 & 2.26161 & 0.196257 & -474.255 & -0.0498344 \\ 0 & -20572.5 & 88.258 & 0 & -2159.22 & -1.94476 \end{pmatrix}$$

$$b = \begin{pmatrix} 0 & 0 & 0 & 32.6768 & -6.37198 & 0 \end{pmatrix}^{T}$$

$$x = \begin{pmatrix} q_1 & \delta_2 & q_3 & \dot{q}_1 & \dot{\delta}_2 & \dot{q}_3 \end{pmatrix}^{T}$$

q_1 :   angle of link.

δ_2 :   displacement of hinge for pendulum.

q_3 :   angle of pendulum.

The state equation of pendulum system without additional dynamics is:

$$A = \begin{pmatrix} 0 & 0 & 1 & 0 \\ 0 & 0 & 0 & 1 \\ 0 & 6.5531 & -0.938621 & -0.144397 \\ 0 & 81.0655 & -1.33818 & -1.78627 \end{pmatrix}$$

$$b = \begin{pmatrix} 0 & 0 & 30.4747 & 43.4473 \end{pmatrix}^{T}$$

$$x = \begin{pmatrix} q_1 & q_2 & \dot{q}_1 & \dot{q}_2 \end{pmatrix}^{T}$$

q_1 :   angle of link.

q_2 :   angle of pendulum.

Based on the state equation of pendulum system without additional dynamics, a discrete-time VS controller with $\bar{P}\bar{R}$-sliding sector is designed. The positive definite symmetric matrix Q is chosen as

$$Q = diag(1, 1, 1, 1)$$

then the solution P of the Riccati equation (22) and matrix R are determined by

$$P = \begin{bmatrix} 1.32132368 & -2.87041542 & 0.372473930 & -0.284276152 \\ -2.87041542 & 33.0669334 & -3.44522372 & 2.89077316 \\ 0.372473930 & -3.44522372 & 0.433959219 & -0.335507650 \\ -0.284276152 & 2.89077316 & -0.335507650 & 0.290947897 \end{bmatrix}$$

$R \;=\; diag(0.5, 0.5, 0.5, 0.5)$

The parameters S and Δ of the PR-sliding sector thus are shown as follows:

$$S \;=\; [\; -1.0 \quad 20.6041295 \quad -1.35212449 \quad 2.41640558 \;]$$
$$\Delta \;=\; diag(0.5, 0.5, 0.5, 0.5)$$

Other parameters of the discrete-time VS controller are chosen as

$$T \;=\; 0.01\text{Second}$$
$$\beta \;=\; 0.5$$
$$K \;=\; 100$$

The simulation results are shown in Figure 3, 4, 5, 6, 7.

Figure 1: Rotational Inverted Pendulum

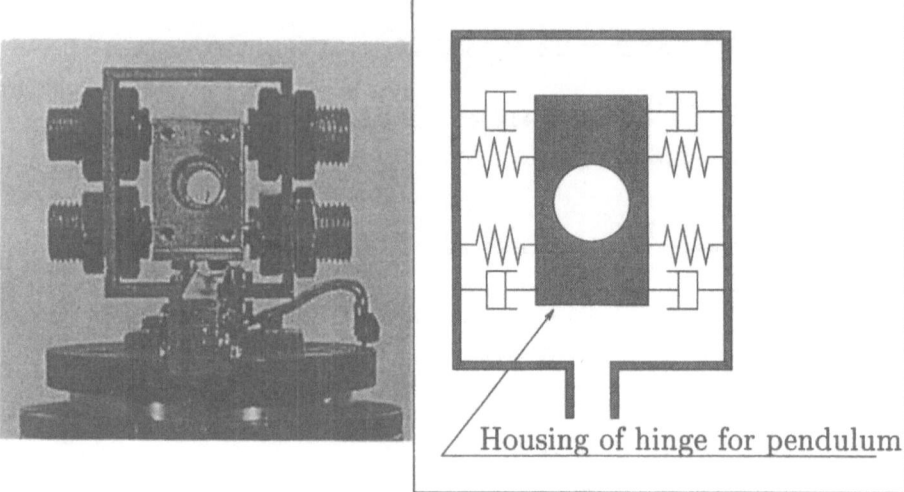

Figure 2: Additional dynamics

5 Conclusion

The continuous-time VS controller with PR-sliding sector is proposed, which quadratically stabilizes the system without chattering phenomena and can be implemented in sampled-data systems. The quadratical stability of the continuous-time VS controller with PR-sliding sector was shown. A discrete-time VS controller with discrete-time $\bar{P}\bar{R}$-sliding sector was designed for sampled-data systems as an extension of the continuous-time VS controller with PR-sliding sector:

1. The discrete-time $\bar{P}\bar{R}$-sliding sector is designed as a subset of the continuous-time PR-sliding sector.

2. The discrete-time VS control law is equal to the continuous-time VS control law at every sampling instant.

It was shown in the paper that the proposed discrete-time VS controller with $\bar{P}\bar{R}$-sliding sector quadratically stabilizes the sampled-data system.

References

DeCarlo, R.A., S.H. Zak and G.P. Matthews (1988). Variable structure control of nonlinear multivariable systems: A tutorial. *Proc. of the IEEE* **76**, 212–232.

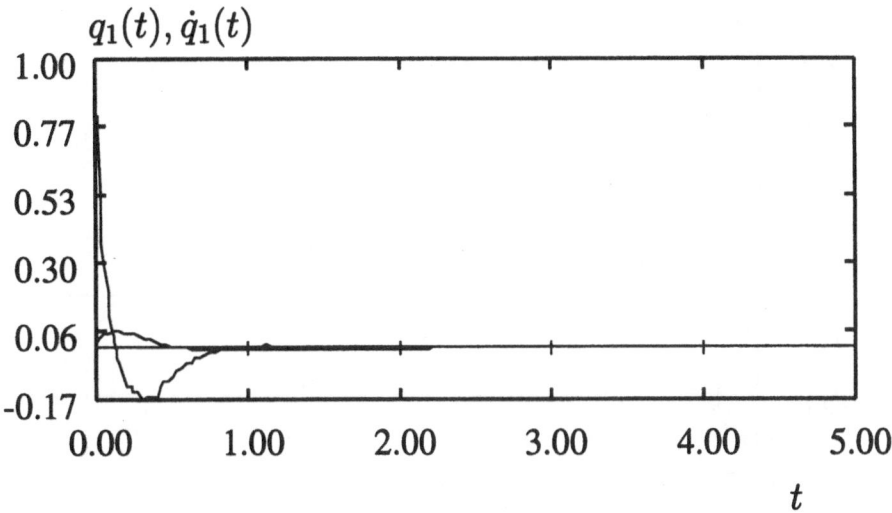

Figure 3: Response with Discrete-time VS Controller

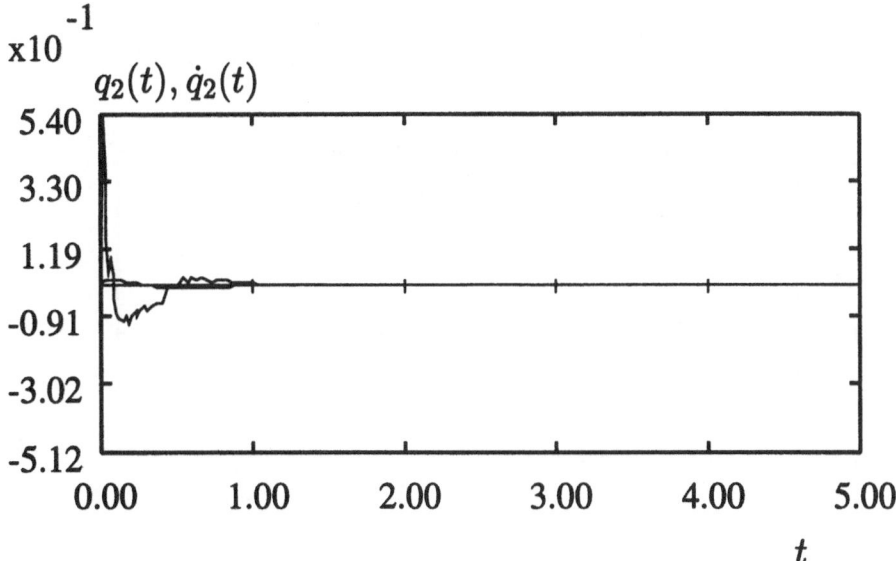

Figure 4: Response with Discrete-time VS Controller

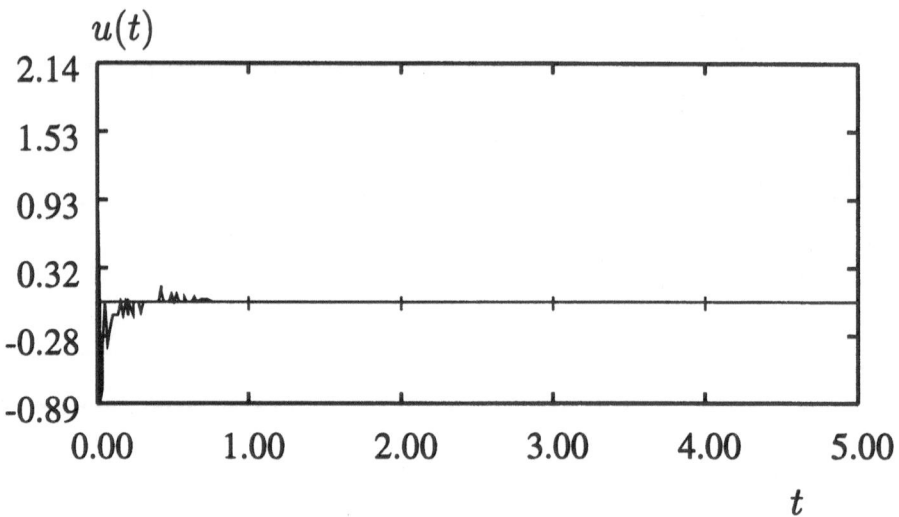

Figure 5: Input with Discrete-time VS Controller

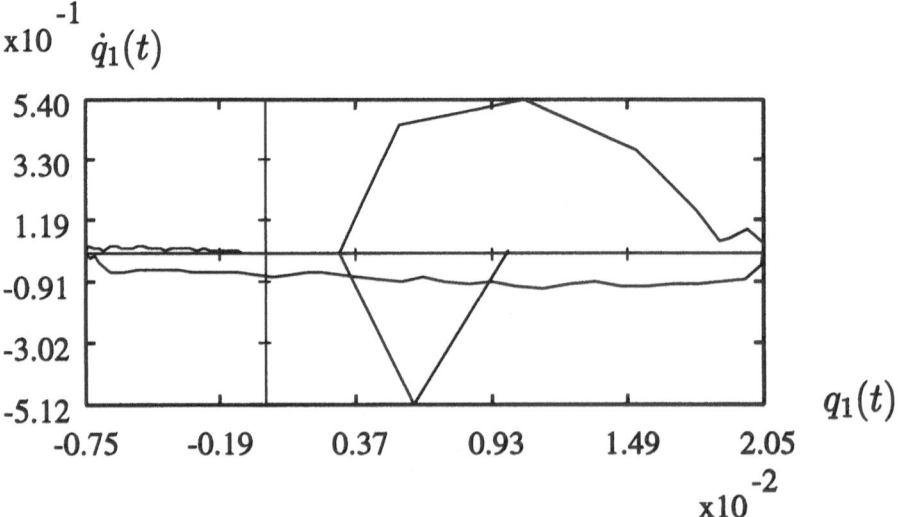

Figure 6: Phase Plane with Discrete-time VS Controller

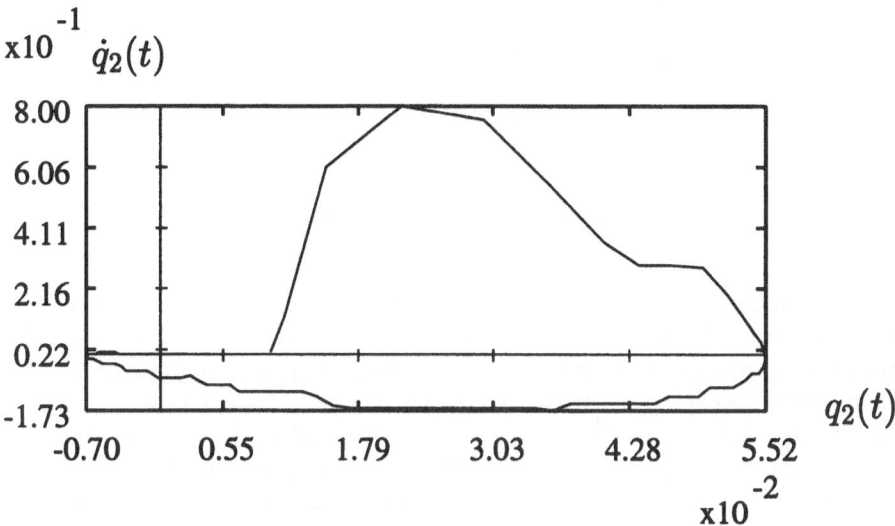

Figure 7: Phase Plane with Discrete-time VS Controller

Furuta, K. (1990). Sliding mode control of a discrete system. *System & Control Letters* **14**, 145–152.

Furuta, K. and Y. Pan (1994). Discrete-time vss control for continuous-time systems. In: *Proc. of the First Asian Control Conference*. Tokyo. pp. 377–180.

Furuta, K. and Y. Pan (1995). A new approach to design a sliding sector for vss controller. In: *Proc. of the American Control Conference*. Seattle. pp. 1304–1308.

Hara, M., K. Furuta, Y. Pan and T. Hoshino (1998). Evaluation of discrete-time vsc on inverted pendulum apparatuswith additional dynamics. *AMCS, to be published.*

Kotta, U. (1989). On the stability of discrete-time sliding mode control system. *IEEE Trans. on Automatic control* **AC-34**, 1021–1022.

Sarpurk, S.Z., Y. Istefanopulos and O. Kaynak (1987). On the stability of discrete-time sliding mode control system. *IEEE Trans. on Automatic Control* **AC-32**, 930–932.

Utkin, V.I. (1964). Variable structure system with sliding mode:a survey. *IEEE Transactions on Automatic Control* **AC-22**, 212–222.

A Control Engineer's Perspective on Fuzzy Control

Stephen Yurkovich*

Control engineers have been slow to embrace fuzzy logic as a common tool in the typical repertoire of useful techniques for solving difficult engineering problems. From its beginnings more than 30 years ago, through its first implementations for control 20 years ago, to its increasingly common appearances within technical programs at serious control meetings throughout the '90s, fuzzy control has indeed come a long way. Along this journey, fuzzy control has gained respect, and is now recognized as having substance beyond its inherent heuristic and intuitive appeal.

This presentation offers the perspective of one "conventional-minded" control engineer on the subject of fuzzy control. By tracing a brief history of fuzzy control from its inception out of fuzzy logic, I would like to explore some of the controversy that has followed its evolution. Without delving deeply into the details of the technology, a little background will be given so that we can discuss current affairs with regard to results for stability and design using fuzzy logic in control. Then I will focus on what initially motivated me to study fuzzy control: applications.

1   Some Conventional Controlist Philosophy

Is fuzzy control overly hyped?

The answer to that question has changed over the years. The answer from the control community many years ago was a re-sounding "yes." That isn't necessarily a commentary on the technology itself, or the state of affairs then, but often rather on the work of numerous individuals who were perceived as "non-controlists," but who had success with fuzzy logic in control applications. And, frankly speaking, there has been a significant amount of sloppiness surrounding the area and proliferating the open literature over the years. The truth is, for these and a variety of other reasons, control engineers developed an early skepticism toward fuzzy control which still lingers today. Many who dabbled and experimented in the technology have recently "come out of the

*Department of Electrical Engineering, The Ohio State University, 2015 Neil Avenue, Columbus OH 43210-1272 USA. *yurkovich@ee.eng.ohio-state.edu*

closet," however, due to some careful (perhaps long overdue) work establishing the limitations and usefulness of fuzzy control as a worthwhile tool.

In a typical design scenario, the control engineer usually follows a predetermined procedure which begins with the need for understanding the process to be controlled as well as the control objectives. Such a procedure might generally proceed by (i) developing a model of the process, (ii) using the (mathematical) model to design a controller (perhaps using a linear model to develop a linear controller with classical techniques), (iii) using the mathematical model of the closed-loop system and mathematical or simulation-based analysis to study its performance (possibly leading to re-design), and (iv) implementing the controller and evaluating its performance in the closed-loop. Although debatable, it can be said that heuristics enter the design process when such a conventional control design process is used; this is especially true when one is concerned with implementation. And, it is fair to say that conventional control engineering approaches using heuristics in tuning have been relatively successful (the vast majority of all controllers currently in operation are conventional PID controllers). One wonders, however, how much of the success can be attributed to the use of the mathematical model and a "conventional" control design approach, and how much should be attributed to heuristic tuning the control engineer uses when implementing the design? That is, how critical a role do heuristics play in the overall design process?

The reason I make this point about conventional control design is to transition directly into our discussion of fuzzy control technology. Simply stated, fuzzy control provides a formal methodology for representing, manipulating, and implementing a human's heuristic knowledge about how to control a system [1, 2]. The design process involves incorporating human expertise on how to control a system into a set of rules. The decision process (inference mechanism) in the fuzzy controller architecture reasons over the information embodied in the rules, measured signals from the process, and the design goals to "decide" what control inputs will achieve the goals when applied to the process. Thus, it could be claimed that the focus in fuzzy control design is on the use of heuristic knowledge to achieve good control, whereas the focus in conventional control design is on the use of a mathematical model for control systems development with subsequent use of heuristics in implementation. Either way, heuristics play an important role.

Along these lines, then, the point which is probably most often raised in discussion of controller *synthesis* using fuzzy logic is that such procedures are usually performed in an *ad hoc* manner, where mechanized synthesis procedures, for the most part, are nonexistent (e.g., it is often not clear exactly how to justify the choices for many controller parameters, such as membership functions, defuzzification strategy, and inference strategy). On the other hand, some mechanized synthesis procedures do exist for particular applications (for example, [3, 4]); typically such procedures arise primarily out of necessity because of system complexity (such as when many inputs and multiple objectives must be achieved). Controller adaptation, in which a form of *automatic controller synthesis* is achieved, is one way of attacking this problem, when no other "direct" synthesis procedure is known.

Another point which raises skepticism among control engineers is the claim that fuzzy control design and implementation are performed in a "model-free" manner. One could argue from experience that most often the control engineer developing a fuzzy control system does have a mathematical model available. While it may not be used directly in controller design, it is often used in simulation to evaluate the performance of the fuzzy controller before it is implemented (and it is often used for rule-base re-design). On the other hand, there are some applications where one can design a fuzzy controller and evaluate its performance directly via an implementation (e.g., for some commercial products where performance is less critical). In such cases there may thus be no need for conducting simulation-based evaluations (requiring a mathematical model) before implementation. In other applications there is the need for a high level of confidence in the reliability of the fuzzy control system before it is implemented.

But even in the case of conventional control design, one must inquire as to how much of the success can be attributed to the use of the mathematical model and design approach, and how much should be attributed to heuristic tuning that the control engineer uses upon implementation. That is, most often conventional control engineering approaches use an approximate mathematical model that is accurate enough to characterize the essential plant behavior, yet simple enough so that the necessary assumptions to apply the analysis and design techniques are satisfied. However, due to the inaccuracy of the model, upon implementation the developed controllers often need to be tuned via the "expertise" of the control engineer. The fuzzy control approach, where explicit characterization and utilization of control expertise is used earlier in the design process, largely avoids the problems with model complexity that are related to design. That is, often fuzzy control system design does not depend on a mathematical model unless it is needed to perform simulations to gain insight into how to choose the rule-base and membership functions.

2   Evolution of a Technology

The issues I have raised to this point (an incomplete list) have been part of the evolutionary process of fuzzy control through the years. It is interesting to explore some of the folklore and history of the technology as well. Unfortunately, it would be futile to attempt to give a broad-sweeping history in this context; I will assume that readers are familiar with some of the well-referenced works, and would like to explore instead some of this folklore. Of course, I cannot imagine discussing the history of fuzzy control, even briefly, without referring to the work and philosophy of Lotfi Zadeh [5].

In the early 1960s Professor Zadeh began sowing the seeds of the concept of fuzziness and "partial truth." He basically conjectured that a problem lay in the fuzziness of concepts which he had attempted to define within the framework of classical mathematics. In a 1962 paper Zadeh expressed his feelings that systems analysis and control were somehow taking a back seat to the development of mathematical theories [6]. Effectively, he was not questioning the power of mathematics, but rather the effectiveness of traditional mathematical methods (which have trouble handling partial truths and uncertainty) in the solution of real-world problems. It was in these early works, it would seem, that the foundation for fuzzy control, based on the idea of fuzzy sets, was being laid; the formulation of the basic concept of fuzzy sets followed and has taken its place in the often-quoted literature (see, for example, [7]).

Zadeh's concept of fuzzy sets drew a mixed reaction, to put it mildly. A few notable scientists and mathematicians (such as Bellman and Moisil) were apparently supportive and enthusiastic; but Zadeh recounts that, for the most part, he mostly experienced skepticism and hostility [5]. Discussion of the folklore would not be complete if I did not mention the now famous comments of Professor Rudy Kalman in 1972 at a meeting in France at which Zadeh first described the concept of a linguistic variable (a variable expressed with words, rather than numbers). What follows is an excerpt taken directly from [5], as a record of Kalman's comments that day:

> proposals could be severely, technical point of view. This would be out of place here. But question remains: Is Professor Zadeh presenting important ideas or is he indulging in wishful "The most serious objection to 'fuzzification' of system analysis is that lack of methods of system analysis is not the principal scientific problem in the 'systems' field. That problem is one of developing basic concepts and deep insight into the nature of 'systems,' perhaps trying to find something akin to the 'laws' of Newton. In my opinion, Professor Zadeh's suggestions have no chance to contribute to the solution of this basic problem.

> "Let me say quite categorically that here is no such thing as a fuzzy concept. We do talk about fuzzy things but they are not scientific concepts. Some people in the past have discovered certain interesting things, formulated their findings in a non-fuzzy way, and therefore we have progressed in science.

> "No doubt Professor Zadeh's enthusiasm for fuzziness has been reinforced by the prevailing political climate in the U.S. — one of unprecedented permissiveness. 'Fuzzification' is a kind of scientific permissiveness; it tends to result in socially appealing slogans unaccompanied by the discipline of hard scientific work and patient observation." alternative for the scientific optimism, 'Wir wollen wissen:

fuzziness is tolerable or formally claimed) that his problems. "In any case, if the 'fuzzification' approach is going to solve any difficult problems, this is yet to be seen." away precise reasoning can solve any nontrivial

Recently control engineers have stepped up open debate on the merits of fuzzy control. An example of this appeared in [8], where prominent members of the control community reacted to claims made in an article which appeared in the February 1993 issue of *IEEE Spectrum* Magazine. Five researchers wrote letters reacting to that article, which, among other things, alluded to the failure of "conventional" adaptive control in critical situations, but implied that adaptive *fuzzy* control would have fared very well. This quite naturally led to a discussion in more than one of the letters about the review process, and even prompted a conjecture that articles on fuzzy control systems are not reviewed as rigorously as other articles on (presumably conventional) control systems research and applications.

I believe it is accurate to say that more recent times (the last three or four years) have seen a bit more careful work in the field of fuzzy control, primarily because more and more control people are venturing into the area. To wit, several new books have appeared (or will soon appear) which focus on fuzzy control, and where once "hype" reigned supreme, now claims are more often reserved and carefully stated. This in itself is an encouraging sign for the future of the technology in control circles.

3   Fuzzy Control Structures

Let's turn our focus now toward the technology itself. We cannot, of course, do justice to the details of fuzzy control in this limited presentation (the reader is referred to [1] for a complete treatment), but we can give a general overview in order to proceed in our discussion to extend on the ideas laid out to this point.

Generally speaking, a fuzzy controller is a static nonlinear map from controller inputs to controller outputs (but often the designer may add, for example, an integrator), whose structure consists of four basic components: a fuzzifier, a knowledge base, an inference mechanism, and a defuzzifier.

The fuzzifier interface essentially converts controller inputs (often functions of measurements from the process) into information on which the inference mechanism can operate. This is where linguistic descriptions become important (more on this in a minute), and where some creativity in names arises (admittedly, some people get carried away in their creativity in this regard). Linguistic variables (such as "error") are defined to have linguistic values or levels (such as "large"). Quantifying the process dynamics with linguistics is not always easy and certainly a better understanding of the process dynamics generally leads to a better linguistic description (recall my previous comments on the need for a process model). Part of this process involves choosing the admissible range of controller inputs and outputs (in the jargon referred to as *universes of discourse*); then, to make the design procedure easier, it is common to normalize these admissible ranges to an interval such as $[-1, 1]$, in which case the normalizing gains become, in themselves, a design variable.

These linguistic descriptions are used in the decision process via the knowledge base. Within the knowledge base (also known as the rule base) are the familiar rules of the form

If <u>Antecedent</u> **Then** <u>Consequent</u>

where the antecedent could represent multiple inputs to the controller (such as "if u_1 is A_1 *and* if u_2 is A_2 *and...*" where the u_i are process inputs and the A_i are linguistic values associated with some fuzzy sets). Note that the antecedent (premise) and consequent (conclusion) are quantified with linguistics in fuzzy sets and membership functions (which specify the degree of certainty that a real variable may be classified heuristically with some prespecified linguistic value) [1]. It is actually the case that for most fuzzy controllers the fuzzification process is so simple that it can be ignored; simply think of it as the act of obtaining a value of an input variable and finding the numeric values of the membership function(s) that are defined for that variable. One can then think of the membership function values as an *encoding* of the fuzzy controller numeric input values.

The rules essentially contain the designer's information on how to control the process. Indeed, the rules represent a very important component with regard to the effectiveness of the controller, and should be chosen so that all pertinent information about control and disturbance rejection are included ("tailored" to the process). The inference mechanism acts on the results of the rules, determining which rules to use, and essentially emulates how control decisions are made. The inference process generally involves two steps: (i) premises of rules are compared to controller inputs to determine which rules apply to the current situation, and (ii) the conclusions (what control actions to take) are determined using the relevant rules for that control instance. To do this, the process of "fuzzy inference" is used, and there are several ways to carry this out. One example is Zadeh's compositional rule of inference [7] which basically states that you can be no more certain about a conclusion than you are about a premise.

Defuzzification converts the conclusion of the inference mechanism into actual inputs for the process. A number of defuzzification strategies exist (indeed, often one can be created to best suit your results); in general, each provides a means to choose a single output based on the results of the inference process (which turn out to be fuzzy sets), each providing a mathematical quantification of the operation of the entire fuzzy system provided that each of the terms in the descriptions are fully defined.

The above description refers to the usual type of fuzzy system encountered most frequently in the control systems literature (sometimes referred to as a "standard" fuzzy system, or even "Mamdani" fuzzy system, such as in the Matlab fuzzy system toolbox). Another very important classification of fuzzy systems, and one that has gained much popularity recently because of its success in functional approximation, is the "functional" fuzzy system, of which the familiar Takagi-Sugeno (TS) fuzzy system [9] is a special case. In the functional fuzzy system, the basic rule takes the form

$$\textbf{If } \underline{\text{Premise}} \textbf{ Then } y_i = 3D f_i(\cdot)$$

where the premise is the same as for the standard fuzzy system, but where the consequent is now a *function* rather than a linguistic term with associated membership function. The choice of the function $f_i(\cdot)$ depends on the application (it can be virtually any function of the process inputs), and does not have an associated membership function.

One way to view the functional fuzzy system is as a nonlinear interpolator between the mappings that are defined by the functions in the consequents of the rules. This is especially appealing to the control engineer if one considers the topical area of gain scheduling, whereby a nonlinear controller is constructed by scheduling linear controllers designed for linear models along an operating line. Thus, in the special case where the consequent takes the form of an affine mapping,

$$f_i(\cdot) = 3D\alpha_{i,0} + \alpha_{i,1}u_1 + \cdots + \alpha_{i,n}u_n \quad,$$

we refer to the representation as a TS fuzzy system. Furthermore, if $\alpha_{i,0} = 3D0$ for all i, then the $f_i(\cdot)$ mappings are linear and the fuzzy system (when inference is performed on the rules) essentially performs a nonlinear interpolation between the linear mappings (in the spirit of the gain scheduling controller). Note that the functional fuzzy system structure can be thought of as being very general, because if we simply choose $f_i = 3D\alpha_{i,0}$, then the resulting fuzzy system is equivalent to the standard fuzzy system described above.

4   Fuzzy Control Design Procedures

With a structure in place (whether a standard or functional fuzzy system), the control engineer turns to specifying the various parameters that make up the fuzzy controller. Simply put, the fuzzy controller design process is nothing more than a heuristic synthesis technique for designing nonlinear controllers. Of course, one could argue that there exist mechanized procedures for carrying out the design (I will mention one in a minute), or in general that this statement overemphasizes the role of heuristics. However, the design process inevitably comes down to choosing (i) linguistic variables (to characterize the physical process); (ii) linguistic values (quantifying the levels associated with the variables); (iii) membership functions (to give the degree of certainty or fuzzy set membership); (iv) control rules (to embody all knowledge about controlling the process); (v) fuzzy inference strategy (for decision making in producing the control action); and, (vi) defuzzification strategy (to interface back to the real world). The choice of these critical components results from our intuitive understanding of the process, not usually (entirely) from a mechanized design process. Often the design process comes down to doing what makes good sense.

Although some mechanized design procedures do exist, there are still many aspects of the control problem which must be delineated based on heuristics (and knowledge of the process). While we cannot hope to present a procedure for design, we can give some basic guidelines [1] for non-adaptive fuzzy controller synthesis:

1. Attempt a conventional design first (unless, of course, a sponsor insists on using fuzzy control — I will comment on this aspect later), such as a typical PID, which is computationally simpler and easy to understand.

2. If you choose to implement a fuzzy controller, carefully assess what inputs (measurements and functions thereof) are needed for the controller (such as proportional, derivative, or integral) using your control engineering intuition. Consider pre-processing the available measured data into a form that would be most useful if you, yourself, would be the controller.

3. Begin with a minimal number of rules and membership functions, and add to them as needed in your design process. Sometimes the addition of more membership functions can help to increase the "resolution" of the control action, improving performance; other times, it may have little or no effect.

4. Understanding exactly what the fuzzy controller is doing (in terms of the nonlinear surface it represents) can help you incorporate higher level knowledge into the controller (shaping the nonlinear mapping/surface).

Some of these guidelines can be found in the case studies illustrated in [1] and in some of the design experiences we overview later in this presentation.

5 Stability Considerations

One of the strongest knocks against fuzzy control for many years has been the lack of convincing stability results, even for special cases of fuzzy system structures. Of course, when fuzzy controllers with "safety nets" are operating our washing machines, vacuum cleaners, or electric razors, we tend to overlook the lack of stability guarantees. On the other hand, a search for fuzzy controllers in critical environments such as a nuclear power plant or an aircraft flight control system would likely come up empty. Where, then, should our confidences lie? If confidence lies in a stability proof, one could make a point that all stability proofs (that is, for particular example systems) are inherently tied to some equation which supposedly represents a physical reality. We may subsequently rationalize that a proof is solid, even though we know that equation must be a bit erroneous. But, then again, are we ready yet to throw away stability theory when working on realistic applications?

Stability results are beginning to appear for fuzzy systems, and it is possible to apply standard tools from nonlinear analysis for some systems. As an example, my colleague Kevin Passino has worked out an interesting example for the familiar simple pendulum system [1]. Consider a mass m attached to a hinged, stiff rod of length ℓ, where the friction at the hinge is proportional to a constant k. Suppose that the angle the pendulum makes with respect to the vertical downward position is θ, and that a control input torque T may be applied at the hinge to balance the pendulum in the inverted (upright) position. Defining phase variables x_1, x_2 and translating the system to the inverted position equilibrium point ($\theta = 3D\pi$, $\dot{\theta} = 3D0$), the equations of motion for this system are easily written as

$$\dot{x}_1 = 3Dx_2$$

$$\dot{x}_2 = 3D\frac{g}{\ell}\sin(x_1) - \frac{k}{m}x_2 + \frac{1}{m\ell^2}T \quad ,$$

where g is the acceleration due to gravity. In order to actuate the pendulum in the upright position, suppose the fuzzy controller receives as inputs x_1 and x_2 and produces as an output the torque; therefore, we write the controller output as $T = 3Dh(x_1, x_2)$ for some unspecified functional relationship embodied in $h(\cdot, \cdot)$, and assume that $h(0,0) = 3D0$. Choosing a Lyapunov function

$$V(x_1, x_2) = 3D\frac{1}{2}(x_1^2 + x_2^2)$$

and forcing $\dot{V} < 0$ results in the relationship

$$h(x_1, x_2) \leq m\ell^2 \left(\left(-\epsilon + \frac{k}{m}\right) x_2 - x_1 - \frac{g}{\ell}\sin(x_1) \right)$$

in some neighborhood $N(\delta)$, for some $\delta > 0$ and some constant $\epsilon > 0$. This, then, provides design restrictions for the fuzzy controller $h(x_1, x_2)$; design would consist of plotting the right-hand side of this inequality and finding δ and ϵ such that the inequality holds, "guaranteeing" asymptotic stability (to the extent that the original equation adequately describes the system). Clearly this is only a local result (valid for initial conditions in $N(\delta)$), and δ may indeed be very small (implying that the initial condition on the pendulum may be very close to the vertical position to assure that the fuzzy controller will stabilize the process).

These rather simple applications of Lyapunov stability are not entirely satisfying since each involves plotting the output of the fuzzy controller in order to verify stability. This is necessary, of course, because of the fact that we do not use an analytic mathematical expression for the fuzzy controller mapping $h(\cdot, \cdot)$. Moreover, for more complicated systems such analysis would be extremely difficult, at best, and perhaps the amount of effort required would be prohibitive. Nonetheless, they serve to illustrate that such analysis is, in fact, possible, relating to familiar control-theoretic stability results. An extension of Lyapunov's stability results, under the realm of absolute stability criteria, is also applicable to fuzzy systems. However, in that case the closed-loop system must be able to be represented as a linear, time-invariant system separable from all nonlinear elements. Presumably, then, the results (which apply the classical circle criterion) are valid for fuzzy controllers applied to a linear plant. Examples which illustrate these stability results are, for the most part, academic in nature. That is, one could naturally question the motivation of designing a fuzzy controller for a linear, time-invariant process. Furthermore, results using the circle criterion typically turn out to be conservative.

One of the more promising areas where stability results have been produced is for systems assuming the TS model form. Briefly, for simplicity suppose we have a TS model with m rules (that is, $i = 3D1, 2, \ldots m$ in our previous expression for the TS structure), and that, in the closed loop, the controller is another TS fuzzy system with m rules. In the simple case that the controller consequent functions (right-hand sides of the functional rules) are linear combinations of the system state, it is often possible (for certain classes of systems in this structure) to specify a quadratic Lyapunov function V in terms of the total system state. A total of m^2 Lyapunov matrix equations result, all with a common matrix P which must be shown to be positive definite for *every* equation in order to guarantee asymptotic stability. Finding this common matrix is not trivial if m and the system order are large. However, linear matrix inequality (LMI) methods can be used to compute P, if it exists.

It should be pointed out that these Lyapunov methods only provide sufficient conditions for stability. Thus, in the TS model result alluded to above, if a P cannot be found, the implication is merely that a different choice for V may actually produce a stabilizing (in terms of these tests) TS controller. Note also that the stability test which specifies conditions on P in no way depends on the membership functions chosen for the plant representation (and ultimately used in the controller). Thus, since the test is essentially valid for any membership functions that could represent the plant, the known nonlinear structure of the plant is in no way being exploited; this further emphasizes the fact that these are conservative results.

Because stability of fuzzy control systems is such an important topical area (and one receiving much attention, from opponents and proponents alike), we should give a few up-to-date comments, and some references, to explore. For example, one should consult [10] for further details on stability results when TS models are used. Design algorithms based on the LMI approach with these stability ideas are possible; an introduction to this segment of the literature may be found in [11, 12]. Other results are surfacing for fuzzy controllers developed using ideas from sliding mode control, while other approaches to the stability analysis of fuzzy control systems include extensions to the circle criterion ideas, Popov's criterion, describing function analysis, and standard analyses using phase portraits and phase plane analyses.

6   Adaptation and Supervision

An obvious requirement in the design and operation of any controller (fuzzy or otherwise) involves questions of system robustness. For instance, performance of a fuzzy controller constructed for some "nominal" plant may degrade if significant and unpredictable plant parameter variations, structural changes, or environmental disturbances occur. Clearly, controller adaptation is one way of overcoming these difficulties, in achieving reliable controller performance in the presence of unmodeled parameter variations and disturbances.

It is well documented that a significant amount and variety of knowledge can be loaded into the rule-base of the fuzzy controller to achieve high performance operation. It is true, however, that while the controller can be designed to capture knowledge about how to directly control the process, there may be additional expertise that can be exploited to enhance performance. The expert may, for example, know how to control the system very well in one set of operating conditions, but if the system switches to another set of operating conditions, the controller may be required to behave differently to again achieve high performance operation. For instance, a PID controller is often designed (tuned) for one set of plant operating conditions, but if the operating conditions change the controller will not be properly tuned. This is such an important problem that the literature abounds with expertise on how to manually and automatically tune PID controllers. Such expertise may be utilized in development of a supervisory fuzzy controller which can observe the performance of a low-level control system and automatically tune controller parameters. That is, expertise that we have on how to tune controllers can be loaded into a rule-base.

Many other examples exist of applications where the control engineer may have a significant amount of knowledge about how to tune a controller. One such example is in aircraft control when controller gains are scheduled based on the operating conditions. Fuzzy supervisory controllers have been suggested as schedulers in such applications. In other applications we may know that conventional or fuzzy controllers need to be switched on based on the operating conditions or a supervisory fuzzy controller may be used to tune an adaptive controller.

It is important to note that I make a subtle distinction between "adaptive" and "supervisory" fuzzy control. By adaptive fuzzy control, discussed in a minute, I refer to the act of adapting the controller parameters on a continuous (or perhaps periodic) basis. That is, controller updates are made based on a limited amount of system knowledge incorporated into some performance measure. On the other hand, while supervisory control techniques also adapt the controller parameters according to characteristics of the process, they may be distinguished from adaptive techniques by the manner in which the operator knowledge is incorporated into higher levels of a hierarchical structure, allowing for instances where, for example, the controller is tuned at the beginning of a control "session." It is therefore this concept of *monitoring* and *supervising* lower-level controllers (possibly fuzzy, possibly conventional) that defines the supervisory control scheme. Indeed, in this sense supervisory and adaptive systems can be described as special cases of one another.

Some would argue that the solution to such problems is always to incorporate more expertise into the rule-base to enhance performance; however, there are several limitations to such a philosophy, including: (i) the difficulties in developing (and characterizing in a rule-base) an accurate intuition about how to best compensate for the unpredictable and significant process variations that can occur for all possible process operating conditions; and (ii) the complexities of constructing a fuzzy controller that potentially has a large number of membership functions and rules. Experience has shown that it is often possible to tune fuzzy controllers to perform very well if the disturbances are known. Hence, the problem does not result from a lack of basic expertise in the rule-base, but from the fact that there is no facility for automatically re-designing (i.e., re-tuning) the fuzzy controller so that it can appropriately react to unforeseen situations as they occur.

There have been many techniques introduced for adaptive fuzzy control. For instance, one adaptive fuzzy control strategy that borrows certain ideas from conventional "model reference adaptive control" (MRAC) is called "fuzzy model reference learning control" (FMRLC) [3]. The FMRLC can automatically synthesize a fuzzy controller for the plant and later tune it if there are significant disturbances or process variations. The FMRLC has been successfully applied to an inverted pendulum, a ship-steering problem [3], anti-skid brakes [4], reconfigurable control for aircraft [13], and in implementation for a flexible-link robot [14]. The work on the FMRLC and subsequent modifications to it tend to follow the main focus in fuzzy control where one seeks to employ heuristics in control. There are other techniques that take an approach that is more like conventional adaptive control in the sense that a mathematical model of the plant and a Lyapunov-type approach is used to construct the adaptation mechanism. Such work is described in [15]. There are many other "direct" and "indirect" adaptive fuzzy control approaches that have been used in a wide variety of applications, too numerous to cite herein, but which represent some of the most promising work in the use of adaptive ideas with fuzzy control [1], [15].

7 Personal Experiences in Applications

Let me now relate the real reasons I entered into this field of fuzzy control, as a long-time "conventional" controlist. My own interest in fuzzy control technology began about seven or eight years ago. And, I readily admit that I was skeptical for a long time as to the role fuzzy control should take in the repertoire of the control engineer. Looking back on this time, I can think of two main facts that prompted me to venture (from my closet, as it were) into the unknown in exploring this technology. First, I had experienced some really challenging problems in my applications research (some control scientists are, in fact, engineers, and therefore work on real systems), where difficulties stemmed not only from the control aspects of the application itself (such as stringent specifications or problem size), but also from the lack of a good mathematical model and/or understanding of the physical processes in these tough problems. Secondly (and I am somewhat embarrassed to admit it), several industrial sponsors I dealt with had a keen interest in exploring fuzzy logic in their research and development; this, of course, translated directly to funding for the research effort of our control group. Naturally, I was motivated to investigate, explore, and compare — and to pay hungry graduate students.

It is difficult to assess exactly what the prevailing attitude in U.S. industry is today toward fuzzy control. One viewpoint is that many companies in the U.S. have investigated the technology but lost enthusiasm because it was not productive. The perceived problem behind this is that many U.S. engineers in industry have been "...led by the technical literature into thinking that using fuzzy rules to implement nonlinear PID control is the prime use of fuzzy logic for control" [16]. That is, after being sent off on an investigation and returning to report little payoff beyond the familiar PID, engineers "...have missed the high value-added, high payoff applications." Despite these perceived notions and problems in the way industry views the technology, the fact remains that many research and development groups are still eager to investigate what fuzzy control can bring to their applications.

So it is, then, that my interest in fuzzy control stemmed from applications-oriented research, in my own research and in that funded by my sponsors. It is accurate to say that a lot of good work is being done in industry and other research organizations in the area, much of which does not get reported to the open literature. For some of us, the most enjoyable research we do involves real industrial control problems, working alongside engineers in industry; the atmosphere is certainly different for an industrial sponsor than for one from a government agency. Proof of concept for the fuzzy control technology, and its ability to add to the solution of real, difficult engineering problems, is the over-riding concern, regardless of what motivated the investigation of the technology in the first place.

Automotive Systems

An important application area for control (in general) and lately for intelligent control techniques (in particular) is within the automotive industry. The particular areas with which I have been interested (and was able to secure funding) are braking systems and powertrain control.

Work in our group on anti-skid braking systems (ABS) was on-going for a period of about five years, and dealt first with conventional control methodologies and then fuzzy control. We have been fortunate in this long-term project to be able to work from analytical modeling, through simulation analysis, into benchtop testing with a production brake system setup, and then finally all the way to the test track to implement and test our braking algorithms on a vehicle loaned to us by Delphi Chassis Division of General Motors. Most of that work involved conventional control design and implementation (including linear, nonlinear and auto-tuning control implementations) [17]. There was little interest from the sponsor, however, in using fuzzy control, primarily because (so we were told) their internal research groups were devoting resources to the effort; still, we were able to investigate (although never actually implement on the vehicle) fuzzy control. In one phase of that project, where we were worked on controlling the torque at the wheel for base braking (non-ABS), we found that fuzzy control offered little over conventional designs we had attempted, and later implemented.

On a related project, we investigated the use of a fuzzy adaptive scheme (the FMRLC discussed previously) for ABS control in a quarter-car (single wheel) model in simulation [4]. The principle objective in that work was to illustrate the FMRLC design methodology for ABS systems with harsh road environments. Having had experience in such systems, we knew we were up against a tough problem; one of the main difficulties in the problem was the uncertainty in the modeling, and the overall complexity (especially once you move from the quarter-car to more complicated systems). Furthermore, it was difficult to gain accurate information about existing control algorithms used in production (our sponsors could not tell us of their algorithms for proprietary reasons). Thus, fuzzy control (particularly the use of the automated synthesis capabilities of the FMRLC approach) were especially appealing for this problem. Our findings were that the FMRLC showed real promise for effective braking (in terms of lateral stability and stopping distance) for various road conditions and transitions in road surfaces, *in simulation*. None of that work has ever made it into a production controller (at least not with our knowledge), so we cannot assess its impact on the field. However, it represented an important step in our own development of this important controller strategy, particularly with respect to a difficult engineering problem such as ABS control. We have since carried out several studies in extension of this work (again, in simulation) for active suspension and braking/steering/suspension integration for automated highway applications. It is my belief that with more complicated applications such as that, the promise of fuzzy control is even greater.

More recently I have been involved with control work in the area of automotive powertrain systems, primarily for internal combustion (IC), spark-ignited engine control. Working with a somewhat simplified model of the IC engine represented with crank-angle as the independent variable (instead of time), we determined that linear controller designs for the idle speed control problem (maintaining smooth idle operation of the IC engine despite encountering torque disturbances such as those due to air conditioning accessory loading) were inadequate when applied to the nonlinear simulation model. This led us to attempt nonlinear designs such as sliding mode control and fuzzy logic, both of which performed quite well (but were essentially equivalent) [18]. These studies began as simulation based, and evolved into full comparative implementation studies in an engine test cell to investigate their effectiveness in application [19].

Moving from this work, we are looking at a special case of the idle speed control problem, that of engine cold start, investigating the use of a fuzzy control scheme in actual implementation on a Ford V8 production engine. In this problem the concern is in controlling the air-to-fuel ratio (AFR) in the first minute or so of operation (thus the term "cold start") before the exhaust gas oxygen (EGO) sensor heats up (which is used for active control in feedback for AFR control in normal operation). Typically in production, this is an open-loop mode because of lack of feedback signals from the EGO sensor. Using an in-cylinder pressure sensor in implementation, we employ a technique which calculates the equivalent heat release duration, and is subsequently used to approximate the AFR (used as an input to the fuzzy controller). This is a problem which readily calls for fuzzy control, as the modeling associated with this process is quite elusive; in fact, I mention the problem only to make the point that this may be the first time I am turning to fuzzy control out of necessity, because the modeling task (for control design purposes) is quite complex. In this case, we are implementing a two-level fuzzy controller consisting of a standard fuzzy controller at the lower level, and a supervisory fuzzy controller at the upper level. Results have been quite promising [20].

Paper Machine Control

Some of the more traditional engineering application areas I have been involved with recently are turbofan engine control and monitoring, glass furnace control, and paper machine control; the first two involve work underway, while the latter is an area I would like to discuss next. I have been fortunate to be involved with good students who were also employees of ABB Industrial Systems, Inc., a world leader in measurement instrumentation for paper machines. Faced with difficult problems associated with controlling the moisture content of the paper in a typical machine, ABB wished to investigate the feasibility of application of fuzzy control technology. We undertook an interesting project in which we developed fuzzy controllers for machine direction (MD) moisture content, and in the end were able to implement the controller in a nearby paper mill in central Ohio, USA [21].

The main function of the paper machine is to form the paper sheet and then remove water from the sheet by means of vacuums, pressing, and evaporation. The paper machine is divided into two main sections referred to as the wet-end and dry-end. The sheet is formed in the wet-end on a continuous synthetic fabric (wire), which also allows water to drain from the forming sheet. Additional water is removed by passing the sheet through the wet press, which compresses a water absorbing fabric against the paper sheet. The dry-end of the machine, which is the area of interest for the work in [21], removes the remaining water in the sheet through evaporation. The paper passes over rotating steel cylinders, which are internally heated via super-heated pressurized steam. As the steam condenses in the drying cylinder, the transfer of heat through the cylinder wall results in an increase in the paper sheet temperature. The temperature of the sheet is varied by adjusting the steam pressure applied to each drying cylinder and is used to control the moisture content of the paper sheet.

The moisture measurement and associated drying section modeling is an active topic of research. When considering control of the MD moisture content the affect of the drying section steam pressure is obviously dependent on the current drying rate. This results in a process response between the drying section steam pressure and moisture content measurement which is a nonlinear function of properties of the paper sheet and machine conditions. Conventional PID control techniques are inadequate when considering the nonlinear effects in the process response and the long process dead time present. Nonetheless, control implementation in the pulp and paper industry is still based mainly on PID (PI) regulation. This approach is widely accepted because of the proven performance obtained for a process with simple dynamic characteristics such as a first order response. There are, however, many process characteristics on the paper machine that are nonlinear functions of measurable and unmeasurable paper machine parameters. An example of this is the moisture content in the paper sheet produced, the model of which includes a substantial amount of dead time and is dependent on the machine operating point.

Model-based approaches (such as the Smith predictor or internal model controller, IMC) have been successfully developed and implemented to compensate for the long process time delay that is inherent to the moisture measurement. However, the control performance of a model-based controller has often been found to be inconsistent for the wider range of process conditions. In general, the model-based controller usually requires rather conservative tuning to prevent aggressive control response at any one given machine operating point.

Fuzzy logic offered an alternative approach for this problem in compensating for effects of nonlinear process responses. As we mentioned previously, the use of fuzzy logic in a supervisory role is not limited to supervising only PI and PID structures at the lower level. In this work, in fact, we used a fuzzy supervisor to automate the tuning of an IMC controller. The moisture response gain as a nonlinear function of the paper machine operating point was identified via extensive system identification where it was found that as the machine operating point varied, the standard IMC controller cannot achieve the desired closed loop to moisture content setpoint changes without modification of the controller parameters. Thus, we were motivated to investigate supervisory structures for automated tuning. =09 The fuzzy supervisor developed for this problem used the machine speed and paper sheet weight inputs to determine the operating point of the machine, resulting in an estimate of the process response from the steam pressure setpoint to moisture content measurement. In [21] the fuzzy controller is compared to the standard IMC controller (fixed control parameters for all machine operating points) by investigating setpoint tracking and disturbance rejection responses. The fuzzy supervisor structure resulted in the desired response for all operating points, while the standard IMC was only adequate at one operating point and too conservative for all other machine operating points. The fuzzy supervisor was implemented on a real paper machine via the ABB Master^{TM} measurement and control system. The existing standard IMC control scheme for the upstream moisture measurement was overlaid with the fuzzy supervisor, which was implemented in a block control language. By using the fuzzy supervisor, the overall setpoint change response time (rise time) was consistently improved for different operating conditions.

Whether or not similar results could have been achieved using other schemes (non-fuzzy), for supervision and controller tuning, was not really the issue for this study. Of course, the inquisitive control engineer surely would be curious in this respect, and I suspect that, indeed, similar results could have been achieved with other schemes, probably rule-based in nature. However, one goal in this work (from the company's point of view) was to introduce a new technology, one that could give the company a competitive edge. Whether this was achieved or not remains to be seen, but good results, with improvements over existing controllers in actual implementation, go a long way towards this goal.

8   Concluding Perspectives

In a brief presentation such as this, we cannot hope to support nor expel predisposed notions of the usefulness and/or advantages of fuzzy control as a technology. We have, however, presented some personal perspectives along these lines.

An important topic which we did not consider here is that of fuzzy identification and estimation. While not falling directly under the area of fuzzy control, function approximation and model building from real data are often first and foremost in solving the overall control problem. Indeed, some of the most promising directions for the use of fuzzy logic for solving real-world engineering control problems (and beyond) are in the realm of approximation and modeling. Rather than go into details about this vast area, or cite the open literature for directions to pursue, I merely direct anyone interested along these lines to Chapter 5 of [1], where construction of fuzzy systems from numerical data is addressed. Many more references related to this important segment of the literature are given there.

Suffice it to say that in choosing tools for design, there is a need for the control engineer to assess what (if any) advantages fuzzy control methods have over conventional methods. This should be done by careful comparative analyses involving modeling, mathematical analysis, simulation, implementation, and a full engineering cost-benefit analysis (which involves issues of cost, reliability, maintainability, flexibility, lead-time to production, etc.). What's more, in addition to possible advantages in using fuzzy control, it makes sense to take a critical look at what possible *disadvantages* there could be to using it. For example, in light of the fact that a fuzzy control system attempts to manipulate a human's knowledge on how to control a system, one might ask if the behaviors observed by a human expert include all situations that can occur due to disturbances, noise, or plant parameter variations? Or, can the human expert realistically and reliably foresee problems that could arise from closed-loop system instabilities? Will the expert be able to effectively incorporate stability criteria and performance objectives (e.g. rise-time, overshoot, and tracking specifications) into a rule-base to ensure that reliable operation can be obtained? With regard to the question of "model-free" approaches, an obvious question arises as to whether or not an effective and widely-used synthesis procedure could essentially be void of mathematical modeling and subsequent use of proven mathematical analysis tools.

The standard control engineering methodology involves repeatedly coordinating the use of modeling, controller (re)design, simulation, mathematical analysis, and experimental evaluations to develop control systems. What is the relevance of this established methodology to the development of fuzzy control systems? Engineering a fuzzy control system uses many ideas from the standard control engineering methodology, except that in fuzzy control it is often said that a formal mathematical model is assumed unavailable so that mathematical analysis is impossible. While it is often the case that it is difficult, impossible, or cost-prohibitive to develop an accurate mathematical model for many processes, it is almost always possible for the control engineer to specify some type of approximate model of the process (after all, we do know what physical object we are trying to control). Indeed, it has been our experience that most often the control engineer developing a fuzzy control system does have a mathematical model available. While it may not be used directly in controller design, it is often used in simulation to evaluate the performance of the fuzzy controller before it is implemented (and it is often used for rule-base re-design). Certainly there are some applications where one can design a fuzzy controller and evaluate its performance directly via an implementation. In such applications one may not be overly concerned with a high performance level of the control system (e.g., for some commercial products such as washing machines or a shaver). In such cases there may thus be no need for conducting simulation-based evaluations (requiring a mathematical model) before implementation. In other applications there is the need for a high level of confidence in the reliability of the fuzzy control system before it is implemented (e.g., in systems where there is a concern for safety).

The current status of the field, as characterized by these limitations, coupled with the importance of nonlinear analysis of fuzzy control systems, make it an open area for investigation that will help establish the necessary foundations for a bridge between the communities of fuzzy control and nonlinear analysis. Clearly a case can be made that says the technology is leading the theory. Practitioners (those implementing control systems on hard, real-world problems) will proceed with the design and implementation of fuzzy control systems without attention to nonlinear analysis. In the mean time, theorists will attempt to develop a mathematical theory for the verification and certification of fuzzy control systems. My hope is that some day soon, the two camps will join together more completely, building on the progress made to this point.

Acknowledgments

It is my pleasure to acknowledge my lengthy collaboration with Professor Kevin M. Passino, who has been involved in the development of a large portion of the philosophy, technical results, and overall attitudes described here with regard to fuzzy control; indeed, a significant portion of what has been cited herein has come from joint contract work, joint publications, joint presentations of workshops and short courses, and endless hours of conversation with Kevin and his students over the last several years. Of course, I would also like to acknowledge the assistance of my students who have learned alongside me and, in many instances, taught me a lot. Finally, several sponsors have supported my research in these and related areas over the last few years, so I would like to acknowledge them: Air Products and Chemicals, Inc., ABB Industrial Systems, Delphi Chassis Division of General Motors, Reliance Electric Co., OMRON, Emerson Electric, General Motors, General Electric Aircraft Engines, Amoco Oil Co., NATO, and the National Science Foundation.

References

[1] K. M. Passino and S. Yurkovich, *Fuzzy Control.* Reading, MA: Addison-Wesley, 1998.

[2] K. M. Passino and S. Yurkovich, "Fuzzy control," in *The Control Handbook* (W. S. Levine, ed.), Boca Raton: CRC Press, 1996.

[3] J. R. Layne and K. M. Passino, "Fuzzy model reference learning control for cargo ship steering," *IEEE Control Systems*, vol. 13, pp. 23–34, Dec. 1993.

[4] J. R. Layne, K. M. Passino, and S. Yurkovich, "Fuzzy learning control for antiskid braking systems," *IEEE Transactions on Control Systems Technology*, vol. 1, pp. 122–129, June 1993.

[5] L. A. Zadeh, "The evolution of systems analysis and control: A personal perspective," *IEEE Control Systems*, vol. 16, pp. 95–98, June 1996.

[6] L. A. Zadeh, "A critical view of our research in automatic control," *IRE Trans. on Automatic Control*, vol. AC-7, p. 74, 1962.

[7] L. A. Zadeh, "Fuzzy sets," *Informat. Control*, vol. 8, pp. 338–353, 1965.

[8] Reader's Forum, "On fuzzy control and fuzzy reviewing...," *IEEE Control Systems*, vol. 13, pp. 5–7, June 1993.

[9] T. Takagi and M. Sugeno, "Fuzzy identification of systems and its applications to modeling and control," *IEEE Transactions on Systems, Man, and Cybernetics*, vol. 15, pp. 116–132, Jan. 1985.

[10] K. Tanaka and M. Sugeno, "Stability analysis and design of fuzzy control systems," *Fuzzy Sets and Systems*, vol. 45, pp. 135–156, 1992.

[11] K. Tanaka, T. Ikeda, and H. O. Wang, "Robust stabilization of a class of uncertain nonlinear systems via fuzzy control: Quadratic stabilizability, h^∞ control theory, and linear matrix inequalities," *IEEE Transactions on Fuzzy Systems*, vol. 4, no. 1, pp. 1–13, 1996.

[12] J. Zhao, R. Gorez, and V. Wertz, "Synthesis of fuzzy control systems based on linear Takagi-Sugeno fuzzy models," in *Multiple Model Approaches to Nonlinear Modeling and Control* (R. Murray-Smith and T. A. Johansen, eds.), UK: Taylor and Francis, 1996.

[13] W. A. Kwong, K. M. Passino, E. G. Laukonen, and S. Yurkovich, "Expert supervision of fuzzy learning systems for fault tolerant aircraft control," *Proceedings of the IEEE*, vol. 83, pp. 466–483, Mar. 1995.

[14] V. G. Moudgal, W. A. Kwong, K. M. Passino, and S. Yurkovich, "Fuzzy learning control for a flexible-link robot," *IEEE Transactions on Fuzzy Systems*, vol. 3, pp. 199–210, May 1995.

[15] L.-X. Wang, *A Course in Fuzzy Systems and Control*. Englewood Cliffs, NJ: Prentice Hall, 1997.

[16] S. Chiu, "Developing commercial applications of intelligent control," *IEEE Control Systems*, vol. 17, pp. 94–97, Apr. 1997.

[17] J. K. Hurtig, S. Yurkovich, K. M. Passino, and D. Littlejohn, "Torque regulation with the General Motors ABS VI electric brake system," in *Proceedings of the American Control Conference*, (Baltimore, MD), pp. 1210 – 1211, 1994.

[18] S. Yurkovich and M. Simpson, "Comparative analysis for idle speed control: A crank-angle domain viewpoint," in *Proceedings of the American Control Conference*, (Albuquerque, NM), 1997.

[19] J. Wills, S. Yurkovich, and G. Rizzoni, "Direct fuzzy control of idle speed in an internal combustion engine," in *IFAC Workshop on Advances in Automotive Control*, (Mohican State Park, OH), Feb. 1998.

[20] W. E. Leisenring and S. Yurkovich, "Using equivalent heat release duration for closed-loop cold start AFR control," in *Proceedings of the American Control Conference*, (Philadelphia, PA), June 1998.

[21] T. F. Murphy, S. Yurkovich, and S.-C. Chen, "Intelligent control for paper machine moisture control," in *Proceedings of the IEEE Conference on Control Applications*, (Dearborn, MI), pp. 826 – 833, 1996.

Part C

Modeling, Identification and Estimation

Identification for Control
– What Is There To Learn?

Lennart Ljung*

Abstract

This paper reviews some issues in system identification that are relevant for building models to be used for control design. We discuss how to concentrate the fit to important frequency ranges, and how to determine which these are. Iterative and adaptive approaches are put into this framework, as well as model validation. Particular attention is paid to the presentation and visualization of the results of residual analysis.

1   Identification for Control

There may of course be several reasons why a model of a dynamical systems is sought. A common one is that the model is needed to design a regulator for the system. It is then important that available design variables are chosen so that the resulting model becomes as appropriate as possible for the control design. Feedback control is both forgiving and demanding: The core property of feedback is that a good closed loop system can be obtained even with very coarse knowledge of the system to be controlled. At the same time, certain aspects of the system have to be known so as to assure stability of the closed loop. In linear systems language: In certain frequency ranges we need reliable information about the system, while in others, a very approximate idea will do fine.

Identification for control purposes therefore naturally should focus on the "important" frequency ranges, and hopefully we should be able to do well with rather simple models. The question is how to achieve this.

What Is There To Learn?

To obtain a model that can be successfully used for control design we need to learn a few things:

---
*Department of Electrical Engineering, Linköping University, S-581 83 Linköping, Sweden. Email: ljung@isy.liu.se   WWW: http://www.control.isy.liu.se

1. What frequency ranges are important

2. A model with fit focused to those ranges

3. If the model structure used is flexible enough to provide relevant information about the remaining ranges

   - Alt: Prior information about other frequency ranges is sufficiently reliable

This covers a whole spectrum of applications from full fledged system identification with sophisticated model validation to very simple techniques.

The Ziegler-Nichols rule for PI-tuning, e.g., fits into this scheme as follows: Solve 1 and 2 simultaneously by increasing the P-gain to the instability limit. This gives the value of the system's frequency function at the phase cross-over frequency (which is the important frequency range). Tune the PI-regulator based on this information. Item 3 is in this case handled by prior information/assumption: The system's frequency function is "nice" (like a monotonically decreasing amplitude); it won't give you any bad surprises at other frequencies. This prior information can also be phrased like this: "We can achieve good control by a PI-regulator". The very successful *autotuners*, e.g. [19], are more sophisticated variants on this theme.

There has been a considerable interest lately in *iterative identification for control* schemes where a succession of experiments (in closed loop) are made in order to iterate between steps 1 and 2 above. See, among many references, e.g. [17],[21], [3] and [11]. These schemes, seemingly, do not address step 3 explicitly.

Iterative control design is closely related to *adaptive control*, which in a sense is the limit as the experiment time decreases down to one sample. See, e.g. [20],[4] and [8] for basic treatments of adaptive control.

Step 3 above concerns *model validation*. This is a classical topic in statistics, but has also been the subject of intense, renewed interest in the control community again due to its importance for identification for control. In particular, several new approaches to deal with the topic in a non-statistical setting have been suggested. See, among many references, e.g.,[16], [18] and [9].

We shall in this paper provide a subjective commentary to issues related to identification for control. In Section 2 we briefly discuss item 1, while identification techniques to achieve step 2 are reviewed in Section 3. The linked, iterative nature of steps 1 and 2 present in both adaptive control and iterative identification for control is commented upon in Section 4. Model validation is then treated in Section 5.

2 What Frequency Ranges Are Important?

For linear systems, the issues of model accuracy are treated by the classical concepts of the *sensitivity function* S and the *complementary sensitivity function* T. See, e.g. [15]. We believe the system is described by the model G and use a regulator $u = -Fy + F_r r$, (r being the reference input) which would give us a nominal closed loop system G_c with nominal output y. If the true system really is given by G_0, we obtain the actual output y_0 which differs from the desired one by

$$\mathrm{E}|y(t) - y_0(t)|^2 = \int |G_0(e^{i\omega}) - G(e^{i\omega})|^2 |S_0(e^{i\omega})|^2 \left|\frac{G_c(e^{i\omega})}{G(e^{i\omega})}\right|^2 \Phi_r(\omega)d\omega \quad (1a)$$

$$S_0 \text{ is the true sensitivity function } 1/(1 + G_0 F) \quad (1b)$$

$$\Phi_r(\omega) \text{ is the spectrum of the reference input} \quad (1c)$$

Here we considered the response from the reference input only, and suppressed the frequency argument in most of the involved functions.

Moreover, to guarantee stability we have

$$|G_0 - G| \cdot \frac{|T|}{|G|} < 1 \quad \text{for all frequencies} \quad (2a)$$

$$T = \frac{FG}{1 + FG} \quad (2b)$$

For a 1-dof regulator ($F_r = F$) we have $G_c = T$, so the two expressions then both tell us that the model needs to be "good" where $T/G = G_c/G$ is large and/or where S_0 is large. Both these things typically happen around the bandwidth of G_c, so this is a rather straightforward message. For 2-dof regulators (such as typically used for pole placement) there may be considerable differences between G_c and T, so the message about which are the important frequency ranges may then be more complicated.

We can turn the question around a bit. Instead of asking what discrepancies we get due to the model error for a fixed regulator, we can ask what is the model's influence on the design of a fixed closed loop system: Suppose we want to achieve $G_c = G_d$ and for a given model G we solve for such a 1-dof regulator $F = F(G)$. That is,

$$G_d = \frac{F(G)G}{1 + F(G)G} \quad (3)$$

The desired output is then $y = G_d r$. We then also have $T = G_d$. The actual closed loop system $F(G)G_0/(1 + F(G)G_0)$ then gives the output y_0, and the discrepancy is still given by (1a).

3   Model Fit

The Method

Given input-output data $Z^N = \{y(1), u(1), \ldots, y(N), u(N)\}$ and a parameterized model structure

$$\hat{y}(t|\theta) = G(q, \theta)u(t) \tag{4}$$

we can estimate the model by the straightforward fit (e.g. [12]):

$$\hat{\theta}_N = \arg\min_\theta V_N(\theta, Z^N) \tag{5a}$$

$$V_N(\theta, Z^N) = \sum_{t=1}^{N} \varepsilon_F^2(t, \theta) \tag{5b}$$

$$\varepsilon_F(t, \theta) = L(q, \theta)(y(t) - G(q, \theta)u(t)) \tag{5c}$$

Here L is a (possibly parameter-dependent) monic prefilter that can be used to enhance certain frequency ranges.

This method can be seen as direct curve-fitting in the frequency domain:

$$V_N(\theta, Z^N) \approx \int |G(e^{i\omega}, \theta) - \hat{G}_N(e^{i\omega})|^2 |L(e^{i\omega}, \theta)|^2 |U_N(\omega)|^2 d\omega \tag{5d}$$

$$\hat{G}_N = \frac{Y_N}{U_N} \quad \text{(the ETFE)} \tag{5e}$$

$$U_N(\omega) = \sum u(t)e^{-i\omega t} \quad Y_N(\omega) = \sum y(t)e^{-i\omega t} \tag{5f}$$

Limit Results

We are interested in what happens as the data sample size, N, increases. To investigate this we assume that the data are subject to

$$y(t) = G_0(q)u(t) + v(t) \tag{6a}$$

$$\Phi_v(\omega) = \lambda_0 |H_0(e^{i\omega})|^2 \quad \text{the spectrum of } v \tag{6b}$$

We also assume that the additive disturbance v is uncorrelated with the reference input r, i.e., that the cross spectrum $\Phi_{rv} = 0$. We can then split the input spectrum Φ_u into that part that originates from r and that part that originates from v:

$$\Phi_u(\omega) = \Phi_u^r(\omega) + \Phi_u^v(\omega) \tag{7}$$

It then follows (Chapter 8 in [12] plus straightforward calculations) that

$$\hat{\theta}_N \to \theta^* = \arg\min_\theta V(\theta) \text{ as } N \to \infty \tag{8a}$$

$$V(\theta) = \int |G(e^{i\omega}, \theta) - G_0(e^{i\omega})|^2 |L(e^{i\omega}, \theta)|^2 \Phi_u^r(\omega) d\omega$$

$$+ \int \left| \frac{1 + G(e^{i\omega}, \theta) F(e^{i\omega})}{1 + G_0(e^{i\omega}) F(e^{i\omega})} \right|^2 |L(e^{i\omega}, \theta)|^2 \Phi_v(\omega) d\omega \quad \text{or} \tag{8b}$$

$$V(\theta) = \int |(G_0(e^{i\omega}) + B(e^{i\omega}, \theta)) - G(e^{i\omega}, \theta)|^2 |L(e^{i\omega}, \theta)|^2 \Phi_u(\omega) d\omega$$

$$+ \int \lambda_0 \left| H_0(e^{i\omega}) - \frac{1}{L(e^{i\omega}, \theta)} \right|^2 |L(e^{i\omega}, \theta)|^2 \frac{\Phi_u^r(\omega)}{\Phi_u(\omega)} d\omega + \lambda_0 \tag{8c}$$

$$|B(e^{i\omega}, \theta)|^2 = \frac{\lambda_0}{\Phi_u(\omega)} \cdot \frac{\Phi_u^v(\omega)}{\Phi_u(\omega)} \cdot |H_0(e^{i\omega}) - 1/L(e^{i\omega}, \theta)|^2 \tag{8d}$$

A number of comments can be made around these results:

1. If the prefilter L and the model G are flexible enough so that for some θ_0, $G(q, \theta_0) = G_0(q)$ and $L(q, \theta_0) = 1/H_0(\theta)$ then $V(\theta_0) = \lambda_0$ so $\theta^* = \theta_0$ (provided this is a unique minimum.) It is thus natural to think of the prefilter as an inverse noise model.

2. If L is θ-independent the second term of (8c) can be omitted and the limit model is the minimum of

$$V(\theta) = \int |(G_0(e^{i\omega}) + B(e^{i\omega})) - G(e^{i\omega}, \theta)|^2 |L(e^{i\omega})|^2 \Phi_u(\omega) d\omega \tag{9}$$

$$|B(e^{i\omega})|^2 = \frac{\lambda_0}{\Phi_u(\omega)} \cdot \frac{\Phi_u^v(\omega)}{\Phi_u(\omega)} \cdot |H_0(e^{i\omega}) - 1/L(e^{i\omega})|^2 \tag{10}$$

3. If, moreover, the system operates in open loop so that $\Phi_u^v = 0$, the "bias-pull"-term $B = 0$. Then the limit model is a clear cut approximation of G_0 in the frequency weighting norm $|L|^2 \Phi_u$.

4. From (8b) we see that a tempting parameterization of L is to use $L(q, \theta) = \tilde{L}(q)/(1 + G(q, \theta) F(q))$. Such a prefilter parameterization corresponds to what is known as *indirect identification* of closed loop systems, [5], [2]. It is the same as identifying the closed loop and then solve for the open loop dynamics, using the (presumed) knowledge of F. The limiting model is then, according to (8b), the minimizing argument of

$$V(\theta) = \int |G(e^{i\omega}, \theta) - G_0(e^{i\omega})|^2 \left| \frac{1}{1 + G(e^{i\omega}, \theta) F(e^{i\omega})} \right|^2 |\tilde{L}(e^{i\omega})|^2 \Phi_u^r(\omega) d\omega \tag{11a}$$

$$= \int |G(e^{i\omega}, \theta) - G_0(e^{i\omega})|^2 \left| \frac{G_c(e^{i\omega})}{G(e^{i\omega}, \theta)} \right|^2 |\tilde{L}(e^{i\omega})|^2 |S_0(e^{i\omega})|^2 \Phi_r(\omega) d\omega \tag{11b}$$

(cf (1a).) This model is a compromise between fitting G to G_0 and making the model sensitivity function $S = 1/(1 + GF)$ small.

Asymptotic variance

As the number of data tends to infinity and the order n of the model G and well as of the prefilter L increases, the asymptotic variance of the frequency function estimate $\hat{G}_N(e^{i\omega}) = G(e^{i\omega}, \hat{\theta}_N)$ as subject to ([12], chapter 9):

$$\operatorname{Var} \hat{G}_N(e^{i\omega}) \approx \frac{n}{N} \frac{\Phi_v(\omega)}{\Phi_u^r(\omega)} \tag{12}$$

Some Design Issues

Based on the above asymptotic results some design problems involving both bias (approximation) aspects and variance can be solved (Chapter 14 in [12]).

We see from (8)-(12) that the properties of the model are only affected by

1. The input spectra Φ_u^r and Φ_u^v, which in turn are consequences of the choices of regulator F_r, F and reference spectrum Φ_r

2. The prefilter L

in addition to the model parameterization $G(q, \theta)$ and the true system's characteristics G_0 and Φ_v.

Suppose now that we would like to choose the experiment design variables so that the weighted,total model error

$$J = \int \operatorname{E}|\hat{G}_N(e^{i\omega}) - G_0(e^{i\omega})|^2 W(\omega) d\omega \tag{13}$$

The model parameterization is given, as is the data length N and we restrict ourselves to parameter-independent prefilters. We also assume that the input power $\operatorname{E}u^2(t)$ is bounded. The total error J contains both the bias error and the variance error. We use the asymptotic variance result (12). The solution is

- Use open loop: $F=0$
- Use the input spectrum $\Phi_u(\omega) \sim \sqrt{W(\omega)\Phi_v(\omega)}$
- Use the prefilter $|L(\omega)|^2 \sim \sqrt{\frac{W(\omega)}{\Phi_v(\omega)}}$

The problem becomes more difficult if the output power is constrained instead. Then the optimal solution will involve closed loop operation, and the double influence of L on B and the weighting function in (8) is more tricky to deal with. The variance contribution to J in (13) is minimized by the following choices:

- Closed loop operation, with F, chosen so that

$$\int |S_0(e^{i\omega})|^2 \Phi_v(\omega) d\omega \rightarrow 0 \qquad (14a)$$

- The reference spectrum

$$\Phi_r \sim \sqrt{W \Phi_v} \cdot \left| \frac{1 + FG_0}{FG_0} \right|^2 |G_0| \qquad (14b)$$

It might of course be difficult to realize this optimal solution, since (14a) requires considerable knowledge of the system.

4 Iterations and Adaptation

Iterative Design

The questions of which frequency range to emphasize and what model/regulator to use (steps 1 and 2 in Section 1) are clearly linked. The regulator determines S and T and hence which ranges are important; these in turn affect the model which gives the regulator, etc.

If we know what bandwidth we are looking for, and we intend to use a design method with full control over the loop shaping aspects, it is fairly safe to focus the model fit to a decade or so around the intended bandwidth.

Sometimes, the possible bandwidth is not known, but part of the information we gain from the system. Then it may be reasonable to make several experiments to gain insight into higher and higher frequencies. (The "windsurfer approach", e.g., [10].) Even if the intended bandwidth is known, it might not be clear in what frequency ranges the model fit has to be good, e.g., due to the design method used (like pole placement). In both these cases, iterative experiments have been suggested along the following lines:

1. Pick a fixed, and typically low order, model structure

2. Pick a design method and a design criterion that uses the model: $F = F(G)$

3. Perform an identification experiment in closed loop with current regulator F_i. Identify the system using indirect identification, giving the model G_i.

4. Compute the regulator $F_{i+1} = F(G_i)$ and go to step 3.

The motivation for the experiment design and method in step 3 is the (formal) similarity between (11b) and (1a). This similarity is somewhat deceptive, though, as we shall see in the convergence analysis below.

To be more specific, assume the we for step 2 choose pole placement, so that $F(G)$ is defined by (3). Let the desired output be $y_d(t) = G_d(q)r(t)$. The actual output is $y_0(t) = y_0(t; F)$, where we marked its dependence on the regulator F. We can denote the model output

$$\hat{y}(t|\theta) = \frac{F(q)G(q, \theta)}{1 + F(q)G(q, \theta)} r(t) = \hat{y}(t; F, G) \tag{15}$$

The criterion (11b) then is

$$J(F, G) = E|y(t; F) - \hat{y}(t; F, G)|^2 \tag{16}$$

and the iterations can be summarized as

$$G_i = \arg \min_G J(F_i, G) \tag{17a}$$

$$F_{i+1} = F(G_i) \tag{17b}$$

Adaptive Control

Adaptive control is the same paradigm as iterative design. Instead of conducting full separate experiments, the model is updated each sample in the direction that the current experiment gives information about. With the above definitions and somewhat symbolic notation the basic update algorithm for adaptive control will be

$$G_{i+1} = G_i - \mu J'_G(F(G_i), G_i) \tag{18}$$

Here we used the notation

$$J'_G(F, G) = \frac{\partial}{\partial G} J(F, G)$$

Convergence Analysis

(The analysis in this subsection has its roots in Section 7.3.2 of [13]. Similar results have been proven by [7] for the identification for control application.) The actual convergence analysis of the iterative and adaptive schemes, (17), (18) is not easy in general. We shall here just comment on the possible convergence points, the fix-points of the schemes. It is clear that (17) and (18) can only converge to a model G^* and corresponding regulator $F^* = F(G*)$ such that

$$J'_G(F(G^*), G^*) = 0 \qquad (19)$$

Is this the right point? The distance of interest in (16) is $\mathrm{E}|y_0(t; F) - y_d(t)|^2$, and since $y_d(t) = \hat{y}(t; F(G), G)$ for all G we have

$$\mathrm{E}|y_0(t; F) - y_d(t)|^2 = \mathrm{E}|y_0(t; F) - \hat{y}(t; F(G), G)|^2 = J(F(G), G) \qquad (20)$$

(This is "the correct interpretation" of (1a) in the case of (3).) Now, the models that are best for control design are those that minimize $J(F(G), G)$ w.r.t. G, that is a model G^* such that

$$0 = \frac{d}{dG} J(F(G), G)_{|G=G^*} = J'_F(F(G^*), G^*) F'_G(G^*) + J'_G(F(G^*), G^*) \quad (21a)$$

Notice the difference between (19) and (20)! They describe the same model(s) G^* if and only if $J'_F = 0$. This means that the criterion of fit $J(F, G)$ shall not depend on the regulator, which in turn (essentially) implies that $G^* = G_0$. The possible convergence points for the iterative/adaptive schemes are thus the desired points only if the model is essentially correct. This brings us directly to the issue of model validation.

5   Model Validation and Model Error Modeling

Recall Step 3 in Section 1: Find out if

- If the model structure used is flexible enough to provide relevant information about the remaining ranges

  - Alt: Prior information about other frequency ranges is sufficiently reliable

Working as in the previous section with a fixed (low order) model structure could – if we are lucky – lead to a model/regulator that is the best we can achieve within the chosen structure. It does not follow that this is "good enough". *Model Validation* is really the topic to find out if what is "best" is also "good enough".

Statistics Over the Residuals

Most of the model validation tests are simply based on the difference between the simulated and measured output:

$$\varepsilon(t) = y(t) - \hat{y}(t) = y(t) - \hat{G}_N(q)u(t) \qquad (22)$$

Filtered version of these residuals are frequently used; we include this case by allowing y and u in the above expression to be prefiltered. Typical model validation tests amount to computing the model residuals and giving some statistics about them. Note that this as such has nothing to do with probability theory. (It is another matter that *statistical model validation* often is complemented with probability theory and model assumptions to make probabilistic statements based on the residual statistics. See, e.g., [1].)

The following statistics for the model residuals are often used:

- The maximal absolute value of the residuals

$$M_N^\varepsilon = \max_{1 \le t \le N} |\varepsilon(t)| \qquad (23)$$

- Mean, Variance and Mean Square of the residuals

$$m_N^\varepsilon = \frac{1}{N} \sum_{t=1}^N \varepsilon(t) \qquad (24)$$

$$V_N^\varepsilon = \frac{1}{N} \sum_{t=1}^N (\varepsilon(t) - m_N^\varepsilon)^2 \qquad (25)$$

$$S_N^\varepsilon = \frac{1}{N} \sum_{t=1}^N \varepsilon(t)^2 = (m_N^\varepsilon)^2 + V_N^\varepsilon \qquad (26)$$

- Correlation between residuals and past inputs.
  Let

$$\varphi(t) = [u(t), u(t-1), \dots, u(t-M+1)]^T \qquad (27)$$

and

$$R_N = \frac{1}{N} \sum_{t=1}^N \varphi(t)\varphi(t)^T \qquad (28)$$

Now form the following scalar measure of the correlation between past inputs (i.e. the vector φ) and the residuals:

$$\tilde{\xi}_N^M = \frac{1}{N} \left\| \sum_{t=1}^N \varphi(t)\varepsilon(t) \right\|_{R_N^{-1}}^2 \qquad (29)$$

Note that this quantity also can be written as

$$\tilde{\xi}_N^M = \hat{r}_{\varepsilon u}^T R_N^{-1} \hat{r}_{\varepsilon u} \qquad (30)$$

where

$$\hat{r}_{\varepsilon u} = [\hat{r}_{\varepsilon u}(0), \dots, \hat{r}_{\varepsilon u}(M-1)]^T \qquad (31)$$

with

$$\hat{r}_{\varepsilon u}(\tau) = \frac{1}{\sqrt{N}} \sum_{t=1}^{N} \varepsilon(t)u(t-\tau) \tag{32}$$

Now, if we were prepared to introduce assumptions about the true system (the measured data Z^N), we could use the above statistical measures to make statements about the relationship between the model and the true system, typically using a probabilistic framework.

If we do not introduce any explicit assumptions about the true system, what is then the value of the statistics (23)-(29)? Well, we are essentially left only with *induction*. That is to say, we take the measures as indications of how the model will behave also in the future:

> "Here is a model. On past data it has never produced a model error larger than 0.5. This indicates that in future data and future applications the error will also be below that value."

This type of induction has a strong intuitive appeal.

In essence, this is the step that motivates the "unknown-but-bounded approach". Then a model or a set of models is sought that allows the preceeding statement with the smallest possible bound, or perhaps a physically reasonable bound.

Note, however, that the induction step is not at all tied to the unknown-but-bounded approach. Suppose we instead select the measure S_N^ε as our primary statistics for describing the model error size. Then the Least Squares (Maximum Likelihood/Prediction Error) identification method emerges as a way to come up with a model that allows the "strongest" possible statement about past behavior.

How reliable is the induction step? It is clear that some sort of invariance assumption is behind all induction. To have some confidence in the induced statement about the future behavior of the model, we thus have to assume that certain things do not change. To look into the invariance of the behavior of ε it is quite useful to reason as follows. (This will bring out the importance of the statistics (29)).

It is very useful to consider two sources for the model residual ε: One source that originates from the input $u(t)$ and one that doesn't. With the (bold) assumption that these two sources are additive and the one that originates from the input is linear, we could write for some transfer function $\tilde{G}$ (*The model error model*)

$$\varepsilon(t) = \tilde{G}(q)u(t) + v(t) \tag{33}$$

Note that the distinction between the contributions to ε is fundamental and has nothing to to with any probabilistic framework. We have not said anything about $v(t)$, except that it would not change, if we changed the input $u(t)$. We refer to (33) as the separation of the model residuals into *Model Error* and *Disturbances*.

The division (33) shows one weakness with induction for measures like M_N^ε and S_N^ε going from one data set to another. The implicit invariance assumption about the properties of ε would require both the input u and the disturbances v to have invariant properties in the two sets. Only if we would have indications that $\tilde{G}$ is of insignificant size, we could allow inductions from one data set to another with different types of input properties. The purpose of the statistics $\tilde{\xi}_N^M$ in (29) is exactly to assess the size of $\tilde{G}$. We shall see this clearly below. (One might add that more sophisticated statistics will be required to assess more complicated contributions from u to ε).

In any case, it is clear that the induction about the size of the model residuals from one data set to another is much more reasonable if the statistics $\tilde{\xi}_N^M$ has given a small value ("small" must be evaluated in comparison with S_N^ε in (26)).

We might add that the assumption (33) is equivalent to assuming that the data Z^N have been generated by a *"true system"*

$$y(t) = G_0(q)u(t) + v(t) \tag{34}$$

where

$$\tilde{G}(q) = G_0(q) - \hat{G}(q) \tag{35}$$

Other Approaches To Characterize the Residuals

Let us turn again to the fundamental relation (33). In connection with robust control design issues there has been recent interest to characterize the model errors in a way that fits new robustness results, see e.g.[16], [18] and [9]. A basic idea is to characterize all $\tilde{G}$ and all bounds on v that are consistent with the model residuals ε and u: In somewhat loose notation, this is the set

$$\{C_G, C_v; |\tilde{G}|_\infty < C_G \,\&\, |v(t)| < C_v \,\&\, \varepsilon = \tilde{G}u + v\} \tag{36}$$

This is in a sense the set of all model error assumptions that are unfalsified by the data and the nominal model. We could, e.g. take $C_v = \max|\varepsilon|$ and $C_G = 0$, saying that there is no model error, just unstructured disturbances with a certain maximum amplitude. Or, we could say that there are no disturbances, but a certain bound on the model error. The idea is then to pick a member in this set of unfalsified models that allows the best, robust control design.

While this approach has many interesting features, it should be remarked that the split in (33) is not entirely up to pure arbitrariness: The data contains information about the split, and the traditional correlation analysis tries to find this information. We stress again that we cannot rely upon any bound C_v unless we believe that v does not contain contributions from u.

Control Oriented Presentation of Residual Analysis

The traditional way to present the result of residual analysis is to compute the cross correlation function (32) and present it and/or the squared sum (29) for inspection and possibly statistical hypothesis tests.

The question now is, what can be said about the model error $\tilde{G}$ based on the information in Z^N.

The procedure will be to form

$$\varepsilon(t) = L(q)(y(t) - \hat{G}(q)u(t))$$

and then $\tilde{\xi}_N^M$ as in (27)-(29). In these calculations replace $u(t)$ outside the interval $[1, N]$ by zero. Assume that $R_N > \delta I$. It is then shown in [14] that

$$\left[\frac{1}{2\pi} \int_{-\pi}^{\pi} \left|\tilde{G}(e^{i\omega})\right|^2 \left|L(e^{i\omega})\right|^2 |U_N(\omega)|^2 \, d\omega \right]^{1/2}$$

$$\leq \quad (1+\eta) \left[\frac{1}{N} \tilde{\xi}_N^M \right]^{1/2} + (1+\eta)x_N + (2+\eta)C_u \sum_{k=M}^{\infty} |\rho_k| \qquad (37)$$

Here

- $x_N = \left| \frac{1}{N} \sum_{t=1}^{N} \tilde{v}(t)\varphi(t) \right|_{R_N^{-1}}$

- $\tilde{v}(t) = L(q)v(t)$

- ρ_k is the impulse response of $L(q)\tilde{G}(q)$

- $|U_N|^2$ is the periodogram (see 5f).

- $\eta = \frac{C_u M}{\sqrt{N\delta}}$

- $C_u = \max_{1 \leq t \leq N} |u(t)|$.

If the input is tapered so that $u(t) = 0$ for $t = N - M + 1, ...N$, the number η can be taken as zero.

Let us make a number of comments:

- The result is really just a statement about the relationship between the sequences $\tilde{v}(t) = L(q)[y(t) - G_0(q)u(t)]$, and $\varepsilon(t) = L(q)[y(t) - \hat{G}(q)u(t)]$ on the one hand and the given transfer functions $L(q), G_0(q), \hat{G}(q)$ together with the given sequences $u(t), y(t)$ on the other hand. There are as yet no stochastic assumptions whatsoever, and no requirement that the "model" $\hat{G}$ may or may not be constructed from the given data.

- By the choice of prefilter $L(q)$ we can probe the size of the model error over arbitrarily small frequency intervals. However, by making this filter very narrow band, we will also typically increase the size of the impulse response tail. (Narrow band filters have slowly decaying impulse responses.)

- In practical use the often erratic periodogram $|U_N|^2$ can be replaced by smoothed variants.

- For the quantities on the right hand side, we note that $\tilde{\xi}_N^M$ is known by the user, as well as η, N and C_u. The tail of the impulse response ρ_k beyond lag M is typically not known. It is an unavoidable term, since no such lag has been tested. The size of this term has to be dealt with by prior assumptions.

- The only essential unknown term is x_N. We call this *"The correlation term"*. The size and the bounds on this term will relate to noise assumptions.

The implications of this result under varying assumptions about the additive disturbance $v(t)$ are discussed in [14].

Visualizing the Result of Residual Analysis: Model Error Models

The result of correlation analysis is traditionally done in standard statistical fashion depicted in figure 1. The information from the cross correlation analysis between ε and u, can also be interpreted as an implicit FIR model for the transfer function $\tilde{G}$ in (33) from u to ε. For control purposes, it is much more effective to present the (amplitude) frequency function of this *model error model*, with uncertainty bounds as in figures 2 – 4. The data used in these figures are simulated from a second order ARMAX model. It is clear that conventional model validation corresponds to increasing the model complexity until the model error model has uncertainty bounds that include zero (as in figure 4), since then there is no clear evidence that $\tilde{G}$ is not zero – the estimated model is then not falsified. But it is also clear that the two plots together; the model and its "sidekick", the model error model, can be used for control design, even if the model is falsified. Look at figure 3. According to the model error model there is significant, but rather small errors in the mid frequency range. The model is thus falsified, but could still very well be used for control design if the information in the lower plot is taken into account.

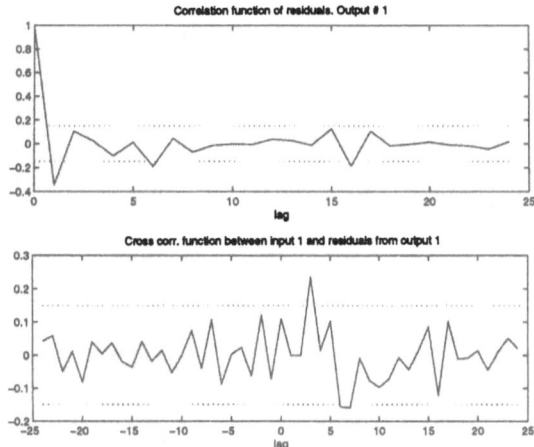

Figure 1: Traditional residual analysis: Auto- and cross-correlation functions with uncertainty regions.

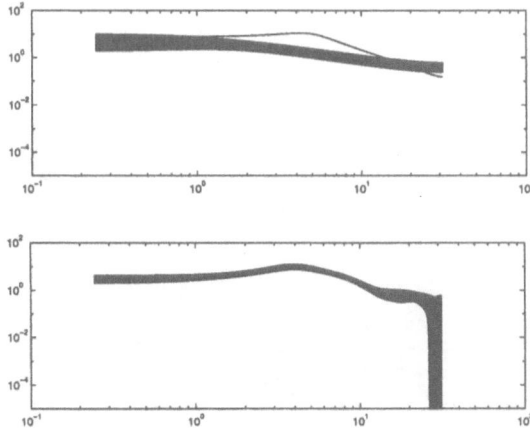

Figure 2: Upper plot: Amplitude Bode plot of a first order model with estimated uncertainty bounds. The true system is also plotted. Lower plot: The model error model computed as a 20:th order ARX model from u to $y - \hat{G}u$

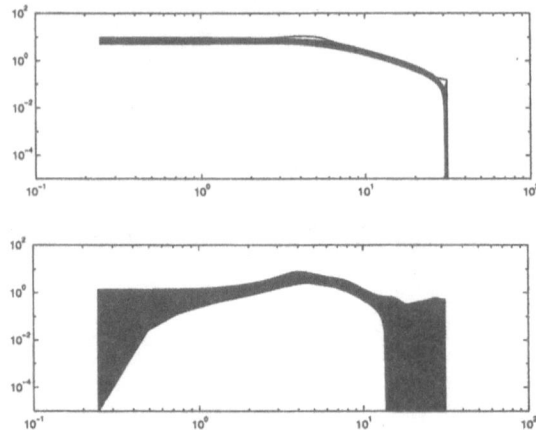

Figure 3: As in previous figure, but second order ARX model

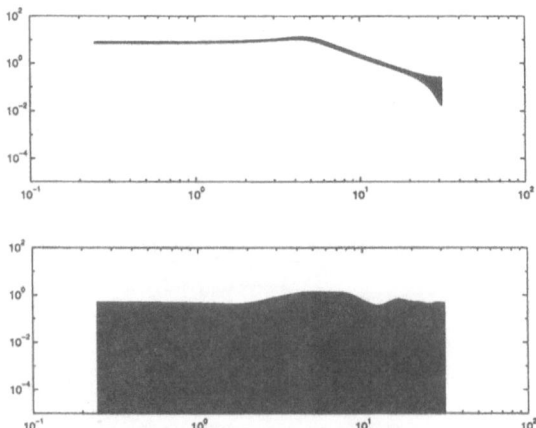

Figure 4: As in previous figure, but second order ARMAX model. Here the model error model contains zero in its uncertainty region, which means that the top model is not falsified.

6   Conclusions: What Is There To Learn?

Identification for control is one of the most important applications of system identification. We have reviewed some basic issues in this area.

First we may note that in most cases it is relatively easy to realize in which frequency range(s) the model need to be accurate – typically around the intended bandwidth of the closed loop system. For the identification experiment we should thus concentrate both input power and prefiltering to such ranges.

Note that if we just consider the bias distribution of the approximating model, there is no need to perform neither iterative experiments nor experiments in closed loop. Any bias weighting can be achieved on the original data set by prefiltering. However, the disturbances acting on the system will also cause variance errors, and to improve on the information in certain frequency ranges new experiments with new input power distribution may be necessary. Also, with constrained output variance, better accuracy can typically be achieved in closed loop experiments. A third situation where closed loop experiments are helpful is when a model of the noise properties as also required for the control design, [6].

There are of course other, practical reasons for making new experiments, such as time-variation, non-linear effects of different operating points, etc.

In general, if the chosen frequency range is small, we can be rather confident that even a quite simple model gives a good fit over this range. The extreme case is formed by the autotuners, which essentially fit a first order model at the phase-cross-over frequency. To use such a model for control design requires however also some insight into the system's properties in other frequency ranges. Often, this can be handled by prior information: "the plant can be reasonably well controlled by a PI-regulator". In situations where automated decisions are not required, there is however no good reason for not performing model validation. Simple analysis of the model residuals gives information about the model error model, that could be instrumental either for the control design or for requiring more accurate models.

There is still a good market for good model structures for model-error-models. The couple model plus model-error-model should work together and effective ways to present and visualize the model-error-model are quite important.

References

[1] N.R. Draper and H. Smith. *Applied Regression Analysis, 2nd ed.* Wiley, New York, 1981.

[2] U. Forsell and L. Ljung. Closed-loop identification revisited. Technical report, Dept of Electrical Engineering, 1997.

[3] M. Gevers. Towards a Joint Design of Identification and Control. In H. L. Trentelman and J. C. Willems, editors, *Essays on Control: Perspectives in the Theory and its Applications*, pages 111–151. Birkhäuser, 1993.

[4] G.C. Goodwin and K.S. Sin. *Adaptive Filtering, Prediction and Control.* Prentice-Hall, Englewood Cliffs, N.J., 1984.

[5] I. Gustavsson, L. Ljung, and T. Söderström. Identification of processes in closed loop – identification and accuracy aspects. *Automatica*, 13:59 – 77, 1977.

[6] H. Hjalmarsson, M. Gevers, and F. De Bruyne. For model-based control design,closed loop identification gives better performance. *Automatica*, 32, 1996.

[7] H. Hjalmarsson, S. Gunnarsson, and M. Gevers. Optimality and sub-optimality of iterative identification and control schemes. In *Proceedings of the American Control Conference*, pages 2559–2563, Seattle, 1995.

[8] P. A. Ioannou and J. Sun. *Robust Adaptive Control.* Prentice-Hall, Upper Saddle River, NJ, 1996.

[9] R.L. Kosut, M.K. Lau, and S.P. Boyd. Set-membership identification of systems with parametric and nonparametric uncertainty. *IEEE Trans. Autom. Control*, 37(7):929–942, 1992.

[10] W. S. Lee, B. D. O. Anderson, I. M. Y. Mareels, and R. L. Kosut. On some key issues in the windsurfer approach to adaptive robust control. *Automatica*, 31(11):1619–1636, 1995.

[11] W. S. Lee, B. D. O. Andersson, R. L. Kosut, and I. M. Y. Mareels. On Robust Performace Improvement through The Windsurfer Approach to Adaptive Robust Control. In *Proceedings of the 32nd IEEE Conference on Decision and Control*, pages 2821–2827, San Antonio, TX, 1993.

[12] L. Ljung. *System Identification - Theory for the User.* Prentice-Hall, Englewood Cliffs, N.J., 1987.

[13] L. Ljung and T. Söderström. *Theory and Practice of Recursive Identification.* MIT press, Cambridge, Mass., 1983.

[14] L.Ljung and L. Guo. The role of model validation for assessing the size of the unmodeled dynamics. *IEEE Trans. Autom Control*, Vol AC-42(9):1230–1239, 1997.

[15] J. M. Maciejowski. *Multivariable Feedback Design.* Electronic Systems Engineering Series. Addison-Wesley, 1989.

[16] K. Poolla, P. P. Khargonekar, A. Tikku, J. Krause, and K.Nagpal. A time-domain approach to model validation. *IEEE Trans. on Automatic Control*, AC-39:951–059, 1994.

[17] R. J. P. Schrama. Accurate models for control design: the necessity of an iterative scheme. *IEEE Transactions on Automatic Control*, 37:991–994, 1992.

[18] R.S. Smith and J.C Doyle. Model invalidation: A connection between robust control and identification. *IEEE Trans. Automatic Control*, 37:942–952, July 1992.

[19] K. J. Åström and T. Hägglund. *Automatic Tuning of PID Regulators*. Instrument Society of America, Triangle Research Park, N.C., 1988.

[20] K.J. Åström and B. Wittenmark. *Adaptive Control*. Addison-Wesley, Reading, MA, 1989.

[21] Z. Zang, R. R. Bitmead, and M. Gevers. Iterative Weighted Least-squares Identification and Weighted LQG Control Design. *Automatica*, 31(11):1577–1594, 1995.

Identification of Complex Systems *

Munther A. Dahleh[†]

Abstract

In this paper, we discuss the problem of identifying a complex system
with a limited-complexity model using finite corrupted data. Complex
systems are ones that cannot be uniformly approximated by a finite dimensional space. Nevertheless, our *prejudice* is represented by selecting
a finitely parameterized set of models from which an estimate of the original system will ultimately be drawn. We will give an account of a new
formulation that shows how such a model should be selected from data.
We will demonstrate this paradigm on the class of linear time-invariant
stable systems and give an overview of the available results concerning
input design, consistency, error bounds, and sample complexity.

1 Introduction

This particular perspective on system identification is motivated from the interest in developing models that are usable for control purposes. In such
applications, only a simplified model of the process is needed as long as it is
accompanied by a "coarse" description of the unmodeled dynamics. A powerful paradigm for robust control has emerged in the last two decades that
enables designers to provide feedback controllers that deliver certain performance specifications utilizing only the simplified models and the description
of the unmodeled dynamics. Nevertheless, the question remains as to how one
can derive such descriptions from finite corrupted data.

Traditional system identification based on the minimum prediction error
paradigm (MPE) provides a powerful framework that has its roots in statistics
[8]. Within this framework, a parametric model is selected to minimize the
prediction error, and it is shown to converge to the actual process asymptotically if the process belongs to the model parametrization and the experiment
satisfies certain richness conditions. In addition, the asymptotic distribution
of the parameter estimates is shown to be Gaussian which aids in estimating
confidence intervals for the parameters. Nevertheless, the approach suffers

---
*This research was supported by NSF Grant number 9157306-ECS, AFOSR F49620-95-
0219, and, Siemens AG.
†Laboratory for Information and Decision Systems, Massachusetts Institute of Technology

from several drawback. First, it focuses largely on problems when exact modeling is possible. Although some partial results exist when the process is not in the model structure, such results are asymptotic and do not constitute a practical tool for identification. In particular, the results lack error bounds as well as necessary conditions on the experimental inputs. Secondly, the prediction error translates into an error bound in terms of the $\mathcal{L}_2$-norm which is not amenable to robust control applications. Thirdly, since the theory is stochastic, assumptions such as stationarity and ergodicity are essential for the results to hold.

Another paradigm for system identification known as set-membership system identification ([11, 12, 13, 14, 16]) was developed to deal with non-stochastic descriptions of noise. Noise is assumed to belong to a prescribed set. This approach can easily be extended to represent undermodeling in a direct fashion. Nevertheless, this approach also suffers from several drawbacks. First, noise descriptions are generally too rich. Secondly, when unmodeled dynamics are present, they are generally described independently of the set of models and hence overlap with the actual model we are trying to identify. Both of these reasons contribute to deriving estimates and error bounds that are quite conservative.

An important concept that emerges from system identification is that of identifiability of certain sets of systems. In particular, given a set of possible systems, a question arises whether or not any system in this set can be identified using finite corrupted data. If a set is identifiable, the next natural question is that of sample complexity; i.e., the time length necessary to identify every single element in this set. Notions such as ϵ-entropy and ϵ-dimension were introduced to capture this notion [23].

Recent work has focused on dealing with some of the drawbacks mentioned above. Analysis of the MPE paradigm in the presence of unmodeled dynamics has appeared in [2]. The problem of approximating the original system as an operator was addressed in several places. For linear time invariant plants, such approximation can be achieved by uniformly approximating the frequency response (in the $\mathcal{H}_\infty$-norm) or the impulse response (in the ℓ_1 norm). In $\mathcal{H}_\infty$ identification, it was shown that robustly convergent algorithms can be furnished, when the available data is in the form of a corrupted frequency response, at a set of points dense on the unit circle [4, 3]. When the topology is induced by the ℓ_1 norm, a complete study of asymptotic identification was given in our past work [18] for arbitrary inputs, and the question of optimal input design was addressed as well. Related work on this problem was also reported in [1, 2, 5, 7, 6, 9, 10, 15, 17]. Although these results relate well to robust control applications, they do not address the issue of undermodeling in a direct fashion.

In here, we describe a new paradigm that attempts to:

1. unify both the minimum prediction error paradigm and the set member-

ship paradigm,

2. incorporate undermodeling as part of the problem formulation,

3. provide a model and error bounds that are amenable to robust control applications, and

4. address identifiability of classes of complex systems.

Finally, it is noted that the formulation and most results in this paper are taken from [19, 20, 21, 22]

2 Problem Formulation

We begin by describing the basic problem of system identification (ID). Generally, system identification is concerned with building a *model* of a specific process that can be used for some objective such as simulation, prediction or control. It is important to emphasize that the resulting model is only an approximation of the real process, and represents our own prejudice in representing it. Hence, it is essential that the outcome of ID contains a description of this approximation error.

2.1 Prior Information

In the sequel, we will assume the following experimental setup: a DT linear time-invariant stable process can be excited by any bounded input, and noisy measurements can be collected for some finite time. We will assume that some *prior* information is available about this setup:

1. The real process T_0 belongs to an infinite dimensional space $\mathcal{T}$ which is equipped with a norm, $\| \cdot \|$.

2. The noise w is an unknown signal that corrupts the output of the system additively. This can be modeled as either a realization of some stochastic process, or simply an arbitrary signal of finite length n that belongs to a prespecified (fully or partially) set $\mathcal{W}_n$ and n denotes the length of the experiment.

3. A model set $\mathcal{G} \subset \mathcal{T}$ is given. This is a finitely parametrized set of systems in $\mathcal{T}$ that represents our prejudice in describing the process. Generally, this set is derived from existing mechanistic models of the process. In this paper, we will assume $\mathcal{G}$ is a subspace of $\mathcal{T}$.

4. A prior bound, γ, on the distance between T_0 and $\mathcal{G}$ is available:

$$\text{dist}(T_0, \mathcal{G}) =: \inf_{G \in \mathcal{G}} \|T_0 - G\| \leq \gamma$$

We will denote by $\mathcal{I}$ the set of all systems T that are within a distance γ from $\mathcal{G}$.

2.2 Data Collection

Given an input experiment $u \in \mathcal{U}$, *Data* is generated via the equation

$$y = T_0 u + w$$

Only a finite set of Data can be collected. Define the set of Data up to time n as

$$Z^n = \{[u(k), y(k)] \mid k \leq n\}$$

So far, u has not been specified. The issue of selecting an appropriate experimental input is a major issue in ID.

2.3 Objective: Mathematical Formulation

If the real process T_0 is known, then the model is simply chosen to be

$$G_0 = \arg\min_{G \in \mathcal{G}} \|T_0 - G\|$$

However, only a finite Data record, Z^n, can be obtained. Let ϕ be an algorithm that maps Z^n to an element $\hat{G}_n \in \mathcal{G}$. We will generally abuse the notation and refer to $\hat{G}_n$ as both the estimate of the model and the algorithm. Define the error function E_n as follows:

$$E_n(\phi, u) = \sup_{T_0 \in \mathcal{I}} \sup_{w \in \mathcal{W}_n} \|G_0 - \hat{G}_n\|$$

E_n is a measure of the maximum possible distance at time n between the estimate provided by ϕ and the best approximation of T_0 in $\mathcal{G}$, namely G_0. Define

$$E(\phi, u) = \limsup_{n \to \infty} E_n(\phi, u)$$

Problem 1 Find an input $u \in \mathcal{U}$ and an algorithm ϕ such that $E(\phi, u)$ is minimized. Equivalently, find u and ϕ that solve the problem

$$\inf_{\phi} \inf_{u \in \mathcal{U}} E(\phi, u) =: E^*$$

An estimate $\hat{G}_n$ is *consistent* if $E^* = 0$.

Another important component of identification is the derivation of rates of convergence to the minimum error. An alternate way of viewing these rates is via the notion of time (sample) complexity. Define the quantity:

$$\mathcal{N}(\eta) = \min\{N \mid |E_n - E^*| \leq \eta\}$$

$\mathcal{N}$ simply indicates the minimum length of the Data record necessary to achieve a near optimal error.

Problem 2 For the choice of u and ϕ in Problem 1, characterize the time complexity function $\mathcal{N}$.

2.4   Remarks

Below, we will make several remarks on the above description:

1. The choice of $\mathcal{T}$ (space and norm) is tied to the general objective behind system identification. For instance, in control applications, it is desired that errors are measured in terms of norms that are usable by robust control techniques. Typically, these norms are operator norms such as $\mathcal{H}_\infty$ or ℓ_1. In general, however, such norms may result in complicated identification algorithms and it may be desirable to work with a Hilbert space structure. It can be shown that a good choice for this space is the Hardy-Sobolev space, $\mathcal{H}$, defined as the set of all stable systems with an inner product

$$< F, G >_{\mathcal{H}} = < F, G > + < F', G' > = \sum_{k \geq 0} (1 + k^2) f_k g_k.$$

   The prior $\mathcal{I}$ indicates some knowledge of the smoothness of the frequency response. Robust controller synthesis with error bounds in terms of this norm can be turned into convex optimization problems that are readily solvable [22]. Of course, weighted versions of this norm can be used as well.

2. The space $\mathcal{T}$ as well as the prior set $\mathcal{I}$ are both complex in the sense that there does not exist a finite dimensional subspace that can uniformly approximate all elements in these sets.

3. The space $\mathcal{G}$ is finitely parametrized and hence represents a more parsimonious description of the process. Most of the results will be derived for the case where $\mathcal{G}$ is a subspace.

4. The prior set $\mathcal{I}$ represents an additive error structure. This is equivalent to saying that the part of the process that will not be modeled can be described by an additive error structure.

5. The expected outcome of identification is the following:

   - An estimate $\hat{G}_n$.
   - A (parametric) estimate of the error $\|G_0 - \hat{G}_n\|$, D_n.
   - An estimate $\hat{\gamma}_n$ of the unmodeled part $\hat{\gamma}_n \leq \gamma$.
   - Additional structure on the unmodeled part, reflected in a set Δ.

- The model

$$\{\hat{G}_n + \Delta \mid \|G - \hat{G}_n\| \leq D_n, \, \Delta \in \mathbf{\Delta}, \, \|\Delta\| \leq \hat{\gamma}_n\}$$

2.5 Structure of the Unmodeled

If $\mathcal{G}$ is a subspace, then every $T_0 \in \mathcal{I}$ can be decomposed as $T_0 = G_0 + \Delta_0$, where G_0 is the best possible approximation of T_0 in $\mathcal{G}$. From standard results in functional analysis, Δ_0 will be aligned with an element in the subspace of annihilators of $\mathcal{G}$, denoted by $\mathcal{G}^{\perp}$. If $\mathcal{T}$ is a Hilbert space $(\ell_2, \mathcal{H})$, then Δ_0 is simply orthogonal to $\mathcal{G}$. This immediately gives an additional structure on the unmodeled part of T_0, namely Δ_0. We will denote the set of such admissible systems as $\mathbf{\Delta}$

Example 2.1 *If $\mathcal{G} \subset \mathcal{H}$ is the set of all FIR systems of order M, then*

$$\mathcal{G}^{\perp} = \{\lambda^{M+1} F(\lambda), \, F \in \mathcal{H}\}.$$

Example 2.2 *Let $\mathcal{T} = \mathcal{H}$, and*

$$\mathcal{G} = \{\theta/1 - a\lambda, \, \theta \in \mathbb{R}, \, a \text{ is fixed}\}$$

then

$$\mathcal{G}^{\perp} = \{\Delta \mid \sum_{k=0}^{\infty} (1 + k^2) a^k \delta_k = 0.\}$$

3 Fundamental Limitations

The fundamental limitations of identification are better captured by introducing the following concepts: sets of unfalsified models and the diameter of uncertainty. These are standard concepts in a field known as Information Based Complexity that provides a natural setting for deterministic learning problems similar to the one defined here.

3.1 Sets of Unfalsified Models

Given a finite data record, we can express the set of all models that can produce this Data record as follows:

$$S_n(Z^n) = \{G \in \mathcal{G} \mid \quad y(k) = Gu(k) + \Delta u(k) + w(k), \, k \leq n,$$
$$\Delta \in \mathbf{\Delta}, \, \|\Delta\| \leq \gamma, \, w \in \mathcal{W}_n\}$$

It is clear that every element in S_n is indistinguishable from G_0 and hence the size of S_n gives a lower bound to the best possible achievable error using any algorithm.

3.2   Diameter of Parametric Uncertainty

For any given experiment, the diameter of $S_n(Z^n)$ measures the maximum distance between any two elements in this set. We can define the maximum possible diameter of uncertainty at time n as

$$D_n(\mathcal{T}, \mathcal{G}, \mathcal{W}_n, u) = \sup_{T \in \mathcal{I}} \sup_{w \in \mathcal{W}_n} \text{diam}(S_n(Z^n))$$

This a measure of the maximum parametric error that can be incurred irrespective of how the Data is generated from the input u. In other words, this is a worst-case type measure. We will often suppress the dependence and just write this expression as $D_n(u)$.

The Asymptotic diameter of uncertainty is defined as

$$D(u) = \limsup_{n \to \infty} D_n(u)$$

It can be easily shown that for any input u, the diameter of uncertainty bounds the optimal asymptotic error:

$$\frac{D(u)}{2} \leq \limsup_{n \to \infty} \inf_{\phi} E_n(\phi, u) \leq D(u)$$

As a result, the identification problem formulated earlier is equivalent to characterizing $D(u)$.

4   Deterministic Noise Models

Noise is assumed to belong to a prespecified set. This set should reflect our understanding of the corrupting noise. Any noise signal in this set is likely to corrupt the output of the process. An alternate approach to modeling noise is by representing it as a realization of a stochastic process. The most used model in this context is a white noise model, or possibly a filtered white noise model. In here, we will show that the latter model can be effectively represented in a set description allowing us to use deterministic tools for deriving estimates and error bounds.

4.1   Unknown-But-Bounded Noise

Consider the set:
$$\mathcal{W}_n = \{w \mid |w(k)| \leq \epsilon, \, k \leq n\}$$

With this description, noise signals are known to be bounded in time, but otherwise arbitrary. Clearly, this is a very rich set of signals and consequently may not be practically viable.

4.2 Low-Correlated Noise

The following set description is motivated from the general properties of white noise. Define the N-window autocorrelation with lag τ as:

$$r_w^N(\tau) = \frac{1}{N} \sum_{i=0}^{N-\tau} w(\tau + i)w(i)$$

Consider the following set:

$$\mathcal{W}_n = \{w \in \mathbb{R}^n \mid \frac{1}{n^2} \sum_{\tau=1}^{n} |r_w^n(\tau)| \leq C\gamma_n\}$$

If γ_n is bounded below from zero, then arbitrary bounded signals will be present in this set. By selecting γ_n so that it approaches zero as n approaches infinity, we get a more restricted set of Noise. Of course, we have to be careful not to end up with a trivial set of signals. This can easily happen if γ_n approaches zero at very fast rate. Below, we will argue that there is a lower bound on the decay rate such that a white noise signal will belong to this set with high probability.

Theorem 1 *If γ_n approaches zero at a rate $\frac{1}{n^\alpha}$, $\alpha < .5$, then this set asymptotically contains white noise with high probability. Equivalently, let $x(0), x(1), x(2), \ldots$ be bounded independent identically distributed random variables with 0 mean. Then*

$$P(\{x(t)\}_{t=0}^{\infty} \cap \mathbb{R}^n \in \mathcal{W}_n) \overset{n \to \infty}{\longrightarrow} 1$$

4.2.1 Cross Correlation Properties

Next, we state a correlation property that will be useful in the future. This result simply shows that the correlations of any admissible noise signal w with a periodic input decays at a polynomial rate to zero.

Lemma 1 *Let u be periodic with period m and $w \in \mathcal{W}_n$, $\gamma_n = n^{-\alpha}$, $\alpha < .5$, and n be the number of measurements then*

$$|r_{uw}^n(\tau)| \leq C\sqrt{m}n^{-\alpha/2}$$
$$\sum_{\tau=0}^{m}(r_{uw}^n(\tau))^2 \leq Cmn^{-\alpha}$$

4.3 Noise Low-Correlated with Sinusoids

There are many ways to describing white noise signals in a deterministic setting. Frequency domain characterizations of such signals are possibly the most powerful for their simplicity and practicality. Consider the set

$$\mathcal{W}_n^F = \left\{ d \in \mathbb{R}^n \mid \frac{1}{\sqrt{n \log n}} \left| \sum_{k=0}^n d_k e^{iwk} \right| \le W(w), \ w \in [0, 2\pi] \right\}$$

This set satisfies the correlation property in Lemma 1 for any Filter W. It can be interpreted as the set of signals that are low correlated with slowly varying sinusoids. In addition, this set is convex. This is an advantage since this simplifies the characterization of the sets of unfalsified models.

To restrict the set of noise further, we can consider signals that are low correlated with fast varying sinusoids. Consider the following set

$$\mathcal{W}_n^L = \left\{ d \in \mathbb{R}^n \mid \sup_{\alpha_1, \alpha_2} \left| \frac{1}{\sqrt{n \log(n)}} \sum_{t=0}^n d_t e^{i(\alpha_1 t^2 + \alpha_2 t)} \right| \le 1 \right\}$$

Again, this set contains white noise signals with high probability. It also satisfies the same correlation property mentioned earlier. In addition, this set is linearly parametrized and hence convex.

5   Results: Exact Modeling

In here, we discuss the case of exact modeling, namely when $\mathcal{T} = \mathcal{G}$ (hence $T_0 = G_0$). For simplicity, we will assume that $\mathcal{G}$ is the space of FIR systems of order M. Since this is a finite dimensional space, the norm we consider is not crucial to the results we obtain. We will also assume that the experimental input u is restricted to have a peak value less than or equal to one.

5.1   Unknown-but-Bounded Noise

It is straightforward to derive a lower bound on the maximum diameter of uncertainty. Since both T_0 and w can vary arbitrarily in their respective sets, then for a given input u one possible experiment can produce the output w_1, $|w_1(k)| \le \epsilon$. A pair (T_0, w) is consistent with this output if

$$w_1 = T_0 u + w$$

In fact for any T_0 with $\|T_0\|_1 \le 2\epsilon$, there exists w and w_1 such that the above equation is satisfied. Hence, the maximum diameter of uncertainty $D(u)$ is at least as large as 2ϵ.

On the other hand, it is possible to derive an experimental input u^* so that $D(u^*) = 2\epsilon$. Define u^* to be the concatenation of all possible sequences of $\{+1, -1\}$ of length M. There are 2^{M-1} possible sequences and so u^* is

$M2^{M-1}$ long. This input has the following persistence of excitation property: For any T_0, FIR of length M,

$$\|T_0 u\|_\infty = \|T_0\|_1$$

One can show, since the set $\mathcal{T}$ is balanced and convex, that the maximum diameter of uncertainty for a given input is equal to the diameter of the set of models that are indistinguishable from 0. Every element in this set satisfies $T_0 u + w = 0$ for some $w \in \mathcal{W}$. When $u = u^*$, the maximum $\|T_0\|_1$ is equal to ϵ. Given that $-T_0$ is also admissible, it follows that

$$\text{diam}(S_n(Z^n)) = 2\epsilon$$

We will summarize these results below [18]:

Theorem 2 *If noise is unknown but bounded by ϵ, then the optimal diameter of uncertainty is equal to 2ϵ. In other words, there does not exist an algorithm that produces a consistent estimate of the process. Nevertheless, there exits an input and an algorithm such that the maximum error satisfies*

$$\epsilon \leq E^* \leq 2\epsilon$$

Any algorithm that picks an element from the set of unfalsified models and any input with the persistence of excitation property (described above) will result in this error bound.

This theorem emphatically shows that this identification setup produces negative results. This can be seen in two accounts: the lack of consistency and the fact that an exponentially long input (in M) is needed for identification. This is not surprising since the noise model is too conservative and does not decouple from the actual system output.

It turns out that the time complexity function $\mathcal{N}(\eta)$ is exponential in both M and η [1, 15].

5.2 Low-Correlated Noise

The properties mentioned earlier for this Noise set suggest that correlation techniques are appropriate to recover the actual process. The input-output relation can be written in the following form:

$$y(t) = \begin{pmatrix} u(t) & u(t-1) & \ldots & u(t-M) \end{pmatrix} \begin{pmatrix} t_0 \\ t_1 \\ \vdots \\ t_M \end{pmatrix} + w(t)$$

Multiplying both sides by $\phi(t) = (\, u(t) \quad u(t-1) \quad \ldots \quad u(t-M)\,)^T$ and summing from $t = 0$ to n we get:

$$\frac{1}{n}\sum_{t=0}^{n}\phi(t)y(t) = \frac{1}{n}\sum_{t=0}^{n}\phi(t)\phi(t)^T \begin{pmatrix} t_0 \\ t_1 \\ \vdots \\ t_M \end{pmatrix} + \frac{1}{n}\sum_{t=0}^{n}\phi(t)w(t)$$

If the input u is chosen to be a periodic input of period $M + 1$ and also persistently exciting (in the traditional sense) of order $M + 1$ (e.g. a PRBS), then

$$\lim_{n\to\infty}\{\frac{1}{n}\sum_{t=0}^{n}\phi(t)\phi(t)^T\}^{-1} \leq C_1 < \infty$$

¿From the properties of w, it follows that

$$\|\frac{1}{n}\sum_{t=0}^{n}\phi(t)w(t)\|_2 \leq C_2 \frac{\sqrt{M}}{n\sqrt{\alpha}}$$

With these properties, it is clear that the standard least squares estimate, namely

$$\hat{G}_n = \{\frac{1}{n}\sum_{t=0}^{n}\phi(t)\phi(t)^T\}^{-1}\frac{1}{n}\sum_{t=0}^{n}\phi(t)y(t)$$

has the property that

$$\|T_0 - \hat{G}_n\|_2 \leq C\frac{\sqrt{M}}{n\sqrt{\alpha}}$$

This can be generalized in the theorem below [20]:

Theorem 3 *For low correlated noise, the least squares estimator provides a consistent estimate with polynomial sample complexity if the input is persistently exciting in the traditional sense.*

It interesting to make the following observations about this result:

1. This result gives a deterministic interpretation of the LS estimator in the presence of noise. In a precise sense, it bridges the gap between deterministic and stochastic formulations.

2. This result does not make stationarity assumptions on the noise signal. The only property needed is the decaying correlations.

3. Uniform convergence is attained with polynomial sample complexity using simple periodic inputs.

4. This result follows for the rest of the noise models discussed earlier. The advantage in using $\mathcal{W}_N^F$ or $\mathcal{W}_N^L$ is that the set of unfalsified models in both cases is convex. Estimating the diameter of this set becomes a much easier task.

5. This result extends for general model structures under some minor assumptions.

6   Results: Explicit Undermodeling

Identification with explicit undermodeling is a much trickier problem than the ones discussed earlier. To illustrate the main issues, we will first consider the following two examples.

6.1   FIR Models of Order Zero

Let $\mathcal{G}$ be the space of FIR models of order zero, i.e., linear scalars. The space of systems can be either $\mathcal{H}$ or ℓ_1. Every process decomposes into $T_0 = g_0 + \Delta_0$ where Δ_0 is in the set

$$\Delta = \{\Delta | \Delta = (0, \delta_1, \delta_2, \ldots), \|\Delta\| \leq \gamma\}$$

The noisy output equation can be written as:

$$y(k) = g_0 u(k) + \sum_{j=1}^{k} \delta_j u(k - j) + w(k)$$

If the output is not noisy, then the best choice for an input is the impulse since g_0 can be recovered exactly from the output. On the other hand, if the output is noisy (we will make this precise later), then a longer input/output relation is needed to enable us to decorrelate the noise signal. As a result, the estimate of g_0 is affected by the unmodeled part.

If we correlate the output equation with the input, we get the following relation:

$$r_{yu}^n(0) = g_0 r_u^n(0) + \sum_{i=1}^{n} \delta_i r_u^n(i) + r_{wu}^n(0)$$

It will possible to recover g_0 if we can construct an input u with the following two properties:

1. $\sup_{1 \leq i \leq n} \left| \frac{r_u^n(i)}{r_u^n(0)} \right| \to 0$ as $n \to \infty$.

2. $\left| \frac{r_{wu}^n(0)}{r_u^n(0)} \right| \to 0$ as $n \to \infty$.

With this input, the estimate is precisely the LS estimate given by

$$\hat{G}_n = \frac{r_{yu}^n(0)}{r_u^n(0)}$$

and satisfies the property that

$$\lim_{n\to\infty} \sup_{T_0 \in \mathcal{I}} \|G_0 - \hat{G}_n\| = 0$$

It remains to see if such an input can be constructed. Notice that an i.i.d. Bernoulli process with zero mean and variance 1 has the property

$$\mathcal{P}\left\{ \max_{1 \le i \le n} r_u^n(i) \ge \alpha \right\} \le n \exp(-nf(\alpha))$$

The input of interest can be viewed as a deterministic realization of such a white signal. It is clear that no periodic input has this property since the auto correlations are nonzero at multiples of the period. This excludes the PRBS as a possible input. It is known, however,that chirps have nice correlation properties. It can be shown that a high order chirp input has uniformly decaying correlations as desired above. An example of such an input is

$$u(t) = \exp(i\alpha t^3), \quad \alpha = \pi(1 + \sqrt{5})$$

which satisfies

$$\max_{1 \le i \le n} \left| \frac{r_u^n(i)}{r_u^n(0)} \right| \le C \frac{\log(n)}{\sqrt{n}}$$

As for the second property of the input, it can be shown that it holds if $w \in \mathcal{W}^L$. The result below generalizes the above discussion [21].

Theorem 4 *Given the following assumption:*

- $\mathcal{T}$ *is* ℓ_1 *or* $\mathcal{H}$.

- $\mathcal{G}$ *is the set of FIR systems of order* M.

- *Noise is in* $\mathcal{W}^L$.

- *Prior* $\mathcal{I}$: *dist*$(T_0, \mathcal{G})$ *is bounded by* γ.

- *Input: High order chirp.*

Then the following holds

- *The least squares estimator converges to* G_0:

$$\sup_{T \in \mathcal{I}} \sup_{w \in \mathcal{W}^L} \|\hat{G}_n - G_0\| \le (C_1\gamma + C_2 M)\frac{\log(n)}{\sqrt{n}}$$

- *The estimator is not tuned to* γ.

- *Complexity is polynomial.*

6.2 One-Pole Models

The situation is slightly more complicated when the model set $\mathcal{G} \in \mathcal{H}$ is not FIR. Consider the following set:

$$\mathcal{G} = \{\frac{\theta}{1 - a\lambda}, \theta \in \mathbb{R}, |a| < 1 \text{ fixed }\}$$

The output equation is given by:

$$y(k) = \theta \tilde{u}(k) + \Delta u + w, \qquad \tilde{u} = \frac{1}{1 - a\lambda} u$$

where Δ is bounded by γ and satisfies:

$$\sum_{k=0}^{\infty} (1 + k^2) a^k \delta_k = 0$$

Notice that the previous correlation technique (or LS) may not produce a consistent estimate since the correlations of $\tilde{u}$ and Δu are not guaranteed to decay to zero ($\delta_0 \neq 0$).

The general approach utilizes the structure of the unmodeled dynamics, namely the existence of a functional that annihilates them. Conceptually, the estimator can be decomposed into two steps:

1. Prefiltering the Data by a filter constructed from the set of annihilators that has the effect of separating the model from the unmodeled part.

2. Applying correlation techniques to recover the model.

For the FIR class of systems, this filter is the identity map. For the one-pole example, the filter is given by:

$$\begin{pmatrix} 1 & 2a & 5a^2 & \cdots & (1 + k^n)a^n \\ \frac{1}{a} & 1 & 2a & \cdots & (1 + k^{n-1})a^{n-1} \\ \vdots & \vdots & \ddots & & \\ & & & \cdots & 1 \end{pmatrix}$$

By applying this filter and following it by the LS estimator, we obtain a consistent estimate. This generalizes to the result below [22].

Theorem 5 *Given the following assumptions:*

- $\mathcal{T} = \mathcal{H}$.

- $\mathcal{G}$: *a finite dimensional subspace of stable systems, with dimension M.*

- *Noise is in $\mathcal{W}^L$.*

- *Prior $\mathcal{I}$: dist($T_0, \mathcal{G}$) is bounded by γ.*

- *Input: High order chirp.*

Then the following hold:

- *There exists a filter $\mathcal{F}$ such that the two-step algorithm converges to G_0*

$$\sup_{T \in \mathcal{I}} \sup_{w \in \mathcal{W}^L} \|\hat{G}_n - G_0\| \leq (C_1 \gamma + C_2 M) \frac{\log(n)}{\sqrt{n}}$$

- *The estimator is not tuned to γ.*

- *Complexity is polynomial.*

7   Identification in Practice

While deriving a paradigm for system identification is important, real world applications do not immediately translate into the appropriate mathematical descriptions required to proceed within this paradigm. In fact, it is usually quite hard to characterize a mathematical linear space to which the real process belongs, and as a consequence, it is hard to verify the validity of the prior information. In practice however, such paradigms are adapted to the application at hand with the hope to create reasonable models of the process. The effectiveness of these paradigms stem from the ability to validate (or invalidate) the set of prior information and simultaneously to provide models with uncertainty descriptions. In this section, we will describe how the error bounds derived in the last section allow us to estimate both the parametric error (error in the space $\mathcal{G}$) and the nonparametric error (an estimate of the prior γ).

Define $z_w(t) = y(t) - Gu(t) - w(t)$ for $t = 0, \ldots n$. Let $\mathbf{z_w}$ be the vector constructed from $z_w(t)$. The set of unfalsified models $S_n(\gamma)$, parametrized by the prior γ, can be written as:

$$S_n(\gamma) = \{G \in \mathcal{G} \mid \mathbf{z_w} = \mathcal{U}\delta, \, < g^i, \Delta > = 0, \, i = 1, \ldots M, \, \|\delta\| \leq \gamma, \, w \in \mathcal{W}_n^L\}$$

$\mathcal{U}$ is the Toeplitz matrix constructed from u and g^i's are the annihilators of Δ. We can eliminate Δ from the above characterization by solving an infinite dimensional quadratic programming problem. The set $S_n(\gamma)$ is then expressed as

$$S_n(\gamma) = \{G \mid \mathbf{z_w}^T V^{-1} \mathbf{z_w} \leq \gamma, \, w \in \mathcal{W}_n^L\}$$

V is the matrix of inner products between u and the g^i's (details omitted). In this form, $S_n(\gamma)$ is a characterized by a finite number of quadratic constraints. An estimate of the parametric error is simply

$$D_n(\gamma) = \mathrm{diam}(S_n(\gamma))$$

Now, we can invoke the error bounds to produce an "optimistic" estimate of γ, $\hat{\gamma}_n$, as follows:

$$\min \gamma$$

$$D_n(\gamma) \leq (C_1\gamma + C_2 M)\frac{\log(n)}{\sqrt{n}}$$

This shows how this proposed paradigm provides a tradeoff between parametric and nonparametric error estimates using measured Data and the noise model. The computations are based on convex analysis.

8   Final Remarks

A framework for identifying simplified models of complex systems has been proposed and some results were presented. In this paper, we have restricted ourselves to deterministic noise models. This, however, is not essential to this framework. The same development can be carried through replacing the error functions by their probabilistic analogs. It should be evident that whenever a result is derived for any of the noise sets containing white noise signals, a similar result will hold if the noise is assumed to be a realization of a white noise process. In such cases, however, it is not possible to characterize the sets of unfalsified models, and hence estimates of the parametric and nonparametric error bounds have to derived differently.

How does the minimum prediction error paradigm fit within this approach? In the exact modeling case, we have already shown that minimizing prediction error (specializes to LS in the setting of this paper) provides consistent estimates for any of the low correlated noise sets. In the undermodeling case, it can be shown that our approach for the class of systems $\mathcal{T} = \ell_2$ (space of finite energy sequences) coincides with the MPE approach. In fact, it has been shown [8] that if w is WN, then for any fixed $T_0 \in \ell_2$, the least sqauares estimator satisfies:

$$\lim_{N\to\infty} \hat{G}_N = \tilde{G} = \min_{G \in \mathcal{G}} \|(T_0 - G)\Phi_u^{\frac{1}{2}}\|_2 \quad w.p.1$$

where Φ_u is the spectral density function of u. Our results strengthen this result in the following ways. First, the convergence is shown to be uniform over the set of prior plants if a high order chirp is used as an input. Secondly, error bounds are derived for the set of prior plants. Thirdly, uniform convergence is shown for the noise set $\mathcal{W}^L$ removing all unnecessary assumptions regarding the noise such as quasi-stationarity. Finally, a systematic approach for trading off parametric and nonparametric errors is proposed using deterministic noise sets.

It is interesting to note that in contrast to the exact modeling case, the specification of the set $\mathcal{T}$ in the undermodeling case plays a major role. If $\mathcal{T} = \ell_2$, it can be shown that the two-step procedure is automatically achieved by minimizing the prediction error. This, however, is not true for the space $\mathcal{H}$, where one can show that no estimate based on any weighted minimum prediction error problem is consistent [19]. The two-step procedure is essential for such problems. We have omitted any detailed discussion on other spaces such as ℓ_1 or $\mathcal{H}_\infty$ [19, 21].

Studying identification in the space $\mathcal{H}$ has many advantages. For one, it is a Hilbert space and hence, as shown earlier, the resulting identification algorithms are computationally attractive. Another reason is that it is well-suited for robust control applications. This stems from the fact that if Δ satisfies $\|\Delta\|_{\mathcal{H}} \leq 1$, then $\Delta(e^{iw})$ can take values in an ellipse in the complex plane. The robustness analysis and synthesis can be turned into convex optimization problems [22]. Finally, priors in terms of this norm include knowledge about the smoothness of the frequency response of the system which, if not known, can be estimated from Data as discussed earlier.

References

[1] M.A. Dahleh, T. Theodosopoulos, and J.N. Tsitsiklis, "The sample complexity of worst-case identification of F.I.R. Linear systems," *System and Control. letters*, Vol 20, No. 3, pp. 157-166, March 1993.

[2] G.C. Goodwin, M. Gevers and B. Ninness, "Quantifying the error in estimated transfer functions with application to model order selection," *IEEE Trans. A-C*, Vol 37, No. 7, July 1992.

[3] G. Gu, P.P. Khargonekar and Y. Li, "Robust convergence of two stage nonlinear algorithms for identification in $\mathcal{H}_\infty$, *Systems and Control Letters*, Vol 18, No. 4, April 1992.

[4] A.J. Helmicki, C.A. Jacobson and C.N. Nett, "Control-oriented System Identification: A Worst-case/deterministic Approach in H_∞," *IEEE Trans. A-C*, Vol 36, No. 10, October 1991.

[5] M.K. Lau, R.L. Kosut, S. Boyd, "Parameter Set Estimation of Systems with Uncertain Nonparametric Dynamics and Disturbances", Proceedings of the 29th Conference on Decision and Control, pp. 3162-3167, 1990.

[6] M. Livstone and M.A. Dahleh. "A framework for robust parametric set membership identification," *IEEE Trans. A-C*, vol 40, pp. 1934-1939, Nov 1995.

[7] L. Lin, L. Wang and G. Zames, "Uncertainty principle and identification n-Widths for LTI and slowly varying systems, " *ACC*, Chicago, IL, 1992.

[8] L. Ljung. *System identification: theory for the user*, Prentice-Hall, Inc, NJ, 1987.

[9] P.M. Makila, "Robust Identification and Galois Sequences", Technical Report 91-1, Process Control Laboratory, Swedish University of Abo, January, 1991.

[10] P.M. Makila and J.R. Partington, "Robust Approximation and Identification in H_∞", Proc. 1991 American Control Conference, June, 1991.

[11] M. Milanese and G. Belforte, "Estimation theory and uncertainty intervals evaluation in the presence of unknown but bounded errors: Linear families of models and estimators", *IEEE Trans. Automatic Control*, AC-27, pp.408-414, 1982.

[12] M. Milanese and R. Tempo, "Optimal algorithm theory for robust estimation and prediction", *IEEE Trans. Automatic Control*, AC-30, pp. 730-738, 1985.

[13] M. Milanese, "Estimation theory and prediction in the presence of unknown and bounded uncertainty: a survey", in *Robustness in Identification and Control*, M. Milanese, R. Tempo, A. Vicino Eds, Plenum Press, 1989.

[14] J.P. Norton, "Identification and application of bounded-parameter models", *Automatica*, vol.23, no.4, pp.497-507, 1987.

[15] K. Poolla and A. Tikku, "On the time complexity of worst–case system identification," IEEE Trans. AC., May 1994.

[16] F. C. Schweppe, "Uncertain Dynamical Systems", Prentice Hall, Englewood Cliffs, NJ, 1973.

[17] R. Smith and M. Dahleh. *The Modeling of Uncertainty in Control Systems*, Springer-Verlag, 1994.

[18] D. Tse, M.A. Dahleh and J.N. Tsitsiklis. Optimal Asymptotic Identification under bounded disturbances. *IEEE Trans. Automat. Contr.*, Vol. 38, No. 8, pp. 1176-1190, August 1993.

[19] S. Venkatesh. Identifcation for Complex Systems. MIT Ph.D thesis No. LIDS-Th-2394, July 1997.

[20] S. Venkatesh and M.A. Dahleh, "Deterministic identification in the presence of unmodeled dynamics", to appear in IEEE Trans. A-C. Dec 1997.

[21] S. Venkatesh and M.A. Dahleh, "Identification of Complex systems with limited-complexity models," Submitted to IEEE Trans. A-C.

[22] S. Venkatesh, A. Megretski and M.A. Dahleh, "A convex parametrization of stabilizing controllers for perturbations belonging to a Hardy-Sobolev Space," submitted.

[23] G. Zames, "On the metric complexity of casual linear systems: ϵ-entropy and ϵ-dimension for continuous-time", *IEEE Trans. on Automatic Control*, Vol. 24, April 1979.

Recent Results on the Analytic Center Approach for Bounded Error Parameter Estimation

Er-Wei Bai*   Roberto Tempo†   Yinyu Ye‡

Abstract

In this paper, we present an overview of some recent work [5] on the so-called *analytic center approach* for bounded error parameter estimation. First, we discuss the optimality properties of well-known algorithms such as the Chebychev center, the projection and the min-max estimates. Subsequently, we propose the analytic center as an alternative algorithm for recursive estimation. We show that the analytic center minimizes the output error and, on the contrary of other estimates like Chebychev, allows for an easy-to-compute sequential algorithm. We argue that the maximum number of Newton iterations required to evaluate a sequence of analytic centers is linear in the number of observed data points and it is comparable to the complexity of off-line algorithms for estimating a single analytic center. Finally, we briefly discuss a number of open problems which are currently under investigation.

1   Problem Statement and Motivations

In this paper, we consider a single input-single output discrete-time system

$$y_i = \phi_i^T \theta + v_i \quad i = 1, 2, \ldots, n$$

where $y_i \in \mathbf{R}$ is the system output, $\phi_i \in \mathbf{R}^m$ the measurable regressor, $\theta \in \mathbf{R}^m$ the unknown parameter vector to be identified and $v_i \in \mathbf{R}$ the noise. In vector form, the system can be re-written more compactly as

$$y = \Phi\theta + v \tag{1}$$

where

$$y = \begin{pmatrix} y_1 \\ y_2 \\ \vdots \\ y_n \end{pmatrix}, \quad \Phi = \begin{pmatrix} \phi_1^T \\ \phi_2^T \\ \vdots \\ \phi_n^T \end{pmatrix} \quad \text{and} \quad v = \begin{pmatrix} v_1 \\ v_2 \\ \vdots \\ v_n \end{pmatrix}.$$

*[1]Department of Electrical and Computer Engineering, University of Iowa, Iowa City, Iowa 52242. erwei@icaen.uiowa.edu

†CENS-CNR, Politecnico di Torino, Torino, Italy. tempo@polito.it

‡Department of Management Science, University of Iowa, Iowa City, Iowa 52242. yyye@dollar.biz.uiowa.edu

The objective of parametric system identification is to compute an estimate $\hat{\theta}$ of the unknown parameter vector θ from available input-output measurements y and Φ. Most of the research on parametric estimation is focused on the so-called *stochastic* framework where the noise is a sequence of random variables with some known probabilistic properties; see, e.g. the classical textbook [11].

An alternative method assumes that the noise v_i is *unknown but bounded* by a constant $\epsilon > 0$

$$|v_i| \le \epsilon \tag{2}$$

for $i = 1, 2, \ldots, n$. Subsequently, we define the membership-set

$$\Omega^n \doteq \bigcap_{i=1}^{n} \{\hat{\theta} \in \mathbf{R}^m : -\epsilon \le y_i - \phi_i^T \hat{\theta} \le \epsilon\}. \tag{3}$$

This is the set of all parameter estimates consistent with equation (1), the input-output data y and Φ and the noise bound ϵ. For specific references on this topic, see e.g. the special issue [1] which is devoted to this line of research. Additional references are also available in [2] and [3].

In the set-membership setting, the main goal is the exact description of Ω^n. This description may reveal how its size can be reduced by a better experimental design or a choice of inputs but, unfortunately, the computational complexity required to compute Ω^n exactly may be very high. This fact follows from the observation that the membership-set can be fairly complex even in the case of linear parametrization since the number of vertices can grow exponentially when the number of observations increases. To reduce the computational burden, approaches of approximating the membership-set via ellipsoidal or orthotopic descriptions have been widely used; see, e.g. [9] and [12]. However, these approaches are often conservative if the shape of the membership-set is very different than the outer bounding sets. In addition, sequential implementations may lead to further conservatism [13].

In the bounded-error context, it is of paramount importance to compute specific estimates within the membership-set enjoying certain optimality properties. One of the most popular optimal estimates is the *Chebyshev center* θ_c of the set Ω^n

$$\theta_c \doteq \arg \min_{\theta \in \Omega^n} \max_{\eta \in \Omega^n} \|\theta - \eta\|. \tag{4}$$

The Chebyshev center is the best worst-case estimate of the true but unknown system parameter vector. For given input-output data, the calculation of θ_c can be performed via linear programming if the norm in (4) is ℓ_∞. It should be pointed out, however, that θ_c is optimal in terms of the parameter estimation error but *not* in terms of the output error

$$e_o \doteq \|y - \Phi \theta_c\|.$$

To avoid this drawback, the *projection estimate* θ_p, also denoted as constrained least squares estimate, is often proposed

$$\theta_p \doteq \arg\min_{\theta \in \Omega^n} \|y - \Phi\theta\|^2. \tag{5}$$

If the norm in (5) is Euclidean, this estimate can be computed with convex programming. The projection estimate minimizes the average error at the expense of a possibly large maximum error. For this reason, the so-called minimax estimate θ_{mm} has been introduced

$$\theta_{mm} \doteq \arg\min_{\theta \in \Omega^n} \max_{1 \leq i \leq n} |y_i - \phi_i^T \theta|$$

This estimate can be also computed with convex programming.

We now make a few remarks regarding these optimal estimates.

1. The Chebyshev center θ_c depends only on the constraints $-\epsilon \leq y_i - \phi_i^T \hat{\theta} \leq \epsilon$ describing the membership-set Ω^n and *not* on the redundant constraints. The projection estimate θ_p also depends on the constraints describing Ω^n and on *all* the input-output data y and Φ which enter into the objective function.

2. It is well-known that the unconstrained least squares estimate is in general *not* in the membership-set. Thus, by the convexity of the set Ω^n and the form of the cost function, we conclude that θ_p is often at the boundary of the membership-set.

3. If the norm in (5) is Euclidean, θ_p is the estimate within Ω^n which minimizes the "average" output error

$$e_o = \sum_{i=1}^{n}(y_i - \phi_i^T \theta)^2$$

or, equivalently, maximizes the complementary average output error

$$e_c = \sum_{i=1}^{n}(\epsilon^2 - (y_i - \phi_i^T \theta)^2).$$

Thus, the projection estimate θ_p can be immediately re-written as

$$\theta_p = \arg\max_{\theta \in \Omega^n} \sum_{i=1}^{n}(\epsilon^2 - (y_i - \phi_i^T \theta)^2).$$

4. The Chebyshev center is a continuous (but *not* necessarily smooth) function of the data y_i and ϕ_i so that the change in θ_c corresponding to new observed data could not be easily followed. We also notice that θ_p is a continuous function of the data y_i and ϕ_i.

5. As previously discussed, the projection estimate θ_p minimizes the "arithmetic average" output error. Generally, the output error does not have to be in the arithmetic form. For instance, the following maximizations

$$\max_{\theta \in \Omega^n} \Pi_{i=1}^n (\epsilon^2 - (y_i - \phi_i^T \theta)^2) \quad \text{or} \quad \max_{\theta \in \Omega^n} \sum_{i=1}^n \ln(\epsilon^2 - (y_i - \phi_i^T \theta)^2)$$

are also indications that the output error is minimized.

In this paper, we present an overview of some recent results on the so-called analytic center (see Section 2 for a precise definition) approach for bounded error parameter estimation. The concept of analytic center was first introduced in [14] and, more recently, it was studied in the context of interior point methods [17]. In [15] and [4], the analytic center has been used for solving model matching problems in control and for developing algorithms for adaptive filtering. In [8], the analytic center of Linear Matrix Inequalities has been studied. The use of the analytic center in identification and, in particular, in bounded error identification is very recent [5] and provides an alternative to more standard estimation tools which are of difficult use when solving on-line problems.

¿From the identification point of view, the analytic center θ_a^n of the set Ω^n enjoys interesting optimality properties. In [5], it has been shown that this estimate maximizes the complementary logarithm weighted average output error

$$\theta_a^n = \arg \max_{\theta \in \Omega^n} \sum_{i=1}^n \ln(\epsilon^2 - (y_i - \phi_i^T \theta)^2).$$

In the same paper, we also demonstrated that the computation of θ_a^n is quite easy and, in addition, two ellipsoids which are centered at the analytic center θ_a^n and inscribe and outscribe the membership-set can be immediately determined. The analytic center is also an interpolatory algorithm in the sense of Information-Based Complexity [16]. Therefore, the worst-case parameter error is guaranteed to be within a factor of two of the parameter error of the Chebyshev center θ_c. That is,

$$\max_{\theta \in \Omega^n} \|\theta - \theta_a^n\| \leq 2 \max_{\theta \in \Omega^n} \|\theta - \theta_c\|.$$

As a second goal, we discuss an easy-to-implement algorithm that generates a *sequence* of analytic centers. Here, by sequential algorithm, we mean the following: The computation of θ_a^i at time i uses only the estimate θ_a^{i-1} at the previous time $(i-1)$ and the current input-output data y_i and ϕ_i. More precisely, if the cost of a non-sequential algorithm for computing θ_a^i at time i is $c(i)$, then a sequential algorithm should produce a sequence of estimates at time $1, 2, \ldots, i$ with a total cost bounded by $O(c(i))$.

The proposed algorithm is based on the Newton method and the number of iterations required to compute a *sequence* of analytic centers up to time

i is linear in i and comparable to the complexity of the batch algorithm for computing a single analytic center at fixed time i; see [5] for details.

2 Definition of the Analytic Center

Consider the system (1) and the noise bound ϵ. Define

$$A^n \doteq (\phi_1, -\phi_1, \phi_2, -\phi_2, \ldots, \phi_n, -\phi_n) \in \mathbf{R}^{m \times 2n}$$

and

$$b^n \doteq \begin{pmatrix} \epsilon + y_1 \\ \epsilon - y_1 \\ \vdots \\ \epsilon + y_n \\ \epsilon - y_n \end{pmatrix} \in \mathbf{R}^{2n}.$$

Then, the membership-set Ω^n can be immediately re-written as

$$\Omega^n = \{\hat{\theta} \in \mathbf{R}^m : (A^n)^T \hat{\theta} \le b^n\}$$

where the vector inequality applies element-wise.

We assume that

- The rank of the matrix A^n is m for $n \ge m$. This is a necessary assumption otherwise the membership-set is unbounded.

- The membership-set Ω^n has a non-empty interior; i.e., there exists some $\rho > 0$ and some $\hat{\theta} \in \Omega^n$ so that

$$-\epsilon + \rho \le y_i - \phi_i^T \hat{\theta} \le \epsilon - \rho \tag{6}$$

for $i = 1, 2, \ldots, n$.

Roughly speaking, the former assumption implies that the system is identifiable and the latter means that the model validation step has been already performed and there exists a non-empty set of parameters which are consistent with input-output data.

Next, let $f(\theta)$ denote the (dual) potential function

$$f(\theta) = \sum_{i=1}^{2n} \ln(b_i - a_i^T \theta). \tag{7}$$

The analytic center θ_a^n of Ω^n is an interior point of Ω^n maximizing the (dual) potential function $f(\theta)$

$$\theta_a^n \doteq \arg \max_{\theta \in \Omega^n} f(\theta).$$

Consequently, we obtain (see [5] for details)

$$\theta_a^n = \arg\max_{\theta\in\Omega^n} \sum_{i=1}^{n} \ln(\epsilon^2 - (y_i - \phi_i^T\theta)^2) = \arg\max_{\theta\in\Omega^n} \Pi_{i=1}^{n}(\epsilon^2 - (y_i - \phi_i^T\theta)^2).$$

The problem we address in the next section can be summarized as follows: How can we efficiently compute the analytic center?

3   An Off-Line Algorithm

In this section, we discuss the calculation of θ_a^n by means of an off-line algorithm. This subject has been analyzed in depth in the paper [5] where technical details of the algorithm are given. Here, we only provide the key ideas behind the computation of θ_a^n. These ideas can be summarized as follows:

1. For a given data set and an arbitrary small error tolerance $\delta > 0$, we aim to calculate an *approximate* analytic center θ_δ^n which is δ-close to the analytic center θ_a^n. This concept can be expressed in several ways. For example, θ_δ^n can be picked so that its potential function $f(\theta_\delta^n)$ is "close" to the potential function $f(\theta_a^n)$ associated to the analytic center θ_a^n. Therefore, the error between $f(\theta_\delta^n)$ and $f(\theta_a^n)$ is small for small δ. This issue has been widely investigated in the literature on interior point methods; see e.g. [10] for technical details.

2. The algorithm for the computation of a δ-approximate center consists of two phases. The objective of Phase I is to find a 1/2-approximate center from the initial conditions and that of Phase II is to compute a δ-approximate center from a 1/2-approximate center.

3. In Phase I, the algorithm proceeds iteratively and, at each iteration, the key operation is the solution of a linear least squares problem. ¿From the computational point of view, this is equivalent to performing one Newton iteration. At the end of Phase I, the algorithm is guaranteed to generate a 1/2-approximate center. If the goal is to produce a δ-approximate center with $\delta < 1/2$, we can perform Phase II, otherwise we terminate.

4. Similarly to Phase I, in Phase II the algorithm proceeds iteratively. In this case, the crucial step is the solution of a linear set of equations, which can be solved via Newton's method. The computational burden is again one Newton iteration.

5. The convergence of the algorithm is guaranteed and a bound on the total number of Newton iterations required can be computed. This bound can be stated as a function of n, δ and ρ, where n is the number of measurements, δ represents the goodness of the δ-approximation and the slack variable ρ is defined in (6). In [5], it has been shown that the number of Newton iterations

required is bounded by

$$O(n \cdot \ln \frac{\epsilon}{\rho} + \ln \ln \frac{2}{\delta}).$$

4   Computation of a Sequence of Analytic Centers

In this section, we study an algorithm for the *sequential* computation of the analytic center. We recall that, for each $i \geq m$, the membership-set is given by

$$\Omega^i = \bigcap_{k=1}^{i} \{\hat{\theta} \in \mathbf{R}^m : -\epsilon \leq y_k - \phi_k^T \hat{\theta} \leq \epsilon\} = \{\hat{\theta} \in \mathbf{R}^m : (A^i)^T \hat{\theta} \leq b^i\}$$

where

$$A^i = (\phi_1, -\phi_1, \phi_2, -\phi_2, \ldots, \phi_i, -\phi_i) \in \mathbf{R}^{m \times 2i}$$

and

$$b^i = \begin{pmatrix} \epsilon + y_1 \\ \epsilon - y_1 \\ \vdots \\ \epsilon + y_i \\ \epsilon - y_i \end{pmatrix} \in \mathbf{R}^{2i}.$$

For identification purposes, we need to compute the analytic center θ_a^i of Ω^i or a δ-approximate center for each $i = m, m+1, \ldots, n$. Clearly, if we apply directly the off-line algorithm of the previous section, the overall complexity may be high. Therefore, in [5] an on-line algorithm for which the calculation of θ_δ^i is based on θ_δ^{i-1} and on the current input-output data has been derived. The advantage is that earlier computations of θ_δ^{i-1} would not be wasted. In particular, in [5] it has been shown that the total number of Newton iterations required to compute a *sequence* of δ-approximate analytic centers

$$\theta_\delta^m, \theta_\delta^{m+1}, \ldots, \theta_\delta^n$$

of the membership-sets

$$\Omega^m, \Omega^{m+1}, \ldots, \Omega^n$$

is bounded by

$$O(n \cdot (\ln \frac{\epsilon}{\rho} + \ln \ln \frac{2}{\delta})).$$

We remark that this number is *linear* in the number of measurements n and, therefore, comparable to the complexity of the off-line algorithm for computing a single δ-approximate analytic center.

5   Conclusions

In this paper, we summarized some recent results on the so-called analytic center approach for bounded error sequential parameter estimation. Two easy-to-compute algorithms have been discussed for both on-line and off-line identification. The computational complexity, in terms of Newton iterations, to estimate a sequence of analytic centers is linear in the number of measurements and it coincides with that of an off-line algorithm for computing a single analytic center. This property distinguishes the analytic center from approaches involving the computation of the Chebyshev center. In fact, the sequential computation of this estimate, or a δ-approximation of it, is presently an open problem; see [7] for details.

A number of open problems are currently under investigation. In particular, we feel that some connections with classical stochastic identification requires attention. Along this direction, conditions for asymptotic convergence and comparisons with recursive least squares need to be established; see [6] for preliminary results. Finally, it is important to characterize classes of noise density functions with bounded support for which the analytic center is a Maximum Likelihood Estimator.

References

[1] (1995). "Special Issue on Bounded Error Estimation (Part II)," *International Journal of Adaptive Control and Signal Processing*, Vol. 9.

[2] (1992). "Special Issue on System Identification for Robust Control Design," *IEEE Transactions on Automatic Control*, Vol. 37.

[3] (1995). "Special Issue on Trends in System Identification," *Automatica*, Vol. 31.

[4] Afkhamie, K. H., Z.-Q. Luo and K. M. Wong (1997). "Interior Point Column Generation Algorithms for Adaptive Filtering," *Technical Report Mc Master University*, Ontario, Canada.

[5] Bai, E.-W., Y. Ye and R. Tempo (1997). "Bounded Error Parameter Estimation: A Sequential Analytic Center Approach," *Proceedings of the IEEE Conference on Decision and Control*, San Diego; also submitted to *IEEE Transactions on Automatic Control*.

[6] Bai, E.-W., M. Fu, R. Tempo and Y. Ye (1997). "Analytic Center Approach to Parameter Estimation: Convergence Analysis," *Technical Report University of Newcastle*, Newcastle, Australia.

[7] Bai, E.-W., R. Tempo and Y. Ye (1997). "Open Problems in Sequential Parametric Estimation," *Technical Report CENS-CNR, Politecnico di Torino*, Torino, Italy.

[8] Boyd, S., L. El Ghaoui, E. Feron and V. Balakrishnan (1994). *Linear Matrix Inequalities in System and Control Theory*, SIAM, Philadelphia, PA.

[9] Fogel, E. and Y. Huang (1982). "On the value of information in system identification-bounded noise case," *Automatica*, Vol. 18, pp. 229-238.

[10] Goffin, J. L., Z. Luo, and Y. Ye (1996). "Complexity analysis of an interior cutting plane method for convex feasibility problems," *SIAM J. Optimization*, Vol. 6, pp. 638-652.

[11] Ljung, L. (1995). "System Identification: Theory for the Users," *Prentice-Hall*, Englewood Cliffs, NJ.

[12] Milanese M. and R. Tempo, "Optimal Algorithms Theory for Robust Estimation and Prediction," *IEEE Transactions on Automatic Control*, vol. AC-30, pp. 730-738, 1985.

[13] Ninness, B. M. and G. C. Goodwin (1995). "Rapproachement between bounded-error and stochastic estimation theory", *International Journal Adaptive Control and Signal Processing*, Vol. 9, pp. 107-132.

[14] Sonnevend, G. (1986). "An analytic center for polyhedrons and new classes of global algorithms for linear programming," *Lecture Notes in Control and Information Science*, Vol. 84, pp. 866-876.

[15] Sonnevend, G. (1987). "A new method for solving a set of linear inequalities and its applications," *Dynamic Modelling and Control of National Economics*, Pergamon, Oxford-New York, pp. 465-471.

[16] Traub, J. F., G. Wasikowski and H. Woźniakowski (1988). *Information-Based Complexity*, Academic, New York.

[17] Ye, Y. (1997). *Interior-Point Algorithm: Theory and Analysis*, Wiley, New York.

Universal Output Prediction and Nonparametric Regression For Arbitrary Data*

S.R. Kulkarni[†]   S.E. Posner[‡]

Abstract

We construct a class of elementary nonparametric output predictors of an unknown discrete-time nonlinear fading memory system. Our algorithms predict asymptotically well for every bounded input sequence, every disturbance sequence in certain classes, and every linear or nonlinear system that is continuous and asymptotically time-invariant, causal, and with fading memory. The predictor is based on k_n-nearest neighbor estimators from nonparametric statistics. It uses only previous input and noisy output data of the system without any knowledge of the structure of the unknown system, the bounds on the input, or the properties of noise. Under additional smoothness conditions we provide rates of convergence for the time-average errors of our scheme. Finally, we apply our results to the special case of stable LTI systems.

1   Introduction

We introduce an elementary algorithm which predicts the output of an unknown nonlinear discrete-time system that satisfies certain generic regularity conditions, such as continuity and approximate time-invariance, causality, and fading memory. The algorithm only uses the past observed input and noisy output data, and works for every bounded input sequence, every system in the class (without parametric and/or structural assumptions), and a wide range of disturbances. In this sense, the algorithm is "universal" in the terminology of information theory and statistics. The algorithm we use to achieve an asymptotically good predictor is an adaptation of the well-known k_n-nearest neighbor algorithm which has been analyzed extensively in the nonparametric statistics, pattern classification, and information theory literature [1, 10, 15].

*This work was partially supported by the National Science Foundation under NYI grant IRI-9457645.

[†]Department of Electrical Engineering, Princeton University, Princeton, NJ 08544. kulkarni@ee.princeton.edu

[‡]ING Baring Securities, Inc., New York, NY 10021. steven.posner@ing-barings.com
This research was completed while Posner was at the Department of Electrical Engineering, Princeton University and the Department of Statistics, University of Toronto.

Most previous work on output prediction has been parametric in nature. This encompasses many important areas in linear systems theory. For example, the Kalman filter uses the parameters of a linear system to construct a predictor in the presence of unknown stochastic disturbances. Similarly, the Luenberger observer uses the state-space matrices to construct an observer of the unknown state. In adaptive control and other schemes, system parameters are estimated and a controller is tuned on-line accordingly. In this paper, we are concerned with *nonparametric* prediction. We construct a predictor of the output of an unknown system assuming only generic conditions, but without any knowledge (or even an estimate) of system parameters.

Our nonparametric approach is in line with the work of several authors. In particular, Greblicki, Pawlak, and Krzyżak (e.g. [8, 9, 12, 13]) consider Hammerstein and Wiener systems which are nonlinear systems composed of linear systems coupled with memoryless nonlinearities. They consider these systems driven by stationary or i.i.d. noise and show that various nonparametric schemes can be used to estimate the nonlinearity. In contrast, we impose only mild regularity assumptions on the system without assuming any particular system structure, and our algorithm works for any bounded input sequence. Surprisingly, we provide a predictor for which we prove that the pointwise prediction errors tend to zero, even with the generality of our setup. The price we pay for this generality is, of course, in the rate of convergence, which is to be expected. In order to make statements about rates of convergence, stronger assumptions must be placed on both the plant and the input, or by making statements about the time-average of the prediction errors, we need only impose stronger conditions on the plant.

The role of prediction is also typically linked to that of system identification. System identification is concerned with using some algorithm to select a model from a model class (generally by selecting the model that best explains the measured data) so that the distance between the model and the true plant is small in some metric. Traditionally, system identification is ultimately used for control. The chosen model is used to design or tune a controller for the underlying system. Some recent work in system identification has focused on the theoretical limits of identification algorithms in a worst-case setting, i.e., in which the output disturbances are only assumed to be bounded (e.g., see [23, 11, 20, 3, 18] and references cited therein). We consider both worst-case and stochastic noise models but in the context of prediction. Our results hold for a broad class of nonlinear systems quite similar to that studied in [2] in the context of worst-case identification. In contrast with identification results which require a "sufficiently rich" input sequence, we show that prediction can be performed for arbitrary input sequences. Also, since we do not provide explicit estimates of the unknown system itself, we do not require any topological structure on the class of systems, which is required from the outset for identification results. However, at present, we make no claims as to how our approach is to be used for the purpose of controller design. Rather, our focus here is simply on output prediction.

In a very different context, a similar scheme is used in [19]. They consider the estimation of conditional probabilities for stationary ergodic time series by looking at similar strings from the past and averaging the next value after each string. In contrast, in this paper, we focus on prediction of the output of a nonlinear dynamical system driven by an arbitrary bounded input. A similar algorithm is also used by Farmer and Sidorowich ([6] and references therein) in the context of predicting chaotic time series, although to our knowledge performance statements such as those presented here have not been shown. We suspect that the results in this paper can be used to make rigorous performance statements of their algorithm as well. Our work is also in the spirit of the work of Feder, Merhav, and Gutman [7] in which they construct a finite-memory predictor of the next outcome of a binary sequence. However, the specific formulations are quite different, in that we are in a systems framework, we have access to the input sequence which provides information about the unknown output, and we focus on different algorithms.

In Section 2 we formulate the problem and precisely define the class of systems, inputs, and noise under consideration. In Section 3, we introduce a class of data-dependent but elementary nonparametric estimators and show (Theorem 1) that with bounded input and noise sequences we can predict *pointwise* asymptotically well to within the level of the noise, and that for stochastic noise, we can get asymptotically zero mean square prediction error. In Section 4, we consider rates of convergence. With only additional regularity conditions on the system (Lipschitz continuity and rates on fading memory), rates of convergence for the time-average of the prediction errors can be obtained for every bounded input sequence (Theorem 2). In order to get rates on the pointwise prediction errors, additional conditions are needed on both the input and the system. We show (Theorem 3) that if we impose independence or stationarity assumptions on the inputs and the additional conditions on the system, then a uniform rate of convergence can be obtained for the pointwise prediction errors. In Section 5 we consider in more detail the special case of stable LTI systems. The complete version of this paper with proofs can be found in [16].

2   Formulation and Preliminaries

We consider the on-line prediction of the output of an unknown discrete-time system based on past inputs and noisy output observations. Suppose an unknown discrete-time system H is driven by an input sequence $u_0, u_1, \ldots$. We consider the following sequential prediction problem. By time n we have observed the past inputs $u_0, \ldots, u_{n-1}$ and corresponding noisy outputs

$$z_i = y_i + e_i \quad i = 1, \ldots, n-1$$

where y_i is the output of H at time i and e_i represents measurement noise. We then observe the input u_n at time n and produce an estimate $\hat{y}_n$ of the uncorrupted output y_n. Our goal is to have small estimation errors as n increases. Precise conditions on the system H, the input $u_0, u_1, \ldots$, and the noise $e_0, e_1, \ldots$ are given below.

We consider systems $H : \mathbb{R}^\infty \to \mathbb{R}^\infty$ where $\mathbb{R}^\infty = \{(u_0, u_1, \ldots) : u_i \in \mathbb{R}, i \geq 0\}$. For a subset $S \subset \{0, 1, 2, \ldots\}$, we define the projection operator, $P_S : \mathbb{R}^\infty \to \mathbb{R}^{|S|}$, in the natural way. We use the notation:

$$\begin{aligned}
\text{for } S = \{m, m+1, \ldots, n\}, \quad & P_{[m,n]}u = (u_m, u_{m+1}, \ldots, u_n) \\
\text{for } S = \{n, n+1, \ldots\}, \quad & P_{[n,\infty)}u = (u_n, u_{n+1}, \ldots) \\
\text{for } S = \{n\}, \quad & P_n u = u_n
\end{aligned}$$

for every $u = (u_0, u_1, \ldots) \in \mathbb{R}^\infty$. Note that in this paper we abuse consistency but conform to standard notation and use some lower case letters to mean vectors in $\mathbb{R}^\infty$ (e.g., e, y, z, u, v) and other lower case letters may be constants or whatever the context dictates. Similarly, upper case letters may be operators, vectors, or constants depending on the context.

Define the closed ball of radius r of a Banach space $(\mathcal{X}, \|\cdot\|)$ as

$$\mathcal{B}_r \mathcal{X} = \{x \in \mathcal{X} : \|x\| \leq r\}.$$

We will mostly deal with the following balls,

$$\mathcal{B}_r \ell_\infty = \{u \in \ell_\infty : \|u\|_\infty \leq r\} \text{ where } \|u\|_\infty = \sup_{i \geq 0} |u_i|,$$

and

$$\mathcal{B}_r \mathbb{R}^L = \{u \in \mathbb{R}^L : \|u\|_\infty \leq r\}, \quad L \geq 1.$$

We consider the on-line prediction of the output of an unknown system that satisfies certain general regularity conditions. The input may be any sequence in $\mathcal{B}_r \ell_\infty$. The measured output is corrupted with an additive disturbance sequence, $e \in \mathbb{R}^\infty$. The system model is

$$y = H(u)$$

$$z = y + e$$

We consider both deterministic and stochastic disturbance classes:

- $e \in \mathcal{B}_\delta \ell_\infty$, with $\delta \geq 0$,

- $e_0, e_1, \ldots$ i.i.d. zero mean and finite variance.

We consider systems that satisfy the following properties, where r is the same parameter as the bound on the allowable input sequences.

(A1) H is continuous (but not necessarily linear) on $\mathcal{B}_r \ell_\infty$, i.e., $\|H(u) - H(v^{(n)})\|_\infty \overset{n \to \infty}{\longrightarrow} 0$ if $\|u - v^{(n)}\|_\infty \overset{n \to \infty}{\longrightarrow} 0$, for $u, v^{(1)}, v^{(2)}, \ldots \in \mathcal{B}_r \ell_\infty$.

(A2) For each $\epsilon > 0$, there exists $T = T(\epsilon) \geq 1$ and $L = L(\epsilon) \geq 1$ such that the output at all times $n, m \geq T$ depends only on the previous L input components to within ϵ, i.e.,

$$|P_n H(u) - P_m H(v)| \leq \epsilon$$

for all $u, v \in \mathcal{B}_r \ell_\infty$ such that $P_{[m-L+1,m]} v = P_{[n-L+1,n]} u$.

Condition (A1) is straightforward. Condition (A2) is an asymptotically time-invariance, causality, and fading memory condition. These conditions are fairly general and contain many classes that are of common interest. For example, any stable linear time-invariant causal discrete-time system satisfies (A1)—(A2) as will be shown in Section 5. Also, Hammerstein and Wiener systems [8, 12, 13] satisfy (A2), and, if the nonlinear element is continuous, (A1).

3 Predictor and Main Result

Our first result is to construct a prediction algorithm that achieves pointwise convergence. We exhibit an algorithm which predicts the output of any unknown system in our class subjected to any bounded input. At time n, we have observed all the past inputs $(u_0, u_1, \ldots, u_{n-1})$ together with u_n, and we have observed noisy outputs $(z_0, z_1, \ldots, z_{n-1})$. We would like to estimate the uncorrupted system output, y_n, using an algorithm that produces an estimate $\hat{y}_n$ so that the prediction errors tend to zero asymptotically.

The algorithm we propose is an adaptation of k_n-nearest neighbor estimators from nonparametric statistics. The basic idea of the algorithm is as follows. Take the most recent L_n inputs (where L_n is a data-dependent parameter specified in detail below) as a nominal vector in $\mathbb{R}^{L_n}$ and find the previous input substring of length L_n that is nearest to it in the sup norm sense. A natural estimate for y_n would be the output associated with the input vector that is nearest the nominal vector. That is, find a time in the past when the most recent L_n inputs were most similar to the current L_n inputs and use the observed output at that time as our prediction. The idea is that by the assumed continuity of the system, nearby inputs should produce nearby outputs. If there were no noise in our observations and if the parameter L_n was chosen wisely we might expect that this algorithm would perform well. However, this basic idea needs to be refined in several ways.

First, since the "order" of the system is assumed unknown it is clear that we will need $L_n \to \infty$ as $n \to \infty$ if we wish to drive prediction errors to zero. The reason we have any hope of driving prediction errors to zero is due to the assumption that the system has fading memory. Actually, the fading memory assumption must be used in another way as well. To avoid the effects of initial conditions, we should not use input strings too close to time 0. Hence, the second refinement is to introduce another parameter T_n which tends to infinity as $n \to \infty$, and only search for nearby input strings that occur after time T_n. The third refinement we need results from the fact that our output observations are noisy. With random noise, the output at the time associated with the nearest input string may not necessarily give a good prediction. To average out the noise in the output observations, we could search for a number of past input strings that are close to the recent string, and average the corresponding outputs to form our prediction. Thus, we introduce a third parameter, k_n, which is the number of "nearest neighbors" in the input string that we search for, and we will need $k_n \to \infty$ as $n \to \infty$. (Actually, in the case of worst-case noise, averaging is not needed, and we can just take $k_n = 1$).

At time n, given $u_0, \ldots, u_n$ and the parameters k_n, L_n, and T_n, let $\{U_j(L_n)\}_{j=T_n+L_n}^n$ be the set of all strings from the past input sequence after time T_n that are of length L_n. That is,

$$U_j(L_n) = P_{[j-L_n+1,j]}u.$$

Note that each $U_j(L_n)$ is a vector in $\mathbb{R}^{L_n}$. Let $m_n^{[i]}$ be the index of the ith nearest neighbor (NN) of $U_n(L_n)$ (which is the most recent string of inputs of length L_n) from the set $\{U_j(L_n)\}_{j=T_n}^{n-1}$. The first NN distance $d_n(1, L_n) = d_n(1, L_n, T_n; u_0, \cdots, u_n)$ satisfies

$$
\begin{aligned}
d_n(1, L_n) &= \min_{T_n \le j < n} \|U_n(L_n) - U_j(L_n)\|_\infty \\
&= \|U_n(L_n) - U_{m_n^{[1]}}(L_n)\|_\infty = \max_{0 \le j < L_n} |u_{n-j} - u_{m_n^{[1]}-j}|.
\end{aligned}
$$

Similarly, $d_n(i, L_n) = d_n(i, L_n, T_n; u_0, \cdots, u_n)$ is the ith smallest distance between $U_n(L_n)$ and $\{U_j(L_n)\}_{j=T_n}^{n-1}$ and equals $\|U_n(L_n) - U_{m_n^{[i]}}(L_n)\|_\infty$. Consider the simple predictor

$$\hat{y}_n = \frac{1}{k_n} \sum_{i=1}^{k_n} z_{m_n^{[i]}}, \tag{1}$$

where $z_{m_n^{[i]}} = P_{m_n^{[i]}} H(u) + e_{m_n^{[i]}}$ is the output observation at time $m_n^{[i]}$.

To complete the specification of the predictor, we need only specify the choice of the parameters k_n, L_n, and T_n as a function of n. Of course, to get asymptotically good predictions, the parameters need to be chosen carefully. In particular, to exploit the continuity of the system, we need the k_n nearest input strings to get closer and closer to the most recent input string as $n \to \infty$. An important quantity is the k_n-th nearest neighbor distance $d_n = d_n(k_n, L_n, T_n; u_0, \ldots, u_n)$ which is the distance between the most recent string of length L_n and the k_n-th nearest neighbor from past strings of length L_n occurring after time T_n. That is,

$$d_n(k_n, L_n, T_n; u_0, \ldots, u_n) = \|U_n(L_n) - U_{m_n^{[k_n]}}(L_n)\|_\infty$$

We need to make sure that $d_n \to 0$ as $n \to \infty$. Boundedness of the input is crucial in this regard. With a boundedness assumption, input strings of any fixed length L belong to a compact subset of $\mathbb{R}^L$, and it is this compactness that allows a suitable choice of parameters to make $d_n \to 0$.

However, it can be shown that with any fixed choice of the parameters k_n, L_n, and T_n, there is always a bounded input $u_0, u_1, \ldots$ for which $d_n(k_n, L_n, T_n; u_0, \ldots, u_n) \not\to 0$. This is typical of data-independent algorithms [15, 21] and is the reason one cannot get pointwise convergence with such algorithms for arbitrary inputs. In such cases, one must resort to making statements about the time-average of the prediction errors. To overcome this problem, we use suitable data-dependent choices of the algorithm parameters as in [14]. By choosing k_n, L_n, and T_n to depend on the observed input $u_0, \ldots, u_n$, we can construct an algorithm for which $k_n, L_n, T_n \to \infty$ and $d_n \to 0$. In this case, we can show that the pointwise prediction errors (and hence also the time-average errors) converge to zero.

Lemma 1 *For every bounded sequence of inputs $u_0, u_1, \ldots$, if $k_n = k_n(u_0, \ldots, u_n)$, $L_n = L_n(u_0, \ldots, u_n)$, and $T_n = T_n(u_0, \ldots, u_n)$, are defined by*

$$(k_n, L_n, T_n) = \mathrm{argmin}_{(k,L,T)} \ \frac{1}{k} + \frac{1}{L} + \frac{1}{T} + d_n(k, L, T; u_0, \ldots, u_n)$$

then we have

   (i) $k_n(u_0, \ldots, u_n) \to_{n\to\infty} \infty$

   (ii) $L_n(u_0, \ldots, u_n) \to_{n\to\infty} \infty$

   (iii) $T_n(u_0, \ldots, u_n) \to_{n\to\infty} \infty$

   (iv) $d_n(k_n, L_n, T_n; u_0, \ldots, u_n) \to_{n\to\infty} 0$

The following theorem, our main result, describes the asymptotic behavior of our data-dependent nonparametric predictor. The algorithm does not need to know any of the parameters used in the assumptions on the input, system, or noise.

Theorem 1 *Consider the predictor $\{\hat{y}_n\}$ given by equation (1) where k_n, L_n, T_n are chosen in a data-dependent manner according to Lemma 1. Then for any $u \in B_r \ell_\infty$ for some $r < \infty$ and any H that satisfies (A1) and (A2), we have that*

1. *for any $e \in \mathcal{B}_\delta \ell_\infty$,*

$$\limsup_{n \to \infty} |y_n - \hat{y}_n| \leq \delta.$$

2. *for any i.i.d. $e_0, e_1, \ldots$ such that $Ee_i = 0$ and $E|e_i|^2 < \infty$,*

$$\lim_{n \to \infty} E(y_n - \hat{y}_n)^2 = 0.$$

Notes

- There is no uniform rate of convergence over the entire input class.

- The parameters used in the algorithm for the proof depends on the actual input sequence, in contrast with Theorem 2 (in the following section) in which the parameters are fixed and independent of the input sequence. Of course, the choice of parameters used in Lemma 1 is not the only one which will work. Many data-dependent schemes can achieve the conclusion of Lemma 1 and hence the result of Theorem 1

- The upper bound for part 1 is clearly tight since errors of at least δ can be forced by the noise sequence each time.

- With arbitrary bounded inputs and without assumption (A1), no asymptotic prediction is possible.

4   Rates of Convergence of Prediction Errors

The result of Theorem 1 shows that an appropriate data-dependent predictor provides estimates of the uncorrupted output such that the estimation errors converge pointwise to zero (for i.i.d. noise). However, it is easy to verify that no uniform rate of convergence is possible. The inability to obtain a uniform convergence rate arises from two distinct and fundamental reasons. One reason is that with arbitrary bounded inputs, one can construct input sequences such that the k_n-th nearest neighbor distance converges to zero arbitrarily slowly. In fact, one can make the 1-NN distance converge to zero arbitrarily slowly. Thus, although the predicted output will be an average of outputs due to nearby inputs, we have no way of bounding how close the inputs will be at any particular time. However, even if we had such a bound, we still could not get a uniform rate of convergence due to a second reason, which involves the regularity of the unknown system. Namely, although continuity implies that nearby inputs result in nearby outputs, we need a stronger assumption such as a Lipschitz condition to have a hope of getting rates. Also, although the system is assumed to have fading memory, we need bounds on the rate at which the memory fades in order to get bounds on the prediction errors.

Thus, to obtain rates of convergence, we need assumptions on the inputs that allow bounding the nearest neighbor distances, and conditions on the system that give stronger versions of assumptions (A1) and (A2). A result of this type is provided in Theorem 3 below. However, first we give a rate result of a different sort. Namely, by considering the time-average of the prediction errors, we can obtain uniform rates on the time-average errors with assumptions only on the system (i.e., that hold for all bounded inputs). The basic ideas of this result will also be used for the pointwise rate result of Theorem 3. Interestingly, the prediction algorithms used in this section are of the same basic form as in Section 3, but with the added simplification that the parameters k_n, L_n, and T_n need only satisfy certain rate conditions, but can be chosen in a data-independent fashion.

The following stronger versions of assumptions (A1) and (A2) will be used to get the rate results in this section.

(A1′) There exists $K, \alpha > 0$ such that for all $u, v \in \mathcal{B}_r \ell_\infty$,

$$\|H(u) - H(v)\|_\infty \leq K\|u - v\|_\infty^\alpha$$

(A2′) There exists $C \geq 0$, $0 < \rho < 1$, and T such that for $n, m > T$,

$$|P_n H(u) - P_m H(v)| \leq C\rho^L$$

for every $u, v \in \mathcal{B}_r \ell_\infty$ such that $P_{[m-L+1,m]} v = P_{[n-L+1,n]} u$.

For example, as we will see in the final section, stable linear systems with a decay rate on the impulse response satisfy (A1′) and (A2′).

Fix nondecreasing sequences $k_n = \log^2 n$, $T_n = \log n$, and $L_n = \sqrt{\log \frac{n-T_n}{k_n}}$. With these data-independent specifications on the parameters k_n, L_n, T_n we use the simple predictor (1).

Theorem 2 *Consider the predictor given by equation (1) where k_n, T_n, L_n are chosen data-independently as above. Then for any $u \in \mathcal{B}_r \ell_\infty$ for some $r < \infty$ and any H that satisfies (A1′)-(A2′), we have that*

1. *for any $e \in \mathcal{B}_\delta \ell_\infty$,*

$$\frac{1}{N} \sum_{n=1}^{N} |y_n - \hat{y}_n| \leq \delta + \frac{\eta_1}{L_N^{2\gamma}} \leq \delta + \frac{2\eta_1}{\log^\gamma N}$$

2. *for any i.i.d. $e_0, e_1, \ldots$ such that $Ee_i = 0$ and $E|e_i|^2 < \infty$,*

$$\frac{1}{N} \sum_{n=1}^{N} E|y_n - \hat{y}_n|^2 \leq \frac{\eta_2}{L_N^{4\gamma}} \leq \frac{2\eta_2}{\log^{2\gamma} N},$$

where $\gamma = \min\{\alpha, -\log \rho\}$ and η_1, η_2 are well-defined constants.

Notes

- It can be shown that with the algorithm used in the proof of Theorem 2, pointwise errors do *not* tend to zero for all bounded input sequences. The problem is that the parameters k_n, L_n, and T_n were chosen at the outset, independent of the inputs observed. In this case, one can always find a system and construct an input sequence for which the pointwise prediction errors do not converge to zero. The construction simply makes sure that the input is chosen so that the distance between the most recent input string of length L_n and its k_n-th nearest neighbor does not converge to zero. This is the same the reason that a time-average criterion was required in [15].

- The algorithm is completely data-independent as well as independent of knowledge of the parameters in (A1') and (A2').

- Interestingly, the same algorithm works regardless of the noise class, although of course the mode of convergence depends on the type of noise.

- The time average nature of the statements in the theorem arises not because of the noise but as a result of the arbitrary bounded inputs that are allowed.

- This algorithm can be readily modified to allow cases in which output data is missing. The only restriction is that the number of omissions must be $o(k_n)$.

Our next result is to obtain a pointwise rate of convergence by imposing stationarity on the input sequences in addition to the mentioned necessary assumptions on the system. We will use a simple modification of the algorithm used in Theorem 2 in order to exploit the stationarity of the input. Specifically, instead of searching for nearest neighbors over all strings $\{U_j(L_n)\}$ of length L_n, we now take only the set of *nonoverlapping* strings of length L_n from the past. I.e., we search for nearest neighbors from the set $\{U_j(L_n) : j = iL_n \text{ and } T_n + L_n \leq j \leq n\}$. the rest of the algorithm is the same. With this modification, we obtain the following result, which is proved in the following section.

Theorem 3 *Consider the predictor given by equation (1) where k_n, T_n, L_n are chosen data-independently and the nearest neighbors are selected from the set of non-overlapping strings as described above. Then for any H that satisfies (A1')-(A2'), we have that for any stationary $u_0, u_1, \ldots \in \mathbb{R}$, and any i.i.d. $e_0, e_1, \ldots$ such that $Ee_i = 0$ and $E|e_i|^2 < \infty$,*

$$E|y_n - \hat{y}_n|^2 = O\left[\left(\frac{1}{\log n}\right)^{2\gamma}\right],$$

where $\gamma = \min\{\alpha, -\log \rho\}$.

Notes

- A similar statement can be made for *independent* inputs $u_0, u_1, \ldots$ with a weaker form of (A1') such as in [10].

5   Linear Systems

In this section we consider a special class of systems that satisfy (A1)–(A2), namely the class of stable causal linear time-invariant discrete-time systems. The system model is

$$
\begin{aligned}
y &= h * u \\
z &= y + e
\end{aligned}
\tag{2}
$$

where $*$ is the convolution operator, z is the measured output, e is a noise sequence, and $h \in \ell_1$.

In the following corollary of Theorem 2 we show that every linear system (2) satisfies (A1)–(A2) and we obtain rates of convergence for various special cases.

Corollary 1 *Using the predictor from Theorem 1 we have that for any $u \in \mathcal{B}_r \ell_\infty$, any $h \in \ell_1$, and for any i.i.d. $e_0, e_1, \ldots$ such that $Ee_i = 0$ and $E|e_i|^2 < \infty$,*

$$
\lim_{n \to \infty} E|y_n - \hat{y}_n|^2 = 0
$$

Using modified algorithms we obtain

$$
\frac{1}{N} \sum_{n=1}^{N} E|y_n - \hat{y}_n|^2 = O[f(N)] \to 0,
$$

with rates of convergence as follows:

- *If $h \in \{g \in \ell_1 : g_k = 0 \ \forall k \geq L\}$ with L known then $f(n) = O(n^{-2/(L+2)})$.*

- *If $h \in \{g \in \ell_1 : \exists L, g_k = 0 \ \forall k \geq L\}$ then $f(n) = O(1/\log^2 n)$.*

- *If $h \in \{g \in \ell_1 : |g_k| \leq B\rho^k\}$ then the $f(n) = O(1/\log^2 n)$.*

6   Summary and Remarks

The prediction algorithm introduced in this paper is in fact representative of a class of alternative but similar algorithms. For simplicity we chose to illustrate our ideas using an adaptation of the nearest neighbor algorithm. However, many of the arguments can be extended easily using other techniques from the consistent nonparametric regression literature such as the kernel and partitioning estimators (e.g. [22, 12, 21]). There has also been related work in the information theory literature on universal prediction for arbitrary sequences, although the results are somewhat different in nature than the results presented here. There, the performance is compared to the best predictor from a given class of predictors. In the present paper, we do not a priori restrict the form of the predictor, but instead we make an assumption on the outputs to be predicted, and try to achieve asymptotically optimal prediction when the system generating the outputs is not given. It would be interesting to compare these approaches and/or attempt to obtain results in the spirit of the information theoretic results in the present formulation.

We now make some comments on the relationship between our prediction schemes to system identification. We first point out that the notion of identification of a system generally requires that some topological structure be imposed on the model class. For example, in the context of stable linear systems it is common to consider the systems to be elements of ℓ_1 (e.g. [23]). In addition, it is known that we require sufficiently rich input sequences (e.g., Galois sequences or persistency of excitation conditions) in order to identify the system (e.g., see [23, 18, 17]). In contrast, our prediction scheme does not require any topological structure, works for nonlinear systems, and works for any bounded input sequence. The difference arises because in identification one needs to "explore the entire state space" whereas for prediction the behavior of the system only along the input sequence needs to be known. In a sense, this idea is captured in Lemma 1 which shows that for any bounded input sequence eventually there is sufficient data for prediction. In fact for non-rich input sequences (that don't excite the system in many ways) prediction should be even easier. On the other hand, it is clear that identification is impossible if data is gathered only in a small region or low-dimensional subspace of all possible inputs.

An obvious issue that needs to be addressed in more detail is the relationship between good identification and good prediction. Certainly good identification implies good prediction. For example, from the work in [23] a system estimate $\hat{h}^{(n)}$ of h is constructed whereby

$$\limsup_{n \to \infty} \|\hat{h}^{(n)} - h\|_1 \leq 2\delta$$

in the case in which the noise is uniformly bounded by δ. A predictor can be extracted from this in the obvious way,

$$\hat{y}_n := P_n(\hat{h}^{(n-1)} * u)$$

leading to a prediction error of

$$\begin{aligned}
|\hat{y}_n - y| &= |P_n(\hat{h}^{(n-1)} * u - h * u)| \leq \|\hat{h}^{(n-1)} * u - h * u\|_\infty \\
&\leq \|\hat{h}^{(n-1)} - h\|_1 \|u\|_\infty \leq r\|\hat{h}^{(n-1)} - h\|_1
\end{aligned}$$

if $u \in \mathcal{B}_r \ell_\infty$. This gives

$$\limsup_{n \to \infty} |\hat{y}_n - y| \leq 2r\delta. \tag{3}$$

In Theorem 1 we obtained a bound of δ on the limsup which improves the result in (3). Although good identification does imply good prediction, our prediction scheme is an improvement and works for every bounded input sequence (independent of the bound).

For the converse, we argue that our predictor when subjected to a "rich" input leads to a notion of identification. We first need to impose a topology on the system class. Take a σ-compact subset $(\mathcal{X}, \rho)$ of nonlinear systems that satisfy (A1) and (A2). Define the class of input sequences $\mathcal{U}$ that have the following "richness" property:

- $\mathcal{U}$ is the set of all bounded inputs $u \in \mathcal{B}_r \ell_\infty$ such that for any $G, H \in (\mathcal{X}, \rho)$, we have that

$$\limsup_{n \to \infty} |P_n G(u) - P_n H(u)| \leq \delta$$

  implies that

$$\rho(G, H) \leq \delta$$

Note that in the case of stable linear systems, $\mathcal{U}$ contains the set of Galois sequences. Now if we apply an input belonging to $\mathcal{U}$ then our prediction algorithm is in fact excluding systems that are 2δ away from the true system. We suspect that an Occam's Razor type algorithm similar to that used by Tse et. al. [23] can be used to construct an identifier for these systems. It seems that the class of "rich" inputs leading to good identification are precisely those inputs for which the prediction algorithm has slow convergence properties. The question of finding interesting nonlinear system classes with interesting "rich" input classes remains to be seen.

A somewhat different question involves the connection between this prediction algorithm and adaptive control. At present no claims are made as to how some version of the present algorithm might be used for control, but one might expect that if one can predict well one should also be able to control well. In fact, one might argue that in control (as with prediction) one is interested in the behavior of the system on the current input rather than on all possible inputs as in system identification. However, the central problem in this case is selecting an appropriate input (to satisfy the control objective) from the family of all admissible inputs. It is precisely this selection problem that seems to require some sort of search over the input space which is more like the identification problem. Nevertheless, it may be possible to devise a strategy that alternates or trades off learning phases with control phases, particularly if suitable assumptions restricting the behavior of the system are imposed. It may be very interesting to pursue such directions.

References

[1] T.M. Cover, "Estimation by the nearest neighbor rule," *IEEE Trans. Information Theory*, vol. IT-14, pp. 50-55, Jan. 1968.

[2] M.A. Dahleh, E.D. Sontag, D.N.C. Tse, and J.N. Tsitsiklis, "Worst-case identification of nonlinear fading memory systems," Automatica, vol. 31, pp. 503-508, 1995.

[3] M.A. Dahleh, T. Theodosopoulos, and J.N. Tsitsiklis, "The sample complexity of worst-case identification for f.i.r. linear systems," *Syst. Contr. Lett.*, vol. 20, pp. 157-166, March 1993.

[4] L. Devroye, "Necessary and sufficient conditions for the pointwise convergence of nearest neighbor regression function estimates," *Z. Wahrscheinlichkeitstheorie verw. Gebiete* vol. 61, pp. 467-481, 1982.

[5] R.M. Dudley, *Real Analysis and Probability*, Chapman & Hall, 1989.

[6] J.D. Farmer and J.J. Sidorowich, "Exploiting chaos to predict the future and reduce noise," *Evolution, Learning, and Cognition*, pp. 265-289, World Scientific, Singapore, 1988.

[7] M. Feder, N. Merhav, and M. Gutman, "Universal prediction of individual sequences," *IEEE Trans. Information Theory*, vol. 38, pp. 1258-1270, July 1992.

[8] W. Greblicki, "Nonparametric identification of Wiener systems by orthogonal series," *IEEE Trans. Automatic Control*, vol. 30, pp. 2077-2086, 1994.

[9] W. Greblicki and M. Pawlak, "Dynamic system identification with order statistics," *IEEE Trans. Information Theory*, vol. 40, pp. 1474-1489, 1994.

[10] L. Györfi, "The rate of convergence of k_n-NN regression estimates and classification rules," *IEEE Trans. Information Theory*, vol. IT-27, pp. 362-364, May 1981.

[11] A.J. Helmicki, C.A. Jacobson, C.N. Nett, "Control Oriented System Identification: A Worst-Case/Deterministic Approach in $\mathcal{H}_\infty$," *IEEE Trans. Automatic Control*, vol. 36, Oct. 1991.

[12] A. Krzyżak, "On estimation of a class of nonlinear systems by the kernel regression estimate," *IEEE Trans. Information Theory*, vol. 36, pp. 141-152, 1990.

[13] A. Krzyżak, "Identification of nonlinear systems by recursive kernel regression estimates," *Int. J. Systems Sci.*, vol. 24, pp. 577-598, 1993.

[14] S.R. Kulkarni, "Data-dependent nearest neighbor and kernel estimators consistent for arbitrary processes," preprint.

[15] S.R. Kulkarni and S.E. Posner, "Rates of convergence of nearest neighbor estimation under arbitrary sampling," *IEEE Trans. Information Theory*, pp. 1028-1039, July 1995.

[16] S.R. Kulkarni and S.E. Posner, "Nonparametric output prediction for nonlinear fading memory systems," to appear *IEEE Trans. Automatic Control*, Nov., 1998.

[17] L. Ljung, *System Identification: Theory for the User*, Prentice-Hall, 1987.

[18] P.M. Mäkilä, "Robust identification and Galois sequences," *Int. J. Contr.*, vol. 54, pp. 1189-1200, 1991.

[19] G. Morvai, S. Yakowitz, and L. Györfi, "Nonparametric inferences for ergodic stationary time series," preprint.

[20] K. Poolla and A. Tikku, "On the time complexity of worst-case system identification," *IEEE Trans. Automatic Control*, vol. 39, pp. 944-950, May 1994.

[21] S.E. Posner, "Nonparametric estimation, regression, and prediction under minimal regularity conditions," Ph.D. Thesis, Department of Electrical Engineering, Princeton University, 1995.

[22] C.J. Stone, "Consistent nonparametric regression," *Ann. Stat.*, vol. 8, pp. 1348-1360, 1977.

[23] D.N.C. Tse, M.A. Dahleh, J.N. Tsitsiklis, "Optimal asymptotic identification under bounded disturbances," *IEEE Trans. Automatic Control*, vol. 38, pp. 1176-1190, Aug. 1993.

Model Reduction for Classes of Uncertain, Multi-Dimensional, Parameter Varying and Non-Linear Systems

K. Glover, P.J. Goddard and Y-C. Chu*

Abstract

This paper surveys recently proposed approaches for the model reduction of certain classes of uncertain, multi-dimensional, parameter varying and non-linear systems. It is shown that each of these systems may be written using a similar formulation. Balanced truncation model reduction, based on the solution of two Linear Matrix Inequalities (LMI's) are discussed for each class of system and the similarities (and differences) highlighted.

This paper is dedicated with warm friendship to Bruce Francis and Mathukumalli Vidyasagar on the occasion of their 50th birthdays.

1 Introduction

A central theme of model-based control theory is the unavoidable trade-off between model accuracy and model simplicity. For although a model must be sufficiently accurate to capture the salient features of a plant, it must remain simple enough to understand and easily manipulate.

There are a wealth of techniques which may be applied for the approximation of finite dimensional, linear time invariant (LTI) models. Arguably the best known of these techniques being balanced truncation, first introduced by Moore ([16]). Balanced truncation is based on the derivation of a similarity transformation for LTI stable systems such that the states of the transformed system are equally as controllable as observable. Since those states which are most weakly controllable and observable contribute least to the system's input-output response, they may be truncated without undue loss of model accuracy. Subsequently Enns ([9]) derived an $\mathcal{H}_\infty$ norm error bound for balanced truncation, and also considered frequency weighting. Other popular methods for approximation of LTI systems include optimal Hankel norm approximation

---

*Department of Engineering, University of Cambridge, Cambridge, CB2 1PZ, UK

([11]), balanced stochastic truncation ([8]) and relative error methods ([23]). These methods also have associated $\mathcal{H}_\infty$ norm error bounds ([11] [13] and [23] respectively).

As our abilities to model systems as being more complex than linear and time-invariant have evolved, so have techniques for their analysis and approximation. In this paper we will review techniques for the approximation of several classes of uncertain, multi-dimensional, parameter varying and nonlinear systems, using methods that are various generalisations of balancing adapted to each situation.

Each of the approaches to be discussed may be formulated in terms of the general interconnection structure shown in Figure 1. Throughout, $M = \begin{bmatrix} A & B \\ C & D \end{bmatrix} \in \mathbb{R}^{(n+p)\times(n+m)}$ is assumed to be a constant matrix, and is interpreted as representing what is known about the system. On the other hand Δ may take on many forms, each of which leads to a different interpretation of the interconnection structure.

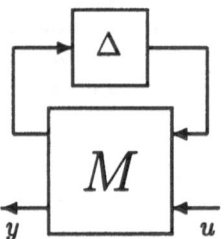

Figure 1: General LFT Interconnection

For Δ in an appropriate set, the input-output mapping for this system is given by

$$y = (\Delta \star M)u$$

where the Redheffer star product is defined as

$$\Delta \star M = D + C\Delta(I - A\Delta)^{-1}B,$$

whenever the inverse is well-defined. Systems of this form will be denoted by the pair (Δ, M).

The most general form of Δ to be considered, is Δ lying in the set,

$$\Delta = \{\mathrm{diag}[\delta_1 I_{n_1}, \dots, \delta_S I_{n_S}, \Delta_1, \dots, \Delta_T]\}, \tag{1}$$

where δ_i and Δ_i are time varying dynamic operators ([17]). For analysis purposes, it is convenient for Δ to lie in the norm-bounded subset of Δ,

$$\mathbf{B_\Delta} = \{\Delta \in \Delta : \|\Delta\|_{i,2} \leq 1\}$$

where $\| \cdot \|_{i,2}$ denotes the induced 2-norm.

For various classes of Δ, we are interested in determining the existence of reduced order systems $(\hat{\Delta}, \hat{M})$, and a bound for the error

$$\|(\Delta \star M) - (\hat{\Delta} \star \hat{M})\|_{i,2} \leq \epsilon. \tag{2}$$

The exact meaning of equation (2), and the notation used, will be developed in subsequent sections.

2 Uncertain and Multi-Dimensional Systems

Perhaps the simplest interpretation of (Δ, M) comes from assuming $\Delta = \delta_1 I = z^{-1} I : l_{2e} \mapsto l_{2e}$, the discrete time transform variable or backward shift operator. Then (Δ, M) may be viewed as a one-dimensional (1D), discrete time system. Several well known methods for state-order reduction of this type of LTI system were discussed in the introduction.

Of more interest for the purposes of this paper are uncertain 1D models. These may be viewed as a (Δ, M) interconnection with $\delta_1 = z^{-1}$ and the remaining elements of Δ as norm bounded repeated scalar and full block perturbations. Model reduction in this context may be aimed at simplifying the uncertainty description in addition to the state-dimension. This is a much more subtle issue than state-order reduction alone.

Alternatively, multi-dimensional (MD) systems, ([14]), may also be written in this framework by viewing the δ_i as different transform variables. In this case, model reduction means reducing the state-order, and as in the 1D case a model which may be reduced without error is called non-minimal.

Balanced truncation reduction for MD systems and 1D uncertain systems where Δ is purely diagonal, has been studied by Wang et al. [20] and Beck et al. [3].

2.1 Balancing (Δ, M)

For 1D systems with no uncertainty, the role of Lyapunov equalities and associated controllability and observability Gramians in determining a unique balanced realisation is well developed [16, 9, 11]. Further, the role of Hankel singular values, $\sigma_1 \geq \sigma_2 \geq \cdots \geq \sigma_n \geq 0$, in determining the *twice the sum of the tail* error bound [9, 11],

$$\|(z^{-1}I_n \star M) - (z^{-1}I_k \star \hat{M})\|_\infty \leq 2 \sum_{i=k+1}^{n} \sigma_i \tag{3}$$

is well understood.

These results have been extended by Wang et al. [20] and Beck et al. [3] for the balancing and model reduction of 1D uncertain systems (with only repeated scalar uncertainty) and MD systems. Assume

$$\Delta = \text{diag}[\delta_1 I_{n_1}, \dots, \delta_S I_{n_S}] : l_{2e} \mapsto l_{2e}, \tag{4}$$

where δ_i are linear time varying uncertain operators and M is partitioned compatibly with Δ, such that,

$$A = \begin{bmatrix} A_{11} & \cdots & A_{1S} \\ \vdots & \ddots & \vdots \\ A_{S1} & \cdots & A_{SS} \end{bmatrix}; B = \begin{bmatrix} B_1 \\ \vdots \\ B_S \end{bmatrix}; C = \begin{bmatrix} C_1 & \cdots & C_S \end{bmatrix}. \tag{5}$$

If the system is stable for all Δ in $\mathbf{B}_\Delta$ then it can be shown [3, 15, 19] there will exist matrices, $X = \text{diag}[X_1, X_2, \dots, X_S] \geq 0$ and $Y = \text{diag}[Y_1, Y_2, \dots, Y_S] \geq 0$, both partitioned compatibly with A, satisfying the Linear Matrix Inequalities:

$$A^T Y A - Y + C^T C \leq 0 \quad \text{and} \quad A X A^T - X + B B^T \leq 0 \tag{6}$$

The matrices X and Y will be called *generalised Gramians* but note that they are not unique and in contrast to the 1-D case there will not generally exist a minimal solution to the LMI's. A matrix of the form $T = \text{diag}[T_1, T_2, \dots, T_S]$ that is nonsingular and partitioned compatibly with Δ will commute with Δ and hence $T^{-1}\Delta T = \Delta$ and T can be used as a state transformation matrix giving,

$$\begin{bmatrix} A & B \\ C & D \end{bmatrix} \rightarrow \begin{bmatrix} T^{-1}AT & T^{-1}B \\ CT & D \end{bmatrix}.$$

Given X and Y satisfying (6) then as in the 1-D case a T is easily found to balance the generalised Gramians. An uncertain system, (Δ, M), will be called balanced if there exist X and Y satisfying (6) such that,

$$X = Y = \Sigma = \text{diag}[\Sigma_1, \dots, \Sigma_S]$$

where $\Sigma_i = \text{diag}[\sigma_{i1} I_{s_{i1}}, \dots, \sigma_{it_i} I_{s_{it_i}}] > 0; \sigma_{i1} \geq \cdots \geq \sigma_{it_i}$ and $\dim(\Sigma_i) = \dim(A_{ii})$ is denoted by $n_i = \sum_{j=1}^{t_i} s_{ij}$.

Beck [2] shows that necessary and sufficient conditions for reducibility of the state dimension of (Δ, M) (with Δ as in (4)) is that there exist singular generalised Gramians. This reducibility result also holds when δ_i are non-commuting indeterminates, however in the MD systems case the δ_i will commute and the theory of minimal realisation is substantially more involved.

Given a balanced realisation, it is possible to use truncation to reduce the model dimension. As with the 1D LTI case, truncation of those rows and columns of Δ and M associated with small generalised singular values results in a small reduction error. However, since generalised singular values are not unique, great care must be taken to ensure that appropriate X and Y have been selected.

2.2 Truncation and Error Bounds

In order to derive the model reduction error for balanced truncation of a 1D uncertain or MD system, M and Δ must be partitioned to separate the sub-blocks which are to be removed. Partition each block of Σ as

$$\Sigma_i = \text{diag}[\hat{\Sigma}_i, \Sigma_{i2}]$$

for $i = 1, \ldots, S$, where

$$\hat{\Sigma}_i = \text{diag}[\sigma_{i1} I_{s_{i1}}, \cdots, \sigma_{ik_i} I_{s_{ik_i}}], \quad k_i \leq t_i$$

and

$$\Sigma_{i2} = \text{diag}[\sigma_{i(k_i+1)} I_{s_{i(k_i+1)}}, \cdots, \sigma_{it_i} I_{s_{it_i}}].$$

Truncate Σ_{i2}, $i = 1, 2, \ldots, S$ and the parameter matrices,

$$A_{ij} = \begin{bmatrix} \hat{A}_{ij} & A_{ij12} \\ A_{ij21} & A_{ij22} \end{bmatrix}, B_i = \begin{bmatrix} \hat{B}_i \\ B_{i2} \end{bmatrix} \quad \text{and} \quad C_j = \begin{bmatrix} \hat{C}_j & C_{j2} \end{bmatrix}$$

to $\hat{A}_{ij}$, $\hat{B}_i$ and $\hat{C}_j$ for each $i, j = 1, \ldots, S$. The resulting system is then,

$$\hat{M} = \begin{bmatrix} \hat{A} & \hat{B} \\ \hat{C} & \hat{D} \end{bmatrix} = \begin{bmatrix} \hat{A}_{11} & \cdots & \hat{A}_{1S} & \hat{B}_1 \\ \vdots & \ddots & \vdots & \vdots \\ \hat{A}_{S1} & \cdots & \hat{A}_{SS} & \hat{B}_S \\ \hat{C}_1 & \cdots & \hat{C}_S & D \end{bmatrix}$$

with the uncertainty structure

$$\hat{\Delta} = \text{diag}[\delta_1 I_{\hat{n}_1}, \cdots, \delta_S I_{\hat{n}_S}]$$

where $\hat{n}_i = \sum_{j=1}^{k_i} s_{ij}$. As in the 1D case, truncation of a balanced system in this way, results in a reduced order system that will be stable for all $\hat{\Delta}$ and is balanced.

Theorem 2.1 ([20]) *Suppose $(\hat{\Delta}, \hat{M})$ is the reduced model obtained from the balanced truncation of (Δ, M). Then*

$$\sup_{\Delta \in \mathbf{B}_\Delta} \|(\Delta \star M) - (\hat{\Delta} \star \hat{M})\|_{i,2} \leq 2 \sum_{i=1}^{S} \sum_{j=k_i+1}^{t_i} \sigma_{ij}. \tag{7}$$

The proof of this result proceeds by forming the state space of the error system and then establishes a state coordinate transformation that makes this realisation a contraction.

Theorem 2.1 gives an upper bound on the induced error that can be arbitrarily conservative because the solutions to the LMI's in (6) are not further specified. In [3] necessary and sufficient conditions are obtained for the reduction error to be less than ϵ for the case when Δ is as in (4), but these conditions are not convex. They do however lead to an alternative reduction method which is shown to satisfy (7) divided by a factor of 2.

3   LPV Systems

Linear parameter-varying systems are a special class of time-varying system where time dependence enters the state equations through one or more exogenous parameters. For a comprehensive discussion of LPV systems and their properties see [4] and the references therein. Conventionally, a continuous time LPV system is represented by the state-space realisation,

$$
\begin{aligned}
\dot{x}(t) &= A(\rho(t))x(t) + B(\rho(t))u(t) \\
y(t) &= C(\rho(t))x(t) + D(\rho(t))u(t)
\end{aligned}
\tag{8}
$$

where $A : \mathbf{R}^s \mapsto \mathbf{R}^{n \times n}, B : \mathbf{R}^s \mapsto \mathbf{R}^{n \times m}, C : \mathbf{R}^s \mapsto \mathbf{R}^{p \times n}$ and $D : \mathbf{R}^s \mapsto \mathbf{R}^{p \times m}$ are continuous functions of the parameter vector $\rho \in F_\rho$. F_ρ defines the set of feasible parameter trajectories which are a subset of all piecewise continuous functions according to

$$
F_\rho \triangleq \{\rho(t) : \mathbf{R} \mapsto \mathbf{R}^s, \rho_{i_{\min}} \leq \rho_i \leq \rho_{i_{\max}}, i = 1, 2 \ldots s\}.
$$

which ensures that for each $\rho(t) \in F_\rho$ the system's transition matrix is unique and continuous.

Note that there is no requirement for the parameter dependence exhibited by the state-space matrices to be linear. However, as in [18], it is often possible to suitably extract the parameter dependence and form a $(\Delta, M) : \mathcal{L}_{2e} \mapsto \mathcal{L}_{2e}$ interconnection of the form of Figure 1 (e.g. when (A, B, C, D) are rational functions of ρ). In this case the model reduction methods outlined in the previous section for uncertain and multi-dimensional systems may be applied to perform state-order reduction and/or modify the parameter dependence. However there may be unnecessary conservatism since $\rho_i(t)$ are scalar time varying gains and not dynamic systems.

An alternative approach, and that considered in the next section, is to define the system $(\Delta, M(\rho))$ where,

$$
M(\rho) = \begin{bmatrix} A(\rho) & B(\rho) \\ C(\rho) & D(\rho) \end{bmatrix}
\tag{9}
$$

and $\Delta = \frac{1}{s}I : \mathcal{L}_{2e} \mapsto \mathcal{L}_{2e}$.

3.1   Balancing

State-order reduction for general systems of the form of equation (9) has been studied in [12, 21, 22]. These approaches rely, as an initial step, on transforming the state space to one which is balanced. States which contribute little to the input-output response are then truncated in an identical manner to LTI balanced truncation.

If a system given by (8) is quadratically stable then there exist X and Y independent of ρ satisfying the LMI's

$$A^T(\rho)Y + YA(\rho) + C^T(\rho)C(\rho) \leq 0 \quad \forall \rho \in F_\rho \qquad (10)$$
$$A(\rho)X + XA^T(\rho) + B(\rho)B^T(\rho) \leq 0 \quad \forall \rho \in F_\rho. \qquad (11)$$

This realisation can then be balanced as in the 1-D case when in these coordinates,

$$X = Y = \Sigma = \text{diag}\,[\sigma_1, \cdots, \sigma_n]$$

where $\sigma_i = \sqrt{\lambda_i(XY)}$.

3.2 Truncation and Error Bounds

The balanced realisation introduced in the preceding section leads immediately to a balanced truncation procedure for parameter-varying systems.

Lemma 3.1 ([12]) *Assume $(\Delta, M_b(\rho))$ is an n-state, quadratically stable, balanced LPV system with $M_b(\rho)$ partitioned as follows*

$$M_b(\rho) = \begin{bmatrix} A_{11}(\rho) & A_{12}(\rho) & B_1(\rho) \\ A_{21}(\rho) & A_{22}(\rho) & B_2(\rho) \\ C_1(\rho) & C_2(\rho) & D(\rho) \end{bmatrix},$$

where $A_{11} \in \mathbf{R}^{k \times k}, A_{12} \in \mathbf{R}^{k \times (n-k)}, A_{21} \in \mathbf{R}^{(n-k) \times k}, A_{22} \in \mathbf{R}^{(n-k) \times (n-k)}, B_1 \in \mathbf{R}^{k \times m}, B_2 \in \mathbf{R}^{(n-k) \times m}, C_1 \in \mathbf{R}^{p \times k}$ and $C_2 \in \mathbf{R}^{p \times (n-k)}$. Then $(\hat{\Delta}, \hat{M}(\rho))$ where

$$\hat{M}(\rho) = \begin{bmatrix} A_{11}(\rho) & B_1(\rho) \\ C_1(\rho) & D(\rho) \end{bmatrix}$$

is a k-state, quadratically stable, balanced approximation, and

$$\|(\frac{1}{s}I_n \star M_b(\rho)) - (\frac{1}{s}I_k \star \hat{M}(\rho))\|_{i,2} \leq 2 \sum_{i=k+1}^{n} \sigma_i.$$

This gives a generalisation of the 1-D result and methods for calculating X and Y have been studied in [21] that are computationally straightforward when M is a linear function of ρ and hence only the extremal plant models need to be considered. However in general ensuring that (10–11) hold for all ρ may require many LMI constraints over a grid for ρ. Also various slightly ad hoc methods can be used for trying to ensure that the neglected σ_i are small.

These results may be very conservative if it is known that $|\dot{\rho}(t)|$ is bounded. In this case it is possible to consider ρ-dependent solutions to the LMI's,

$$\dot{Y} + A^T(\rho)Y(\rho) + Y(\rho)A(\rho) + C^T(\rho)C(\rho) \leq 0 \quad \forall \rho \in F_\rho, |\dot{\rho}| < \dot{\rho}_{imax}$$
$$-\dot{X} + A(\rho)X(\rho) + X(\rho)A(\rho) + B(\rho)B^T(\rho) \leq 0 \quad \forall \rho \in F_\rho, |\dot{\rho}| < \dot{\rho}_{imax}$$

These solutions can then be balanced with a ρ-dependent state transformation although care must be taken to ensure that such a transformation is a differentiable function of ρ. The resulting transformed state space will now be a function of ρ and $\dot{\rho}$ (although when ρ is scalar Wood [21] shows in principle how to remove this dependency on $\dot{\rho}$). Truncation error bounds are apparently no longer valid [21] however the matrices $X(\rho)$ and $Y(\rho)$ may be substantially smaller than when they are required to be independent of ρ.

In [12, 22], Goddard et al. propose an alternative approach for model reduction of LPV systems which is a generalisation of the LTI optimal Hankel norm approximation approach. However, unlike its LTI counterpart, an interpretation of this scheme being optimal in any sense is no longer clear.

4   Non-Linear Systems

Consider the class of discrete time non-linear systems described by (Δ, M) where

$$M = \begin{bmatrix} A & B \\ C & D \end{bmatrix}, \quad \Delta = \mathrm{diag}[\phi, \phi, \dots, \phi] \circ z^{-1} I_n, \tag{12}$$

$\circ$ denotes the composition and $\phi : \mathrm{R} \mapsto \mathrm{R}$ is a non-linear function which is possibly uncertain but known to be odd and 1-Lipschitz:

$$\phi(-s) = -\phi(s), \quad |\phi(s) - \phi(t)| \leq |s - t|. \tag{13}$$

This model is similar to certain recurrent neural networks as for example studied in [1], and it can be shown that certain universality results hold and also that in general the only state space coordinate transformations giving equivalent systems are permutations and sign changes. Chu et al. [5, 6, 7] show that by using Lyapunov functions and storage functions that are diagonally dominant rather accurate (but still conservative) stability tests and induced 2-norm calculations result. Such diagonally dominant matrices exploit the assumptions on the non-linear function, ϕ.

4.1   Reduction and Error Bounds

The lack of state space realisations in the non-linear case comes from the fact that Δ in (12) does not generally commute with a matrix T, even if T is diagonal. That is, $T^{-1} \circ \Delta \circ T \neq \Delta$. So in contrast to the 1D uncertain systems, the non-linear systems cannot be balanced via similarity transformations and then truncated even when the solutions X, Y to the Lyapunov inequalities exist. However, stability and induced norm results are obtained using the property that $T^{-1} \circ \Delta \circ T$ is a contraction if $(TT^T)^{-1}$ is diagonally dominant (need not be diagonal). Hence a reduced order model with an error bound can be derived with the same spirit as the balanced truncation of linear systems.

Lemma 4.1 ([7]) *Assume (Δ, M) is an n-state system given by (12–13). If there exist $X > 0$ and $Y > 0$ satisfying*

$$A^T Y A - Y + C^T C / \sigma \leq 0 \quad \text{and} \quad A X A^T - X + B B^T / \sigma \leq 0 \qquad (14)$$

such that

$$\begin{bmatrix} Y + X^{-1} & Y - X^{-1} \\ Y - X^{-1} & Y + X^{-1} \end{bmatrix} \qquad (15)$$

is diagonally dominant and $n - \hat{n}$ rows and columns of $Y - X^{-1}$ are zero, then there exists a $\hat{n}$-state model $(\hat{\Delta}, \hat{M})$ with

$$\hat{\Delta} = diag[\phi, \phi, \dots, \phi] \circ z^{-1} I_{\hat{n}} \qquad (16)$$

such that

$$\|(\Delta \star M) - (\hat{\Delta} \star \hat{M})\|_{i,2} \leq 2\sigma.$$

The reduced order model can be obtained as follows. Assume without loss of generality that the last $n - \hat{n}$ rows and columns of $Y - X^{-1}$ are zero. The matrix T that diagonalizes XY can be chosen such that

$$T = \begin{bmatrix} \hat{T} & 0 \\ T_{21} & T_{22} \end{bmatrix}$$

where $\hat{T} \in \mathrm{R}^{\hat{n} \times \hat{n}}$. Let

$$M_b = \begin{bmatrix} A_b & B_b \\ C_b & D \end{bmatrix} = \begin{bmatrix} T^{-1} A T & T^{-1} B \\ C T & D \end{bmatrix}$$

and truncate A_b, B_b C_b to $\hat{A}_b \in \mathrm{R}^{\hat{n} \times \hat{n}}$, $\hat{B}_b \in \mathrm{R}^{\hat{n} \times m}$ and $\hat{C}_b \in \mathrm{R}^{p \times \hat{n}}$. The reduced order model $(\hat{\Delta}, \hat{M})$ is then given by (16) and

$$\hat{M} = \begin{bmatrix} \hat{T} \hat{A}_b \hat{T}^{-1} & \hat{T} \hat{B}_b \\ \hat{C}_b \hat{T}^{-1} & D \end{bmatrix}. \qquad (17)$$

The non-linear system (Δ, M) is not equivalent to the associated balanced system (Δ, M_b) and therefore cannot be approximated by simply truncating the latter. It is also shown [7] that if there exists a non-linear system $(\hat{\Delta}, \hat{M})$ such that $\|(\Delta \star M) - (\hat{\Delta} \star \hat{M})\|_{i,2} < \sigma$ is provable by a diagonally dominant storage function then there will exist $X > 0$, $Y > 0$ solving the Lyapunov inequalities (14) such that (15) is diagonally dominant and rank$(Y - X^{-1}) \leq \hat{n}$. It is seen that Lemma 4.1 further restricts the null space to a coordinate subspace.

As in the 1-D uncertain case, there do not generally exist a minimal solution to the Lyapunov inequalities and the solutions X, Y are not unique. Minimising σ subject to the conditions in Lemma 4.1 is a non-convex problem, though it can be recast as a LMI problem with an additional deficient-rank constraint [5, 7] which is widely studied in the literature but no completely satisfactory solution is yet available. An alternative approach is proposed in [5, 7] which is slightly ad hoc but seems to be more practical and has a Hankel norm interpretation.

With the diagonally dominant matrices replaced by block-diagonal diagonally dominant matrices it is easily seen that the results generalise to (Δ, M) with

$$\Delta = \text{diag}[\phi_1, \phi_1, \ldots, \phi_1, \phi_2, \phi_2, \ldots, \phi_2, \ldots, \phi_S, \phi_S, \ldots, \phi_S] \circ z^{-1} I_{n_1 + n_2 + \cdots + n_S}. \tag{18}$$

The procedure of matrix partition is similar to the 1-D uncertain case in Section 2.

5 Conclusion

In this paper, approximation methods for certain classes of uncertain, multidimensional, parameter varying and non-linear systems have been proposed. These methods are based on the balanced truncation approach first introduced by Moore [16] for linear, time-invariant systems. However, each method has its idiosyncrasies and care is required in their application.

References

[1] F. Albertini and E. D. Sontag, "For neural networks, function determines form," *Neural Networks*, vol. 6, pp. 975–990, 1993.

[2] C. Beck. "Model Reduction and Minimality for Uncertain Systems", *PhD Dissertation*, California Institute of Technology, 1995.

[3] C. Beck, J.C. Doyle and K. Glover. "Model Reduction of Multi-Dimensional and Uncertain Systems", *IEEE Trans. Auto. Cont.*, Vol. AC-41, pp. 1466-1477, 1996.

[4] S. Boyd, L. El Ghaoui , E. Feron and V. Balakrishnan. "Linear Matrix Inequalities in System and Control Theory", SIAM Studies in Applied Mathematics, 1994.

[5] Y-C. Chu. "Control of Systems with Repeated Scalar Nonlinearities", *PhD Dissertation*, University of Cambridge, 1996.

[6] Y-C. Chu and K. Glover. "Bounds of the Induced Norm and Model Reduction Errors for a Class of Nonlinear Systems", *Proceedings of the 35th IEEE CDC*, Kobe, Japan, pp. 4288-4293, 1996.

[7] Y-C. Chu and K. Glover. "Bounds of the Induced Norm and Model Reduction Errors for Systems with Repeated Scalar Nonlinearities", To appear *IEEE Trans. Auto. Cont.*, 1998.

[8] U.B. Desai and D. Pal. "A Transformation Approach to Stochastic Model Reduction", *IEEE Trans. Auto. Cont.*, AC-29, pp. 1097-1100, 1984

[9] D. Enns. "Model Reduction for Control Systems Design", *PhD Dissertation*, Stanford University, 1984.

[10] A. Feintuch, "The Gap Metric for Time-Varying Systems", *Systems and Control Letters*, Vol. 16, pp. 277-279, 1991.

[11] K. Glover. " All Optimal Hankel Norm Approximations of Linear Multivariable Systems and their L_∞ error bounds", *Int. J. Cont.*, Vol. AC-39, No. 6, pp. 1115-1193, 1984.

[12] P.J. Goddard. "Performance-Preserving Controller Approximation", *PhD Dissertation*, University of Cambridge, 1995.

[13] M. Green. "A relative error bound for balanced stochastic truncation," *IEEE Trans. Auto. Cont.*, AC-33, pp. 961-965, 1988.

[14] IEEE Proceedings, Special Issue on Multi-Dimensional Systems, 1977.

[15] A. Megretskii. "Necessary and sufficient conditions of stability: A multiloop generalisation of the circle criterion", *IEEE Trans. Auto. Cont.*, AC-38, pp. 753-756, 1993.

[16] B.C. Moore. "Principal Component Analysis of Linear Systems: Controllability, Observability and Model Reduction", *IEEE Tans. Auto. Cont.*, AC-26, pp. 17-32, 1981

[17] A. Packard and J. Doyle "The Complex Structured Singular Value", *Automatica*, Vol. 29, No. 1, 1993.

[18] A. Packard. "Gain Scheduling via Linear Fractional Transformations", *Systems and Control Letters*, Vol. 22, pp. 79-92, 1994.

[19] J. Shamma. "Robust Stability with Time-Varying Structured Uncertainty", *IEEE Trans. Auto. Cont.*, AC-39, pp. 714-724, 1994.

[20] W. Wang, J.C. Doyle, C. Beck and K. Glover. "Model Reduction of LFT Systems", *Proceedings of the 30th IEEE CDC*, Brighton, 1991.

[21] G.D. Wood. "Control of Parameter-Dependent Mechanical Systems", *PhD Dissertation*, University of Cambridge, 1996.

[22] G.D. Wood, P.J. Goddard and K. Glover. "Approximation of Linear Parameter-Varying Systems", *Proceedings of the 35th IEEE CDC*, Kobe, 1996.

[23] K. Zhou, J.C. Doyle and K. Glover. Robust and Optimal Control, Prentice-Hall, 1995.

Information and Control: Witsenhausen Revisited

Sanjoy Mitter   Anant Sahai *†

Dedicated to Bruce Francis and Mathukumalli Vidyasgar
on the occasion of their fiftieth birthday.

Abstract

The role of information in the context of control is a deep issue. To get at this, we review Witsenhausen's notions of *information patterns* for control problems. While staying in that basic framework, we then use ideas from traditional information theory as we re-examine Witsenhausen's famous "counterexample". In the process, we construct a family of nonlinear "quantizing" control laws that can perform infinitely better than the best linear policies.

1 Introduction

In traditional information theory, a technical notion of information is developed that is independent from the actual use of that information. Aside from its considerable aesthetic appeal, this body of ideas has proven itself to be quite useful in the context of information transmission. However, fundamental to most of the results in information theory is the use of long block lengths and letting sequence lengths tend to infinity as a way of getting the laws of large numbers to work to reduce uncertainty. In a control context, the focus is on the present. While there is a sense in which all of feedback control is about trying to reduce uncertainty, a control action must be applied now and we can not afford to wait forever.

In this report, we will attempt to get a handle on the role of information in control by revisiting two classic papers. The first of these is Witsenhausen's 1971 survey paper [4] on the "Separation of Estimation and Control for Discrete Time Systems." Here, we will give the essentials of Witsenhausen's

*This research supported by U.S. Army Grant PAAL03-92-G-0115, Center for Intelligent Control Systems.

†Department of Electrical Engineering and Computer Science and Laboratory for Information and Decision Systems, Massachusetts Institute of Technology. mitter@lids.mit.edu, sahai@mit.edu

framework for talking about stochastic control problems. The key idea is that of *information patterns* — a formal way of talking about the issue of "who knows what and when do they know it." Using this, we will restate his main assertions on the various forms of separation between estimation and control. Though the language is general, we will quickly find ourselves talking about linear systems with quadratic costs and Gaussian distributions for primitive random variables — the LQG problem.

With the basics behind us, we next consider Witsenhausen's 1968 "Counterexample In Stochastic Optimum Control" [3] which shows how important the information pattern really is to the control problem. It details a deceptively simple 2-stage LQG problem and shows that when you restrict to memoryless control, affine[1] controllers are no longer sufficient to minimize cost. The paper does this by computing the best affine controller and then exhibiting a nonlinear control law which does better.

We then present a simpler family of nonlinear control laws and use them to get something much stronger — a demonstration that the ratio of the cost of the best affine controller and a nonlinear controller can go to infinity! Then, we try to use ideas from information theory to give some intuition as to why the affine controllers are suboptimal. At its heart, the problem seems to boil down to one of communication between stages 1 and 2. We argue that the restriction to affine controllers is suboptimal because it forces a tension between the complexity of the message and the reliability of its transmission. We show how the nonlinear controller is able to circumvent this tension, achieving better performance.

2   Separation of Estimation and Control

In Witsenhausen's classic survey paper [4], he sets out to elucidate the relationships between estimation and control for discrete time, Bayesian[2] systems. The fundamental issue stems from the distinction between the control *laws* and the actual realizations of the control *variables* applied to the system. The designer chooses the *laws* to fulfill some objective, and until that choice is made, the control *variables* are still "random variables to be of yet uncertain status."

2.1   Problem Framework

Witsenhausen considers a general finite-horizon distributed discrete time control problem. Time goes from 1 to T, there are M observation posts[3] , and K control stations[4] The causal sequence is as follows:

---

[1] Linear plus constant
[2] All "uncertainty" in the system is modeled probabilistically
[3] For example, consider geographically distributed sensors
[4] These usually represent distributed controllers

1. Generation of random initial state x_0.

2. Observations of outputs $y_1^1, \cdots, y_1^M = (g_1^1(x_0, w_1^1), \cdots, g_1^M(x_0, w_1^M))$

3. Application of controls $u_1^1, \cdots, u_1^K$

4. Transition to state $x_1 = f_1(x_0, v_1, u_1^1, \cdots, u_1^K)$

and then this continues until the final state x_T is reached.

The uncertainty in the system is modeled by a basic set of independent primitive random variables: $x_0; v_t, w_t^m (t = 1, \cdots, T; m = 1, \cdots, M)$. The v_t enter into the state transition functions f_t and the w_t^m into the observation functions g_t^m in the obvious ways.

Finally, the preferences between outcomes are expressed consistently through an additive cost function on the state and the controls: $\sum_{t=1}^{T} h_t(x_t, u_t^1, \cdots, u_t^K)$. The goal of the designer is to pick a *design* γ specifying control *laws* γ_t^k that select the u_t^k to minimize the expected cost. Furthermore, once all the γ_t^k are selected, all the variables in the closed loop system become well defined random variables. More technically, given a complete design γ and a pair of sets of values for some arbitrary sets of the output and control variables, Y and U, we have a clearly defined σ-field $\mathcal{F}(Y, U; \gamma)$ in probability space and thus conditional distributions[5] for all the variables in the system[6].

2.2   Information Patterns

As stated above, the problem is still incompletely specified. We need to know the sets from which we are allowed to pick the functions γ_t^k. Stated informally, the key questions are "who knows what when" and "what are they allowed to do with that information?". To formalize the first of these questions, the notion of *information pattern* is defined. This assigns to every control variable u_t^k, two sets $Y_{t,k}$ and $U_{t,k}$ of pairs of indices specifying which observation variables y_τ^μ and control variables u_θ^κ the control law γ_t^k has access to[7]. Generally, no restriction is put on the functional form or range of γ_t^k, except the trivial one of saying that it should be measurable over the σ-field generated by its arguments. However, sometimes it is interesting to restrict attention to jointly affine γ_t^k.

For the idea of *information pattern* to be useful, we need a notion of equivalence over it. So, patterns $(Y_{t,k}, U_{t,k})$ and $(\tilde{Y}_{t,k}, \tilde{U}_{t,k})$ are *equivalent* if for any design γ feasible with the first, there is a design $\tilde{\gamma}$ feasible with the second

---

[5] The underlying probability space and measure are determined by the primitive random variables.

[6] For example, the conditional probability $P(y_3^2 \in [-1, 1] | y_2^4 = 7, y_3^3 = 5, u_4^2 = 0.5, \gamma)$ should be defined and make sense

[7] To be precise, γ_t^k takes as arguments all the y_τ^μ and u_θ^κ where $(\tau, \mu) \in Y_{t,k}$ and $(\theta, \kappa) \in U_{t,k}$

such that every system variable agrees under the two designs almost surely.[8]
Witsenhausen next defines some classifications of *information patterns*. A pattern is said to have *perfect recall* if $Y_{t,k} \subseteq Y_{t+1,k}$ and $U_{t,k} \subseteq U_{t+1,k}$. A pattern is said to be *classical* if it has *perfect recall* and moreover $Y_{t,k}$ and $U_{t,k}$ are independent of k.[9] We define two related terms that will also be useful. A pattern is said to be *perfectly classical* if every station has knowledge of all past outputs and controls. For the common case when the observation posts have a natural identification with the control stations[10] , a pattern is said to be *locally classical* if every station can remember all of its past inputs and outputs.

Now, the point of these definitions is to begin to get at the notion that as long as we have information about the relevant past control *variables* and outputs, we might not need to know all the control *laws* in order to have well defined random variables. Let L be a set of indices (θ, k). We use γ_L to refer to the restriction of design γ to just the laws γ_θ^k. Now, call a triple (Y, U, L) a *field basis* if for any two designs $\gamma, \hat{\gamma}$, $\gamma_L = \hat{\gamma}_L$ implies $\mathcal{F}(Y, U; \gamma) = \mathcal{F}(Y, U; \hat{\gamma})$. So, knowledge of the values of these particular Y and U together with knowing the laws γ_L is sufficient to understand the underlying probability space.[11]

2.3   Results

With these definitions in hand, Witsenhausen proceeds to state 11 distinct "Assertions" in the paper. Rather than going through all of them, we restate 4 of them that seem most important.

This first assertion is perhaps the most fundamental, and is the basis for many of the separation results for linear systems.

Assertion 1 *If, for every (t,k), $(Y_{t,k}, U_{t,k}, \emptyset)$ is a field basis, then the given feedback control problem is equivalent to a feedforward control problem.*

A feedforward control problem is defined as one in which the observation functions depend only on the primitive random variables, and not on the actual control variables applied. Let $(\check{x}, \check{y}, \check{u}, \check{f}, \check{g})$ be the suitably constructed

---

[8] With respect to the probability measure defined by the basic set of independent random variables.

[9] Independence of k means that all the control laws at any given time have access to the same information.

[10] In block diagrams for example, for each block there is a natural identification of the input arrows with the output ones.

[11] This is not enough to know all the conditional distributions for all the random variables in the system. To understand this, consider the following example. For a simple single-input single-output scalar system, suppose $Y = (1,1)$, $U = (1,1)$, $L = \emptyset$. Now this is a field basis because knowledge of the control law does not tell us anything more about the underlying probability space than what we already know by seeing y_1, u_1. However, unless we have a control law in hand, we can not talk about the conditional distribution of u_2.

feedforward control problem depending on the *same primitive random variables* as the original problem. The systems are equivalent if $\forall\gamma\exists\check{\gamma}$ such that $P(u = \check{u}) = 1$ and similarly $\forall\check{\gamma}\exists\gamma$ such that $P(u = \check{u}) = 1$.

Assertion 2 *Consider a problem with perfectly classical information pattern. Let F_t be the conditional distribution for x_{t-1} given all the past outputs and applied controls. Then, there is no loss if we restrict our control laws γ_t to be of the form: $\gamma_t = \phi_t(F_t)$ where ϕ_t is a function defined over the (possibly infinite dimensional) space of distributions for x_{t-1}.*

This second assertion states that the conditional law for the state is a sufficient statistic for the purpose of control. Thus, for a perfectly classical information pattern, a clear separation exists between filtering (estimating F_t) and control. Although Witsenhausen does not point this out, it is important to note that this assertion rests on the assumption that the primitive random variables are all independent. Without that, we must first explicitly augment the state to capture the dependence before this most basic separation can hold.

It is also important to notice that any nontrivial distributed system will not have a perfectly classical information pattern. This will be brought out sharply in the discussion of the "counterexample" in the next section.

Assertion 3 *For a perfectly classical linear Gaussian system, the conditional distribution of x_{t-1} has a Gaussian version with covariance independent of the data and mean affine in the data.*

The above assertion tells us that in the case of linear Gaussian systems, the filtering problem can be solved (since Gaussian random variables have their distributions parameterized by the mean and covariance) even if we restrict ourselves to time-varying affine functions to do the filtering. However, notice that no assertion is made about the form of the control law ϕ_t. For that, we need some extra assumption on the cost function.

Assertion 4 *For a perfectly classical linear system with quadratic cost criteria[12] consider the same system, except with perfect state observation and setting all the primitive random variables v_t to their mean values.[13] Let $\phi_t^*(x_{t-1})$ be the (obviously affine) optimal control law for this simpler system except thinking of it as starting at time t with the initial distribution for the state x_{t-1} being*

---

[12] By quadratic cost we mean that the incremental cost functions h_t should be quadratic in state x_t and in the individual controls u_t^i.

[13] Witsenhausen states this assertion subtly incorrectly in his paper. He says to use the same system except "*fixing all the primitive random variables at their mean values.*" This is too much of a restriction. To see this, suppose that all the primitive random variables, which includes x_0, had zero-mean. Then, identically zero control laws ϕ^* would be optimal for this system since everything would be zero. Clearly, this need not be optimal for the original problem!

a point mass at x_{t-1}. Then, $\gamma_t = \phi_t^*(\bar{F}_t)$ is an optimal control law for the original system where $\bar{F}_t$ is the mean of the conditional distribution for x_{t-1}.

This assertion represents a phenomenon often called "certainty equivalence". Here, the mean of the conditional distribution is sufficient to determine the optimal control action. The variance just contributes to the expected cost. Notice that here, only the ϕ^* part is affine. But for LQG problems with perfectly classical information patterns, we can combine this assertion with the previous one, and so both the ϕ_t^* and the $\bar{F}_t$ are affine. Thus, so are the optimal γ_t. This is the separation result that we are all most familiar with.

3   Counterexample

The natural question that arises is whether Witsenhausen is being overly conservative in his separation assertions. For affine control laws to be optimum, do we really need all four of the properties: linear systems, Gaussian primitive variables, quadratic cost, and perfectly classical information patterns? That the LQG part is critical seems clear, but one may have a doubt when it comes to the perfectly classical information patterns. To see this, we consider Witsenhausen's famous "counterexample" [3].

3.1   Problem

The problem is deceptively simple. Stated using the notation above, let us consider the problem (k, σ) as:

- $T = 2$

- x is a scalar, with x_0 Gaussian - zero mean, variance σ^2

- The state transition functions:[14] $x_1 = f_1(x_0, u_1) = x_0 + u_1$ and $x_2 = f_2(x_1, u_2) = x_1 - u_2$

- The output equations: $y_1 = g_1(x_0) = x_0$ and $y_2 = g_2(x_1) = x_1 + w$ where w is a zero mean, unit variance Gaussian random variable.

- The cost expressions: $h_1(x, u) = k^2 u^2$, $h_2(x, u) = x^2$

- The information patterns: memoryless[15] : $Y_1 = \{y_1\}; U_1 = \emptyset$ $Y_2 = \{y_2\}; U_2 = \emptyset$

---

[14] We follow Witsenhausen's notation here.
[15] Recall that the perfectly classical information patterns for this system would have been: $Y_1 = \{y_1\}; U_1 = \emptyset$ and $Y_2 = \{y_1, y_2\}; U_2 = \{u_1\}$

Before we proceed to analyze the problem as given, consider what would happen if we had a perfectly classical information pattern. In that case, we could take advantage of the given cost function and achieve zero cost with the following affine control laws: $\gamma_1(y_1) = 0$ and $\gamma_2(y_1, y_2, u_1) = y_1$.

3.2 Affine Controls

We want to now find the best possible affine control laws under the specified information pattern. By inspection, it is clear that since everything has zero-mean, they will be linear.

Let $\gamma_1(y_1) = ay_1 = ax_0$ and $\gamma_2(y_2) = by_2$. Clearly, x_1 will be Gaussian, with zero-mean and variance $(1+a)^2\sigma^2$. So, since h_2 is just x_2^2, it is clear that the optimal $\gamma_2 = \hat{x}_1 = E(x_1|y_2)$. So, using the familiar properties of sums of Gaussian random variables, $b = \frac{(1+a)^2\sigma^2}{1+(1+a)^2\sigma^2}$. We can also compute $E(h_2) = E(x_2^2) = E((x_1 - \hat{x}_1)^2) = \frac{(1+a)^2\sigma^2}{1+(1+a)^2\sigma^2}$ Now, we have an expression for the expected total cost:

$$k^2a^2\sigma^2 + \frac{(1+a)^2\sigma^2}{1+(1+a)^2\sigma^2} \tag{1}$$

To find the minimum of this expression with respect to a, we take its derivative and set it equal to zero. After some simplification, we get the equation:

$$2k^2\sigma^2 a(1 + \sigma^2(1+a)^2)^2 + 2\sigma^2(1+a) = 0 \tag{2}$$

We divide through by $2k^2\sigma^2$ and following Witsenhausen, we let $t = \sigma(1+a)$ to get:

$$(t - \sigma)(1 + t^2)^2 + \frac{t}{k^2} = 0 \tag{3}$$

Which we can rewrite as

$$\frac{t}{(1 + t^2)^2} = k^2(\sigma - t) \tag{4}$$

Now, let us compute them for the case $k = 0.1$, $\sigma = 10$. We can see graphically where the solutions will be in Figures 1 and 2. Numerically, we find that the optimal value for t is 9.899 which results in $a = -0.0101$ and total cost $= 0.99$.

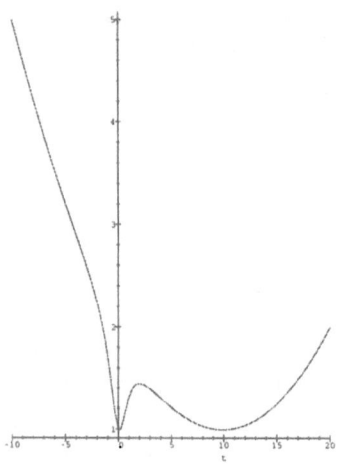

Figure 1: Expected cost vs t parameter

3.3   Nonlinear Controls

As an alternative, Witsenhausen suggests that we try the nonlinear controllers:[16]

$$\gamma_1(y_1) = -y_1 + \sigma\mathrm{sgn}(y_1) \tag{5}$$

So, at the end of the first stage, x_1 is a two-point distribution at $\pm\sigma$ depending on the sign of x_0.

$$\gamma_2(y_2) = \sigma\tanh(\sigma y_2). \tag{6}$$

We analyze the resulting expected costs, term by term. $E(h_1) = k^2 E((x_0 - \sigma\mathrm{sgn}(x_0))^2)$. Simplifying this, we get $2k^2\sigma^2(1 - E(|\frac{x_0}{\sigma}|))$. But since $\frac{x_0}{\sigma}$ is just a unit-variance Gaussian, $E(h_1) = 2k^2\sigma^2(1 - \sqrt{\frac{2}{\pi}})$. The second term, $E(h_2) = E(x_2^2)$, can not be evaluated symbolically. But, after some simplifications:[17]

$$E(h_2) = \sigma^2 \int_{-\infty}^{+\infty} \frac{(1 - \tanh(\sigma^2 + \sigma w))^2}{\sqrt{2\pi}} e^{-\frac{w^2}{2}} dw \tag{7}$$

Setting $k = 0.1$ and $\sigma = 10$ as before, we compute numerically that the expected cost is: 0.404. Compare this with the best value possible with affine

---

[16] Witsenhausen motivates these controllers by showing that this form (with σ replaced with an adjustable parameter a) is optimal if x_0 had been chosen as being $\pm\sigma$ with probability $\frac{1}{2}$ each. In this case, the first control pushes out the state, and the second control is the optimum response to the resulting two-point distribution.

[17] Witsenhausen simplifies this further, but since we were going to integrate it numerically anyway, there was no point in getting bogged down in additional unnecessary manipulations.

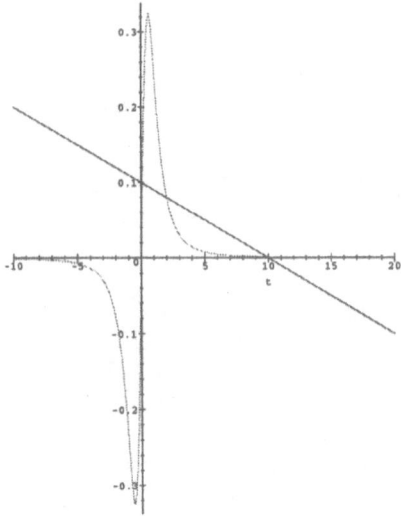

Figure 2: Graphical setting of the first derivative to zero

controllers, 0.99! The nonlinear controller is more than twice as good as the best affine control law.[18]

3.4 "Quantizing" Controllers

We would like to point out that Witsenhausen's example non-linear controllers are unnecessarily confusing — the integrals and hyperbolic functions obfuscate the essential simplicity of what is going on. Consider the following controller pair that is much clearer and still close to Witsenhausen's pair:

$$\gamma_1'(y_1) = -y_1 + \sigma\text{sgn}(y_1) \tag{8}$$

$$\gamma_2'(y_2) = \sigma\text{sgn}(y_2) \tag{9}$$

We can think of this as a 1-bit quantizer, followed by simple ML decoding. Now, by close inspection we can see that for large σ, the expected cost at the second stage is nearly zero since it is equal to $4\sigma^2 P_e(\sigma)$ where P_e is the

---

[18] No claim is made for the optimality of this nonlinear controller. In fact Witsenhausen says that we can numerically construct even better nonlinear controllers for this problem.

probability of decoding error at the second stage. But P_e obviously dies off as $e^{-\frac{x^2}{2}}$ since it is the integral of a tail of a Gaussian random variable. *No integrals need to be computed.* Furthermore, we see that we only needed 1 simple nonlinear element (the sgn function — a comparator) for each controller, making the practical significance of these results clearer. This phenomenon is not something that we need "complicated" nonlinearities to take advantage of.

Building on the intuition given above, consider the following family of "quantizing" controllers, parametrized by a single number B. [19]

$$\gamma_1^B(y_1) = -y_1 + B\lfloor \frac{y_1}{B} + \frac{1}{2}\rfloor \tag{10}$$

$$\gamma_2^B(y_2) = B\lfloor \frac{y_2}{B} + \frac{1}{2}\rfloor \tag{11}$$

The first stage takes the input and "quantizes" it into bins of size B. The decoder then just looks to see which bin the value is in. Consider now a series of problems $(k, \sigma)_n$ and non-linear controllers as follows:

$$k_n = \frac{1}{n^2} \tag{12}$$

$$\sigma_n = n^2 \tag{13}$$

$$B_n = n \tag{14}$$

For our purposes, the analysis of the performance of these controllers is also simple. The first stage cost is $k^2 E((\gamma_1^B(x_0))^2)$ which by inspection can certainly be bounded by $\frac{k^2 B^2}{4}$ since the absolute value of the control is clearly bounded above by $\frac{B}{2}$. Since, $k_n^2 B_n^2 = \frac{1}{n^2}$, the first stage cost tends to zero in this sequence.

For the second stage, we notice that since the bin size B grows as n while the variance of the observation noise w stays fixed at 1, that the second stage cost is zero, unless the noise w has magnitude greater than $\frac{B}{2} = \frac{n}{2}$. But since w is Gaussian, this tail event happens with a probability that tends to zero as $e^{-\frac{n^2}{8}}$. So, in the limit of large n, the second stage cost is zero as well. Thus:

$$\lim_{n\to\infty} E(J_n|\gamma^{B_n}) = 0 \tag{15}$$

---

[19] This family has an important role to play in another situation as well. Consider the paramaterized pair $(\alpha * \gamma_1^B(y), \beta * (y - \gamma_1^B(y)))$. It can be shown [1] that based on appropriate choices of (α, β, B) this pair of joint source-channel encoders, together with suitable decoders, can achieve higher end-to-end distortion meeting a given power constraint for a 2 dimensional AWGN channel than is possible with the best linear encoding. In fact, as power tends to infinity, the non-linear encoder/decoder's distortion tends to zero faster than the best linear encoder/decoder's distortion.

But what happens to the affine cost? Examining Equation 1, and substituting, we have:

$$E(J_n|\gamma_{\text{affine}}) = a^2 + \frac{(1+a)^2}{\frac{1}{n^4} + (1+a)^2} \tag{16}$$

Clearly,

$$\lim_{n \to \infty} E(J_n|\gamma_{\text{affine}}) = a^2 + 1 \tag{17}$$

And so, we can see that the minimum cost is achieved by setting a to zero, giving us:

$$\lim_{n \to \infty} E(J_n|\gamma_{\text{bestaffine}}) = 1 \tag{18}$$

So, the ratio $\frac{E(J_n|\gamma_{\text{bestaffine}})}{E(J_n|\gamma^{B_n})}$ tends to infinity!

3.5 Discussion

We have seen that in the case of this particular information pattern, a nonlinear controller can be superior to the best linear one. Can we get any intuition as to why this situation arose?

It seems that since the cost of control in stage 2 is zero, all that mattered at the second stage was how well it could predict x_1. Also, by not penalizing the state and keeping the cost of control in stage 1 low, we were effectively giving the first stage a lot of freedom in setting x_1 and a strong incentive to view the output x_1 purely as a way to communicate over a Gaussian channel with the second stage about the state. This coincidence of the message[20] and the messenger[21] is what is causing this seemingly strange behavior.

Ideally, what we would like is for the message to be simple (*ie* low entropy = informative prior[22]) so that there is less-information for the decoder to try and extract from the signal. However, to get the message across intact, we would like the messenger to have high-energy so that the signal-to-noise ratio is favorable (high mutual information = informative likelihoods[23]). Unfortunately, when we restrict ourselves to affine controllers for this problem, *these two objectives are in direct opposition*. An affine controller implies Gaussian state and for a Gaussian random variable, high energy implies high entropy and low entropy implies low energy. If you look at the plot in Figure 1, you will see that the two minima correspond to exactly these two cases. In the one

---

[20] x_1 is exactly what we want to communicate to the second stage.

[21] x_1 is also the input to the "channel"

[22] The intuition involved is that low entropy implies less unpredictability. Less unpredictability means that our prior knowledge is quite strong.

[23] The intuition for the case of signalling is that we want to reduce the effect of the noise. We do this by having a large mutual information between the input and output of the channel. Using the terms of hypothesis-testing, this means that we would like our "likelihood" terms to be strongly discriminating.

near $t = 0$, the entropy of x_1 is low. In the other one near $t = \sigma = 10$, the power in x_1 is high.

The nonlinear controllers have no such tension and they try to achieve the best of both. The resulting x_1 has differential entropy equal to zero[24] , and still manages to have significant power — allowing the messenger to be decoded over the noise with a low probability of error. So, the cost can be driven all the way to zero.

4   Conclusion

Fundamentally, we can now say that even through the general stochastic control problem formulation gives us a single cost function for the control objective, there seem to be intrinsically three distinct things going on naturally in the closed-loop system.

1. The first and most obvious is the overt control-objective itself. We want to use information in order to keep the state and control small in some sense.

2. The second is estimation. The system needs to have good estimates of the true state to be able to act. This can be viewed as aggregating information.

3. The third is communication. Different parts of the system need to share information.

The importance of the first two is widely recognized (Dual control, etc.), but the Witsenhausen counterexample effectively shows how a problem with non-classical information pattern really has a strong communication aspect to it. It also showed by example that the class of affine functions may not have sufficient freedom to do a good job in balancing the various factors involved and hence will not lead to optimal solutions. We are currently looking at control problems that explicitly contain a communications channel[2].

References

[1] Anant Sahai. Sending 1 signal over 2 channels. Unpublished Work, 1998.

[2] Sekhar Tatikonda, Anant Sahai, and Sanjoy Mitter. LQG control under communication constraints. Submitted to the 1998 CDC.

---

[24] We realize that differential entropy is not the best thing to look at in this case, however no matter how you look at it, the signal x_1 constructed by the first stage is very simple — effectively a discrete random variable.

[3] H. S. Witsenhausen. A counterexample in stochastic optimum control. *SIAM Journal of Control*, 6(1):131–147, 1968.

[4] H. S. Witsenhausen. Separation of estimation and control for discrete time systems. *Proceedings of the IEEE*, 59(11):1557–1566, 1971.

A Control Problem for Gaussian States

Masahiro Yanagisawa and Hidenori Kimura*

Abstract

This paper is concerned with an analysis of physical limitations of macroscopic control due to the laws of quantum mechanics. A control problem for a class of quantum states, called Gaussian, which are widely used in quantum optics is considered. Our interest is focused on controlling the mean value and the variance of two obsevables, coordinate and momentum, with respect to Gaussian states. A necessary and sufficient condition for the mean value to be controllable is obtained. Furthermore it is shown that a problem to attain the smallest possible variance under the condition that its mean value is stable is reduced to an optimal regulator problem.

1 Introduction

The revolutionary change in our understanding of microscopic phenomena that took place early in the twentieth century had never precedented in the history of natural sciences. We were confronted with severe limitations in the validity of classical physics, but we found the alternative theory that replaced the classical physics with far richer scope and range of applicability.

Some sort of microscopic effects arise in some fields of engineering, e.g. communication, signal detection, etc. about one hundred years after the birth of quantum mechanics. Many innovational technologies which utilize characteristics of photon, e.g. optical computer, optical switch, photo electronics etc. are devised in physics and engineering. New coding systems, which are perfectly protected from tapping, are realized by quantum optics, and new calculation algorithms are exploited to overcome the limitations of the present algorithms [8] [24]. Photo communication plays now important roles in the information processing. The communication theory for photon based on the quantum mechanical description, that is called the *quantum communication theory*, was initiated by Helstrom in 1967 as the quantum mechanical analog to the classical communication theory by Shannon [10]. He formulated a

*Department of Mathematical Engineering and Information Physics, The University of Tokyo, Bunkyo-ku, Tokyo 113–8656, Japan.

measurement process from a quantum-mechanical point of view which is essentially different from the conventional ones. On the other hand, the entropy theory has been formulated as a method of analyzing and constructing communication system based on the quantum mechanics. Entropy was considered by von Neumann in 1932 from statistical physics, and later, it was extended to relative entropy of quantum mechanics by Ohya et.al. [19]. A quantum channel model which is one of the important components of communication was also introduced [20]. Now, new theory that deals with *quantum information* emerges in a variety of research fields in engineering and physics.

In general the smaller the scale of the experiment becomes, the larger the interaction between the system to be measured and the measuring system. Some disturbances could be ignored in the macroscopic phenomena, but those become serious in the microscopic cases. For example, continuous application of simultaneous measurements of coordinate and momentum to a flying rocket do not cause any disturbances to the rocket. But when a experimenter applies such a measurement to a microscopic particle, the particle must be changed depending on the results of the measurement. Most experiments such as quantum calculation etc. do not succeed yet since the environmental systems destroy part of the quantum state, sometimes whole experiment itself. If a theorist wants to realize the quantum calculation then a experimenter must be able to produce the quantum states as required exactly. At that time control theorists will be expected to solve the problem, how to choose inputs to quantum system in order to attain a desired state of the system against the disturbance.

The following state

$$S = \int \frac{d^2\zeta}{\pi N} |\zeta\rangle\langle\zeta| e^{-\frac{|\zeta-z|^2}{N}}, \tag{1.1}$$

where $|\zeta\rangle$ is a coherent vector in a Hilbert space and $z \in C$, $N \in R$, and their analogs for many degrees of freedom are widely used in quantum optics for the description of radiation fields, both chaotic as natural light and coherent as those generated by lasers. Moreover the state (1.1) possess certain attractive analytical properties which are interesting as physical models, as well as from a mathematical point of view. Since these properties are essentially due to the Gaussian character of the states, it is convenient to look (1.1) from a more general point of view. The purpose of this paper is to study a control problem associated with such a state for making the quantum communication and information theory more realistic.

Notations:

R	The set of real numbers.
C	The set of complex numbers.
M^T	Transpose of a matrix M.

Tr S Trace of an operator S.

$S^\dagger$ Self-adjoint an operator of S.

2   Gaussian States and Canonical Measurements

2.1   States and measurements

Denote by $\mathcal{H}$ a Hilbert space whose inner product is denoted by $\langle\psi|\varphi\rangle$. We assume that $\mathcal{H}$ has countable complete orthonormal system of vectors.

A quantum state S is defined as a Hermitian operator in $\mathcal{H}$ satisfying

$$S \geq 0, \qquad \text{Tr } S = 1. \tag{2.1}$$

Denote the set of states by S, which is a convex set. The extreme points of S are called *pure* whose matrix representation has rank 1.

Let U be a measurable space of outcomes of the measurement. We call a quantum measurement with values in U an affine map $S \rightarrow \mu_S(du)$ of the convex set of quantum states of $\mathcal{H}$ into the set of all probability distributions on U. $\mu_S(du)$ is interpreted as the probability distribution of the results of the measurement in the state S. The measurements are described in terms of resolution of identity of Hermitian operators $M = \{M(B)|B \in \mathcal{A}(U)\}$ in $\mathcal{H}$, where $\mathcal{A}(B)$ means the Borel σ-field of the set B, satisfying

$$M(\emptyset) = 0, \qquad M(U) = I, \qquad M(B) = M^\dagger(B) \geq 0 \tag{2.2}$$

and for any at most countable decomposition $\{B_j\}$ of $B \in \mathcal{A}(U)$

$$M(B) = \sum_j M(B_j), \tag{2.3}$$

where the series is weakly convergent. A particularly important class called the *simple measurements* $E = \{E(B)\}$ which constitute an orthogonal resolution of identity, satisfying the additional requirement

$$E(B_1)E(B_2) = 0 \qquad \text{if} \quad B_1 \cap B_2 = \phi. \tag{2.4}$$

Any simple measurement is an extreme point of the convex set M, but the converse is not true.

Any measurement with values in the space U is described by an affine map of the set of states S into the set of probability distributions on U. The structure of any such map is described by the following theorem [11].

Theorem 2.1 *Let $S \rightarrow \mu_S$ be a U-measurement. Then there exists a unique resolution of identity $M = \{M(B)|B \in \mathcal{A}(U)\}$ in $\mathcal{H}$ such that for any state S*

$$\mu_S(B) = \text{Tr } SM(B). \tag{2.5}$$

Conversely, any resolution of identity defines an U-measurement by (2.5).

A non-orthogonal resolution of identity $\{M(B)\}$ in $\mathcal{H}$ arises as a restriction of an orthogonal resolution of identity $\{E(B)\}$ in a larger Hilbert space $\tilde{\mathcal{H}}$. The converse statement was proved by [1].

Theorem 2.2 *Any resolution of identity $\{M(B)\}$ in $\mathcal{H}$ can be dilated to an orthogonal resolution of identity in $\{E(B)\}$ in a larger Hilbert space $\tilde{\mathcal{H}}$ so that the following equation holds.*

$$M(B) = \tilde{E} E(B) \tilde{E}, \qquad (2.6)$$

where $\tilde{E}$ is the projection from $\tilde{\mathcal{H}}$ onto $\mathcal{H}$.

The measurements E and M are statistically equivalent in the sense that they have the same probability distributions for any state S [11].

Theorem 2.3 *For any measurement $M = \{M(B)\}$ in $\mathcal{H}$ there are Hilbert space $\mathcal{H}_0$, a pure state S_0 in $\mathcal{H}_0$ and a simple measurement $E = \{E(B)\}$ in $\mathcal{H} \otimes \mathcal{H}_0$ such that*

$$\mu_S^M(B) = \mu_{S \otimes S_0}^E(B) \qquad (2.7)$$

for any state S in $\mathcal{H}$. Conversely, any such $(\mathcal{H}_0, S_0, E)$ gives rise to the unique measurement M in $\mathcal{H}$ satisfying (2.7).

2.2 $\mathcal{L}^2$ spaces associated with a quantum state

Let us introduce a non-commutative analog of the Hilbert space of random variables with finite second moment [11] [12] [14]. A symmetric operator X in $\mathcal{H}$ such that $\langle X\psi|\varphi \rangle = \langle \psi|X\varphi \rangle$ for $\psi, \varphi \in \mathcal{D}(X)$, where $\mathcal{D}(X)$ means the domain of a operator X, is called *square summable* with respect to the state S, if X satisfies

$$\sum_j s_j \|X\psi_j\|^2 < \infty, \qquad (2.8)$$

where $S = \sum_j s_j |\psi_j\rangle\langle\psi_j|$. Let us denote by $\mathcal{L}^2(S)$ a Hilbert space of the square summable operators with respect to the state S in $\mathcal{H}$. The inner product in $\mathcal{L}^2(S)$ is defined for $X, Y \in \mathcal{L}^2(S)$ by

$$\langle Y, X \rangle_S = \operatorname{Tr} \frac{1}{2}((X\sqrt{S})\sqrt{S} + \sqrt{S}(X\sqrt{S})^\dagger)Y^\dagger, \qquad (2.9)$$

and the bilinear skew symmetric form by

$$[Y, X]_S = i\operatorname{Tr}\ ((X\sqrt{S})\sqrt{S} - \sqrt{S}(X\sqrt{S})^\dagger)Y^\dagger. \qquad (2.10)$$

These forms are related by the following inequalities:

$$\langle X_1, X_1 \rangle_S + \langle X_2, X_2 \rangle_S \ \geq\ [X_1, X_2]_S \qquad X_1, X_2 \in \mathcal{L}^2(S), \qquad (2.11)$$

$$\langle X_1, X_1 \rangle_S \langle X_2, X_2 \rangle_S \ \geq\ \frac{1}{4}[X_1, X_2]_S^2 \qquad X_1, X_2 \in \mathcal{L}^2(S). \qquad (2.12)$$

2.3   Canonical conjugate relation

Let us denote by (Z, α) a Euclidean space, i.e. Z is an arbitrary real linear space with dimension $2s$ and by $\alpha(z, z')$ an inner product on Z. Furthermore, denote by (Z, Δ) a symplectic space, i.e. Z is an arbitrary real linear space with dimension $2s$ and $\Delta(z, z')$ a bilinear skew symmetric form on Z. (Z can be of infinite dimension, but we confine ourselves to a finite case for simplicity.) Assume that it is nondegenerate, i.e. if $\Delta(z, z') = 0$ for all $z \in Z$, then $z' = 0$. A basis $\{e_j, h_j | j = 1 \cdots s\}$ in (Z, Δ) is supposed to be symplectic, i.e.

$$\Delta(e_k, h_l) = \delta_{kl} \qquad \Delta(e_k, e_l) = \Delta(h_k, h_l) = 0 \qquad (2.13)$$

The family of unitary operators $\{V(z) | z \in Z\}$ in a Hilbert space $\mathcal{H}$ is a representation of the *canonical conjugate relation* (CCR) with s degrees of freedom [22] if

$$V(z)V(z') = e^{\frac{i}{2}\Delta(z,z')}V(z + z'). \qquad (2.14)$$

By Stone's theorem the following self-adjoint operators $\{R(z) | z \in Z\}$ called *canonical observables* exist

$$V(tz) = e^{itR(z)}. \qquad (2.15)$$

Here it is convenient to introduce the hyper operator instead of commutator. For operators A, B in a Hilbert space $\mathcal{H}$, the hyper operator δ and δ^+ are defined as

$$\delta_A B = AB - BA \qquad \delta_A^+ B = AB + BA. \qquad (2.16)$$

Since $R(z)$ depends on z linearly, the following properties can be derived easily.

$$\nabla_z V(z) = \frac{i}{2}\delta_x^+ V(z), \qquad (2.17)$$

$$\delta_x V(z) = HzV(z), \qquad (2.18)$$

where $x = [\, R(e_1)\ R(h_1)\ R(e_2)\ R(h_2)\ \cdots\,]^T$ and H is skew symmetric matrix. For example in the case of one degree, i.e. $s = 1$, $H = \begin{bmatrix} 0 & -1 \\ 1 & 0 \end{bmatrix}$.

The representations of CCR satisfy the following complete relation [3] [9].

Theorem 2.4 *Let $z \rightarrow V(z)$ be an irreducible representation of the CCR in a Hilbert space $\mathcal{H}$. Then the matrix elements $\langle \psi | V(z)\varphi \rangle$ are square integrable functions of z. If $\{e_j\}$ is an orthonormal basis in $\mathcal{H}$ then the functions*

$$\left\{ \sqrt{\frac{1}{2\pi}} \langle e_j | V(z)e_k \rangle \right\} \qquad (2.19)$$

form an orthonormal basis in the space $\mathcal{L}^2(Z)$ of complex square integrable functions of Z, so that the following orthogonality relations hold:

$$\int \frac{d^{2s}z}{(2\pi)^s} \overline{\langle e_j | V(z)e_k \rangle} \langle e_l | V(z)e_m \rangle = \delta_{jl}\delta_{km}. \qquad (2.20)$$

A non-commutative analog of Fourier transformation is introduced for a *trace-class* operator T in the representation space

$$\mathcal{F}_z[T] = \text{Tr } TV(z). \tag{2.21}$$

We call the transformation $\mathcal{F}_z[S]$ the *characteristic function* of the state S [23] [18]. Let $E_z(d\lambda)$ be the spectral measure of $R(z)$ such that $R(z) = \int \lambda E_z(d\lambda)$. The characteristic function is written as

$$\mathcal{F}_{tz}[S] = \int e^{it\lambda} \mu_S^{Ez}(d\lambda). \tag{2.22}$$

Assume that the nth absolute moment of the distribution $\mu_S^{Ez}(d\lambda)$ is finite. Then the function $\mathcal{F}_{tz}[S]$ is n times differentiable and the nth moment of $\mu_S^{Ez}(d\lambda)$ is equal to

$$m_n(z) = \frac{1}{i^n} \frac{d^n}{dt^n} \mathcal{F}_{tz}[S]\Big|_{t=0}. \tag{2.23}$$

The mean value of the state S is given by

$$m(z) \equiv m_1(z) = E_S(R(z)) = \int \lambda \mu_S^{Ez}(d\lambda) \tag{2.24}$$

which is linear in z. Introducing the real symmetric bilinear form,

$$m_2(z, z') = -\frac{\partial^2}{\partial t \partial s} \mathcal{F}_{tz+sz'}[S]\Big|_{t=s=0}, \tag{2.25}$$

the correlation function of the state S is defined as

$$\alpha(z, z') = m_2(z, z') - m(z)m(z'). \tag{2.26}$$

We call S the *state with finite second moments* if $m_2(z) < \infty$ for all $z \in Z$. Since $m_2(z) < \infty$ implies $R(z) \in \mathcal{L}^2(S)$, the mean value and the correlation function are written as

$$m(z) = \langle I, R(z) \rangle_S, \tag{2.27}$$
$$\alpha(z, z') = \langle R(z) - m(z), R(z') - m(z') \rangle_S. \tag{2.28}$$

Therefore the variance of the state S is represented by

$$\alpha(z, z) = D_S(R(z)) = \langle R(z) - m(z), R(z) - m(z) \rangle_S. \tag{2.29}$$

2.4   Commutation relation and uncertainty relation

Using the notations introduced in the previous sections, the rigorous Heisenberg commutation relation is derived.

Theorem 2.5

$$[R(z), R(z')]_S = \Delta(z, z'). \tag{2.30}$$

Moreover the correlation function of a state with finite second moments satisfies the following equivalent inequalities.

Theorem 2.6

$$\alpha(z, z)\alpha(z', z') \;\geq\; \frac{1}{4}\Delta(z, z')^2, \tag{2.31}$$

$$\alpha(z, z) + \alpha(z', z') \;\geq\; \Delta(z, z'), \tag{2.32}$$

$$[\alpha(z_j, z_k)] \;\geq\; \pm\frac{1}{2}i\,[\Delta(z_j, z_k)]. \tag{2.33}$$

Here $[\alpha(z_j, z_k)]$ denotes a matrix whose (j, k) element is $\alpha(z_j, z_k)$. The relation (2.31) is just the *Heisenberg uncertainty relation* for the canonical observables $R(z)$ and $R(z')$.

The uncertainty relation is easily obtained for the pure state [21]. Consider a pair of observables X_1, X_2 and a pure state $S_\psi = |\psi\rangle\langle\psi|$ such that $\psi \in \mathcal{D}(X_1) \cap \mathcal{D}(X_2)$ assuming such a $\psi \neq 0$ exists. For any real σ

$$0 \;\leq\; |(X_1 - E_S(X_1)\psi - i\sigma^2(X_2 - E_S(X_2)\psi|^2 \tag{2.34}$$

$$=\; D_{S_\psi}(X_1) + 2\sigma^2\mathrm{Im}\langle X_1\psi|X_2\psi\rangle + \sigma^4 D_{S_\psi}(X_2), \tag{2.35}$$

from which the following uncertainty relation is derived:

$$D_{S_\psi}(X_1)D_{S_\psi}(X_2) \geq |\mathrm{Im}\langle X_1\psi|X_2\psi\rangle|^2, \tag{2.36}$$

where the equality holds if and only if

$$[(X_1 - E_S(X_1) + i\sigma^2(X_2 - E_S(X_2)]\psi = 0. \tag{2.37}$$

Such a pure state is called *minimum uncertainty state* and denoted by $|\bar{q}, \bar{p}; \sigma^2\rangle$,

$$[(q - \bar{q}) + i\sigma^2(p - \bar{p})]\,|\bar{q}, \bar{p}; \sigma^2\rangle = 0, \tag{2.38}$$

where q and p denote the operators of coordinate and momentum respectively and $\bar{q}$ and $\bar{p}$ are scalars. For each σ this equation has a solution in $\mathcal{L}^2(\mathbf{R})$ which is unique up to a multiplicative scalar factor. Indeed, in the coordinate representation we have

$$\left[(\xi - \bar{q}) + \sigma^2\left(\frac{d}{d\xi} - i\bar{p}\right)\right]\langle\xi|\bar{q}, \bar{p}; \sigma^2\rangle = 0, \tag{2.39}$$

whence

$$\langle\xi|\bar{q}, \bar{p}; \sigma^2\rangle = k e^{i\bar{p}\xi - \frac{1}{2\sigma^2}(\xi-\bar{q})^2} \tag{2.40}$$

with $|k| = 1$. In the momentum representation,

$$\langle\eta|\bar{q}, \bar{p}; \sigma^2\rangle = k e^{-i(\eta-\bar{p})\bar{q} - \frac{\sigma^2}{2}(\eta-\bar{p})}. \tag{2.41}$$

The vector $|\bar{q}, \bar{p}; \sigma^2\rangle$ is obtained from the vacuum state through the action of the representation of CCR

$$|\bar{q}, \bar{p}; \sigma^2\rangle = W_{\bar{q}, \bar{p}} |0, 0; \sigma^2\rangle, \tag{2.42}$$

where $W_{\bar{q}, \bar{p}}$ is the representation of CCR with one degree of freedom $W_{\bar{q}, \bar{p}} = V(\bar{z})$ with $\bar{z} = [-\bar{q}\ \bar{p}]^T$. The minimum uncertainty states satisfy the complete relation by Theorem 2.4, i.e.,

$$\int \frac{d^2\bar{z}}{2\pi} |\bar{q}, \bar{p}; \sigma^2\rangle \langle \sigma^2; \bar{p}, \bar{q}| = I. \tag{2.43}$$

2.5 Gaussian states

Let $z \to V(z)$ be an irreducible representation of the CCR on a symplectic space (Z, Δ). The state S in $\mathcal{H}$ is called *Gaussian* if its characteristic function has the form

$$\mathcal{F}_Z[S] = e^{im(z) - \frac{1}{2}\alpha(z, z)} \tag{2.44}$$

where $m(z)$ is a linear functional and $\alpha(z, z')$ is a bilinear symmetric form, which are the mean value and the correlation function of S respectively [16] [15] [9]. Gaussian states are widely used in the quantum optics and deeply related to the uncertainty relation.

Theorem 2.7 $\mathcal{F}_Z[S]$ *given by (2.44) is the characteristic function of a quantum state, if and only if $\alpha(z, z')$ satisfies one of the uncertainty relations of Theorem 2.6*

Let $\{e_j, h_j\}$ be a symplectic basis in which the vector z has the components $[\ x_1\ y_1\ x_2\ y_2\ \cdots\ ]$ and

$$q_j = R(e_j) \qquad p_j = R(h_j), \tag{2.45}$$

where R is a canonical observable. Then the characteristic function of a Gaussian state takes the form

$$\mathcal{F}_Z[S] = \prod_j e^{i(\bar{p}_j x_j + \bar{q}_j y_j) - \frac{a_j}{2}(x_j^2 + y_j^2)}, \tag{2.46}$$

where $\bar{p}_j = E_S(p_j)$, $\bar{q}_j = E_S(q_j)$, $a_j = D_S(q_j) = D_S(p_j)$. The fact that the characteristic function (2.46) can be decomposed into the factors corresponding to mutually commuting pairs of the canonical observables $\{p_j, q_j\}$ means that the space $\mathcal{H}$ of the irreducible representation $z \to V(z)$ can be represented as the tensor products

$$\mathcal{H} = \otimes \mathcal{H}_j \qquad S = \otimes S_j \tag{2.47}$$

where S_j is a Gaussian state in $\mathcal{H}_j$. Therefore, we restrict our consideration to the case of degree one in the sequel.

2.6   Parametric symmetry groups and covariant measurements

Let G be a parametric group of transformations of a set Θ and $g \to U_g$ be a continuous projective unitary representation of G in a Hilbert space $\mathcal{H}$. G acts as a group of automorphism of S such that $S \to U_g S U_g^\dagger$. Let $M(d\theta)$ be a measurement with values in Θ. The measurement $M(d\theta)$ is *covariant* with respect to the representation $g \to U_g$ if

$$U_g^\dagger M(B) U_g = M(B_{g^{-1}}) \tag{2.48}$$

for any $B \in \mathcal{A}(\Theta)$, where $B_g = \{\theta | \theta = g\theta_0, \theta_0 \in B\}$ is the image of the set B under the transformation g [25] [5].

Assume that θ is a parameter describing some aspects of the state denoted by S_θ and S_0 the basic state corresponding to the value θ_0. The transformation g results in the creation of the new state $S_\theta = U_g S_0 U_g^\dagger$, where $\theta = g\theta_0$. If the state and the measurement are transformed simultaneously, the whole experimental set-up remains relatively unchanged, i.e.,

$$\mu_{gS}^{gM} = \mu_{S_0}^M. \tag{2.49}$$

Hence the covariant measurements are physically admissible.

We denote by $\mu(dg)$ a measure on the σ-field $\mathcal{A}(G)$ of Borel subsets of G, by $\nu(d\theta)$ a measure on $\mathcal{A}(\Theta)$ and by G_0 the stationary subgroup. Let μ be normalized such that for an arbitrary state S

$$\int U_g S U_g^\dagger \mu(dg) = I. \tag{2.50}$$

The measure ν can be explicitly constructed from μ by the relation $\nu(B) = \mu\left(\theta^{-1}(B)\right)$, where $\theta^{-1}(B) = \{g | g\theta_0 \in B\}$ is the pre-image of $B \in \mathcal{A}(\Theta)$. The following theorem shows the structure of the covariant measurements [13] [5] [6].

Theorem 2.8 *Let $g \to U_g$ be an irreducible representation of a group G of transformations of the set Θ. The relation*

$$M(B) = \int_B U_g S_0 U_g^\dagger \nu(d\theta) \qquad B \in \mathcal{A}(\Theta) \qquad \theta = g\theta_0 \tag{2.51}$$

establishes a one to one affine correspondence between the covariant measurements M and the states S_0 commuting with $\{U_g | g \in G_0\}$.

Consider the group G of shift $(\xi, \eta) \to (\xi + \bar{q}, \eta + \bar{p})$ of the plane $\Theta = \mathbf{R}^2$. The unitary representation $(\bar{q}, \bar{p}) \to W_{\bar{q}, \bar{p}}$ is square integrable and the orthogonality relation holds by Theorem 2.4 corresponding to the invariant measure $\frac{d\bar{q}d\bar{p}}{2\pi}$. Theorem 2.8 implies that any measurement $M(d\bar{q}d\bar{p})$ covariant with respect to the representation $(\bar{q},\bar{p}) \to W_{\bar{q},\bar{p}}$ has the form

$$M(d\bar{q}d\bar{p}) = W_{\bar{q},\bar{p}}S_0 W_{\bar{q},\bar{p}}^\dagger \frac{d\bar{q}d\bar{p}}{2\pi}. \qquad (2.52)$$

2.7 Canonical measurements

Let us define the mean value in the case of multidimensional parameter

$$E_q\{M\} \equiv \int \bar{q}\mu_{S_{q,p}}^M(d\bar{q}d\bar{p}) \qquad E_p\{M\} \equiv \int \bar{p}\mu_{S_{q,p}}^M(d\bar{q}d\bar{p}) \qquad (2.53)$$

and the marginal variances

$$D_q\{M\} \equiv \int (\bar{q}-E_q\{M\})^2 \mu_{S_{q,p}}^M(d\bar{q}d\bar{p}) \quad D_p\{M\} \equiv \int (\bar{p}-E_p\{M\})^2 \mu_{S_{q,p}}^M(d\bar{q}d\bar{p}). \qquad (2.54)$$

We introduce the measure of accuracy of joint measurement as

$$\mathcal{R}\{M\} = g_q D_q\{M\} + g_p D_p\{M\}, \qquad (2.55)$$

where g_q and g_p are positive constants. $\mathcal{R}\{M\}$ given by (2.55) is an affine functional of a measurement M, which is finite if M has finite second moments and achieves its minimum at an extreme point of the convex set of covariant measurements $M(d\bar{q}d\bar{p})$ having finite second moments. Therefore we must take a pure state as S_0 in (2.52) to construct the optimal covariant measurement minimizing (2.55), which is expressed by

$$M(d\bar{q}d\bar{p}) = W_{\bar{q},\bar{p}}|\psi\rangle\langle\psi|W_{\bar{q},\bar{p}} \frac{d\bar{q}d\bar{p}}{2\pi}, \qquad (2.56)$$

where ψ is a unit vector of $\mathcal{H}$. This means that for a Borel B

$$\langle\varphi|M(B)\varphi\rangle = \int_B \langle\varphi|W_{\bar{q},\bar{p}}\psi\rangle\langle\psi|W_{\bar{q},\bar{p}}\varphi\rangle \frac{d\bar{q}d\bar{p}}{2\pi}, \qquad (2.57)$$

the integral is guaranteed to converge by Theorem 2.4.

According to Theorem 2.3, we can obtain a procedure to construct the realization of the measurement (2.56). In addition to $\mathcal{H}$ with q and p, we take the identical space $\mathcal{H}_0$ with q_0 and p_0. Consider the operators in the tensor product $\mathcal{H} \otimes \mathcal{H}_0$

$$p_t = p \otimes I_0 + I \otimes p_0 \qquad q_t = q \otimes I_0 - I \otimes q_0, \qquad (2.58)$$

where I_0 is the unit operator in $\mathcal{H}_0$. These operators are infinitesimal generators of the unitary groups

$$e^{i\xi q_t} = e^{i\xi q} \otimes e^{-i\xi q_0} \qquad e^{i\eta p_t} = e^{i\eta p} \otimes e^{i\eta p_0} \qquad (2.59)$$

which commute due to CCR. Thus, q_t and p_t are commuting observables. It follows that q_t and p_t are compatible and admit the joint measurement $E(d\bar{q}d\bar{p})$.

Let us take $S_0 = |\bar{\psi}\rangle\langle\bar{\psi}|$ as the auxiliary state of (2.6), where the vector $\bar{\psi}$ is complex conjugate to ψ of (2.56). Then it follows that

$$\min_M \mathcal{R}\{M\} = g_q D_S(q) + g_p D_S(p) + \sqrt{g_q g_p}, \tag{2.60}$$

where the minimum is achieved for the unique optimal covariant measurement

$$M_*(d\bar{q}d\bar{p}) = W_{\bar{q},\bar{p}}|0,0;\sigma^2\rangle\langle\sigma^2;0,0|W_{\bar{q},\bar{p}}^\dagger \frac{d\bar{q}d\bar{p}}{2\pi} \tag{2.61}$$

with σ^2 being equal to $\frac{1}{2}\sqrt{\frac{g_q}{g_p}}$. The measurement (2.61) is called *the canonical measurement* [10].

3   Quantum Dynamics Involving Measurements

The Schrödinger equation represents the dynamics of the closed quantum systems which possess a limited class of external forces and converges to the corresponding classical theory as Plank constant goes to zero. On the other hand, the state equation of the control theory due to the classical mechanics represents the open systems which possess various inputs to manipulate the variables of the system and interacts with other systems to be controlled. Hence it becomes naturally dissipative and non-conservative. If we consider a control problem for the quantum system we need to have a description of the quantum open system for the purpose of control.

Here we consider the time evolution of the quantum state interacting with the measuring apparatus which is represented by the canonical measurement of coordinate and momentum (2.61).

3.1   Iterative measurements

Consider first a set of simple measurements E generating discrete data. When we apply the measurements to a state S and observe the results, the measuring apparatus interacts with the state in order to extract the information ,which inevitably cause the change of the state, called *reduction*. One simple measurement reduces the state to

$$S_1 = \frac{E_1 S E_1}{\text{Tr } E_1 S E_1}. \tag{3.1}$$

Since the denominator is re-written using the orthogonality of simple measurements as

$$\text{Tr } ESE = \text{Tr } SE \tag{3.2}$$

which is a resulting probability distribution given by Theorem 2.1, (3.1) represents a conditional probability distribution. The iterative simple measurement E_2 reduces the state to

$$S_2 = \frac{E_2 S_1 E_2}{\text{Tr } E_2 S_1 E_2} = \frac{E_2 E_1 S E_1 E_2}{\text{Tr } E_2 E_1 S E_1 E_2}. \tag{3.3}$$

Therefore N times simple measurements change the state S to S_N given by

$$S_N = \frac{E_N \cdots E_1 S E_1 \cdots E_N}{\text{Tr } E_N \cdots E_1 S E_1 \cdots E_N}. \tag{3.4}$$

In a similar way the reduction of a measurement, not simple measurement, is expressed in a same form as (3.1)

$$S_1 = \frac{M_1^{\frac{1}{2}} S M_1^{\frac{1}{2}}}{\text{Tr } M_1^{\frac{1}{2}} S M_1^{\frac{1}{2}}}, \tag{3.5}$$

and the reduction of the iterative measurements in a form

$$S_N = \frac{M_N^{\frac{1}{2}} \cdots M_1^{\frac{1}{2}} S M_1^{\frac{1}{2}} \cdots M_N^{\frac{1}{2}}}{\text{Tr } M_N^{\frac{1}{2}} \cdots M_1^{\frac{1}{2}} S M_1^{\frac{1}{2}} \cdots M_N^{\frac{1}{2}}}. \tag{3.6}$$

Note that the reduction is a statistical process, not a dynamical one. In the previous equations measurements are applied to a static quantum state which does not evolve if the measurement is absent. According to quantum mechanics, if the quantum state does not interact with any other systems, then its time evolution is expressed by unitary operator $U(t)$. Applying iterative measurements to the evolving state, the probability as the measurements obtain the outcomes of measurements $(d\bar{q}_N, d\bar{p}_N), \cdots, (d\bar{q}_1, d\bar{p}_1)$ is written as

$$
\begin{aligned}
&\text{Pr}\left((d\bar{q}_N, d\bar{p}_N), \cdots, (d\bar{q}_1, d\bar{p}_1)\right) \\
&= \text{Tr } U(\epsilon) \, M^{\frac{1}{2}}(d\bar{q}_N, d\bar{p}_N) \, U(\epsilon) \, M^{\frac{1}{2}}(d\bar{q}_{N-1}, d\bar{p}_{N-1}) \cdots M^{\frac{1}{2}}(d\bar{q}_1, d\bar{p}_1) \, U(\epsilon) \, S \\
&\qquad U^{\dagger}(\epsilon) \, M^{\frac{1}{2}}(d\bar{q}_1, d\bar{p}_1) \cdots U^{\dagger}(\epsilon) \, M^{\frac{1}{2}}(d\bar{q}_N, d\bar{p}_N) \, U(\epsilon), \tag{3.7}
\end{aligned}
$$

where ϵ denotes infinitesimal interval of time.

3.2 Continuous canonical measurement

Let us introduce the canonical measurement of coordinate and momentum (2.61) as M in (3.7). Inserting the identity between the measurement and the unitary evolution of (3.7) we obtain

$$
\begin{aligned}
&\text{Pr}\left((\bar{q}_N, \bar{p}_N), \cdots, (\bar{q}_1, \bar{p}_1)\right) \\
&= \text{Tr } \int dq_N \, dq_{N-1} \cdots dq_0 \, dq_0' \, dq_1' \cdots dq_N'
\end{aligned}
$$

$$U(\epsilon) \, |q_N\rangle\langle q_N| \, M^{\frac{1}{2}}(d\bar{q}_N, d\bar{p}_N) \, U(\epsilon) \, |q_{N-1}\rangle\langle q_{N-1}| \cdots$$
$$\cdots |q_1\rangle\langle q_1| \, M^{\frac{1}{2}}(d\bar{q}_1, d\bar{p}_1) \, U(\epsilon) \, |q_0\rangle\langle q_0| \, S \, |q_0'\rangle\langle q_0'| U^\dagger(\epsilon)$$
$$M^{\frac{1}{2}}(d\bar{q}_1, d\bar{p}_1) \, |q_1'\rangle\langle q_1'| \cdots$$
$$\cdots |q_{N-1}'\rangle\langle q_{N-1}'| \, U^\dagger(\epsilon) \, M^{\frac{1}{2}}(d\bar{q}_N, d\bar{p}_N) \, |q_N'\rangle\langle q_N'| \, U(\epsilon). \qquad (3.8)$$

The transfer function from the state previous to the measurement to the one after the measurement is calculated by path integral [7].

$$G(q, p; q', p') = \int_{q_0, q_0'}^{q, q'} d[q] \, d[q']$$
$$\exp\left[\int d\tau \, \{i(p\dot{q} - H)\} + i(p - \bar{p})(q - \bar{q}) - \frac{1}{2\sigma^2}(q - \bar{q})^2 - \frac{\sigma^2}{2}(p - \bar{p})^2 \right.$$
$$\left. \int d\tau \, \{-i(p'\dot{q}' - H')\} - i(p' - \bar{p})(q' - \bar{q}) - \frac{1}{2\sigma^2}(q' - \bar{q})^2 - \frac{\sigma^2}{2}(p' - \bar{p})^2 \right],$$
$$(3.9)$$

where the integration means the sum over all possible paths from (q_0, q_0') to (q, q').

Let us replace the parameter such that $\dfrac{1}{\sigma^2} = \rho^2 \epsilon$, which shifts the measurement temporally. The measurement results do not become transitive without this replacement [4] [17].

3.3 Introduction of filter

Assume that the outcomes of measurements $\tilde{q}$ and $\tilde{p}$ are obtained through a linear system generated by the kernel f such that

$$\begin{bmatrix} \tilde{q}(t) \\ \tilde{p}(t) \end{bmatrix} = \int_0^t ds \, f(t - s) \begin{bmatrix} \tilde{q}(s) \\ \tilde{p}(s) \end{bmatrix} \qquad (3.10)$$

Denote the final results of measurement $\tilde{q}$ and $\tilde{p}$ by $\bar{q}$ and $\bar{p}$ in the sequel.

In order to calculate the transfer function which depends on the results of measurement obtained through the linear system, we need the following lemma.

Lemma 3.1

$$\int d[x] \, \delta\left(\bar{x} - \int_0^\epsilon d\tau f(\epsilon - \tau)x(\tau)\right) \exp\left[-\alpha \int_0^\epsilon d\tau(x(\tau) - u(\tau))^2\right]$$
$$= \left(\frac{\alpha}{\pi \int_0^\epsilon d\tau f^2(\tau)}\right)^{\frac{1}{2}} \exp\left[-\frac{\alpha}{\int_0^\epsilon d\tau f^2(\tau)}\left(\bar{x} - \int_0^\epsilon d\tau f(\epsilon - \tau)u(\tau)\right)^2\right].$$
$$(3.11)$$

Using Lemma 3.1, we can rewrite (3.9) to

$$
\frac{\rho^2 \sigma^2}{\pi \int d\tau f^2(\tau)} \exp\Bigg[\int d\tau \{ i(p\dot{q} - H) - i(p'\dot{q}' - H') \}
$$

$$
\frac{\rho^2}{\int d\tau f^2(\tau)} \left(\bar{q} - \int d\tau f(t - \tau) \frac{A}{2} \right)^2 + \rho^2 \frac{A^2}{4} - \frac{\rho^2}{2}(q^2 + q'^2)
$$

$$
\frac{\sigma^4 \rho^2}{\int d\tau f^2(\tau)} \left(\bar{p} - \int d\tau f(t - \tau) \frac{B}{2} \right)^2 + \sigma^4 \rho^2 \frac{B^2}{4} - \frac{\sigma^4 \rho^2}{2}(p^2 + p'^2)
$$

$$
+ i\sigma^2 \rho^2 (\dot{q}p - q'p') \Bigg]
\tag{3.12}
$$

with

$$
A = (q + q') - i\sigma^2 (p - p') \qquad B = (p + p') - \frac{i}{\sigma^2}(q - q').
\tag{3.13}
$$

3.4 Time evolution equation under the canonical measurement

In order to represent the effects generated by observing the results of measurements, let us perform Fourier transform with respect to the results of measurement $\int d\bar{x} e^{-i\bar{x}\cdot\zeta}$ with $\bar{x} = [\ \bar{q}\ \bar{p}\ ]^T$, $\zeta = [\ \xi\ \eta\ ]^T$. The expression (3.12) yields the transfer function written as

$$
G(q, p; q', p') = \int_{q_0, q_0'}^{q, q'} d[q]\, d[q']
$$

$$
\exp\Bigg[\int d\tau \left\{ i(p\dot{q} - H) - i(p'\dot{q}' - H') - \frac{\rho^2}{2}(q - q')^2 - \frac{\sigma^4 \rho^2}{2}(p - p')^2 \right\}
$$

$$
\xi^2 \frac{\int d\tau f^2(\tau)}{4\rho^2} - i\xi \int d\tau f(t - \tau) \frac{A}{2} + \eta^2 \frac{\int d\tau f^2(\tau)}{4\sigma^4 \rho^2} - i\eta \int d\tau f(t - \tau) \frac{B}{2} \Bigg]
\tag{3.14}
$$

Differentiation of the transfer function (3.14) with respect to time leads to the time evolution equation of a quantum state under the canonical measurement of coordinate and momentum as follows:

$$
\begin{aligned}
\dot{S} =\ & -i[H, S] - \frac{\rho^2}{2}[q, [q, S]] - \frac{\sigma^4 \rho^2}{2}[p, [p, S]] \\
& - \xi^2 \frac{f^2(0)}{4\rho^2} S - i\xi \frac{f(0)}{2} \left(\{q, S\} - i\sigma^2 [p, S] \right) \\
& - \eta^2 \frac{f^2(0)}{4\sigma^4 \rho^2} S - i\eta \frac{f(0)}{2} \left(\{p, S\} - \frac{i}{\sigma^2}[q, S] \right).
\end{aligned}
\tag{3.15}
$$

The first term of (3.15) is the unitary evolution by Hamiltonian. The process such that the measurements are done but nobody observes the results means just an interaction between the quantum state and the measuring apparatus. Such a process is called *non-selective measurement*, which corresponds to integration with respect to the resultant value of measurement, i.e. putting $\zeta = 0$. Therefore the second and third terms of (3.15) expresses the disturbance from the measuring apparatus to the quantum state due to the interaction. On the contrary, other terms containing ζ are interpreted as denoting reactions to the state caused by the generation of the results by measuring apparatus. This process is called *selective measurement*.

4   Control Problem for Gaussian States

In the case of the macroscopic system, we can manipulate the system during measurements. The manipulation is the dual notion of the measurement. Since the quantum mechanics contains the classical theory in the sense that the classical dynamics is the limit case of the quantum dynamics, all phenomena must be analyzed quantum mechanically. There will be a process which converges to such a classical process, manipulation during measurement, as the limit case. In the preceding section, the dynamics involving the measurement process is obtained as a kind of expression of interaction between the quantum state and its environment. Here we consider the manipulation process. All quantities obtained by measurement are macroscopic in the case of the quantum system as well as classical one. In fact the measurement to the quantum state gives us the macroscopic concept of coordinate and momentum with respect to the state that leads us to understanding of microscopic phenomena. Even in the case of the quantum system, only macroscopic observables are allowed to manipulate states. Hence, assume that we can manipulate the quantum state by going through the measurement process inversely. This means that ζ, which is the outcomes of the measurement, also plays a role of macroscopic inputs to manipulate the quantum system.

4.1   Generalization of time evolution equation

Let us introduce some generalizations of (3.15). The basis $\{e_j, h_j\}$ such as (2.45) is used in the preceding section. If we make a linear invertible transformation to the basis, the structure of the system changes, but none of the properties of the probability distribution obtained by measurement is lost. It is desirable to make the form (3.15) invariant with respect to such a transformation.

The unitary evolution parts, that is generated by a Hamiltonian of the objective system itself, is written generally as

$$\dot{S} = -i\left[\sum_{ij} a_{ij}\delta^+_{x_i}\delta_{x_j} + \sum_i f_i\delta_{x_i}\right]S, \tag{4.1}$$

where δ and δ^+ represent commutator and anti-commutator defined in (2.16), respectively, and $x_1 = q$, $x_2 = p$. In the case of the most general quadratic form of one dimensional Lagrangian such as a harmonic oscillator under a gravitational force proportional to its coordinate, Hamiltonian is given by

$$H = \frac{1}{2m}p^2 + \frac{c}{2}q^2 + eq. \tag{4.2}$$

Then $A \equiv [a_{ij}]$ and $f \equiv [f_i]$ in (4.1) are written as

$$A = \begin{bmatrix} \frac{c}{2} & 0 \\ 0 & \frac{1}{2m} \end{bmatrix} \qquad f = \begin{bmatrix} e \\ 0 \end{bmatrix} \tag{4.3}$$

The term which does not contain ξ and η in (3.15), disturbance from a measuring apparatus, is generalized to

$$-\sum_{ij} d_{ij}\delta_{x_i}\delta_{x_j} \tag{4.4}$$

with $[d_{ij}] \equiv D = D^T$ satisfying uncertainty relation $\det D \geq \frac{1}{4}$.

The terms containing ζ that is input as we assumed is generalized as

$$-\sum_{ij} \xi_i[D^{-1}]_{ij}\xi_j - i\sum_{ij} \xi_i b_{ij}\delta^+_{x_j} - \sum_{ij} \xi_i c_{ij}\delta_{x_j}. \tag{4.5}$$

Summing up the above generalizations, the state for any time is represented as $S(t) = \mathbf{X}S(0)$, where

$$\dot{\mathbf{X}} = \left[-i\delta^+_x A\delta_x - if \cdot \delta_x - \delta_x D\delta_x - \zeta D^{-1}\zeta - i\zeta B\delta^+_x - \zeta C\delta_x\right]\mathbf{X}. \tag{4.6}$$

4.2 Time evolution of Gaussians

Let us consider the time evolution of a Gaussian state. The characteristic function of $S(t)$ is also defined in the same way as (2.21) in which a operator in Hilbert space $\mathcal{H}$ is transformed to a scalar function by the inner product in $\mathcal{H}$, trace.

$$\mathcal{F}_z[S] = \text{Tr } S(t)V(z) = \text{Tr } S(0)\mathbf{X}^\dagger V(z) \tag{4.7}$$

Define the complex conjugate operator $\mathbf{X}^\dagger$ in usual way, e.g. for operators A, B in Hilbert space

$$(\delta_A B)^\dagger = -\delta_{A^\dagger}B^\dagger \qquad (\delta^+_A B)^\dagger = \delta^+_{A^\dagger}B^\dagger \tag{4.8}$$

and so on. In order to calculate (4.7) we have a scalar function $J(t)$ such that $\mathbf{X}^\dagger V(\mathbf{z}) = V(\mathbf{z}(t))e^{J(t)}$. ¿From time differentiation of the both sides and (2.17) (2.18), we obtain

$$\frac{i}{2}\delta_{\boldsymbol{x}}^+ \cdot \dot{\boldsymbol{z}}\, V + \dot{J}\, V$$

$$= \; [-i\delta_{\boldsymbol{x}}^+ A\boldsymbol{z} - if H\boldsymbol{z} - \boldsymbol{z}H^\dagger D H\boldsymbol{z} - \zeta D^{-1}\zeta + i\delta_{\boldsymbol{x}}^+ B\zeta - \zeta CH\boldsymbol{z}]\, V \quad (4.9)$$

Taking replacement such that $AH \to A$, $CH \to C$,

$$\dot{\boldsymbol{z}} \;=\; A\boldsymbol{z} + B\zeta, \qquad\qquad\qquad\qquad\qquad (4.10)$$

$$\dot{J} \;=\; -if \cdot \boldsymbol{z} - \boldsymbol{z}D\boldsymbol{z} - \zeta D^{-1}\zeta - \zeta C\boldsymbol{z}. \qquad\qquad (4.11)$$

Now, we take the initial state $S(0)$ to be Gaussian such that

$$\mathcal{F}_{\boldsymbol{z}}[S(0)] = e^{i\boldsymbol{\theta}\cdot\boldsymbol{z} - \frac{1}{2}\boldsymbol{z}K\boldsymbol{z}}, \qquad\qquad\qquad (4.12)$$

where $\boldsymbol{\theta}$ represents the mean values of coordinate and momentum and K is the covariance matrix. The characteristic function of the state for arbitrary time is finally calculated to be

$$\mathcal{F}_{\boldsymbol{z}}[S(t)] = e^{i\boldsymbol{\theta}\cdot\boldsymbol{z}(t) - \frac{1}{2}\boldsymbol{z}(t)K\boldsymbol{z}(t) + J(t)}$$

$$= \; \exp\left[i\{\boldsymbol{\theta}\cdot\boldsymbol{z}(t) - \int_0^t ds f \cdot \boldsymbol{z}(s)\} \right.$$

$$\left. -\frac{1}{2}\boldsymbol{z}(t)K\boldsymbol{z}(t) - \int_0^t ds\left\{\boldsymbol{z}(s)D\boldsymbol{z}(s) + \zeta(s)D^{-1}\zeta(s) + \zeta(s)C\boldsymbol{z}(s)\right\}\right],$$

$$(4.13)$$

where $\dot{\boldsymbol{z}} = A\boldsymbol{z} + B\zeta$.

Our purpose is to decide the input ζ to control the mean value and the variance in the state expressed through (4.13).

4.3 Control problem of mean value

Here we are concerned with only $\boldsymbol{z}$, $\int ds\boldsymbol{z}(s)$ satisfying (4.10) since the mean value of the Gaussian state is related to the term of the first order of $\boldsymbol{z}$ in (4.13).

Theorem 4.9

$$\textit{Mean value is controllable} \;\Leftrightarrow\; |B| \neq 0. \qquad\qquad (4.14)$$

The controllability of the mean value is similar to the classical case. This fact is expected by the principle of quantum mechanics, that is, mean value corresponds to the classical observable.

4.4   Minimizing variance

Producing a quantum state for some expriments, it would be desirable that the arbitrary accuracy of its variance is attainable. But it is not sure whether such a desire is necessarily ensured. The necessary and sufficient condition for the meán value to be controllable is obtained in the preceding section. Here we consider to minimize the variance as small as possible. If the system $\dot{z} = Az + B\zeta$ is stable then the value $\int dtz$ is also stable. Hence considering to stabilize the mean value which consists of z and $\int dtz$, it is sufficient to concentrate our interests on stabilizing the value z. Since the microscopic phenomena evolve so fast, our consideration is restricted to the the case of infinite time.

The condition $\lim_{t\to\infty} z(t) = 0$ leads minimizing the variance of the state to

$$\min \int_0^\infty dsw(s), \tag{4.15}$$

where $w(s) \equiv \zeta D^{-1}\zeta + 2\zeta Cz + zDz$. $|B| \neq 0$ assures the controllability of the system $\dot{z} = Az + B\zeta$, so it is shown that minimizing the variance with controlling the mean values of coordinate and momentum with respect to the Gaussian state reduced to a least squares optimal control problem.

Theorem 4.10 *Assume that $|B| \neq 0$ in the system $\dot{z} = Az + B\zeta$.*

$$\min \int_0^\infty dsw(s) \quad subject\ to \quad \lim_{t\to\infty} z(t) = 0 \tag{4.16}$$

with $w(s) = \zeta D^{-1}\zeta + 2\zeta Cz + zDz$ is uniquely attained by the feedback law

$$\zeta = -D(BP^+ + C)z \tag{4.17}$$

where P^+ is a stabilizing solution to the algebraic Riccati equation

$$A^T P + PA - (PB + C^T)D(B^T P + C) + D = 0, \tag{4.18}$$

moreover the attainable covariance matrix of Gaussian (4.13) is given by P^+.

The attainable values of variance of coordinate and momentum are decided by the structure of manipulator D which is related to the uncertainty relation.

5   Conclusion

This paper contains two major results. The first is the derivation of a new evolution equation which governs the quantum interactions involved in the measurement process. The equation has been obtained based on the pass integral for the quantum state evolution subject to continuous canonical measurements. It contains unitary evolution by Hamiltonian, interaction with the measuring apparatus represented by non-selective measurement and reaction by selective measurement. The second is the application of the evolution equation derived in the first part to Gaussian states. We have formulated a control problem for the Gaussian states to achieve the minimum variance. It has been shown that the control problem is reduced to an optimal regulator problem, and the attainable minimum variance of Gaussian states depends on the uncertain relation indirectly.

Someone might think that the results are similar to the LQG, in which the uncertainty of the outputs results from the uncertainty of inputs. In the case of the quantum system, no matter how certain the inputs are, the outputs certainly contain the uncertainty which results from the non-commutativity of the physical parameters inside the system. The property, non-commutativity of the observables, is the essence of the microscopic phenomena and absolutely crucial difference between the classical and the quantum mechanics.

References

[1] N. I. Akhiezer and I. M. Glazman. *Theory of Linear Operators in Hilbert Space.* Ungar, New York, 1963.

[2] A. Barchielli, L. Lanz and G. M. Prosperi. A model for the macroscopic description and continual observation in quantum mechanics. *Nuovo Cimento*, B72 (1982) 79–121.

[3] V. Bargmann. On a Hilbert space of analytic functions and an associated integral transform. *Comm. Pure Appl. Math.*, 14(3) (1961) 187–214.

[4] C. M. Caves. Quantum mechanics of measurement distributed in time. *Phys. Rev. D*, 33 (1986) 1643–1665.

[5] E. B. Davies. On repeated measurements of continuous observables in quantum mechanics. *J. Funct. Anal.*, 6 (1970) 318–346.

[6] N. Dunford and J. T. Schwartz. *Linear Operators. Part I.* Interscience, New York, 1958.

[7] R. P. Fynman. Space-time approach to non-relativistic quantum mechanics. *Rev. Mod. Phys.*, 20 (1948) 367–387.

[8] R. P. Fynman. Quantum mechanical computers. *Found. Phys.*, 16 (1986) 507–531.

[9] R. J. Glauber. The quantum theory of optical coherence. *Phys. Rev.*, 130 (1963) 2529–2539.

[10] C. W. Helstrom. *Quantum Detection and Estimation Theory.* Academic Press, New York, 1976.

[11] A. S. Holevo. Statistical decision theory for quantum systems. *J. Multivariate Anal.*, 3(4) (1973) 337–394.

[12] A. S. Holevo. Commutation superoperator of a state and its applications in the noncommutative statistics. *Rep. Math. Phys.*, 12(2) (1977) 251–271.

[13] A. C. Holevo. Covariant measurements and uncertainty relations. *Rep. Math. Phys.*, 16(3) (1979) 385–400.

[14] K. Kraus and J. Schröter. Expectation values of unbounded observables. *Internat. J. Theoret. Phys.*, 7(6) (1973) 431–442.

[15] W. Louisell. *Radiation and Noise in Quantum Elictronics.* McGraw-Hill, New York, 1964.

[16] J. Manuceau and A. Verbeure. Quasifree states of the CCR. *Comm. Math. Phys.*, 9(4) (1968) 293–302.

[17] M. B. Mensky. Group-theoretical structure of quantum continuous measurements. *Phys. Lett.*, 150A (1990) 331–336.

[18] J. E. Moyal. Quantum mechanics as a statistical theory. *Proc. Camb. Phil. Soc.*, 45 (1949) 99–124.

[19] M. Ohya. Quantum Ergodic Channels in Operator Algebras. *J. Math. Anal. Appl.*, 84(2) (1981) 318–327.

[20] M. Ohya. On compound state and mutual information in quantum information theory. *IEEE. Tras. Inf. Theory*, 29 (1983) 770–774.

[21] H. P. Robertson. The uncertainty principle. *Phys. Rev.*, 34(1) (1929) 163–164.

[22] I. Segal. *Mathematical Problems of Relativistic Physics.* AMS, Providence, RI, 1963.

[23] I. Segal. A generating functional for the states of linear Boson field. *Canad. J. Math.*, 13(1) (1961) 1–18.

[24] P. W. Shor. Proceeding of the 35th Annual Symposium on Foundation of Computer Science. *IEEE Computer Society Press*, Los Alamis, CA, (1994) 116–123.

[25] V. S. Varadarajan. *Geometry of Quantum Theory, II. Quantum Theory of Covariant Systems.* Van Nostrand, New York, 1970.

Part D

Robust Control

Frequency Domain Solution of the $\mathcal{H}_\infty$ Problem for Descriptor Systems

Huibert Kwakernaak*

Abstract

The standard $\mathcal{H}_\infty$ problem for descriptor systems is solved by frequency domain techniques relying on spectral factorization. An algorithm of Clements for the spectral factorization of rational matrices is adapted to the descriptor problem. It leads to a compact and efficient solution that involves transforming two suitable matrix pencils to Clements form.

1 Introduction

The celebrated standard $\mathcal{H}_\infty$-problem [3] is represented by the block diagram of Fig. 1. G is the transfer matrix of the generalized plant. The input w is an external driving signal, the signal u is the control input, the output signal z has the significance of error signal, and the output signal y, finally, is measured and available for feedback through the compensator with transfer matrix K. The open-loop transfer matrix of the plant may be partitioned as

$$\begin{bmatrix} z \\ y \end{bmatrix} = \underbrace{\begin{bmatrix} G_{11} & G_{12} \\ G_{21} & G_{22} \end{bmatrix}}_{G} \begin{bmatrix} w \\ u \end{bmatrix} \tag{1}$$

The dimensions are $\dim w = k_1$, $\dim u = k_2$, $\dim z = m_1$, and $\dim y = m_2$. The closed-loop transfer matrix from the external signal w to the error signal z is

$$H = G_{11} + G_{12}(I - KG_{22})^{-1}KG_{12} \tag{2}$$

The $\mathcal{H}_\infty$-problem consists of determining the compensator K that minimizes the ∞-norm $\|H\|_\infty$ of the closed-loop transfer matrix.

*Systems and Control Group, Faculty of Applied Mathematics, University of Twente, P.O. Box 217, 7500 AE Enschede, The Netherlands. E-mail h.kwakernaak@math.utwente.nl

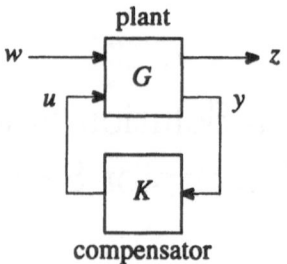

Figure 1: The standard $\mathcal{H}_\infty$ problem

Frequency domain solution. In a series of papers (summarized in [6]) the frequency domain solution of the continuous-time $\mathcal{H}_\infty$-problem has been developed. We briefly recapitulate it. Let γ be a given nonnegative real number. Given the transfer matrix G define the para-Hermitian rational matrix Π_γ whose inverse is

$$\Pi_\gamma^{-1} = \begin{bmatrix} G_{\widetilde{12}} & G_{\widetilde{22}} \\ 0 & I \end{bmatrix} \begin{bmatrix} I - \frac{1}{\gamma^2}G_{11}G_{\widetilde{11}} & -\frac{1}{\gamma^2}G_{11}G_{\widetilde{21}} \\ -\frac{1}{\gamma^2}G_{21}G_{\widetilde{11}} & -\frac{1}{\gamma^2}G_{21}G_{\widetilde{21}} \end{bmatrix}^{-1} \begin{bmatrix} G_{12} & 0 \\ G_{22} & I \end{bmatrix} \tag{3}$$

Determine a spectral factorization of Π_γ^{-1}, if any exists, of the form

$$\Pi_\gamma^{-1} = M_\gamma^{\sim} J M_\gamma \tag{4}$$

We denote $M_\gamma^{\sim}(s) = M_\gamma^T(-s)$. The spectral factor M_γ is a square rational matrix such that both M_γ and its inverse M_γ^{-1} have all their poles in the open left-half complex plane. The constant matrix J is a signature matrix of the form $J = \mathrm{diag}(I_{k_2}, -I_{m_2})$, where the subscript of a unit matrix I indicates its dimension.

Given the spectral factorization (4) we may characterize the transfer matrix of all compensators that achieve $\|H\|_\infty \leq \gamma$ (if any exist) as

$$K_\gamma = (M_{\gamma,11} + UM_{\gamma,21})^{-1}(M_{\gamma,12} + UM_{\gamma,22}) \tag{5}$$

where U is an arbitrary stable rational matrix such that $\|U\|_\infty \leq 1$, and the matrices $M_{\gamma,ij}$ follow by suitably partitioning the spectral factor M_γ.

Solution of the spectral factorization problem. In the mainstream solution of the $\mathcal{H}_\infty$-problem the plant G is represented in state form. The factorization of Π_γ^{-1} may then be reduced to the solution of a pair of algebraic matrix

Riccati equations [4]. The compensator also emerges in state form. Effective algorithms exist for the solution of matrix Riccati equations. To some extent the state formulation is restrictive, however. In particular, plants with non-proper transfer matrices cannot be handled directly. Such plants arise when nonproper weighting functions are introduced in the mixed sensitivity problem and other versions of the $\mathcal{H}_\infty$-problem in order to achieve a prescribed degree of high-frequency roll-off.

In [6] and earlier work it is explained how the matrix Π_γ^{-1} may be factored if the plant G is represented in polynomial matrix fraction form. This approach removes the need for G to be proper and leads to factorization algorithms that involve polynomial matrices. Although such algorithms are available [7] they are fairly cumbersome.

More flexible than the state representation is the descriptor representation of the generalized plant. Nonproper plants can be handled without any difficulty. This is why in the present paper we explore the frequency domain solution of the $\mathcal{H}_\infty$-problem for descriptor systems. In particular, we assume that the generalized plant (1) is characterized by the descriptor equations

$$
\begin{aligned}
E\dot{x} &= Ax + B_1 w + B_2 u \\
z &= C_1 x + D_{11} w + D_{12} u \\
y &= C_2 x + D_{21} w + D_{22} u
\end{aligned}
\tag{6}
$$

We aim at algorithms that involve constant matrix manipulations as provided in the MATLAB core.

A solution to the $\mathcal{H}_\infty$ problem for descriptor systems based on J-spectral factorization through generalized algebraic Riccati equations is presented in [9]. It generalizes the celebrated two Riccati equation solution of the standard state $\mathcal{H}_\infty$ problem [4].

A well-known reference on descriptor systems is the seminal paper [11]. Interesting and useful material on descriptor systems may be found in [1].

2   Descriptor representation of Π_γ^{-1}

In this paper we occasionally use the device of representing a system, or related objects such as transfer matrices, in *signal form*. This representation is connected to J. C. Willems' behavioral approach to system theory [8].

The first application of the idea is the signal representation of the matrix Π_γ^{-1} as given by (3). We interpret the rational matrix as a transfer matrix and write

$$
\begin{bmatrix} y_1 \\ y_2 \end{bmatrix} = \Pi_\gamma^{-1} \begin{bmatrix} u_1 \\ u_2 \end{bmatrix}
\tag{7}
$$

This equation imposes a relation between the signals u_1, u_2, y_1 and y_2. By introducing several ancillary signals we may find an alternative signal representation for the matrix Π_γ^{-1}. Define two new signals v_1 and v_2 by

$$\begin{bmatrix} v_1 \\ v_2 \end{bmatrix} = \begin{bmatrix} I - \frac{1}{\gamma^2}G_{11}G_{\tilde{11}} & -\frac{1}{\gamma^2}G_{11}G_{\tilde{21}} \\ -\frac{1}{\gamma^2}G_{21}G_{\tilde{11}} & -\frac{1}{\gamma^2}G_{21}G_{\tilde{21}} \end{bmatrix}^{-1} \begin{bmatrix} G_{12} & 0 \\ G_{22} & I \end{bmatrix} \begin{bmatrix} u_1 \\ u_2 \end{bmatrix} \qquad (8)$$

Then from (3) and (8) we see that equivalently to (7) we have

$$\begin{bmatrix} y_1 \\ y_2 \end{bmatrix} = \begin{bmatrix} G_{\tilde{12}} & G_{\tilde{22}} \\ 0 & I \end{bmatrix} \begin{bmatrix} v_1 \\ v_2 \end{bmatrix},$$

$$\begin{bmatrix} I - \frac{1}{\gamma^2}G_{11}G_{\tilde{11}} & -\frac{1}{\gamma^2}G_{11}G_{\tilde{21}} \\ -\frac{1}{\gamma^2}G_{21}G_{\tilde{11}} & -\frac{1}{\gamma^2}G_{21}G_{\tilde{21}} \end{bmatrix} \begin{bmatrix} v_1 \\ v_2 \end{bmatrix} = \begin{bmatrix} G_{12} & 0 \\ G_{22} & I \end{bmatrix} \begin{bmatrix} u_1 \\ u_2 \end{bmatrix} \qquad (9)$$

After defining the additional signal $w = G_{\tilde{11}}v_1 + G_{\tilde{21}}v_2$, eliminating v_2 and renaming v_1 to v it easily follows that

$$\begin{bmatrix} v \\ -u_2 \end{bmatrix} = G \begin{bmatrix} \frac{1}{\gamma^2}w \\ u_1 \end{bmatrix}, \qquad \begin{bmatrix} w \\ y_1 \end{bmatrix} = G^\sim \begin{bmatrix} v \\ y_2 \end{bmatrix} \qquad (10)$$

These equations define a relation between the signals y_1, y_2, u_1 and u_2 that is equivalent to (7).

Descriptor representation of Π_γ^{-1}. Application of the descriptor equations (6) to (10) yields

$$\begin{aligned}
E\dot{x}_1 &= Ax_1 + \tfrac{1}{\gamma^2}B_1w + B_2u_1 \\
v &= C_1x_1 + \tfrac{1}{\gamma^2}D_{11}w + D_{12}u_1 \\
-u_2 &= C_2x_1 + \tfrac{1}{\gamma^2}D_{21}w + D_{22}u_1 \\
-E^T\dot{x}_2 &= A^Tx_2 + C_1^Tv + C_2^Ty_2 \\
w &= B_1^Tx_2 + D_{11}^Tv + D_{21}^Ty_2 \\
y_1 &= B_2^Tx_2 + D_{12}^Tv + D_{22}^Ty_2
\end{aligned} \qquad (11)$$

Rewriting the second and fifth of these equations as

$$\begin{bmatrix} I & -\frac{1}{\gamma^2}D_{11} \\ -D_{11}^T & I \end{bmatrix} \begin{bmatrix} v \\ w \end{bmatrix} = \begin{bmatrix} C_1x_1 + D_{12}u_1 \\ B_1^Tx_2 + D_{21}^Ty_2 \end{bmatrix} \qquad (12)$$

we may solve for v and w and obtain

$$\begin{aligned}
v &= (I - \tfrac{1}{\gamma^2}D_{11}D_{11}^T)^{-1} \\
&\quad \cdot (C_1x_1 + D_{12}u_1 + \tfrac{1}{\gamma^2}D_{11}B_1^Tx_2 + \tfrac{1}{\gamma^2}D_{11}D_{21}^Ty_2), \\
w &= (I - \tfrac{1}{\gamma^2}D_{11}^TD_{11})^{-1}(B_1^Tx_2 + D_{21}^Ty_2 + D_{11}^TC_1x_1 + D_{11}^TD_{12}u_1)
\end{aligned} \qquad (13)$$

Substitution into the remaining equations yields with the introduction of a third pseudo state variable $x_3 = y_2$

$$
\begin{aligned}
E\dot{x}_1 &= Ax_1 + \tfrac{1}{\gamma^2}B_1(I - \tfrac{1}{\gamma^2}D_{11}^T D_{11})^{-1} \\
&\quad \cdot (B_1^T x_2 + D_{21}^T x_3 + D_{11}^T C_1 x_1 + D_{11}^T D_{12}u_1) + B_2 u_1, \\
-E^T\dot{x}_2 &= A^T x_2 + C_1^T(I - \tfrac{1}{\gamma^2}D_{11}D_{11}^T)^{-1} \\
&\quad \cdot (C_{11}x_1 + D_{12}u_1 + \tfrac{1}{\gamma^2}D_{11}B_1^T x_2 + \tfrac{1}{\gamma^2}D_{11}D_{21}^T x_3) + C_2^T x_3, \\
-u_2 &= C_2 x_1 + \tfrac{1}{\gamma^2}D_{21}(I - \tfrac{1}{\gamma^2}D_{11}^T D_{11})^{-1} \\
&\quad \cdot (B_1^T x_2 + D_{21}^T x_3 + D_{11}^T C_1 x_1 + D_{11}^T D_{12}u_1) + D_{22}u_1, \\
y_1 &= B_2^T x_2 + D_{12}^T(I - \tfrac{1}{\gamma^2}D_{11}D_{11}^T)^{-1} \\
&\quad \cdot (C_{11}x_1 + D_{12}u_1 + \tfrac{1}{\gamma^2}D_{11}B_1^T x_2 + \tfrac{1}{\gamma^2}D_{11}D_{21}^T x_3) + D_{22}^T x_3, \\
y_2 &= x_3
\end{aligned}
\tag{14}
$$

Inspection shows that after rearrangement these equations imply that

$$
\begin{aligned}
&\Pi_\gamma^{-1}(s) \\
&= \begin{bmatrix} b_2^T & d_{12}^T & d_2^T \\ 0 & 0 & I \end{bmatrix}
\begin{bmatrix} -\tfrac{1}{\gamma^2}r_1 & se - a & -\tfrac{1}{\gamma^2}d_{21}^T \\ -se^T - a^T & -r_2 & -c_2^T \\ -\tfrac{1}{\gamma^2}d_{21} & -c_2 & -\tfrac{1}{\gamma^2}r_3 \end{bmatrix}^{-1}
\begin{bmatrix} b_2 & 0 \\ d_{12} & 0 \\ d_{22} & I \end{bmatrix} \\
&\qquad\qquad\qquad\qquad\qquad\qquad\qquad + \begin{bmatrix} r_0 & 0 \\ 0 & 0 \end{bmatrix}
\end{aligned}
\tag{15}
$$

where

$$
\begin{aligned}
d_{12} &= C_1^T(I - \tfrac{1}{\gamma^2}D_{11}D_{11}^T)^{-1}D_{12} \\
b_2 &= B_2 + \tfrac{1}{\gamma^2}B_1(I - \tfrac{1}{\gamma^2}D_{11}^T D_{11})^{-1}D_{11}^T D_{12} \\
d_{22} &= D_{22} + \tfrac{1}{\gamma^2}D_{21}(I - \tfrac{1}{\gamma^2}D_{11}^T D_{11})^{-1}D_{11}^T D_{12} \\
r_1 &= B_1(I - \tfrac{1}{\gamma^2}D_{11}^T D_{11})^{-1}B_1^T \\
e &= E \\
a &= A + \tfrac{1}{\gamma^2}B_1(I - \tfrac{1}{\gamma^2}D_{11}^T D_{11})^{-1}D_{11}^T C_1 \\
c_2 &= C_2 + \tfrac{1}{\gamma^2}D_{21}(I - \tfrac{1}{\gamma^2}D_{11}^T D_{11})^{-1}D_{11}^T C_1 \\
r_2 &= C_1^T(I - \tfrac{1}{\gamma^2}D_{11}D_{11}^T)^{-1}C_1 \\
d_{21} &= D_{21}(I - \tfrac{1}{\gamma^2}D_{11}^T D_{11})^{-1}B_1^T \\
r_3 &= D_{21}(I - \tfrac{1}{\gamma^2}D_{11}^T D_{11})^{-1}D_{21}^T \\
r_0 &= D_{12}^T(I - \tfrac{1}{\gamma^2}D_{11}D_{11}^T)^{-1}D_{12}
\end{aligned}
\tag{16}
$$

Π_γ^{-1} is in the descriptor form $\Pi_\gamma(s) = R + P^T(sF - Q)^{-1}P$ with R and Q symmetric and F skew-symmetric.

Takaba *et al.* [9] list a number of assumptions on the generalized plant (6). The assumptions are sufficient (but not necessary) to guarantee that for γ large enough both Π_γ and its inverse exist and have no poles on the imaginary axis. We discuss the existence of a solution to the $\mathcal{H}_\infty$ problem in § 7.

3   Clements factorization algorithm

The algorithm for the factorization of a rational matrix in descriptor form that we present in § 4 relies on an algorithm of Clements [2]. Clements' algorithm, in turn, is based on the orthogonal transformation of a suitably defined para-Hermitian real matrix pencil $sE - A$ to what we call here its *Clements form*

$$\begin{bmatrix} 0 & 0 & sE_1 - A_1 \\ 0 & -A_2 & sE_3 - A_3 \\ -sE_1^T - A_1^T & -sE_3^T - A_3^T & sE_4 - A_4 \end{bmatrix} \tag{17}$$

By assumption the pencil $sE - A$ has no finite zeros with zero real part. The finite zeros of the pencil $sE_1 - A_1$ either all have negative real part or all have positive real part, depending on what is required. These zeros are precisely the finite zeros of $sE - A$ with negative or with positive real parts.

The Clements form (17) is obtained in a finite number of deflation steps. We review the procedure as presented in [2]. The appendix of the present paper provides details that are not found in [2].

1. *Nonfinite deflation.* If the size of the pencil $sE - A$ is denoted as n then the pencil has n zeros. At least $n - \text{rank}\, E$ of these are infinite. Some of the remaining zeros may also be infinite. Corresponding to each of these remaining infinite zeros there exists a vector x such that $Ex = 0$, $Ax \neq 0$ and $x^T Ax = 0$. ¿From this vector x an orthogonal matrix U may be constructed such that

$$U(sE - A)U^T = \begin{bmatrix} 0 & 0 & a \\ 0 & sE_o - A_o & \times \\ a^T & \times & \times \end{bmatrix} \tag{18}$$

   where a is a real number. The precise form of the entries marked $\times$ is unimportant. Both the size and the rank of E_o are two less than the size and rank of E.

   Setting $E := E_o$ and $A_o := A$ the deflation step is repeated until nonfinite deflation is no longer possible. Upon termination an orthogonal matrix U has been found such that the original pencil $sE - A$ is transformed according to (18), where a now is a real matrix. The rank of E_o equals the number of *finite* zeros of the pencil $sE_o - A_o$.

2. *Finite deflation.* Given a finite root λ of the pencil $sE - A$ finite deflation amounts to constructing a unitary matrix U such that

$$U(sE - A)U^H = \begin{bmatrix} 0 & 0 & se - a \\ 0 & sE_o - A_o & \times \\ -s\bar{e}^T - \bar{a}^T & \times & \times \end{bmatrix} \tag{19}$$

where e and a are complex numbers such that $\lambda = a/e$. The overbar denotes the complex conjugate. The pencil $sE_o - A_o$ has two finite zeros fewer than $sE - A$.

After repeating this step half as many times as there are finite zeros a unitary matrix U is found such that (19) holds, where the quantities e and a now are matrices such that the zeros of $se - a$ together with those of $-s\bar{e}^T - \bar{a}^T$ constitute the finite zeros of $sE - A$. The pencil $sE_o - A_o$ has infinite zeros only.

If the deflation steps are suitably selected then U is real orthogonal and the pencil $se - a$ is real such that all its zeros have either strictly negative or strictly positive real parts.

3. *Completion.* If first all nonfinite deflations are completed and after this all finite deflations then the rank of the matrix E_o has been reduced to zero, that is, E_o is the zero matrix. Thus, after completion of the nonfinite and finite deflations we have an orthogonal matrix U such that

$$U(sE - A)U^H = \begin{bmatrix} 0 & 0 & sE_1 - A_1 \\ 0 & -A_o & sE_3 - A_3 \\ -sE_1^T - A_1^T & -sE_3^T - A_3^T & sE_4 - A_4 \end{bmatrix}$$

$$(20)$$

A_o is symmetric and typically indefinite. The final step of the algorithm is to construct an orthogonal transformation to bring A_o into the form

$$\begin{bmatrix} 0 & 0 & a \\ 0 & A_2 & \times \\ a^T & \times & \times \end{bmatrix} \qquad (21)$$

where A_2 is positive-definite or indefinite, depending on the situation.

Clements' factorization algorithm   Clements' algorithm [2] for the spectral factorization of a rational matrix Φ requires it to be in *separated descriptor form*. Φ is in separated descriptor form if it may be written as

$$\begin{aligned} \Phi(s) = \ & D + B^T(-sE^T - A^T)^{-1}C + C^T(sE - A)^{-1}B \\ & + B^T(-sE^T - A^T)^{-1}Q(sE - A)^{-1}B \end{aligned} \qquad (22)$$

To find the desired factorization define the para-Hermitian pencil

$$s\mathcal{E} - \mathcal{A} = \begin{bmatrix} 0 & sE - A & -B \\ -sE^T - A^T & -Q & -C \\ -B^T & -C^T & -D \end{bmatrix} \qquad (23)$$

If we assume that Φ has no zeros on the imaginary axis then neither has the pencil $s\mathcal{E} - A$. Transform $s\mathcal{E} - A$ to Clements form

$$U(s\mathcal{E} - A)U^T = \begin{bmatrix} 0 & 0 & s\mathcal{E}_1 - A_1 \\ 0 & -A_2 & s\mathcal{E}_3 - A_3 \\ -s\mathcal{E}_1^T - A_1^T & -s\mathcal{E}_3^T - A_3^T & s\mathcal{E}_4 - A_4 \end{bmatrix} \tag{24}$$

such that all the zeros of the pencil $s\mathcal{E}_1 - A_1$ are in the open right-half plane and the size of A_2 equals that of Φ. Suppose that Φ is $m \times m$ and that A and E are $n \times n$, and partition

$$U = \begin{bmatrix} U_{11} & U_{12} \\ U_{21} & U_{22} \end{bmatrix} \tag{25}$$

where U_{11} is $(n+m) \times n$, U_{12} is $(n+m) \times (n+m)$, U_{21} is $n \times n$, and U_{22} is $n \times (n+m)$. Factor

$$U_{12}^{-1} \begin{bmatrix} 0 & 0 \\ 0 & A_2 \end{bmatrix} (U_{12}^{-1})^T = \begin{bmatrix} H^T \\ L^T \end{bmatrix} R [\, H \quad L \,] \tag{26}$$

where R is $m \times m$. In particular, we may take

$$R = A_2, \qquad \begin{bmatrix} H^T \\ L^T \end{bmatrix} = U_{12}^{-1} \begin{bmatrix} 0 \\ I \end{bmatrix} \tag{27}$$

Then

$$\Phi(s) = M^T(-s)RM(s), \quad M(s) = L + H(sE - A)^{-1}B \tag{28}$$

The zeros of M are the mirror images of the zeros of the pencil $s\mathcal{E}_1 - A_1$ with respect to the imaginary axis and, hence, all have negative real parts.

If needed R may be transformed into a signature matrix using Schur decomposition.

Clements [2] only discusses the case that Φ is positive-definite on the imaginary axis. Then A_2 in (24) is positive-definite and without loss of generality R may be taken equal to the unit matrix. If Φ is indefinite on the imaginary axis (but with constant inertia) then A_2 is nonsingular indefinite with the same inertia. R may now be taken to be a signature matrix.

4   Factorization of a descriptor form rational matrix

In this section we consider the factorization of a para-Hermitian rational matrix in real descriptor form

$$\Phi(s) = D + B^T(sE - A)^{-1}B \qquad (29)$$

with D and A symmetric and E skew-symmetric. In § 5 we specialize the results to the factorization of Π_γ^{-1} as given by (15).

Transformation to separated form.   We first discuss how (29) is brought into separated form. The pencil $sE - A$ may be transformed to Clements form according to

$$V(sE - A)V^T = \begin{bmatrix} 0 & 0 & sE_1 - A_1 \\ 0 & -A_2 & sE_3 - A_3 \\ -sE_1^T - A_1^T & -sE_3^T - A_3^T & sE_4 - A_4 \end{bmatrix} \qquad (30)$$

where $sE_1 - A_1$ has all its zeros in the open left-half plane. The size of A_2 is unimportant here so that in the algorithm of § 3 the third step may as well be omitted. Write

$$\begin{bmatrix} 0 & 0 & sE_1 - A_1 \\ 0 & -A_2 & sE_3 - A_3 \\ -sE_1^T - A_1^T & -sE_3^T - A_3^T & sE_4 - A_4 \end{bmatrix} = (se - a)Q(-se^T - a^T) \qquad (31)$$

where

$$se - a = \begin{bmatrix} sE_1 - A_1 & 0 & 0 \\ sE_3 - A_3 & -A_2 & 0 \\ (sE_4 - A_4)/2 & 0 & I \end{bmatrix}, \quad Q = \begin{bmatrix} 0 & 0 & I \\ 0 & -A_2^{-1} & 0 \\ I & 0 & 0 \end{bmatrix} \qquad (32)$$

Then we have Φ in the separated form

$$\Phi(s) = d + b^T(-se^T - a^T)^{-1}q(se - a)^{-1}b \qquad (33)$$

where $d = D$, $b = VB$, and $q = Q^{-1}$. Φ is now in a form where we may straightforwardly apply Clements' factorization algorithm.

Reduction.   Rather than leaving the algorithm at this we exploit the structure of the pencil $se - a$ to reduce the dimension of the pencil (23) whose

Clements form needs to be computed for the factorization of (33). Moreover, this computation may be decoupled from that of the Clements form of $sE - A$.

Partitioning b in an obvious way it follows easily from (33) that the pencil (23) is given by

$$
\begin{bmatrix}
0 & 0 & 0 & sE_1 - A_1 & 0 & 0 & -b_1 \\
0 & 0 & 0 & sE_3 - A_3 & -A_2 & 0 & -b_2 \\
0 & 0 & 0 & (sE_4 - A_4)/2 & 0 & I & -b_3 \\
-sE_1^T - A_1^T & -sE_3^T - A_3^T & (-sE_4^T - A_4^T)/2 & 0 & 0 & -I & 0 \\
0 & -A_2 & 0 & 0 & A_2 & 0 & 0 \\
0 & 0 & I & -I & 0 & 0 & 0 \\
-b_1^T & -b_2^T & -b_3^T & 0 & 0 & 0 & -D
\end{bmatrix}
\tag{34}
$$

By suitable symmetric elementary row and column operations this pencil may be transformed into the more structured form

$$
\begin{bmatrix}
0 & 0 & 0 & 0 & 0 & 0 & -I \\
0 & 0 & 0 & sE_1 - A_1 & -b_1 & 0 & sE_1 - A_1 \\
0 & 0 & -A_2 & sE_3 - A_3 & -b_2 & 0 & sE_3 - A_3 \\
0 & -sE_1^T - A_1^T & -sE_3^T - A_3^T & sE_4 - A_4 & -b_3 & 0 & (sE_4 - A_4)/2 \\
0 & -b_1^T & -b_2^T & -b_3^T & -D & 0 & 0 \\
0 & 0 & 0 & 0 & 0 & A_2 & 0 \\
-I & -sE_1^T - A_1^T & -sE_3^T - A_3^T & (-sE_4^T - A_4^T)/2 & 0 & 0 & 0
\end{bmatrix}
\tag{35}
$$

Inspection shows that multiplication on the left of the second row of blocks by V^T and of the second column of blocks on the right by V results in

$$
\begin{bmatrix}
0 & 0 & 0 & 0 & -I \\
0 & sE - A & -B & 0 & \times \\
0 & -B^T & -D & 0 & \times \\
0 & 0 & 0 & A_2 & 0 \\
-I & \times & \times & 0 & 0
\end{bmatrix}
\tag{36}
$$

Further inspection reveals that the computation of the Clements form of the pencil (34) actually has been reduced to the computation of the Clements form of the pencil

$$
\begin{bmatrix}
sE - A & -B \\
-B^T & -D
\end{bmatrix}
\tag{37}
$$

Note that this pencil corresponds to the "denominator pencil" of the inverse of Φ. Suppose that the pencil (37) is transformed to Clements form by the orthogonal matrix W so that after suitable partitioning

$$
\underbrace{\begin{bmatrix} W_{11} & W_{12} \\ W_{21} & W_{22} \\ W_{31} & W_{32} \end{bmatrix}}_{W} \begin{bmatrix} sE - A & -B \\ -B^T & -D \end{bmatrix} W^T
$$

$$
= \begin{bmatrix} 0 & 0 & s\bar{E}_1 - \bar{A}_1 \\ 0 & -\bar{A}_2 & s\bar{E}_3 - \bar{A}_3 \\ -s\bar{E}_1^T - \bar{A}_1^T & -s\bar{E}_3^T - \bar{A}_3^T & s\bar{E}_4 - \bar{A}_4 \end{bmatrix} \tag{38}
$$

Again, generally the third step of the algorithm of § 3 may be omitted so that $\bar{A}_2$ has maximal size.

It may now be checked that given W and partitioning

$$
V = \begin{bmatrix} V_1 \\ V_2 \\ V_3 \end{bmatrix} \tag{39}
$$

the pencil (34) is transformed into the Clements form

$$
\left[\begin{array}{cc|cc|cc} 0 & 0 & 0 & 0 & 0 & -I \\ 0 & 0 & 0 & 0 & s\bar{E}_1 - \bar{A}_1 & \times \\ 0 & 0 & -\bar{A}_2 & 0 & \times & \times \\ 0 & 0 & 0 & A_2 & 0 & 0 \\ \hline 0 & -s\bar{E}_1^T - \bar{A}_1^T & \times & 0 & \times & \times \\ -I & \times & \times & 0 & \times & 0 \end{array} \right] \tag{40}
$$

by the (nonorthogonal) transformation U

$$
U = \left[\begin{array}{ccc|ccc|c} 0 & 0 & 0 & 0 & 0 & I & 0 \\ W_{11}V_1^T & W_{11}V_2^T & W_{11}V_3^T & W_{11}V_3^T & W_{11}V_2^T & 0 & W_{12} \\ W_{21}V_1^T & W_{21}V_2^T & W_{21}V_3^T & W_{21}V_3^T & W_{21}V_2^T & 0 & W_{22} \\ 0 & 0 & 0 & 0 & I & 0 & 0 \\ \hline W_{31}V_1^T & W_{31}V_2^T & W_{31}V_3^T & W_{31}V_3^T & W_{31}V_2^T & 0 & W_{32} \\ 0 & 0 & 0 & I & 0 & 0 & 0 \end{array} \right] \tag{41}
$$

If the "center block"

$$
\begin{bmatrix} -\bar{A}_2 & 0 \\ 0 & A_2 \end{bmatrix} \tag{42}
$$

of the pencil (40) does not have the required dimensions $m \times m$ then Step 3 of the algorithm of § 3 may be applied until the block has the required dimensions.

If on the other hand the center block happens to have the required size then it may be checked that the correct partitioning of U needed to complete the factorization according to Clements' algorithm is

$$U = \begin{bmatrix} U_{11} & U_{12} \\ U_{21} & U_{22} \end{bmatrix}$$

$$= \left[\begin{array}{ccc|cccc} 0 & 0 & 0 & 0 & 0 & I & 0 \\ W_{11}V_1^T & W_{11}V_2^T & W_{11}V_3^T & W_{11}V_3^T & W_{11}V_2^T & 0 & W_{12} \\ W_{21}V_1^T & W_{21}V_2^T & W_{21}V_3^T & W_{21}V_3^T & W_{21}V_2^T & 0 & W_{22} \\ 0 & 0 & 0 & 0 & I & 0 & 0 \\ \hline W_{31}V_1^T & W_{31}V_2^T & W_{31}V_3^T & W_{31}V_3^T & W_{31}V_2^T & 0 & W_{32} \\ 0 & 0 & 0 & I & 0 & 0 & 0 \end{array} \right]$$

$$(43)$$

5   Factorization of Π_γ^{-1}

In this section we apply the algorithm of § 4 to the factorization of Π_γ^{-1} as given by (15).

Transformation of the "denominator."   We first transform the denominator

$$\begin{bmatrix} -\frac{1}{\gamma^2}r_1 & se-a & -\frac{1}{\gamma^2}d_{21}^T \\ -se^T - a^T & -r_2 & -c_2^T \\ -\frac{1}{\gamma^2}d_{21} & -c_2 & -\frac{1}{\gamma^2}r_3 \end{bmatrix} \tag{44}$$

$(sE - A$ in § 4) of Π_γ^{-1} to the Clements form

$$\begin{bmatrix} 0 & 0 & sE_1 - A_1 \\ 0 & -A_2 & sE_3 - A_3 \\ -sE_1^T - A_1^T & -sE_3^T - A_3^T & sE_4 - A_4 \end{bmatrix} \tag{45}$$

such that $sE_1 - A_1$ has all its zeros in the open left-half plane. Denoting the size of a as n the pencil (45) has size $2n + m_2$. The transformation to Clements form is completed as soon as its center block A_2 has size m_2. The transformation is accomplished by the orthogonal matrix

$$V = \begin{bmatrix} V_1 \\ V_2 \\ V_3 \end{bmatrix} = \begin{bmatrix} V_{11} & V_{12} \\ V_{21} & V_{22} \\ V_{31} & V_{32} \end{bmatrix} \tag{46}$$

where the rows have heights n, m_2 and n, respectively, and the columns on the rightmost size successively have widths $2n$ and m_2.

Transformation of the "numerator." We next consider the transformation of the "numerator pencil," denoted

$$
\begin{bmatrix} sE - A & -B \\ -B^T & -D \end{bmatrix}
\tag{47}
$$

in § 4, to Clements form. In the factorization of Π_γ^{-1} this pencil is

$$
\left[
\begin{array}{ccc|cc}
-\frac{1}{\gamma^2}r_1 & se - a & -\frac{1}{\gamma^2}d_{21}^T & -b_2 & 0 \\
-se^T - a^T & -r_2 & -c_2^T & -d_{12} & 0 \\
-\frac{1}{\gamma^2}d_{21} & -c_2 & -\frac{1}{\gamma^2}r_3 & -d_{22} & -I \\
\hline
-b_2^T & -d_{12}^T & -d_{22}^T & -r_0 & 0 \\
0 & 0 & -I & 0 & 0
\end{array}
\right]
\tag{48}
$$

The unit matrices in the (3, 5) and (5, 3) positions of (48) may be used to clear the third row and third column, respectively. The result is that finding the Clements form of the numerator pencil amounts to transforming the $(2n + k_2) \times (2n + k_2)$ pencil

$$
\begin{bmatrix}
-\frac{1}{\gamma^2}r_1 & se - a & -b_2 \\
-se^T - a^T & -r_2 & -d_{12} \\
-b_2^T & -d_{12}^T & -r_0
\end{bmatrix}
\tag{49}
$$

to the Clements form

$$
\begin{bmatrix}
0 & 0 & s\bar{E}_1 - \bar{A}_1 \\
0 & -\bar{A}_2 & s\bar{E}_3 - \bar{A}_3 \\
-s\bar{E}_1^T - \bar{A}_1^T & -s\bar{E}_3^T - \bar{A}_3^T & s\bar{E}_4 - \bar{A}_4
\end{bmatrix}
\tag{50}
$$

such that all the zeros of $s\bar{E}_1 - \bar{A}_1$ are in the open right-half plane and $\bar{A}_2$ has size k_2. Suppose that the orthogonal matrix

$$
W = \begin{bmatrix} W_{11} & W_{12} \\ W_{21} & W_{22} \\ W_{31} & W_{32} \end{bmatrix}
\tag{51}
$$

with row heights n, k_2 and n and column widths $2n$ and k_2, accomplishes this. Then it may be checked that

$$
\left[
\begin{array}{cc|cc}
0 & 0 & 0 & I \\
\hline
W_{11} & 0 & W_{12} & 0 \\
W_{21} & 0 & W_{22} & 0 \\
W_{31} & 0 & W_{32} & 0 \\
\hline
0 & I & 0 & 0
\end{array}
\right]
\tag{52}
$$

transforms (48) to the Clements form

$$
\begin{bmatrix}
0 & 0 & 0 & 0 & -I \\
0 & 0 & 0 & s\bar{E}_1 - \bar{A}_1 & \times \\
0 & 0 & -\bar{A}_2 & s\bar{E}_3 - \bar{A}_3 & \times \\
0 & -s\bar{E}_1^T - \bar{A}_1^T & -s\bar{E}_3^T - \bar{A}^T & s\bar{E}_4 - \bar{A}_4 & \times \\
-I & \times & \times & \times & -r_3
\end{bmatrix}
\tag{53}
$$

Completion of the factorization. We use the results of § 4 to complete the factorization. Since the center block (42) has precisely the required size $k_2 + m_2$ we may partition U as in (43). With appropriate substitutions it follows that the factorization of Π_γ^{-1} is given by $\Pi_\gamma^{-1} = M^\sim R M$, where

$$
M(s) = L + H \begin{bmatrix}
sE_1 - A_1 & 0 & 0 \\
sE_3 - A_3 & -A_2 & 0 \\
(sE_4 - A_4)/2 & 0 & I
\end{bmatrix}^{-1}
\begin{bmatrix}
G_1 \\
G_2 \\
G_3
\end{bmatrix},
\tag{54}
$$

$$
\begin{bmatrix}
G_1 \\
G_2 \\
G_3
\end{bmatrix} = V \begin{bmatrix}
b_2 & 0 \\
d_{12} & 0 \\
d_{22} & I
\end{bmatrix}, \qquad
R = \begin{bmatrix}
\bar{A}_2 & 0 \\
0 & -A_2
\end{bmatrix}
\tag{55}
$$

H and L are given by

$$
H = \begin{bmatrix}
H_{11} & 0 & 0 \\
H_{21} & I & 0
\end{bmatrix}, \qquad
L = \begin{bmatrix}
L_{11} & 0 \\
L_{21} & 0
\end{bmatrix}
\tag{56}
$$

where

$$
\begin{bmatrix}
W_{11}V_{31}^T & W_{12} \\
W_{21}V_{31}^T & W_{22}
\end{bmatrix}
\begin{bmatrix}
H_{11}^T & H_{21}^T \\
L_{11}^T & L_{21}^T
\end{bmatrix} =
\begin{bmatrix}
0 & -W_{11}V_{21}^T \\
I & -W_{21}V_{21}^T
\end{bmatrix}
\tag{57}
$$

Inspection of (54) and (56) shows that the spectral factor M may be simplified to

$$
M(s) = \begin{bmatrix}
L_{11} & 0 \\
L_{21} & 0
\end{bmatrix} +
\begin{bmatrix}
H_{11} & 0 \\
H_{21} & I
\end{bmatrix}
\begin{bmatrix}
sE_1 - A_1 & 0 \\
sE_3 - A_3 & -A_2
\end{bmatrix}^{-1}
\begin{bmatrix}
G_1 \\
G_2
\end{bmatrix}
\tag{58}
$$

Using Schur decomposition R may be made into a signature matrix.

6   Computation of the compensator

Using the algorithm of § 5 we obtain the spectral factorization $\Pi_\gamma^{-1} = M_\gamma^\sim J M_\gamma$, where M_γ is in the descriptor form

$$
M_\gamma(s) = L + H(sN - F)^{-1}G
\tag{59}
$$

and $J = \mathrm{diag}(I_{m_2}, -I_{k_2})$. With this spectral factor suboptimal compensators may be computed from (5).

Derivation of the compensator.   We omit the subscript γ in what follows. The compensator (5) may be characterized by the signal representation $u = Ky$, or

$$u = (M_{11} + UM_{21})^{-1}(M_{12} + UM_{22})y \qquad (60)$$

It follows that

$$[\, I \ \ U \,]M \begin{bmatrix} -u \\ y \end{bmatrix} = 0 \qquad (61)$$

Setting

$$x_1 = (sN - F)^{-1}H \begin{bmatrix} -u \\ y \end{bmatrix}, \qquad x_2 = u \qquad (62)$$

and partitioning

$$L = \begin{bmatrix} L_{11} & L_{12} \\ L_{21} & L_{22} \end{bmatrix}, \qquad H = \begin{bmatrix} H_1 \\ H_2 \end{bmatrix}, \qquad G = [\, G_1 \ G_2 \,] \qquad (63)$$

we obtain with (59) the relations

$$\begin{aligned} (sN - F)x_1 + G_1 x_2 &= G_2 y \\ (H_1 + UH_2)x_1 - (L_{11} + UL_{21})x_2 &= -(L_{12} + UL_{22})y \\ u &= x_2 \end{aligned} \qquad (64)$$

These equations define the compensator as a descriptor system with transfer matrix

$$K(s) = [\, 0 \ \ I \,]\begin{bmatrix} sN - F & G_1 \\ H_1 + UH_2 & -L_{11} - UL_{21} \end{bmatrix}^{-1}\begin{bmatrix} G_2 \\ -L_{12} - UL_{22} \end{bmatrix} \qquad (65)$$

If we take in particular $U = 0$ then we obtain the central compensator

$$K(s) = [\, 0 \ \ I \,]\begin{bmatrix} sN - F & G_1 \\ H_1 & -L_{11} \end{bmatrix}^{-1}\begin{bmatrix} G_2 \\ -L_{12} \end{bmatrix} \qquad (66)$$

The compensator (65) is a lower fractional transformation corresponding to the block diagram of Fig. 2. The descriptor system K is defined by

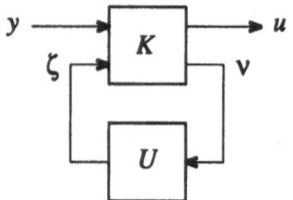

Figure 2: All compensators

$$\begin{bmatrix} N & 0 \\ 0 & 0 \end{bmatrix} \dot{\xi} = \begin{bmatrix} F & -G_1 \\ -H_1 & L_{11} \end{bmatrix} \xi + \begin{bmatrix} G_2 & 0 \\ -L_{12} & I \end{bmatrix} \begin{bmatrix} y \\ \zeta \end{bmatrix} \qquad (67)$$

$$\begin{bmatrix} u \\ v \end{bmatrix} = \begin{bmatrix} 0 & I \\ H_2 & -L_{21} \end{bmatrix} \xi + \begin{bmatrix} 0 & 0 \\ 0 & -L_{22} \end{bmatrix} \qquad (68)$$

7   Conclusions

In this paper the solution of the standard $\mathcal{H}_\infty$ problem is studied for descriptor systems of the form

$$\begin{aligned} E\dot{x} &= Ax + B_1 w + B_2 u \\ z &= C_1 x + D_{11} w + D_{12} u \\ y &= C_2 x + D_{21} w + D_{22} u \end{aligned} \qquad (69)$$

Descriptor representations are flexible tools for defining practical $\mathcal{H}_\infty$ problems that arise in control system design, including problems with nonproper weighting functions.

The solution that is presented relies on the transformation of two matrix pencils to Clements form. The solution is described in Sections 5 and 6 and may straightforwardly be coded for MATLAB.

No doubt the present solution may be used to rederive the solution of Takaba *et al.* [9] that is phrased in terms of generalized algebraic Riccati equations. These GAREs are numerically resolved by computing the generalized eigenstructure defined by matrix pencils that are closely related if not identical to the matrix pencils (44) and (49) that are introduced in § 5. This approach does not appear to have numerical advantages over the compact solution presented in the present paper.

The conditions imposed on the descriptor system (69) in [9] guarantee that the pencils (44) and (49) of § 5 are nonsingular. The pencils have no zeros on the imaginary axis if γ is sufficiently large [6]. To check whether the compensators obtained according to § 6 actually are solutions of the $\mathcal{H}_\infty$ problem it is enough to check whether any of the compensators (for instance the central solution) stabilizes the generalized plant [6]. More general conditions on the descriptor system (69) than those of [9] that guarantee that the pencils (44) and (49) are nonsingular and have no zeros on the imaginary axis for γ large enough remain to be investigated. Until this problem has been resolved the most practical solution is to check the pencils themselves before proceeding with the algorithm.

Even if the pencils (44) and (49) are nonsingular and have no zeros on the imaginary axis the algorithm may fail because the coefficient matrix on the left-hand side of (57) is singular. This may happen in particular if γ equals the minimal value of the ∞-norm [6]. By continuity, numerical ill-conditioning arises when γ is close to this value.

It is not difficult to modify the algorithm so that ill-conditioning is completely avoided, and, indeed, *optimal* (as opposed to sub-optimal) compensators may be computed. For reasons of space the procedure is not included in this paper. Also the accompanying pole-zero cancellation problem has been resolved.

Appendix: Details of Clements' transformation

In this appendix we provide the details of the implementation — in terms of basic MATLAB operations — of the three steps of the transformation to Clements form described in § 3.

Nonfinite deflation. Let the columns of the matrix N be orthogonal and span the null space of E. Determine a nontrivial vector ξ such that $N^T A N \xi = 0$. Then $x = N\xi$. Let V_1 be equal to x after normalization such that $V_1^T V_1 = 1$. Construct the matrix V_2 such that its columns are orthogonal and orthogonal to V_1 and AV_1. V_2 may be found by computing the null space of $[V_1 \ AV_1]^T$. Next, construct the matrix V_3 such that its columns are orthogonal and orthogonal to the columns of V_1 and V_1. This is done by computing the null space of $[V_1 \ V_2]^T$. Then $V = [V_1 \ V_2 \ V_3]$ is an orthogonal matrix and in (18) we may take $U = V^T$.

Finite deflation. The finite deflations may be completed in one step. Using the ordered QZ algorithm described at the end of this appendix unitary matrices q and z may be computed such that

$$q(sE - A)z = \begin{bmatrix} se_o - a_o & \times \\ 0 & \times \end{bmatrix} \tag{70}$$

where the zeros of the (complex-valued) square pencil $se_o - a_o$ are the zeros of $sE - A$ with positive real part. Correspondingly, q and z may be partitioned as

$$q = \begin{bmatrix} q_1 \\ q_2 \end{bmatrix}, \quad z = [\, z_1 \ z_2 \,] \tag{71}$$

Construct a real matrix V_1 whose columns are orthogonal and span the subspace spanned by the columns of z_1. Similarly, construct a real matrix W whose columns are orthogonal and span the subspace spanned by the columns of q_1^H. Let V_2 be a real matrix whose columns are orthogonal and orthogonal to the columns of V_1 and W. Finally, let the columns of V_3 be orthogonal and orthogonal to the columns of V_1 and V_2. Then in (19) U may be taken as $U = V^T$ where $V = [V_1 \ V_2 \ V_3]$.

In the application of the ordered QZ algorithm the separation of the finite and infinite zeros sometimes is delicate because numerically it may be difficult to distinguish between zeros that are large but finite and zeros that are infinite. This problem may be avoided by first using singular value decomposition of E to transform the pencil $sE - A$ orthogonally to the form

$$\begin{bmatrix} sE_{11} - A_{11} & -A_{12} \\ -A_{21} & -A_{22} \end{bmatrix} \tag{72}$$

followed by singular value decomposition of $[\, A_{21} \ A_{22} \,]$ to arrange the pencil as

$$\begin{bmatrix} sE_{11} - A_{11} & \times \\ 0 & -A_{22} \end{bmatrix} \tag{73}$$

The pencil $sE_{11} - A_{11}$ has finite zeros only. Application of the ordered QZ transformation to $sE_{11} - A_{11}$ yields (70).

Completion. The desired transformation in the third step of the algorithm may be performed by first bringing A_o into Schur form

$$A_o = \mathrm{diag}(-\alpha_{m_1}, \ -\alpha_{m_1-1}, \ \cdots, \ -\alpha_1, \ \beta_1, \ \beta_2, \ \cdots, \ \beta_{m_2}) \tag{74}$$

where α_i, $i = 1, 2, \cdots, m_1$, and β_i, $i = 1, 2, \cdots, m_2$, are increasing sequences of nonnegative numbers. If $m_1 > 0$ and $m_2 > 0$ then define

$$V = \begin{bmatrix} c & 0 & -s \\ 0 & I & 0 \\ s & 0 & c \end{bmatrix} \tag{75}$$

where

$$c = \sqrt{\frac{\beta_{m_2}}{\alpha_{m_1} + \beta_{m_2}}}, \quad s = \sqrt{\frac{\alpha_{m_1}}{\alpha_{m_1} + \beta_{m_2}}} \tag{76}$$

V is orthogonal and

$$V^T A_o V = \begin{bmatrix} 0 & 0 & \sqrt{\alpha_{m_1}\beta_{m_2}} \\ 0 & \bar{A}_o & 0 \\ \sqrt{\alpha_{m_1}\beta_{m_2}} & 0 & \times \end{bmatrix} \tag{77}$$

with $\bar{A}_o = \mathrm{diag}(-\alpha_{m_1-1}, -\alpha_{m_1-2}, \cdots, -\alpha_1, \beta_1, \beta_2, \cdots, \beta_{m_2-1})$. This transformation step is repeated as often as needed until the dimension of A_2 in (21) reaches the desired value.

The ordered QZ transformation. The QZ algorithm [5] is a generalization of the Schur transformation. Given two complex valued constant matrices A and B, the QZ algorithm produces unitary matrices Q and Z and upper triangular matrices U and V such that $QAZ = U$ and $QBZ = V$. Denoting the diagonal entries of U as u_{ii} and those of V as v_{ii}, the numbers $\lambda_i := v_{ii}/u_{ii}$ are the generalized eigenvalues of the matrix pair (A, B). They are also the zeros of the pencil $A - sB$.

For certain applications it is useful to have an ordering on the diagonal entries of U and V. The ordering typically is on the real parts of the zeros. The usual QZ algorithm, as implemented in MATLAB, does not impose any order of this type. Ordering, if required, may be accomplished by a series of diagonal interchanges that may be achieved by a double application of the complex Givens rotation.

If U and V are 2×2 upper triangular matrices then it is not difficult to determine complex Givens rotations G_1 and G_2 such that $G_1 U G_2^H$ and $G_1 V G_2^H$ are again upper triangular but the order of the zeros as determined by the diagonal entries (including infinite zeros) is reversed as compared to the original matrix pair (U, V).

This generalizes the well-known result that given a 2×2 upper triangular matrix U a complex Givens rotation G may be determined such that GUG^H is again upper triangular but with the diagonal entries interchanged.

If U and V are upper triangular of dimension greater than 2 then by application of suitable Givens rotations to the columns and rows defined by adjacent diagonal entries of U and V the order of the zeros defined by those diagonal entries may be reversed while preserving the upper triangularity of U and V.

By repeated application of suitable Givens rotations the zeros of (U, V) may be sorted to appear in any desired order along the diagonal. Because Givens rotations are unitary transformations the accumulated transformation is also unitary.

The procedure is similar to that described by Van Dooren [10] except that Van Dooren proposes the use of real Givens transformations.

References

[1] J. D. Aplevich. *Implicit linear systems*, volume 152 of *Lecture Notes in Control and Information Sciences*. Springer-Verlag, Berlin, etc., 1991.

[2] D. J. Clements. Rational spectral factorization using state-space methods. *Systems & Control Letters*, 20:335–343, 1993.

[3] J. C. Doyle. Lecture Notes, ONR/Honeywell Workshop on Advances in Multivariable Control, Minneapolis, Minn., 1984.

[4] J. C. Doyle, K. Glover, P. P. Khargonekar, and B. A. Francis. State-space solutions to standard $\mathcal{H}_2$ and $\mathcal{H}_\infty$ control problems. *IEEE Trans. Aut. Control*, 34:831–847, 1989.

[5] G. M. Golub and C. Van Loan. *Matrix Computations*. The Johns Hopkins University Press, Baltimore, Maryland, 1983.

[6] H. Kwakernaak. Frequency domain solution of the standard H_∞ problem. In M. J. Grimble and V. Kučera, editor, *Polynomial Methods for Control Systems Design*, chapter 2, pages 1741–1746. Springer, London, etc., 1996.

[7] H. Kwakernaak and M. Šebek. Polynomial J-spectral factorization. *IEEE Trans. Aut. Control*, 39(2):315–328, 1994.

[8] J. W. Polderman and J. C. Willems. *Introduction to mathematical systems theory — A behavioral approach*. Springer, New York, etc., 1997.

[9] K. Takaba, N. Morihira, and T. Katayama. H_∞ Control for descriptor systems — A J-spectral factorization approach. In *Proc. 33rd IEEE Conf. Decision & Control*, pages 2251–2256, Lake Buena Vista, FL, 1994.

[10] P. Van Dooren. A generalized eigenvalue approach for solving Riccati equations. *SIAM J. Sci. Stat. Comput.*, 2(2):121–135, 1981.

[11] G. C. Verghese, B. Lévy, and Th. Kailath. A generalized state-space for singular systems. *IEEE Trans. Aut. Control*, 26(4):811–831, 1981.

A Unified Approach to H_∞ Control by J-causal-anticausal Factorization

S. Hosoe*

1 Introduction

H_∞ control theory has attracted a great deal of attention of researchers right after its birth and nowadays there are considerable results for its solutions. The results cover, for instance, the standard H_∞ problems for linear continuous time systems [1], time-varying systems [2], discrete time systems and sampled data systems [3],[4],[5]. There are also a variety of solving techniques such as all-pass embedding, LQG like method, J-spectral factorization approach, a conjugation approach and others (see [6],[7],[9] and references therein).

In this paper, we consider once again the well known and already classical problem of obtaining all suboptimal controllers. This is accomplished by considering J-factorization of a transfer matrix or an operator described by

$$ H = \begin{bmatrix} 0 & 0 & I_{m_2} \\ 0 & G_{21} & G_{22} \end{bmatrix} \begin{bmatrix} -I_{p_1} & G_{11} & G_{12} \\ G_{11}^* & -I_{m_1} & 0 \\ G_{12*} & 0 & 0 \end{bmatrix}^{-1} \begin{bmatrix} 0 & 0 \\ 0 & G_{21}^* \\ I_{m_2} & G_{22}^* \end{bmatrix}, \quad (1) $$

where * denotes the conjugate transpose of the transfer matrices or the adjoint for the operators. This form of matrix has first appeared in the Kwakernaak's paper [8] in the framework of polynomial matrices and has been used to obtain solutions for time-invariant continuous-time plants. Independently, the present author arrived at the same factorization representation in the framework of RH_∞ [10]. Later, the idea was extended to the operator theoretic setting and was used to obtain solutions for sampled-data systems [11]. In this paper, these results are re-arranged in a unified framework. Controller's expressions and their existence conditions are given both in the transfer matrices or operators setting and also in the state space form. The Riccati (algebraic, differential, difference) equations are naturally defined in the process of carrying out the factorization of the operators. The final forms for the most results reported here are already known in the literature, except perhaps for time-varying sampled-data systems with varying sampling interval. Thus the

*Department of Electronic-Mechanical Engineering, Nagoya University, Furo-cho, Chikusa-ku, Nagoya 464-01, Japan.

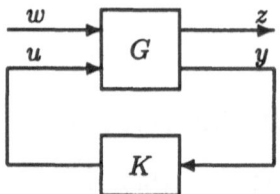

Figure 1: Standard Problem

present paper has the tutorial character. However we would like to emphasize the generality of our approach. Actually, although only continuous time systems and sampled data systems are described here in detail, the readers will be convinced that the procedure given here can naturally be extended to discrete-time systems, multi-rate sampled-data systems [12], descriptor form systems and others.

The paper is organized as follows. After the introduction, Section 2 presents the preliminaries and problem formulation. Our main results will be stated in section 3 where a general algorithm is presented to generate all suboptimal controllers. Also state space calculation is performed for linear time-varying systems and sampled-data systems. Section 4 will give the proofs for the theorems. Section 5 contains the conclusions.

2 Problem Statement

We consider the feedback configuration of Fig. 1, where $z \in R^{p_1}, y \in R^{p_2}, w \in R^{m_1}, y \in R^{m_2}$ are respectively the controlled error, the observed output, the exogenous input and the control input.

When discussing a time-invariant plant in the infinite horizon case, G denotes a transfer function matrix. On the other hand, for time-varying systems or sampled-data systems, and in finite horizon cases, G is regarded as a linear operator. In any cases, it is assumed that we are given the state-space realization of G. Specifically, the continuous time linear time-varying (CLTV) plant and the linear sampled-data (LSD) plant with a zero-order hold and an ideal sampler have the following realizations.

CLTV system:

$$
\begin{aligned}
\dot{x}(t) &= A(t)x(t) + B_1(t)w(t) + B_2(t)u(t), \quad x(t_0) = 0 \\
z(t) &= C_1(t)x(t) + D_{12}(t)u(t) \\
y(t) &= C_2(t)x(t) + D_{21}(t)w(t).
\end{aligned}
\tag{2}
$$

LSD system:

$$
\begin{aligned}
\dot{x}(t) &= A(t)x(t) + B_1(t)w_c(t) + B_2(t)x_h(t), \quad x(t_0) = 0, \\
\dot{x}_h(t) &= 0, \\
x_h(t_i) &= u[t_i], \quad x_h(t_0) = 0 \\
z(t) &= C_1(t)x(t) + D_{12}(t)x_h(t), \\
y[t_i] &= C_{2i}x(t_i) + D_{21i}w_d[t_i],
\end{aligned}
\tag{3}
$$

where $t'_i s$ are the sampling instants and x_h denotes the state of the hold. The coefficient matrices are all assumed to be bounded functions of t and t_j. The control interval for the finite horizon case is denoted by (t_o, t_f) for CLTV plant and $\{t_o = t_0, t_1, t_2, ..., t_N = t_f\}$ $(t_0 < t_1 < t_2 < \cdots < t_N)$ for LSD plant. For the notational convenience, the same expression will also be used for the infinite interval by admitting t_o to take minus infinity and t_f and N plus infinity.

The following simplifying assumptions are made:

(A1) $D_{12}^T(t)D_{12}(t) = I, D_{21}(t)D_{21}^T(t) = I, D_{21i}D_{21i}^T = I$.

(A2) (In the infinite horizon case) Systems(2) ,(3) are stabilizable from u and detectable from y.

Note that with system (2) operator G for the CLTV plant can be explicitly written as

$$G = \begin{bmatrix} G_{11} & G_{12} \\ G_{21} & G_{22} \end{bmatrix} : \mathcal{L}_2^{m_1}(t_0, t_f) \times \mathcal{L}_2^{m_2}(t_0, t_f) \longrightarrow \mathcal{L}_2^{m_1}(t_0, t_f) \times \mathcal{L}_2^{m_2}(t_0, t_f) \quad (4)$$

$$\begin{bmatrix} z \\ y \end{bmatrix}(t) = \int_{t_0}^t \begin{bmatrix} C_1 \\ C_2 \end{bmatrix}(t)\Phi(t, \tau)(B_1 w(\tau) + B_2 u(\tau)d\tau + \begin{bmatrix} D_{12}(t)u(t) \\ D_{21}(t)w(t) \end{bmatrix}(t)$$
$$(5)$$

where $\Phi(t, \tau)$ is the state transition matrix of $\dot{x}(t) = A(t)x(t)$. The operator G corresponding to the LSD plant is similarly be defined.

In the following, in addition to systems (2) and (3) we need the state space representations of G^*, the adjoint operator of G. It can be easily seen that the adjoint for CLTV plant is given by

$$\begin{aligned} -\dot{\bar{x}}(t) &= A^T(t)\bar{x}(t) + C_1^T(t)\bar{z}(t) + C_2^T(t)\bar{y}(t), \ x(t_f) = 0 \\ \bar{w}(t) &= B_1^T(t)\bar{x}(t) + D_{21}^T(t)\bar{y}(t) \\ \bar{u}(t) &= B_2^T(t)\bar{x}(t) + D_{12}^T(t)\bar{z}(t). \end{aligned} \quad (6)$$

To describe LSD plant (3) and its adjoint in a compact form, denote

$$\bar{A}(t) = \begin{bmatrix} A & B_2 \\ 0 & 0 \end{bmatrix}(t), \quad \bar{B}_1(t) = \begin{bmatrix} B_1(t) \\ 0 \end{bmatrix}, \quad J_1 = \begin{bmatrix} I_n & 0 \\ 0 & 0 \end{bmatrix},$$
$$J_2 = \begin{bmatrix} 0 \\ I_{m_2} \end{bmatrix}, \quad \bar{C}_1(t) = [\ C_1(t) \quad D_{12}(t)\ ], \quad \bar{C}_{2i} = [\ C_{2i} \quad 0\ ].$$
$$(7)$$

Let $x_e = [\ x^T \quad x_h^T\ ]^T$. Then (3) can be rewritten as

$$\begin{aligned} \dot{x}_e(t) &= \bar{A}(t)x_e(t) + \bar{B}_1(t)w_e(t) \\ x_e(t_i) &= J_1 x_e(t_{i-}) + J_2 u[t_i], \ x_e(t_{0-}) = 0 \\ z(t) &= \bar{C}_1 x_e(t) \\ y[t_i] &= \bar{C}_{2i}x_e(t_i) + D_{21i}w_d[t_i] \end{aligned} \quad (8)$$

with $x_e(t_{i-}) = \lim_{t\uparrow t_i} x(t)$.
Its adjoint is obtained as

$$
\begin{array}{rcl}
-\dot{\bar{x}}_e(t) & = & \overline{A}^T(t)\bar{x}_e(t) + \overline{C}_1^T(t)\bar{z}(t), \\
\bar{x}_e(t_{i-}) & = & J_1^T\bar{x}_e(t_i) + \overline{C}_{2i}^T\bar{y}[t_i], \bar{x}_e(t_f) = 0, \\
w_c(t) & = & \overline{B}_1^T(t)\bar{x}_e(t) \\
w_d[t_i] & = & D_{21i}^T\bar{y}[t_i] \\
\bar{u}[t_i] & = & J_2^T\bar{x}_e(t_i).
\end{array}
\tag{9}
$$

Denote by T_{zw} the operator mapping w to z of the closed loop system, which with operator expression K for the controller is given by

$$
T_{zw} = G_{11} + G_{12}K(I - G_{22}K)^{-1}G_{21}.
$$

Its induced norm is defined by

$$
\|T_{zw}\|_{\infty,t_0,t_f} = sup_{\|u\|\neq 0}(\|z\|/\|w\|).
\tag{10}
$$

with the norms of $\mathcal{L}_2(t_0, t_f)$ for continuous-time functions $z(t), w(t)$ and $\ell_2(t_0, t_f)$ for discrete-time signals.

Now the standard H_∞ control problem is stated as follows. Find a finite dimensional time-varying continuous-time (discrete-time, resp.) controller K for CLTV plant (2) (LSD plant (3), resp.) such that it internally stabilizes the closed-loop system, and for a given real number γ, it makes

$$
\|T_{zw}\| < \gamma.
\tag{11}
$$

In the following, it is assumed without loss of generality that $\gamma = 1$.

3   Main Results

The general procedure for the construction of suboptimal H_∞ controllers is described as follows.

step 1 Obtain the state space representations Σ_H and $\Sigma_{H^{-1}}$ for H and H^{-1}, where H is given by (1).

step 2 To carry out the factorization of $\Sigma_H, \Sigma_{H^{-1}}$, Riccati matrices X(t) and Y(t) are defined, for which matrix solutions of the homogeneous parts of Σ_H *and* $\Sigma_{H^{-1}}$ are used.

step 3 Execute a coordinate transformation to Σ_H. Then it brings the "A matrices" of $\Sigma_H, \Sigma_{H^{-1}}$ into a triangular form, meaning that H is decomposed into the product of a causal and a anti-causal operators as

$$
H = \hat{K}J\hat{K}^*, \quad J = diag[I_{m_2}, -I_{p_2}]
\tag{12}
$$

where both $\hat{K}$ and $\hat{K}^{-1}$ are causal.

step 4 Controllers are determined from the causal part $\hat{K} = [\hat{K}_{ij}]_{i,j=1,2}$ as

$$K = (\hat{K}_{11}Q + \hat{K}_{12})(\hat{K}_{21}Q + \hat{K}_{22})^{-1} \qquad (13)$$

where **Q** is an arbitrary causal bounded map with norm less than $\gamma(=1)$.

Now, let us demonstrate the procedure in cases of CLTV and LSD systems. First, to derive the state space representations Σ_H and $\Sigma_{H^{-1}}$, let us express the functional relation of (1) as

$$\begin{bmatrix} u \\ y \end{bmatrix} = H \begin{bmatrix} \bar{u} \\ \bar{y} \end{bmatrix}. \qquad (14)$$

Then by introducing the intermediate variables z and w ($w = \begin{bmatrix} w_c^T & w_d^T \end{bmatrix}^T$ for LSD system), it can be rewritten as

$$\begin{bmatrix} z \\ w \\ u \end{bmatrix} = \begin{bmatrix} -I & G_{11} & G_{12} \\ G_{11}^* & -1 & 0 \\ G_{12}^* & 0 & 0 \end{bmatrix}^{-1} \begin{bmatrix} 0 & 0 \\ 0 & G_{21}^* \\ I & G_{22}^* \end{bmatrix} \begin{bmatrix} \bar{u} \\ \bar{y} \end{bmatrix}$$

$$\begin{bmatrix} u \\ y \end{bmatrix} = \begin{bmatrix} 0 & 0 & I \\ 0 & G_{21} & G_{22} \end{bmatrix} \begin{bmatrix} z \\ w \\ u \end{bmatrix} \qquad (15)$$

and from which

$$\begin{bmatrix} z \\ y \end{bmatrix} = \begin{bmatrix} G_{11} & G_{12} \\ G_{21} & G_{22} \end{bmatrix} \begin{bmatrix} w \\ u \end{bmatrix}, \qquad (16)$$

$$\begin{bmatrix} w \\ \bar{u} \end{bmatrix} = \begin{bmatrix} G_{11}^* & G_{21}^* \\ G_{12}^* & G_{22}^* \end{bmatrix} \begin{bmatrix} z \\ -\bar{y} \end{bmatrix}. \qquad (17)$$

Observe that these are the original plant and its adjoint. Thus, from (2), (6) for CLTV plant and (8), (9) for LSD system, we can get the following state-space representations Σ_H of H.

Realization of H for CLTV plant:

$$\begin{bmatrix} \dot{x} \\ \dot{\bar{x}} \end{bmatrix}(t) = \begin{bmatrix} A - B_2 D_{12}^T C_1 & B_1 B_1^T - B_2 B_2^T \\ C_1^T(D_{12}D_{12}^T - I)C_1 & -(A - B_2 D_{12}^T C_1)^T \end{bmatrix}(t) \begin{bmatrix} x \\ \bar{x} \end{bmatrix}(t)$$
$$+ \begin{bmatrix} B_2 & -B_1 D_{21}^T \\ -C_1^T D_{12} & C_2^T \end{bmatrix}(t) \begin{bmatrix} \bar{u} \\ \bar{y} \end{bmatrix}_c$$

$$\begin{bmatrix} u \\ y \end{bmatrix} = \begin{bmatrix} -D_{12}^T C_1 & -B_2^T \\ C_2 & D_{21}B_1^T \end{bmatrix}(t) \begin{bmatrix} x \\ \bar{x} \end{bmatrix}(t) + \begin{bmatrix} 1 & 0 \\ 0 & -1 \end{bmatrix} \begin{bmatrix} \bar{u} \\ \bar{y} \end{bmatrix}(t). \qquad (18)$$

Realization of H for LSD system:

$$\begin{bmatrix} \dot{x}_e \\ \dot{\overline{x}}_e \end{bmatrix}(t) = \begin{bmatrix} \overline{A} & \overline{B}_1\overline{B}_1^T \\ -\overline{C}_1^T\overline{C}_1 & -\overline{A}^T \end{bmatrix}(t)\begin{bmatrix} x_e \\ \overline{x}_e \end{bmatrix}(t)$$

$$\begin{aligned} x_e(t_i) &= J_1 x_e(t_{i-}) + J_2 J_2^T x_e(t_i), \\ \overline{x}_e(t_{i-}) &= J_1^T \overline{x}_e(t_i) - \overline{C}_{2i}^T \overline{y}[t_i], \\ J_2^T \overline{x}_e(t_i) &= \overline{u}[t_i]. \\ u[t_i] &= J_2^T x_e(t_i), \\ y[t_i] &= \overline{C}_{2i} x_e(t_i) - \overline{y}[t_i]. \end{aligned} \tag{19}$$

The state-space realization $\Sigma_{H^{-1}}$ can be obtained similarly by considering

$$H^{-1} = \begin{bmatrix} 0 & G_{12}^* & G_{22}^* \\ 0 & 0 & -I \end{bmatrix}\begin{bmatrix} I & -G_{11}^* & -G_{12}^* \\ -G_{11} & 1 & 0 \\ -G_{21} & 0 & 0 \end{bmatrix}^{-1}\begin{bmatrix} 0 & 0 \\ G_{12} & 0 \\ G_{22} & -I \end{bmatrix}. \tag{20}$$

The result is as follows.

Realization of H^{-1} for CLTV plant:

$$\begin{aligned} \begin{bmatrix} \dot{x} \\ \dot{\overline{x}} \end{bmatrix}(t) &= \begin{bmatrix} A - B_1 D_{21}^T C_2 & B_1(I - D_{21}^T D_{21})B_1^T \\ C_2^T C_2 - C_1^T C_1 & -(A - B_1 D_{21}^T C_2)^T \end{bmatrix}(t)\begin{bmatrix} x \\ \overline{x} \end{bmatrix}(t) \\ &\quad + \begin{bmatrix} B_2 & B_1 D_{21}^T \\ -C_1^T D_{12} & -C_2^T \end{bmatrix}(t)\begin{bmatrix} \overline{u} \\ \overline{y} \end{bmatrix}(t) \\ \begin{bmatrix} u \\ y \end{bmatrix}(t) &= \begin{bmatrix} D_{12}^T C_1 & B_2^T \\ C_2 & D_{21}B_1^T \end{bmatrix}(t)\begin{bmatrix} x \\ \overline{x} \end{bmatrix}(t) + \begin{bmatrix} 1 & 0 \\ 0 & -1 \end{bmatrix} \\ &\quad \begin{bmatrix} \overline{u} \\ \overline{y} \end{bmatrix}(t). \end{aligned} \tag{21}$$

Realization of H^{-1} for LSD system:

$$\begin{bmatrix} \dot{x}_e \\ \dot{\overline{x}}_e \end{bmatrix}(t) = \begin{bmatrix} \overline{A} & \overline{B}_1\overline{B}_1^T \\ -\overline{C}_1^T\overline{C}_1 & -\overline{A}^T \end{bmatrix}(t)\begin{bmatrix} x_e \\ \overline{x}_e \end{bmatrix}(t),$$

$$\begin{aligned} x_e(t_i) &= J_1 x_e(t_i) + J_2 u[t_i], \\ \overline{x}_e(ti-) &= J_1^T \overline{x}_e(t_i) - \overline{C}_{2i}^T \overline{C}_{2i} x_e(t_i) + \overline{C}_{2i}^T y[t_i], \\ \overline{u}[t_i] &= J_2^T \overline{x}_e(t_i), \\ \overline{y}[t_i] &= \overline{C}_{2i} x_e(t-i) - \overline{y}[t_i]. \end{aligned} \tag{22}$$

Riccati matrices X and Y in step 2 are determined as follows. Consider the homogeneous part of (18)

$$\begin{bmatrix} \dot{x} \\ \dot{\overline{x}} \end{bmatrix}(t) = \begin{bmatrix} A - B_2 D_{12}^T C_1 & B_1 B_1^T - B_2 B_2^T \\ C_1^T(D_{12}D_{12}^T - I)C_1 & -(A - B_2 D_{12}^T C_1)^T \end{bmatrix}(t)\begin{bmatrix} x \\ \overline{x} \end{bmatrix}(t). \tag{23}$$

and denote by $\begin{bmatrix} \Phi_{1,t_f}^T(t) & \Phi_{2,t_f}^T(t) \end{bmatrix}^T$ its matrix solution with the terminal condition

$$\begin{bmatrix} \Phi_{1,t_f}^T(t_f) & \Phi_{2,t_f}^T(t_f) \end{bmatrix}^T = \begin{bmatrix} I & 0 \end{bmatrix}^T.$$

Then Riccati matrix $X_{t_f}(t)$ is defined by

$$X_{t_f}(t) = \Phi_{2,t_f}(t)\Phi_{1,t_f}(t)^{-1}. \tag{24}$$

The Riccati matrix $Y_{t_f}(t)$ is obtained by considering the matrix solution $\begin{bmatrix} \Psi_{1,t_0}^T(t)\Psi_{2,t_0}^T(t) \end{bmatrix}^T$ for the homogeneous part of (21) with the initial condition

$$\begin{bmatrix} \Psi_{1,t_0}^T(t_0) & \Psi_{2,t_0}^T(t_0) \end{bmatrix}^T = \begin{bmatrix} 0 & I \end{bmatrix}^T.$$

Then $Y_{t_f}(t)$ is given by

$$Y_{t_0}(t) = \Psi_{1,t_0}(t)\Psi_{2,t_0}(t)^{-1}. \tag{25}$$

The derivation of the Riccati matrices for LSD system is almost the same. For instance, Riccati matrix X_{e,t_f} follows from the homogeneous part of (19)

$$\begin{bmatrix} \dot{x}_e \\ \dot{\overline{x}}_e \end{bmatrix}(t) = \begin{bmatrix} \overline{A} & \overline{B}_1\overline{B}_1^T \\ -\overline{C}_1^T\overline{C}_1 & -\overline{A}^T \end{bmatrix}(t)\begin{bmatrix} x_e \\ \overline{x}_e \end{bmatrix}(t),$$
$$\begin{aligned} x_e(t_i) &= J_1 x_e(t_{i-}) + J_2 J_2^T \overline{x}_e(t_i), \\ \overline{x}_e(t_{i-}) &= J_1^T \overline{x}_e(t_i), \\ J_2^T \overline{x}_e(t_i) &= 0. \end{aligned} \tag{26}$$

Now, using the above definition, it will be straightforward to confirm that $X_{t_f}(t)$ and $Y_{t_f}(t)$ satisfy the following well known differential equations.

Riccati Differential Equation for CLTV system:

$$\begin{aligned} -\dot{X}_{t_f}(t) &= X_{t_f}(t)(A - B_2 D_{12}^T C_1)(t) + (A - B_2 D_{12}^T C_1)^T(t)X_{t_f}(t) \\ &\quad + X_{t_f}(t)(B_1 B_1^T - B_2 B_2^T)(t)X_{t_f}(t) + C_1^T(I - D_{12}D_{12}^T)C_1(t), \\ X_{t_f}(t_f) &= 0. \end{aligned} \tag{27}$$

$$\begin{aligned} \dot{Y}_{t_0}(t) &= (A - B_1 D_{21}^T C_2)Y_{t_0}(t) + Y_{t_0}(t)(A - B_1 D_{21}^T C_2)^T(t) \\ &\quad + B_1(I - D_{21}^T D_{21})B_1^T(t) + Y_{t_0}(t)(C_1^T C_1 - C_2^T C_2)(t)Y_{t_0}(t), \\ Y_{t_0}(t_0) &= 0. \end{aligned} \tag{28}$$

Riccati Differential Equation with Jumps for LSD system:

$$\begin{aligned} -\dot{X}_{e,t_f}(t) &= X_{e,t_f}(t)\overline{A}(t) + \overline{A}^T(t)X_{e,t_f}(t) \\ &\quad + X_{e,t_f}(t)\overline{B}_1\overline{B}_1^T(t)X_{e,t_f}(t) + \overline{C}_1^T\overline{C}_1(t), \\ X_{e,t_f}(t_{i-}) &= X_{e,t_f}(t_i)\{I - \overline{J}_2(\overline{J}_2^T X_{e,t_f}(t_i)\overline{J}_2)^{-1}\overline{J}_2^T X_{e,t_f}(t_i)\} \\ X_{e,t_f}(t_f) &= 0. \end{aligned} \tag{29}$$

$$\begin{aligned}
\dot{Y}_{e,t_0}(t) &= \overline{A}(t)Y_{e,t_0}(t) + Y_{e,t_0}(t)\overline{A}^T(t) \\
&\quad + Y_{e,t_0}(t)\overline{C}_1^T\overline{C}_1 Y_{e,t_0}(t) + \overline{B}_1\overline{B}_1^T, \\
\overline{Y}_{e,t_0}(t_i) &= Y_{e,t_0}(t_{i-})\{I + \overline{C}_{2i}^T\overline{C}_{2i}Y_{e,t_0}(t_{i-})\}^{-1}, \\
Y_{e,t_0}(t_0) &= 0.
\end{aligned} \tag{30}$$

Next, let us proceed to step 3. The coordinate transformation is given by

$$\begin{bmatrix} \xi \\ \dot{\xi} \end{bmatrix}(t) = \begin{bmatrix} I & -YZ \\ X & -Z \end{bmatrix}^{-1} \begin{bmatrix} x \\ \overline{x} \end{bmatrix}(t) \tag{31}$$

where $Z = (I - XY)^{-1}$. The subscripts are dropped for convenience of expression. Notice that the transformation matrix for CLTV and LSD systems is the same.

With this choice of the new coordinates, (18) becomes

$$\begin{aligned}
\begin{bmatrix} \dot{\xi} \\ \dot{\overline{\xi}} \end{bmatrix}(t) &= \begin{bmatrix} A_F & B_{F11}B_{F11}^T - B_{F12}B_{F12}^T \\ 0 & -A_F^T \end{bmatrix}(t) \begin{bmatrix} \xi \\ \overline{\xi} \end{bmatrix}(t) \\
&\quad + \begin{bmatrix} B_{F11} & B_{F12} \\ B_{F21} & B_{F22} \end{bmatrix}(t) \begin{bmatrix} \overline{u} \\ \overline{y} \end{bmatrix}(t) \\
\begin{bmatrix} u \\ y \end{bmatrix}(t) &= \begin{bmatrix} -B_{F21}^T & B_{F11}^T \\ -B_{F22}^T & B_{F12}^T \end{bmatrix}(t) \begin{bmatrix} x \\ \overline{x} \end{bmatrix}(t) + \begin{bmatrix} 1 & 0 \\ 0 & -1 \end{bmatrix} \begin{bmatrix} \overline{u} \\ \overline{y} \end{bmatrix}(t)
\end{aligned} \tag{32}$$

where

$$\begin{aligned}
A_F(t) &= \{A - B_2 D_{12}^T C_1 + (B_1 B_1^T - B_2 B_2^T)X\}(t), \\
\begin{bmatrix} B_{F11} & B_{F12} \\ B_{F21} & B_{F22} \end{bmatrix}(t) &= \begin{bmatrix} Z^T(B_2 + YC_1^T D_{12}) & -Z^T(B_1 D_{21}^T + YC_2^T) \\ XB_2 + C_1^T D_{12} & -XB_1 D_{21}^T - C_2^T \end{bmatrix}(t).
\end{aligned} \tag{33}$$

This result shows that operator H has the factorization of the form (12) and the state-apace realization of $\hat{K}$ is given by

$$\hat{K}: \quad \begin{aligned} \dot{\xi}(t) &= A_F(t)\xi(t) + B_{F11}(t)s(t) + B_{F12}(t)t(t) \\ \begin{bmatrix} u \\ y \end{bmatrix}(t) &= \begin{bmatrix} -B_{F21}^T \\ -B_{F22}^T \end{bmatrix}(t)\xi + \begin{bmatrix} 1 & 0 \\ 0 & -1 \end{bmatrix} \begin{bmatrix} s \\ t \end{bmatrix}(t). \end{aligned} \tag{34}$$

Working similarly with (19), (29)-(31), we can obtain the causal map $\hat{K}$ of the factorization of H in the case of LSD system as

$$\hat{K}: \quad \begin{aligned} \dot{\xi}_e(t) &= (\overline{A}(t) + \overline{B}_1(t)\overline{B}_1^T(t)X_{e,t_f}(t))\xi_e(t) \\ \xi_e(t_i) &= K(t_i)\xi_e(t_{i-}) + L_1(t_i)s[t_i] + L_2(t_i)t[t_i] \\ \begin{bmatrix} u \\ y \end{bmatrix}[t] &= \begin{bmatrix} J_2^T \\ \overline{C}_{2i} \end{bmatrix}(t_i)\xi_e(t_i) + \begin{bmatrix} 0 & 0 \\ N_1 & N_2 \end{bmatrix}(t_i) \begin{bmatrix} s \\ t \end{bmatrix}[t], \end{aligned} \tag{35}$$

where

$$K(t_i) = I - J_2 \Delta^{-1}(t_i)J_2^T X_{e,t_f}(t_i),$$

$$[\ L_1 \quad L_2\ ](t_i) = \Big[cZ_e^T J_2 \Delta^{-1/2} R^{1/2}$$
$$(I - J_2 \Delta^{-1} J_2^T X_{e,t_f}) Q_- \overline{C}_{2i}^T (I + \overline{C}_{2i} Q_- \overline{C}_{2i}^T)^{-1/2} \Big] (t_i),$$
$$[\ N_1 \quad N_2\ ](t_i) = \Big[\ -(I + \overline{C}_{2i} Q_- \overline{C}_{2i}^T)^{-1} \overline{C}_{2i} \quad (I + \overline{C}_{2i} Q_- \overline{C}_{2i}^T))^{-1/2}\ \Big] (t_i), \tag{36}$$

with

$$\Delta(t_i) = J_2^T X_{e,t_f}(t_i) J_2,$$
$$Q(t_i) = Y_{e,t_0}(t_i) Z_e(t_i), Q_-(t_i) = Y_{e,t_0}(t_{i-}) Z_e(t_{i-}),$$
$$\tilde{C}(t_i) = \Delta^{-1/2}(t_i) J_2^T X_{e,t_f}(t_i), R = I - \tilde{C}(t_i)(I + Q_- \overline{C}_{2i}^T \overline{C}_{2i})^{-1} Q_- \tilde{C}(t_i).$$

Now we can state our main result.

Theorem 1 *Finite horizon case:*

(i) The following conditions are equivalent.

(a) The finite horizon H_∞ control problem stated in section 2 is solvable.

(b) The restricted maps $G_{11}|Null(G_{21})$ and $G_{11}^*|Null(G_{12}^*)$ satisfy the norm conditions

$$\| G_{11}|Null(G_{21})\ \|_{\infty,t_0,t_f} < 1, \tag{37}$$
$$\| G_{11}^*|Null(G_{12}^*)\ \|_{\infty,t_0,t_f} < 1. \tag{38}$$

and operator H admits a causal-anticausal factorization

$$H = \hat{K} J \hat{K}^*, \quad J = diag[I_{m_2}, -I_{p_2}] \tag{39}$$

where $\hat{K}$ and $\hat{K}^{-1}$ are causal.

(c) Riccati matrices $X_{t_f}(t)(, X_{e,t_f}(t))$ and $Y_{t_0}(t)(, Y_{e,t_0}(t))$ exist for all $t \in (t_0, t_f)$ and $det(I - X_{t_f} Y_{t_0})(t)(, det(I - X_{e,t_f} Y_{e,t_0})(t)) \neq 0$.

(ii) When the conditions in (i) hold, all the controllers achieving the norm condition are given by (13) with (34) for LTV plant and (35) for LSD system respectively. The state equations of the central solutions (i.e., $Q = 0$ in (13)) take the forms:

$$\dot{\xi}(t) = (A_F - B_{F12} B_{F22}^T)(t)\xi(t) - B_{F12}(t)y(t),$$
$$u(t) = B_{F21}^T(t)\xi(t) \tag{40}$$

for CLTV system, and

$$\dot{\xi}_e(t) = (\overline{A}(t) + \overline{B}_1(t)\overline{B}_1^T(t)X_{e,t_f}(t))\xi_e(t)$$
$$\xi_e(t_i) = (I - J_2 \Delta^{-1}(t_i) J_2^T X_{e,t_f}(t_i))(I + Q_-(t_i)\overline{C}_{2i}^T \overline{C}_{2i})^{-1}$$
$$\cdot\{\xi_e(t_{i-}) + Q_-(t_i)\overline{C}_{2i}^T y[t_i]\}$$
$$u[t_{i-}] = J_2^T \xi_e(t_i) \tag{41}$$

for LSD System.

Theorem 2 *Infinite horizon case:*

(i) The following conditions are equivalent.

(a) The infinite horizon H_∞ control problem over (t_0, ∞) is solvable.

(b) Norm conditions (37), (38) hold uniformly with respect to t_f, i.e., there exists sufficiently small $\epsilon > 0$, independentl of t_f, such that

$$\| G_{11} | Null(G_{21}) \|_{\infty, t_0, t_f} < \gamma - \epsilon, \tag{42}$$

$$\| G_{11}^* | Null(G_{12}^*) \|_{\infty, t_0, t_f} < \gamma - \epsilon \tag{43}$$

for all $t_f \in (t_0, \infty)$
and H admits a causal-anticausal factorization of the form (39) in which $\hat{K}$, $\hat{K}^{-1}$, π_{22}^{-1}, $\pi_{22}^{-1} G_{22}$ are all stable with $\pi_{22} = \hat{K}_{22} - G_{22} \hat{K}_{12}$.

(c-1) Riccati matrix $X_\infty(X_{e,\infty}$, resp.) exists for all $t \in (t_0, \infty)$, that satsfies (27) ((29), resp.) without the terminal constraint. It is bounded, non-negative definite and such that $\dot{x}(t) = (A - B_2 D_{12}^T C_1 + (B_1 B_1^T - B_2 B_2^T) X_\infty)(t) x(t)$ ($\dot{x}_e(t) = (\overline{A} + \overline{B}_1 \overline{B}_1^T X_{e,\infty}(t) x_e(t)$, $x_e(t_{i-}) = \{I - X_{e,\infty}(t_i) \overline{J}_2 (\overline{J}_2^T X_{e,\infty}(t_i) \overline{J}_2)^{-1} \overline{J}_2^T) x_e(t_i)\}$, $resp.$) is exponentially stable.

(c-1) Riccati matrix $Y_{t_0}(Y_{e,t_0}$, resp.) exists for all $t \in (t_0, \infty)$ that satisfies (28)((30), resp.) with the zero initial condition. It is bounded, non-negative definite and such that $\dot{y}(t) = (A - B_1 D_{21}^T C_2 + Y_{t_0}(C_1^T C_1 - C_2^T C_2))(t) y(t)$ $(\dot{y}_e(t) = (\overline{A} + Y_{e,t_0} \overline{C}_1^T \overline{C}_1)(t) y_e(t)$, $y_{e,t_0}(t_i) = \{I + Y_{e,t_0}(t_{i-}) \overline{C}_{2i}^T \overline{C}_{2i}^{-1}\} y_e(t_{i-})$, $resp.$) is exponentially stable.

(c-3) $det(I - X_\infty) Y_{t_0})(t)(det(I - X_{e,\infty} Y_{e,t_0})$, $resp.) \neq 0$ for all $t \in (t_0, \infty)$.

(ii) This part is the same as in (2) of Theorem 1 with using $X_\infty(, X_{e,\infty})$ instead of $X_{t_f}(, X_{e,t_f})$.

4   Proof of the Theorem

Only the proof for the case of CLTV system will be given. The proof for LSD system can similarly be treated. The infinite horizon case can be considered by taking the limit (see [2]).

Proof of Theorem 1.

(a) $\Longrightarrow$ (c). Suppose that a controller K exists satisfying the norm condition. We will show the existence of $Y_{t_0}(t)$. The existence of $X_{t_f}(t)$ then follows by considering the adjoint expression of the generalized plant G and norm condition (11).

Now, by (25), to prove the existence of $Y_{t_0}(t)$ it is sufficient to show that $det(\Psi_{2,t_0}(t) \neq 0$ for all $t \in [t_0, t_f)$. Assume, on the contrary, that there is a time T and a vector $\bar{x}_T \neq 0$ such that $\Psi_{2,t_0}(T)\bar{x}_T = 0$. This implies that the solution of equation

$$\begin{bmatrix} \dot{x} \\ \dot{\bar{x}} \end{bmatrix}(t) = \begin{bmatrix} A - B_1 D_{21}^T C_2 & B_1(I - D_{21}^T D_{21})B_1^T \\ C_2^T C_2 - C_1^T C_1 & -(A - B_1 D_{21}^T C_2)^T \end{bmatrix}(t) \begin{bmatrix} x \\ \bar{x} \end{bmatrix}(t) \quad (44)$$

with the initial condition

$$\begin{bmatrix} x \\ \bar{x} \end{bmatrix}(t_0) = \begin{bmatrix} 0 \\ \bar{x}_T \end{bmatrix} \quad (45)$$

satisfies

$$x(t_0) = 0, \qquad \bar{x}(T) = \Psi_{2,t_0}(T)\bar{x}_T = 0. \quad (46)$$

Rewrite the first equation of (44) as

$$\begin{aligned} \dot{x} &= Ax + B_1 w \\ w &= (I - D_{21}^T D_{21})B_1^T \bar{x} - D_{21}^T C_2 x. \end{aligned} \quad (47)$$

Regarding the above w as an external disturbance to the plant (2), we have $y(t) = C_2 x(t) + D_{21}w(t) \equiv 0$ and thus $u \equiv 0$ since the output of the causal controller K to zero input is identically zero. Therefore we get

$$\begin{aligned} z^T z - w^T w &= x^T(C_1 - C_2)x - \bar{x}^T(B_1(I - D_{21}^T D_{21})B_1^T \bar{x} \\ &= -\frac{d}{dt}\bar{x}^T x \end{aligned} \quad (48)$$

Integrating this with using (46) gives

$$\|z\|^2 - \|w\|^2 = 0$$

which clearly contradicts (11). This proves that $det(\Psi_{2,t_0}(t) \neq 0$ for all $t \in [t_0, t_f)$.

Next, we show that $det(I - X_{t_f} Y_{t_0})(t)$ $(det(I - X_{e,t_f} Y_{e,t_0})(t)$, resp.$) \neq 0$. Again, on the contrary to what we want to prove, let us suppose that there is a time T and a vector $\bar{x}_T \neq 0$ such that

$$X_{t_f}(T)Y_{t_0}(T)\bar{x}_T = \bar{x}_T. \quad (49)$$

Consider the differential equation

$$\dot{\bar{x}}(t) = -((C_1^T C_1 - (C_2^T C_2)(t))Y_{t_0}(t) - (A + B_1 D_{21}^T C_2)(t)^T)\bar{x}(t), \qquad t \in [t_0, T]$$

with the terminal condition $\bar{x}(T) = \bar{x}_T$. Define

$$x(t) = Y_{t_0}(t)\bar{x}(t).$$

Then using (28) it is straightforward to see that $x(t)$ satisfies

$$\dot{x} = Ax + B_1 w$$

where

$$w = (I - D_{21}^T D_{21})B_1^T \bar{x} - D_{21}C_2 x.$$

Note that $y(t) = C_2 x + D_{21}w \equiv 0$ and thus $u \equiv 0$ for any causal controller K. Also, by direct computation, we have

$$\frac{d}{dt}\bar{x}^T(t)Y_{t_0}(t)\bar{x}(t) = w^T(t)w(t) - z^T(t)z(t).$$

Consequently, by integrating over $[t_0, T]$, we get

$$\bar{x}_T^T Y_{t_0}(T)\bar{x}_T = \|w\|_{[t_0,T]}^2 - \|z\|_{[t_0,T]}^2. \tag{50}$$

On the other hand, on the time interval $[T, t_f]$, consider the system

$$\begin{aligned} \dot{x} &= A(t)x + B_1(t)w + B_2(t)u, \\ w &= B_1^T X_{t_f} x \end{aligned} \tag{51}$$

with the initial condition

$$x(T) = Y_{t_0}(T)\bar{x}_T \tag{52}$$

To evaluate the norm, differentiate $x^T X_{t_f} x$ along the trajectory of (51) to get

$$\frac{d}{dt}x^t X_{t_f}x = (Ax + B_1 w + B_2 u)^T X_{t_f}x + x^t X_{t_f}(Ax + B_1 w + B_2 u) + x^t \dot{X}_{t_f}x.$$

Substituting (27) and (51) gives

$$\frac{d}{dt}x^t X_{t_f}x = w^T w - z^T z + (u + (D_{12}^T C_1 + B_2^T X_{t_f})x)^T(u + (D_{12}^T C_1 + B_2^T X_{t_f})x).$$

Therefore, with integrating over $[T, t_f]$ we get

$$-x^T(T)X_{t_f}(T)x(T) = \|w\|_{[T,t_f]}^2 - \|z\|_{[T,t_f]}^2 + \|u + (D_{12}^T C_1 + B_2^T X_{t_f})x\|_{[T,t_f]}^2. \tag{53}$$

Notice that this holds for any u.

Now, combining (49) and (52) gives

$$x^T(T)X_{t_f}(T)x(T) = \bar{x}_T^T Y_{t_0}(T)X_{t_f}(T)Y_{t_0}(T)\bar{x}_T = \bar{x}_T^T Y_{t_0}(T)\bar{x}_T.$$

So, summing the both sides of (50) and (53) gives

$$0 = \|w\|_{[t_0,t_f]}^2 - \|z\|_{[t_0,t_f]}^2 + \|u + (D_{12}^T C_1 + B_2^T X_{t_f})x\|_{[T,t_f]}^2.$$

This implies that for any choice of the causal controller K we have

$$\|w\|_{[t_0,t_f]}^2 \leq \|z\|_{[t_0,t_f]}^2.$$

This contradicts the assumption that there exists a controller K satisfying (11).

(c) $\implies$ (b). When there exist $X_{t_f}(t)$ and $Y_{t_0}(t)$ satisfying $det(I - X_{t_f} Y_{t_0})(t) \neq 0$, it has been already shown that operator H admits a causal-anticausal factorization of the form (39) with $\hat{K}$ being given by (34). So, to complete the proof, we need only to show (37) ((38) can be proved by considering the adjoint). For this, notice that the adjoint operator of $G_{11}|Null(G_{21})$ is represented by $(G_{11}|Null(G_{21}))^* = (I - G_{21}^*(G_{21}G_{21}^*)^{-1}G_{21})G_{11}^*$. Thus the condition (37) can equivalently be rewritten as

$$
\begin{aligned}
I &- (G_{11}|Null(G_{21}))(G_{11}|Null(G_{21}))^* \\
&= I - G_{11}G_{11}^* + G_{11}G_{21}^*(G_{21}G_{21}^*)^{-1}G_{21} \\
&> 0,
\end{aligned}
\tag{54}
$$

where $M > 0$ implies that operator M is self-adjoint and positive. Next, let us observe that analogously to the decomposition of H we have the factorization of the operator

$$
\begin{bmatrix} I - G_{11}G_{11}^* & -G_{11}G_{21}^* \\ -G_{21}G_{11}^* & -G_{21}G_{21}^* \end{bmatrix} = LJL^*
\tag{55}
$$

with L and L^{-1} causal. In fact, representing the operator in the left hand-side by the state-space representation

$$
\begin{aligned}
\begin{bmatrix} \dot{x} \\ \dot{\bar{x}} \\ z \\ y \end{bmatrix}
&=
\begin{bmatrix} A & B_1 B_1^T \\ 0 & -A^T \end{bmatrix}
\begin{bmatrix} x \\ \bar{x} \end{bmatrix}
+
\begin{bmatrix} 0 & B_1 D_{21}^T \\ -C_1^T & -C_2^T \end{bmatrix}
\begin{bmatrix} \bar{z} \\ \bar{y} \end{bmatrix} \\
&=
\begin{bmatrix} -C1 & 0 \\ -C1 & -B_1 B_1^T \end{bmatrix}
\begin{bmatrix} x \\ \bar{x} \end{bmatrix}
+
\begin{bmatrix} I & 0 \\ 0 & -I \end{bmatrix}
\begin{bmatrix} \bar{z} \\ \bar{y} \end{bmatrix}.
\end{aligned}
\tag{56}
$$

and applying the coordinate change $\begin{bmatrix} \xi \\ \bar{\xi} \end{bmatrix} = \begin{bmatrix} I & Y \\ 0 & I \end{bmatrix} \begin{bmatrix} x \\ \bar{x} \end{bmatrix}$, the state-space representation of L is obtained as

$$
\dot{\xi} = A\xi + \begin{bmatrix} YC_1 & -B_1 D_{21}^T - YC_2^T \end{bmatrix} \begin{bmatrix} \eta \\ \bar{\eta} \end{bmatrix},
\tag{57}
$$

$$
\begin{bmatrix} z \\ y \end{bmatrix} = - \begin{bmatrix} C_1 \\ C_2 \end{bmatrix} + \begin{bmatrix} \eta \\ \bar{\eta} \end{bmatrix}.
\tag{58}
$$

Thus, representing L as $L = \begin{bmatrix} L_1 & L_2 \\ L_3 & L_4 \end{bmatrix}$, (55) gives

$$
I - G_{11}G_{11}^* = L_1 L_1^* - L_1 L_1^*, \quad -G_{11}G_{21}^* = L_1 L_3^* - L_2 L_4^*, \quad -G_{21}G_{21}^* = L_3 L_3^* - L_4 L_4^*.
$$

So that

$$
\begin{aligned}
I &- G_{11}G_{11}^* + G_{11}G_{21}^*(G_{21}G_{21}^*)^{-1}G_{21} \\
&= (L_1 - L_2 L_4^{-1} L_3)(I - L_3^* L_4^{-*} L_4^{-1} L_3)^{-1}(L_1^* - L_3^* L_4^{-*} L_2^*) > 0.
\end{aligned}
$$

This proves (54), as desired. The proof of (38) can be performed exactly in the same way.

(b) $\implies$ (a) and (ii). Consider the controller K given by (13) with $\hat{K}$ as in (34). The state space realization of (13) in the form of standard linear fractional transformation can be obtained by rewriting (34) as u and s to be the inputs and y and t the outputs and putting $s = Qt$. When $Q = 0$, it is given by (40). This proves (ii).

It remains therefore to show that under the assumption of (b), K satisfies the norm condition

$$\|G_{11} + G_{12}K(I - G_{22}K)^{-1}G_{21}\|_{\infty,t_0,t_f} < 1. \tag{59}$$

Rewrite (39) as

$$\hat{K}^* H^{-1} \hat{K} = J$$

where H^{-1} is given by (20). Multiplying the both sides by $\begin{bmatrix} Q \\ I \end{bmatrix} (\hat{K}_{21}Q +$ $\hat{K}_{22})^{-1}$ from the right and its adjoint from the left, we have

$$\begin{bmatrix} K^* & I \end{bmatrix} H^{-1} \begin{bmatrix} K \\ I \end{bmatrix} = (\hat{K}_{21}Q + \hat{K}_{22})^{-*}(Q^*Q - I)(\hat{K}_{21}Q + \hat{K}_{22})^{-1} < 0. \tag{60}$$

Now recall that (37) is equivalent to (54). Hence there exists invertible operator M such that $I - G_{11}G_{11}^* + G_{11}G_{21}^*(G_{21}G_{21}^*)^{-1}G_{21} = MM^*$. Thus, H^{-1} is computed as

$$\begin{aligned} H^{-1} &= \begin{bmatrix} G_{12} & 0 \\ G_{22} & I \end{bmatrix}^* \begin{bmatrix} I - G_{11}G_{11}^* & -G_{11}G_{21}^* \\ -G_{21}G_{11}^* & -G_{21}G_{21}^* \end{bmatrix}^{-1} \begin{bmatrix} G_{12} & 0 \\ G_{22} & I \end{bmatrix} \\ &= \begin{bmatrix} I & -G_{11}G_{21}(G_{21}G_{21}^*)^{-1} \\ 0 & I \end{bmatrix}^* \\ &\quad \times \begin{bmatrix} M^{-*}M^{-1} & 0 \\ 0 & -(G_{21}G_{21}^*)^{-1} \end{bmatrix} \begin{bmatrix} I & -G_{11}G_{21}(G_{21}G_{21}^*)^{-1} \\ 0 & I \end{bmatrix}. \end{aligned}$$

Substituting into (60), we can get

$$\| M^{-1}\{G_{11}G_{21}^*(G_{21}G_{21}^*)^{-1} + G_{12}K(I - G_{22}K)^{-1}\}(G_{21}G_{21}^*)^{1/2} \|_{\infty,t_0,t_f} < I$$

$$\Longleftrightarrow$$

$$\begin{aligned} &\{G_{11}G_{21}^*(G_{21}G_{21}^*)^{-1} + G_{12}K(I - G_{22}K)^{-1}\} \\ &\times (G_{21}G_{21}^*)\{G_{11}G_{21}^*(G_{21}G_{21}^*)^{-1} + G_{12}K(I - G_{22}K)^{-1}\} < MM^* \end{aligned}$$

$\Longleftrightarrow$ (59).

This proves the theorem.

5 Conclusion

In this paper, H_∞ control problem has been discussed for linear continuous time-varying and also sampled-data systems in a unified framework. Controller characterization achieving suboptimal control has been given both in state-space and also in operator theoretic framework.

Acknowledgment The author wishes to express his thanks to Professors Y. Hayakawa, Nagoya University and F. Zhang, Tokyo University of Mercantile Marine for their helpful discussion.

References

[1] J.C. Doyle, K. Glover, P.P. Khargonekar, and B. Francis, State-space solutions to standard H_2 and H_∞ control problems, IEEE Trans. Automat. Control, 34(1989), pp. 831-847.

[2] R. Ravi, K.M. Nagpal, and P.P. Khargonekar, H_∞ control problem of linear time-varying systems: A state space approach, SIAM J. Control and Optimization, 29-6(1991), pp. 1394-1413.

[3] B. Bamieh and J.B. Pearson, A generalized framework for linear periodic systems with application to H_∞ sampled-data control, IEEE Trans. Automat. Control, 37(1992), pp. 418-435.

[4] W. Sun, K.M. Nagpal and P.P. Khargonekar, H_∞ control and filtering for sampled-data systems, IEEE Trans. Automat. Control, 38(1993), pp. 1162-1175

[5] Y. Hayakawa, S. Hara, Y. Yamamoto, H_∞ type problem for sampled-data control systems—A solution via minimum energy characterization, IEEE Trans. Automat. Control, 38(1994), pp. 2278-2284.

[6] M. Green, K. Glover, D. Limebeer, and J. Doyle, A J-spectral factorization approach to H_∞ control, SIAM J. Control Optim., 28 (1990), pp.1350-1371

[7] H. Kwakernaak: Robust control and H_∞ -optimization-Tutorial paper, Automatica, 29-2 (1993), pp.255-273

[8] H. Kwakernaak: The polynomial approach to H_∞-control, Lecture Notes in Mathematics 1496, Springer-Verlag (1990), pp.141-221

[9] H.Kimura: Chain-scattering representation, J-lossless factorization and H_∞ control, Lecture Notes on Robust Control, Seaou National University, 1993

[10] S. Hosoe: Solving H_∞ control problem by transfer functions, Lecture notes for SICE seminar (1991), pp.15-34 (in Japanese)

[11] S.Hosoe: Derivation of H_∞ sampled data suboptimal controllers by making use of J-unitary operators, Proc. of the Seoul Workshop on Robust Control (1993), 183-195

[12] S.Hosoe, and T.Kato: Derivation of H_∞ multirate sampled data suboptimal controllers by making use of J-unitary factorization, Proc. of American Control Conference (1997)

H_∞ in Flight

Ian Postlethwaite   Alex Smerlas   Daniel J. Walker*

Abstract

This paper is about the flight testing of an H_∞-controlled helicopter, the Bell 205 Airborne Simulator, carried out in Ottawa in July 1997. The tests are part of a major research activity between the University of Leicester (supported by the UK Engineering and Physical Sciences Research Council), the UK Defense and Evaluation Research Agency (Bedford), the Canadian NRC Flight Research Laboratory (Ottawa), and GKN Westland Helicopters. This is thought to be the first time that an H_∞ controller has been used on a helicopter in flight.

Introduction

Current trends in Fly-By-Wire (FBW) systems are towards the application of higher feedback augmentation levels to decrease pilot workload and increase mission effectiveness. However, in helicopters, FBW developments have lagged noticeably behind their counterparts on fixed wing programmes, one of the major hurdles being the high levels of dynamic uncertainty and cross-axis coupling present in helicopters. From the mid 1980s a significant research effort has been devoted to the development of H_∞-optimisation as a means of addressing multi-axis performance and robustness requirements. Theoretical developments have produced attractive control design methods but no rotary design examples have reached the maturity of a flight test until now. DERA (Bedford) and the University of Leicester have worked together for many years to close the gap between theory and practice and have conducted a number of successful ground-based piloted simulations ([1],[2],[3]). In the last two years this research activity has concentrated on the design of H_∞-control laws for the Canadian Bell 205 variable stability helicopter ([4],[5]). The teaming with the NRC Flight Research Laboratory (FRL) has provided an excellent opportunity for in-flight demonstration of modern multivariable control design methods. The paper is organised as follows:

Section 2 deals with the helicopter control problem. Insight into aerodynamic behaviour and flight control law requirements play a key role in the develop-

*Department of Engineering, University of Leicester, Leicester LE1 7RH UK

ment of this FBW technology demonstration. The benefits of using multi-variable control theory are marked, especially in an H_∞ loop shaping context where all the basic performance requirements can be embodied into the design methodology.

Section 3 describes the control law architecture and how the design method was used to provide task tailored handling qualities. In principle, the design philosophy is similar to the pioneering Advanced Digital Optical/Control System (ADOCS) program, but the new element included here is the explicit robustness specifications through the use of H_∞ optimisation. An important conclusion drawn from this part is that the design methodology adopted presents no restrictions on the candidate architectures (eg. Rate Command Attitude Hold) for a helicopter stability augmentation system.

In section 4, we will describe some of the preliminary work carried out at GKN Westland Helicopters' simulator facilities and the Large Motion Simulator (LMS) at DERA, Bedford. As with the design of any primary flight control system, ground-based simulations are an essential ingredient of a research program to ensure safety and inceptor functionality.

In section 5, we will present some of the flight test results supplemented by handling qualities ratings (HQRs). HQRs are awarded by pilots for an aircraft flying a Mission Task Element (MTE), and are determined following the Cooper-Harper scale [6]. A short video of the flight tests will be shown at the workshop. The final pilot rating depends on the task performance achieved and the pilot workload expended.

Finally, section 6 presents some conclusions drawn from the tests and shows future research directions relevant to the application of multivariable control methods to FBW technology.

Section 2

Helicopters have a broad spectrum of tasks to perform in bad weather and low level environments. In order to achieve the performance required during these tasks, helicopters need to be agile yet easy to control. However, typical rotor-crafts are naturally unstable and exhibit inter-axis coupling. This coupling is inherent to the vehicle and is due to the asymmetric forces and moments produced by the main and tail rotor systems. Therefore, even simple tasks require complicated multivariable control inputs, which imply high pilot workload and the possibility of degraded handling qualities. Consider, for example, a side-step task, where the pilot has to laterally translate the rotorcraft mainly using rolling commands. Roll inputs will dominate the response, but at the same time the pilot needs to coordinate heading angle (with the pedals), height (with the collective lever) and keep track over the ground (using longitudinal inputs). Heading excursions are caused by the vertical offset of the tail rotor thrust and the vertical fin sideforce from the centre of mass. Height does not remain constant as the main rotor thrust is directed sideways and lift no longer counter-balances the vertically directed rotorcraft's weight. Also, any lateral demands cause longitudinal motion as the thrust is not always normal to the main rotor disk, and the inclination of the lift vectors on individual blade sections contributes a significant amount of off-axis inputs. The level of cross coupling depends on the cross inertias and the particular characteristics of the rotor disk (teetering, hingeless etc.).

From the side-step task example it is evident that a multivariable strategy is appropriate for the primary flight control system of a helicopter; the cross axis effects are not neglected, and coupling is handled in a natural way. In general, there are three translational and three rotational degrees of freedom that the pilot closes with the controls. These modes may be characterised by three velocities (u, v, w) and three angular rates (q, p, r). Here, u, v, and w stand for forward, lateral and vertical velocities, while q, p, and r are the pitch, roll and yaw rates.

The Bell 205 helicopter model used for the work described in this paper can be found in [7]. It includes the standard six degrees of freedom (6DOF) stability and control derivatives calculated with the standard Bell stabiliser bar on. As shown in [8], the dynamic response of the helicopter appears to be reduced compared with the actual aircraft. The 6DOF motion can be considered adequate primarily for low-moderate frequency handling qualities analysis. The rationale behind the neglected dynamics is that the higher rotor and inflow phenomena are much faster than the fuselage motions and have enough time to reach their steady state within the time constraints of the airframe response modes. However, the need for high bandwidth control systems makes the unmodelled dynamics significant contributing factors to model uncertainty, especially at frequencies above $10 rad/sec$. Additionally, considerable time delays between inputs and fuselage responses (due to the teetering rotor system), coupled with high frequency uncertainty, can excite under-damped rotor modes which are of primary importance for control design considerations. Therefore, the appropriate control design methodology should employ frequency domain design criteria to ensure adequate disturbance rejection at high frequencies, and it must be able to tolerate model uncertainty, especially around the desired closed loop bandwidth. From this point of view H_∞ optimisation is attractive for helicopter applications; bandwidth requirements are specified by means of frequency dependent weighting functions, and trade-off between performance and robustness is easy for the control law designer to achieve. Several background simulation studies ([1], [2]), have confirmed the above statements, and the Loop Shaping Design Procedure (LSDP) ([9],[10]) has emerged as the most applicable design methodology for the helicopter control problem. The simple solution to the robust stabilisation problem in the LSDP, alongside the observer-based implementation of [11], were instrumental in the implementation of the control law architecture described in the following section.

Section 3

The control law structure is, and should be, independent of the design method used, and tailored to the desired task mode of the vehicle. Low speed and hovering conditions are challenging control problems imposing high workload on the pilot whose attention is primarily directed to the position of the vehicle relative to the ground. In this case, an attitude-command-attitude-hold (ACAH) type of response is what is typically required from the control system, where longitudinal and lateral stick deflections result in direct control of pitch (θ) and roll (ϕ) attitudes. Yaw rate (r) was chosen as the primary controlled variable in the heading loop, while the heave loop was left unaugmented as the direct drive from the collective lever is highly desired by pilots. A schematic representation of the control structure can be found in figure 1. The pilot's demands d are shown entering the prefilter K_1, which is used to define the desired vehicle characteristics for the task tailored mode, according to the latest Aeronautical Design Standard for military rotorcraft [12]. The desired dynamics, in familiar control terms, are specified by diagonal transfer function models which the closed loop responses are forced to track ([13]). The feedback

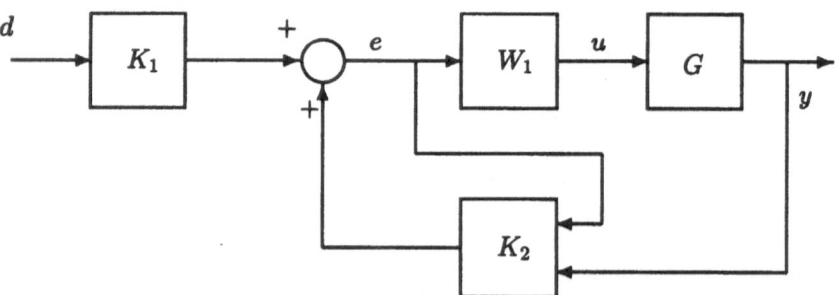

Figure 1: Controller Structure

signals y are derived from the sensors. These include the primary controlled variables (θ, ϕ, r) and two additional measurements, pitch and roll rates (q and p respectively) that are known to increase damping in the closed loop system. In this way, when the designer requires high crossover frequencies (in a loop shaping context), the low frequency dynamics are largely cancelled by the zeros of the open loop transfer function and lead compensation (which is provided by simultaneously feeding back rates and their associated attitudes).

From the above discussion it can be seen that the control law can be tailored to the vehicle task requirements without limiting the designer freedom in terms of mode specification. Although the controller provides an ACAH type of response, other response types such as rate-command-attitude-hold (RCAH) or translational-rate-command-attitude-hold (TRCAH) are also possible. Ideally, if the number of flight control command modes was not a restriction (due to the piloting techniques and mode initialisation), future control laws for rotorcraft would incorporate all the above specifications.

Section 4

As mentioned above in the introduction, considerable time was dedicated to ground simulation and testing of the controller. The objectives of these simulations with the pilot in the loop can be summarised as follows:

- Investigate various add-on features to the basic ACAH controller such as prefiltering functions and additional nonlinear compensation.

- Ensure that the controller performs as intended for a wide range of operating conditions.

- Expose the test pilot to control law functionality similar to the characteristics expected in the flight tests.

- Code the controller and explore any computational savings the observer architecture may offer.

The first flight of any flight control system always has an element of risk. Although all the necessary procedures / precautions ensuring crew safety and vehicle integrity have had many years of successful implementation on the NRC Bell 205 helicopter, ground-based simulations were very important to get pilot clearance for the flight test. In this respect, two flight simulations were performed, one at GKN Westland Helicopters and one at DERA, Bedford. In brief, the following experiments were conducted:

The controller stability and performance over the whole rotorcraft operating envelope were explored until the linear controller robustness margins were "exhausted". The operational envelope of the Bell 205 helicopter extends up to 120 knots IAS (Indicated Airspeed). However, in practice the speed envelope is limited to 90 knots as the teetering rotor system induces significant fuselage vibrations which result in increased stressing loads on the airframe. It was felt that a conservative approach would be the most appropriate in this case, and therefore it was decided that in-flight evaluation should present no problem for the hover regime and possibly moderate speeds.

Investigation of command path shaping functions was carried out. Different response types require different command path shaping schemes. For example, a rate command system requires quadratic types of shaping functions appropriate to the feel system used for the experiment. This results in fast short-term demands and slower large amplitude inputs during full stick throws, appropriate to aircraft's attitudes. Here, it has to be noted that the conclusions were given more pedagogical value rather than precise aims to arrive at a suitable design for the flight test.[1]

It was felt necessary to give the pilot an idea of "what the controller does", subject to simulator constraints. Although it is generally considered routine for the test pilots to fly a novel control law there is no substitute for recent experience, familiarity and understanding when conducting flight experiments.

Finally, coding was found to be easy to perform since the controller was implemented as state-space blocks. The controller was integrated using Euler type integration methods and the computational loads on the on-board computer were estimated not to exceed the computational capabilities of the Bell 205. Had the controller been too demanding, discrete time synthesis and finite difference equations would have been employed for the design and implementation respectively.

Section 5

As with any augmented vehicle, establishing flying qualities requires a combination of quantitative criteria that define the designer's aim at the given time, and subjective (pilot) opinion of how well the aircraft behaved. In the ADS-33 standard, the flying qualities are task-oriented and they are judged on a combination of results from clinical open-loop and closed-loop test manoeuvres. Qualitative criteria for the different levels of coupling between the axes (grades 1 up to 3 with level 1 being the best) are also defined. In addition, these criteria vary for different response types and they depend on the control axis being evaluated. Further qualitative feedback from the flight test is gained from the evaluation methodology in different frequency ranges. In this respect, the tasks defined in [12] are using large and moderate amplitude criteria (to assess control power and agility) as well as short term frequency measures to expose any possible stability problems in the closed loop system.

Overall, the specific aim of the flight trial was to check the flying qualities with the pilot-in-the-loop, levels of workload, task performance and agility, and to provide important data for comparison with predictions from off-line simulations.

---

[1] Both ground-based simulations implemented a displacement centre-stick while the flight test was conducted using a force side-stick.

The test manoeuvres included, primarily, low speed tasks. These were the sidestep, precision hover, spot turn, quick-hop and pirouette manoeuvres. The side-step and quick-hop are essentially hover re-positioning manoeuvres in minimum time. Spot-turn is a yaw axis task in order to assess the turn-to-target capability of the vehicle. Precision hover is a typical task for rescue missions and pirouette is a multi-axis manoeuvre giving the pilot information about the overall handling of the helicopter.

The task performance is classified into "desired" and "adequate" levels. The levels of aggression are associated with the capability to achieve maximum attitude, angular and translational rates of the aircraft in the primary axis response. The following paragraphs summarise some of the handling qualities' evaluation results:

- Forward flight up to 80 knots.

  The pilot flew a few circles around the Flight Research Laboratory area to gain general information about the overall handling of the aircraft. The cyclic response was predictable with roll responses well damped and little coupling between pitch and roll. Flight path control was easy. The hold capability was not adequate and turn-coordination was difficult due to the non-centering pedals.

- ADS-33 Mission Task Elements.

  Side-step: The sidestep task requires the pilot to laterally translate the aircraft as fast as possible acquiring maximum acceleration while compensating for heading, height and pitch transients. The primary task objectives are to assess the lateral/directional handling qualities of the vehicle for aggressive manoeuvring, to judge the interaxis coupling and to check the coordination of bank and collective to hold constant altitude. In all axes (except yaw) desired performance was achieved using maximum aggression tasks. Commanded angles were up to 30 degrees.

  Quick-hop: This task requires the pilot to pitch down the aircraft up to -30 degrees, accelerate, acquire adequate speed and rapidly decelerate by pitching up $+30$ degrees. Similarly to the side-step, handling is being analysed for pitch and heave axes, and coupling to the other axes is judged. The desired performance was achieved but the pilot workload was high.

  Precision Hover: The scenario required the aircraft to approach a given altitude and maintain hover within certain tolerances in all axes. The evaluation pilot found it very easy to achieve the performance desired even in strong cross winds.

Pirouette: This task can be described as the execution of an up-side down cone with the centre of gravity of the helicopter. Considerable pilot workload was necessary to compensate for strong cross winds but the desired performance was achieved.

Turn-to-target: This manoeuvre was initiated from a downwind hover with a yaw demand through 180 degrees back into the wind while maintaining horizontal position. The primary response was outstanding with little coupling into the other loops.

The on-board sensors of the Bell 205 helicopter made it possible to measure aircraft attitudes and rates as well as actuator and control positions. Figure 2 shows some of the roll axis' recorded time histories, namely, pitch attitude and rate, and lateral cyclic input and acceleration. Comparison with the

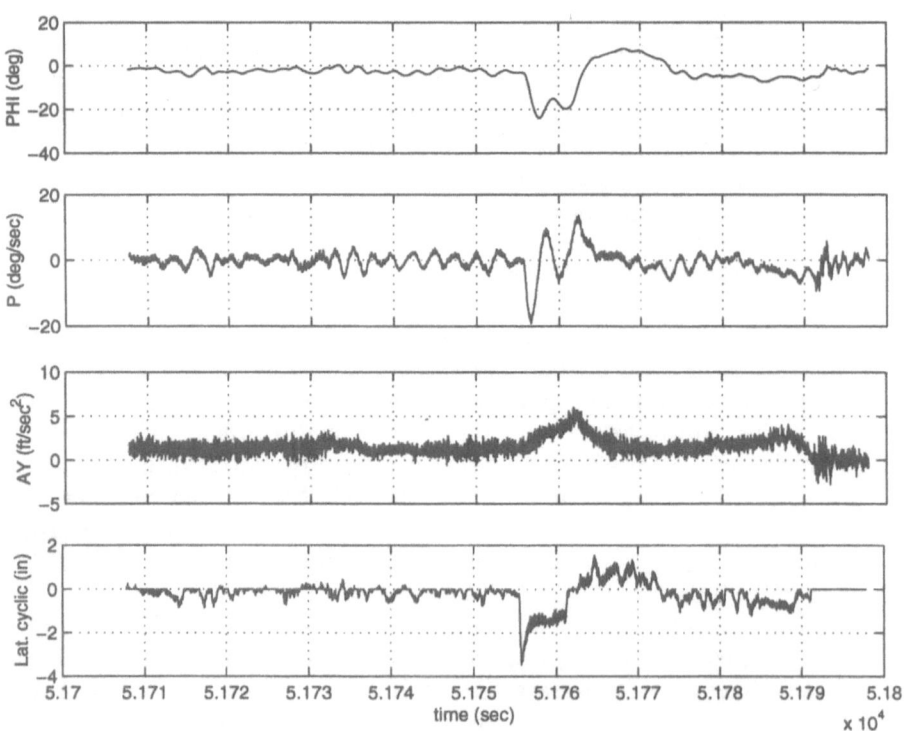

Figure 2: Recorder time histories, roll angle (ϕ), rate (p), acceleration (A_y) and Lateral cyclic input.

off-line simulations is currently under way and will concentrate on spectral analysis of the obtained results. The estimation of linear frequency responses can provide important information to assess how well the specified loop gains and bandwidths were achieved in flight. For example, figure 3 shows the closed loop frequency gain and phase responses from the stick input (after some signal conditioning on the force sensing side-stick) to yaw rate response. A coherence function is also plotted indicating the linear relationship between input and output. A value of 1 indicates direct linear relationship and a value of 0 shows that the input and output are uncorrelated. The successful evolution

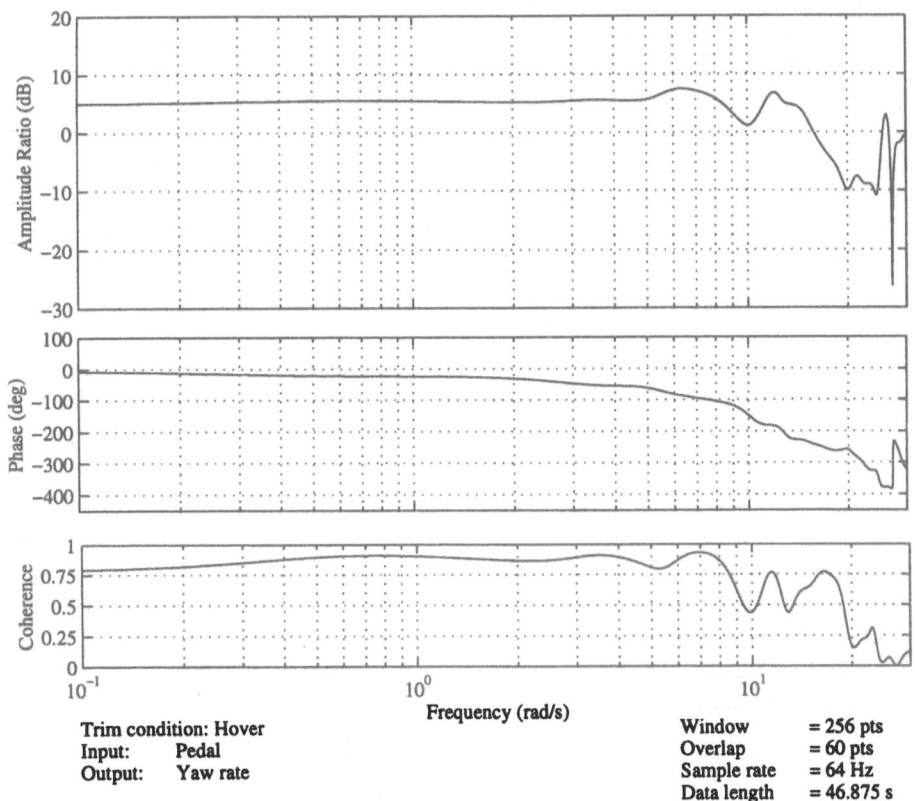

Figure 3: Frequency sweep - Yaw loop

of a new control law methodology is largely dependent on the improvements the designer is able to make after the first flight. In our H_∞ flight trials improvements are sought in the couplings identified by the pilots and confirmed by the flight test data in different frequency ranges. Other aspects of the design can also be improved such as noise attenuation and validation of existing linear models can be carried out using the open loop results.

Section 6

This paper has given a brief description of the overall program of the design and evaluation of an H_∞ control law on the Bell 205 airborne simulator. The flight trials took place at the Flight Research Laboratory of the National Research Council of Canada. The program proved to be an excellent demonstration of cooperation between all the participating organisations, and ground simulations were crucial to the preliminary assessment of the control system. The results drawn from the first analyses of the flight test data and pilot comments are most encouraging. Although only one test pilot has flown the H_∞-based control law the following conclusions can be drawn from the experiment:

The pilot's perception of the primary helicopter responses were clear and predictable. Couplings were present but considering the adverse weather conditions (10-15 knots) and the poor linear models that have been used for design purposes (almost 20 years old) the deficiencies were judged as fair. A number of mechanical problems masked some of the benefits provided by the controller but its functionality was as intended. Further research is being conducted to incorporate gain schedules into the design methodology as well as to use the information gathered during the flight tests to select better weighting functions. This will enable us in the next flight trials scheduled for the summer of 1998 to obtain better matching between specified and achieved designer specifications.

Acknowledgments

The authors would like to acknowledge the contributions of Jeremy Howitt and Michael Strange of DERA Bedford (for modelling and simulation work), Cdr. Robert I. Horton of DERA who was the test pilot, Arthur Gubbels and Stewart Baillie of the Canadian NRC Flight Research Laboratory (for arranging and managing the flight tests), Robert Erdos and Stephan Carignan also of the Flight Research Laboratory (who were the safety pilots) and Adrian Alford of GKN Westland Helicopters who organised and made simulation facilities available prior to the simulation studies at Bedford. In addition we would like to thank the UK Engineering and Physical Sciences Research Council for financial support.

References

[1] A. Yue and I. Postlethwaite. Improvement of Helicopter Handling Qualities Using H_∞ Optimisation. *IEE Proceedings*, D:115–129, 1990.

[2] D.J. Walker, I.Postlethwaite, J.Howitt, and N.P.Foster. Rotorcraft Flying Qualities Improvement Using Advanced Control. *American Helicopter*

Society/NASA Conference on Flying Qualities and Human Factors, page No.2.3.1, 1993.

[3] D.J.Walker and I.Postlethwaite. Advanced Helicopter Flight Control Using Two-Degrees-of-Freedom Optimal Control. *Journal of Guidance Control and Dynamics*, 19(2):461–468, 1996.

[4] A.J. Smerlas, I.Postlethwaite, and D.J.Walker. Full Envelope Robust Control Law for the Bell-205 Helicopter. *Proceedings. of the 22nd European Rotorcraft Forum, 17-19 Sept., Brighton, UK*, pages 105.1–105.6, September 1996.

[5] A.J. Smerlas, I.Postlethwaite, and D.J.Walker. Robust Gain Scheduling in Helicopter Control. *Proceedings of the 23rd European Rotorcraft Forum, 16-18 Sept., Treff Hotel, Dresden, Germany*, pages 84.1–84.5, 1997.

[6] G. Cooper and R. Harper. The use of pilot rating in the evaluation of aircraft handling qualities. Technical Report D-5153, NASA, USA, 1986.

[7] R.K. Heffley, W.F. Jewell, J.M. Lehman, and R.A. Van Winkle. A Compilation and Analysis of Helicopter Handling Qualities Data. Contractor Report 3144, NASA, 1979.

[8] S.W. Baillie, J.M.Morgan, and K.R.Goheen. Practical Experiences in Control Systems Design using the NCR Bell 205 Airborne Simulator. *Flight Mechanics Panel Symposium*, pages 27.1–27.12, January 1994.

[9] D. McFarlane and K. Glover. An $\mathcal{H}_\infty$ Design Procedure Using Robust Stabilization of Normalized Coprime Factors. *Proceedings of the 27th Conference on Decision and Control*, pages 1343–1348, December 1988.

[10] D. McFarlane and K. Glover. *Robust Controller Design Using Normalized Coprime Factor Plant Descriptions*. Information and Control Sciences. Springer-Verlag, 1990.

[11] J. Sefton and K. Glover. Pole-Zero Cancellations in the General $\mathcal{H}_\infty$ Problem with Reference to a Two Block Design. *Systems & Control Letters*, 14:295–306, 1990.

[12] Anonymous. Handling Qualities Requirements for Military Rotorcraft, Aeronautical Design Standard ADS-33C. Technical report, US Army AVSCOM, St. Louis, Missouri, 1989.

[13] D. Hoyle, R. Hyde, and D.J.N. Limebeer. An $\mathcal{H}^\infty$ Approach to Two-Degree-Of-Freedom Design. *Proceedings of the IEEE CDC*, pages 1581–1585, December 1991.

[14] G.D.Padfield, M.T.Charlton, Maj. T.Mace, and Maj. R. Morton. Flying Qualities Evaluation of the UK Attack Helicopter Contenders Using the ADS-33 Methodology; Clinical Criteria and Piloted Simulation Trials.

21st European Rotorcraft Forum, Saint Petersburg, Russia, 2:IV–2.1 – IV2.20, Aug. 30 - Sept. 1 1995.

μ-Analysis and Synthesis based on Parameter Dependent Multipliers

Gan Chen*   Toshiharu Sugie†

Abstract

It is well known that the structured singular value μ plays a crucial role in analysis and synthesis of the systems having structured uncertainties. In this paper, first, we propose new upper bounds of μ based on the parameter dependent multipliers, which are less conservative than former ones. Second, using this type of multipliers, we give a necessary and sufficient condition for dynamical systems to have μ less than a specified value γ for every frequency, which requires neither frequency sweep nor higher order multipliers. Third, based on this result, we solve a state feedback μ synthesis problem via LMI's. Numerical examples demonstrate the effectiveness of the proposed method.

1 Introduction

It is well known that the structured singular value μ plays a crucial role in analysis and synthesis of the systems having structured uncertainties [1, 9]. Unfortunately, because it is difficult to compute the true value of μ, its upper bound is used instead of μ. One of the well known upper bounds is given by Fan et al.[1] which is based on scaled small gain theorem and implemented in CADs. This is quite effective when we treat the complex block uncertainties. However it is not very good at taking care of real repeated uncertainties such as physical parameter variations.

Recently, new improved upper bounds of μ were found out in [2, 3]. Their upper bounds are proved to be less conservative than the conventional ones. In particular, their methods are effective for real repeated uncertainties. However, there still remains a (possibly big) gap between the true value and their upper bounds of μ. In addition, as for controller design, it is not clear how to use them for μ synthesis. For example, the well known D-K iteration can not be applied to their new upper bounds.

---
*Comp. & Sys. Sci. Eng., Osaka Prefecture University, Sakai, Osaka 593, Japan. kan@cs.osakafu-u.ac.jp

†Dept. of Applied Systems Science,Kyoto University, Uji, Kyoto 611, Japan. sugie@robot.kuass.kyoto-u.ac.jp

On the other hand, Safonov et al.[10, 7], derived other type of μ upper bounds which are convenient for μ analysis of dynamical systems, and they have shown how to obtain μ suboptimal controllers in the framework of BMI (Biaffine Matrix Inequality). However it is not clear whether their μ upper bounds are less conservative compared to the conventional one [1]. In addition, they need higher order multipliers in order to obtain less conservative μ upper bounds.

In this paper, first, we propose new upper bounds of μ based on parameter dependent multiplies, which are less conservative than those of [2, 3]. Second, using the multiplier technique, we give a necessary and sufficient condition for dynamical systems to have μ less than a specified value γ for every frequency, which requires neither frequency sweep nor higher order multipliers. Third, based on this result, we solve the state feedback μ synthesis problem via LMI's.

2   Preliminaries

Consider the closed loop system which consists of the constant matrix $G \in \mathbf{C}^{n \times n}$ and the time invariant uncertainty Δ_s/γ shown in Fig. 1. The uncer-

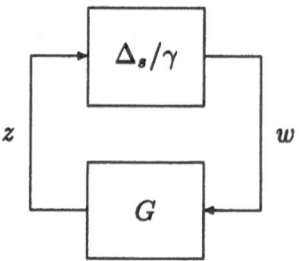

Figure 1: Feedback connection of G and Δ_s

tainty Δ_s belongs to the set $\mathcal{U}$ which is defined as follows:

$$\mathcal{U} := \{\mathrm{diag}\{\Delta_R, \Delta_C\} \in \mathbf{C}^{n \times n} \mid \Delta_R \in \mathcal{R}, \Delta_C \in \mathcal{C}\} \qquad (1)$$

$$\mathcal{R} := \{\mathrm{diag}\{r_1 I, \cdots, r_L I\} \mid r_i \in [-1, 1]\} \qquad (2)$$

$$\mathcal{C} := \{\mathrm{diag}\{\Delta_{c1}, \cdots, \Delta_{c\ell}\} \mid \bar{\sigma}(\Delta_i) \leq 1\} \qquad (3)$$

where, I denotes the identity matrix with appropriate size, Δ_i's are complex square matrices, and $\bar{\sigma}(\cdot)$ denotes the maximum singular value.

For simplicity, we use the following notation

$$\mathrm{Herm}\{M\} := M + M^* \qquad (4)$$

for given constant matrix M, where M^* is the complex conjugate transpose of M.

For given G and $\mathcal{U}$, the structured singular value $\mu(G)$ is defined as follows.

Definition 1

$$\mu(G) := \left\{ \inf \gamma \geq 0 \mid \det \left(I - G \frac{\Delta_s}{\gamma} \right) \neq 0, \ \forall \Delta_s \in \mathcal{U} \right\} \tag{5}$$

This implies that $\mu(G) < \gamma$ holds iff

$$F(\Delta_s) := I - G(\Delta_s/\gamma) \tag{6}$$

is nonsingular. As for the non-singularity of matrices, we have the following lemma which plays the key role in this paper.

Lemma 1 *[2] A matrix $F \in \mathbf{C}^{n \times n}$ is nonsingular if and only if there exists an $M \in \mathbf{C}^{n \times n}$ such that*

$$Herm\{FM\} < 0 \tag{7}$$

holds, where M is called a multiplier.

Also, In the following sections, we use the following lemmas.

Lemma 2 (GSPR Lemma) *[10, 8] Let A, B, C, D be state space matrices of the transfer function $G(s)$, i.e. $G(s) = C(sI - A)^{-1}B + D$. Then,*

$$Herm\{G(j\omega)\} < 0, \ \forall \omega \in \mathbf{R}, \tag{8}$$

holds, if there exists a matrix $P = P^*$ such that

$$Herm\left\{ \begin{bmatrix} A & B \\ C & D \end{bmatrix} \begin{bmatrix} P & 0 \\ 0 & I \end{bmatrix} \right\} < 0 \tag{9}$$

holds. Here, we permit A, B, C, D be complex matrices.

The above lemma is known to be generalized Strict Positive Real Lemma.

Lemma 3 *[5, 6] Let the parameter vector $p := [p_1, p_2, \cdots]$ belongs to given polytope set $\mathcal{P}$. Consider a parameter dependent LMI $F(x, p) < 0$ with respect to the variable x. Then*

$$F(x, p) < 0 \quad \forall p \in \mathcal{P} \tag{10}$$

holds if the following two conditions are satisfied.

$$\frac{\partial^2}{\partial p_i^2} F(p, x) \geq 0, \quad \forall i \tag{11}$$

$$F(x, p) < 0 \quad \forall p \in \mathcal{V}_p \tag{12}$$

where the set $\mathcal{V}_p$ is defined by $\{p \mid vertex \ of \ \mathcal{P}\}$

The lemma implies that we can transform the non-convex problem with respect to p to the convex one by adding the constraint (11).

Lemma 4 (S-procedure) *[8, 2] Let $Y = Y^* < 0$, M, N be suitable size matrices, and Δ be any uncertainty such that $\bar{\sigma}(\Delta) \leq 1$. Then*

$$Herm\{M\Delta N\} + Y < 0 \qquad (13)$$

holds, if there exists a matrix $Q > 0$ such that

$$MQM^* + N^*Q^{-1}N + Y < 0 \qquad (14)$$
$$Q\Delta = \Delta Q \qquad (15)$$

holds.

3 Computation of new upper bounds of μ

In this section we propose a computation method of improved upper bounds of μ based on parameter dependent multipliers.

3.1 Basic idea

By using Lemma 1 and the definition of μ, it was shown in [2] that $\mu(G) < \gamma$ holds if there exists a $M \in \mathbf{C}^{n \times n}$ such that

$$Herm\{F(\Delta_s)M\} < 0 \quad \forall \Delta_s \in \mathcal{U} \qquad (16)$$

holds. It was also shown (i) that the above condition can be reduced to a finite set of LMI's by using S-procedure and (ii) that the upper bounds are less conservative than that of [1]. However, it is not necessary to restrict the multiplier M be constant and it would introduce conservativeness to use the single multiplier for every Δ_s.

Therefore, we allow the multiplier M to be dependent on Δ_s in this paper. By doing this, we obtain the following lemma.

Lemma 5 *$\mu(G) < \gamma$ holds if and only if there exists an $M(\Delta_s)$ such that the following LMI holds.*

$$L(\Delta_s) := Herm\{F(\Delta_s)M(\Delta_s)\} \quad \forall \Delta_s \in \mathcal{U} \qquad (17)$$

Since the problem of finding such an $M(\Delta_s)$ contains the square of the term of Δ_s and is not convex with respect to Δ_s, it is not convenient to solve the problem directly. Therefore we try to obtain a sufficient condition by modifying the problem to more tractable one, which provides us a tighter upper bounds of μ.

3.2   Real repeated scalar uncertainty

First, we consider the case of $\mathcal{U} = \mathcal{R}$, that is, the real repeated scalar uncertainty case. In this case, $F(\Delta_s)$ is of the form

$$F(\Delta_s) = F_0 + \sum_{i=1}^{L} r_i F_i \tag{18}$$

where F_i's are appropriate matrices determined from G and the uncertainty structure of $\mathcal{R}$. Corresponding to the above $F(\Delta_s)$, we employ the multiplier of the form

$$M(\Delta_s) = M_0 + \sum_{i=1}^{L} r_i M_i \tag{19}$$

without loss of generality, where M_i's are parameters to seek. Then, simple calculation shows that

$$\frac{\partial^2}{\partial r_i^2} L(\Delta_s) = \mathrm{Herm}\{F_i M_i\} \quad \forall i \geq 1 \tag{20}$$

Therefore, by applying Lemma 3, we obtain the following result.

Theorem 1 *Suppose $\mathcal{U} = \mathcal{R}$. Then $\mu(G) < \gamma$ if there exist M_i's such that*

$$\mathrm{Herm}\{F(\Delta_v)M(\Delta_v)\} < 0, \quad \forall \Delta_v \in \mathcal{V}_R \tag{21}$$
$$\mathcal{V}_R := \{\Delta_v : \text{ vertex of } \mathcal{R}\} \tag{22}$$
$$\mathrm{Herm}\{F_i M_i\} \geq 0, \quad i = 1, 2, \cdots, L \tag{23}$$

hold.

Though the obtained upper bound could be conservative because of the additional convexity constraint (23), the main point of the theorem is that it is computable via LMI's. In addition, if we choose $M_i = 0$ ($i \geq 1$), the above condition can be reduced to Fu's result [2]. Therefore, it is guaranteed that the upper bounds of μ via Theorem 1 is less conservative than the existing ones. The cost we have to pay is more computation burden.

[**Example 1**] Suppose the matrix G and the set $\mathcal{U}$ in Fig.1 are given by

$$G = \begin{bmatrix} 0 & 1 & 0 & 1 \\ 0.5 & 0 & 0.5 & 0 \\ 2a & 0 & a & 0 \\ 0 & -2a & 0 & -a \end{bmatrix} \tag{24}$$
$$\mathcal{U} = \{\mathrm{diag}\{r_1 I_2, r_2 I_2\} \mid r_i \in [-1, 1]\} \tag{25}$$

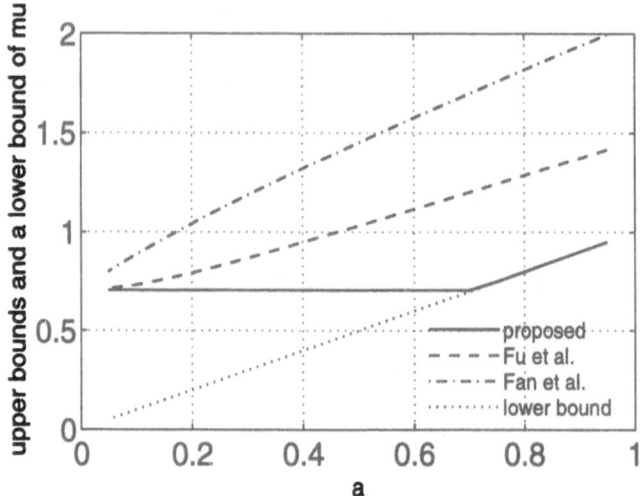

Figure 2: Upper bounds of μ for real repeated scalar uncertainty

For this system, we compute the upper bounds of μ for various a's via Theorem 1 and former ones. In Fig. 2, the dashed line shows the upper bound of μ proposed by [2]. This upper bound is the same as that proposed by [3]. The solid line shows the proposed method. The dash-and-dotted line shows the upper bound [1] of μ used in MATLAB. This figure shows that the upper bounds obtained by former methods are conservative. Furthermore, in this particular example, it can be shown that our upper bound happens to be the exact value of μ.

3.3   Mixed uncertainty

Next, we consider the case where the uncertainty set $\mathcal{U}$ is given by (1), that is, the mixed uncertainty case. In this case, it is difficult to describe the full block uncertainty in a less conservative way by using the polytope type uncertainty like Theorem 1. One of the most reasonable ways to treat this type of uncertainty structure could be to use the parameter dependent multipliers only for the real repeated uncertainty $\Delta_R \in \mathcal{R}$ and to use S-procedure for the full block uncertainty $\Delta_C \in \mathcal{C}$. In what follows we show how to do it.

Now we have $\Delta_s = \mathrm{diag}\{\Delta_R, \Delta_C\}$. Define the following variables.

$$F_R(\Delta_R) := I - G \begin{bmatrix} \Delta_R & 0 \\ 0 & 0 \end{bmatrix} / \gamma \qquad (26)$$

$$T_C := \begin{bmatrix} 0 \\ I \end{bmatrix} \qquad (27)$$

Also, we restrict the multiplier $M(\Delta)$ as $M(\Delta_R)$ depending only on the real repeated scalar uncertainty. Then, $L(\Delta_s) = \text{Herm}\{F(\Delta_s)M(\Delta_s)\}$ is written by

$$L(\Delta_s) = \text{Herm}\{F_R(\Delta_R)M(\Delta_R) + G/\gamma T_C \Delta_C T_C^* M(\Delta_R)\} \tag{28}$$

Applying Lemma 4, $L(\Delta_s) < 0$ holds, if there exists a $D > 0$ satisfying

$$L_a(\Delta_s) \quad := \quad Y + M(\Delta_R)^* T_C D^{-1} T_C^* M(\Delta_R) < 0 \tag{29}$$

$$D\Delta_C \quad = \quad \Delta_C D \tag{30}$$

$$Y \quad := \quad \text{Herm}\{F_R(\Delta_R)M(\Delta_R)\} + G/\gamma T_C D T_C^* G^*/\gamma. \tag{31}$$

Using Schur compliment, (29) can be replaced by

$$L_b(\Delta_R) := \begin{bmatrix} Y & M(\Delta_R)^* T_C \\ T_C^* M(\Delta_R) & -D \end{bmatrix} < 0 \tag{32}$$

Since $F_R(\Delta_R)$ is of the form

$$F_R(\Delta_R) = F_{R0} + \sum_{i=1}^{L} r_i F_{Ri} \tag{33}$$

we use the multiplier $M(\Delta_R)$ defined by

$$M(\Delta_R) := M_0 + \sum_{i=1}^{L} r_i M_i, \tag{34}$$

Then, we obtain the following theorem.

Theorem 2 *Suppose $\mathcal{U}$ is given by (1), then $\mu(G) < \gamma$ holds if there exist M_i's and $D >$ such that*

$$L_b(\Delta_v) < 0, \quad \forall \Delta_v \in \{\text{ vertex of } \mathcal{R}\} \tag{35}$$

$$\text{Herm}\{F_{Ri}M_i\} \geq 0, \quad i = 1, 2, \cdots, L \tag{36}$$

$$D\Delta_C = \Delta_C D, \quad \forall \Delta \in \mathcal{C} \tag{37}$$

hold.

The above theorem gives us a method to compute an upper bound of μ via LMI's.

[Example 2]

Suppose the matrix G and the uncertainty set $\mathcal{U}$ are given by

$$G = \begin{bmatrix} 0 & 1 & 0 & a & 0 & 1-a \\ 0.5 & 0 & a/2 & 0 & (1-a)/2 & 0 \\ 2a & 0 & a & 0 & 0 & 0 \\ 0 & -2a & 0 & -a & 0 & 0 \\ (1-a)/2 & 0 & 0 & 0 & 1-a & 0 \\ 0 & (a-1)/2 & 0 & 0 & 0 & a-1 \end{bmatrix} \tag{38}$$

$$\mathcal{U} = \{\mathrm{diag}\{\delta_r I_2, \delta_c I_2, \Delta_2\} \mid \delta_r \in [-1,1], \quad \delta_c \in \mathbf{C}, \ |\delta_c| \le 1,$$
$$\Delta_2 \in C^{2\times 2}, \ \bar{\sigma}(\Delta_2) \le 1\} \quad (39)$$

We compute the upper bounds of μ for various a's. In Fig. 3, the solid line, the dotted line and the dash-and-dotted line show the proposed upper bound, the upper bound proposed by [2] and the upper bound proposed by [1], respectively. Using the parameter dependent multiplier, we obtain the less conservative result than the upper bound via S-procedure[2]. The upper bound via S-procedure[2] is almost the same as the upper bound proposed by [1] for all a. The proposed upper bound is the same as former ones if $a \approx 0$ where the full block uncertainty is dominant. However, it is less conservative than former ones if $a \approx 1$ where the repeated scalar uncertainty is dominant. This figure shows that the proposed method is effective compared to S-procedure[2]. Note that we cover the complex repeated scalar uncertainty by square polytope in the complex plane, and used the parameter dependent multiplier technique.

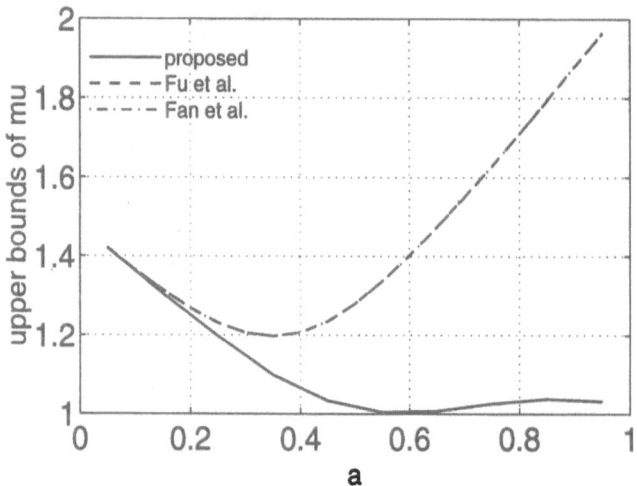

Figure 3: Upper bounds of μ for mixed uncertainty

4 μ Analysis based on dynamic multipliers

In this section, we consider the μ of dynamical systems rather than constant matrices.

Suppose G in Fig.1 is given by $G(s)$ whose state-space realization is given by

$$\dot{x} = Ax + Bw \quad (40)$$
$$z = Cx + Dw \quad (41)$$

with A stable. For the purpose of μ analysis, we have to compute γ such that

$$\sup_{\omega \in \mathbf{R}} \mu(G(j\omega)) < \gamma \tag{42}$$

holds. Note that (42) holds if there exists a dynamic multiplier $M(s)$ such that

$$\text{herm}\{(\gamma I - G(j\omega)\Delta_s)M(j\omega)\} < 0, \quad \forall \Delta_s \in \mathcal{U} \tag{43}$$

holds for every ω. However, it is not an easy task to check the above condition for every frequency even if $M(s)$ is given in advance. In order to avoid the difficulty, Kawanishi and Sugie [8] used Lemma 2 to obtain the following sufficient condition.

Lemma 6 *[8]* $\sup_\omega \mu(G(j\omega)) < \gamma$ *holds if there exist $P_w = P_w^*$ and dynamic multiplier $M(s)$ such that*

$$L_w := Herm\left\{\begin{bmatrix} A_w & B_w \\ C_w & D_w \end{bmatrix}\begin{bmatrix} P_w & 0 \\ 0 & I \end{bmatrix}\right\} < 0, \quad \forall \Delta_s \in \mathcal{U} \tag{44}$$

holds, where $\{A_w, B_w, C_w\ D_w\}$ is a state-space realization of $(\gamma I - G(s)\Delta_s)M(s)$.

In order to find such an $M(s)$ via LMI's, we have to define the order and the basis of the multiplier in advance [8], which is similar to [10]. Therefore, in order to obtain a less conservative upper bound, we have to adopt the higher order multiplier.

Furthermore, though P_w and $M(s)$ are fixed ones in the above lemmas, it is not necessary to restrict P_w and the multiplier $M(s)$ be fixed. Therefore, to obtain a less conservative upper bound, we employ the parameter-dependent multiplier $M(s, \Delta_s)$ of the form

$$M(s, \Delta_s) := C_m(\Delta_s)(sI - A_m(\Delta_s))^{-1}B_m(\Delta_s) + D_m(\Delta_s) \tag{45}$$

Then we obtain the following theorem, where the size of $A_m(\Delta_s)$ is the same as A.

Theorem 3 $\sup_\omega \mu(G(j\omega)) < \gamma$ *holds if and only if there exist $C_s(\Delta_s)$, $D_m(\Delta_s)$ and $P(\Delta_s) = P^*(\Delta_s)$ such that*

$$Herm\left\{\begin{bmatrix} AP(\Delta_s) - B\Delta_s C_s(\Delta_s) & CP(\Delta_s) + (\gamma I - D\Delta_s)C_s(\Delta_s) \\ -B\Delta_s D_m(\Delta_s) & (\gamma I - D\Delta_s)D_m(\Delta_s) \end{bmatrix}\right\} < 0, \tag{46}$$

holds for any $\Delta_s \in \mathcal{U}$.

proof *(Sufficiency)* In Lemma 6, choose the multiplier $M(s, \Delta_s)$ as in (45), and let $P_w = P_w^*$ be of the form

$$P_w = \begin{bmatrix} P(\Delta_s) & S(\Delta_s) \\ S(\Delta_s) & S(\Delta_s) \end{bmatrix} \tag{47}$$

Also, define T, $C_s(\Delta_s)$, $L_{21}(\Delta_s)$, and $L_{23}(\Delta_s)$ as follows.

$$T := \begin{bmatrix} I & 0 & 0 \\ I & -I & 0 \\ 0 & 0 & I \end{bmatrix} \tag{48}$$

$$C_s(\Delta_s) := C_m(\Delta_s)S(\Delta_s) \tag{49}$$

$$L_{21}(\Delta_s) := AP(\Delta_s) - B\Delta_s C_s(\Delta_s) - A_m(\Delta_s)S(\Delta_s)$$
$$+ (P(\Delta_s) - S(\Delta_s))A^* \tag{50}$$

$$L_{23}(\Delta_s) := -B\Delta_s D_m(\Delta_s) - B_m(\Delta_s) + (P(\Delta_s) - S(\Delta_s))C^* \tag{51}$$

Then, from Lemma 6, $\sup_\omega \mu(G(j\omega)) < \gamma$ holds if

$$TL_w T^* = \text{Herm}\{ \begin{bmatrix} AP(\Delta_s) - B\Delta_s C_s(\Delta_s) \\ L_{21}(\Delta_s) \\ CP(\Delta_s) + (\gamma I - D\Delta_s)C_s(\Delta_s) \end{bmatrix}$$

$$\begin{matrix} 0 & -B\Delta_s D_m(\Delta_s) \\ A(P(\Delta_s) - S(\Delta_s)) & L_{23}(\Delta_s) \\ 0 & (\gamma I - D\Delta_s)D_m(\Delta_s) \end{bmatrix} \}$$

$$< 0 \tag{52}$$

is satisfied. Since $A_m(\Delta_s)$ and $B_m(\Delta_s)$ appear only in $L_{21}(\Delta_s)$, $L_{23}(\Delta_s)$, respectively, we can always let $L_{21}(\Delta_s) \equiv 0$ and $L_{23}(\Delta_s) \equiv 0$ by choosing $A_m(\Delta_s)$ and $B_m(\Delta_s)$ as follows.

$$A_m(\Delta_s) := \{AP(\Delta_s) - B\Delta_s C_s(\Delta_s)$$
$$+ (P(\Delta_s) - S(\Delta_s))A^*\}S(\Delta_s)^{-1} \tag{53}$$

$$B_m(\Delta_s) := -B\Delta_s D_m(\Delta_s)$$
$$+ (P(\Delta_s) - S(\Delta_s))C^* \tag{54}$$

And there always exists a nonsingular $S(\Delta_s)$ such that $\text{Herm}\{A(P(\Delta_s) - S(\Delta_s))\} < 0$ holds. Therefore, (52) is satisfied if (46) holds.

(Necessity) Suppose $\sup_\omega \mu(G(j\omega)) < \gamma$ holds. Then, the closed system shown in Fig. 1 is internally stable. Therefore, $I - D\Delta_s/\gamma$ is nonsingular and the closed loop system

$$\dot{x}_{cl} = \{A + B\Delta_s/\gamma(I - D\Delta_s/\gamma)^{-1}C\}x_{cl} \tag{55}$$

is stable for any $\Delta_s \in \mathcal{U}$. This implies that there exist $D_m(\Delta_s)$ and $P(\Delta_s) > 0$ such that

$$\text{Herm}\{(\gamma I - D\Delta_s)D_m(\Delta_s)\} < 0 \tag{56}$$

$$\text{Herm}\{(A + B\Delta_s(\gamma I - D\Delta_s)^{-1}C)P(\Delta_s)\} < 0 \tag{57}$$

hold. Suppose we choose $C_s(\Delta_s)$ as

$$C_s(\Delta_s) := -(\gamma I - D\Delta_s)^{-1}CP(\Delta_s) \tag{58}$$

and $D_m(\Delta_s)$ be sufficiently small. Then (56), (57) and (58) yields (46). This proves the theorem.   Q.E.D.

The theorem implies that the full order multiplier is enough for computing μ without conservativeness, if we allow the multiplier to be parameter dependent. Also, the LMI condition does not contain A and B matrices of the multiplier, the size of LMI remains small in spite of employing the multiplier. The latter property is useful for μ synthesis as shown later.

Since it is difficult to check the LMI conditions in Theorem 3 for every Δ_s. We derive a sufficient condition in a similar way to the previous section.

Corollary 1 $\sup_\omega \mu(G(j\omega)) < \gamma$ *holds if there exist* $C_s(\Delta_R)$, $D_m(\Delta_R)$, $P(\Delta_R) = P(\Delta_R)^*$, $U_i \geq 0$, *and* $Q(\Delta_R) > 0$ *such that*

$$Y_u(\Delta_R) \quad := \quad \begin{bmatrix} X_j(\Delta_R) + \sum_{i=1}^{L} r_i^2 U_i & \begin{bmatrix} C_{sc}(\Delta_R)^* \\ D_{mc}(\Delta_R)^* \end{bmatrix} \\ [C_{sc}(\Delta_R) \quad D_{mc}(\Delta_R)] & -Q(\Delta_R) \end{bmatrix} < 0, \forall r_i = \{-1, 1\} \tag{59}$$

$$X_{ci} \quad := \quad -Herm\left\{ \begin{bmatrix} B_r \\ D_r \end{bmatrix} J_i \begin{bmatrix} C_{sri} & D_{mri} \end{bmatrix} \right\} + U_i \geq 0 \quad \forall i = 1 \sim L. \tag{60}$$

hold, where,

$$\Delta_R \quad = \quad diag\{r_1 I, r_2 I, \cdots, r_L I\} \tag{61}$$

$$X_j(\Delta_R) \quad := \quad X_r(\Delta_R) + \begin{bmatrix} B_c \\ D_c \end{bmatrix} Q(\Delta_R) \begin{bmatrix} B_c \\ D_c \end{bmatrix}^* \tag{62}$$

$$X_r(\Delta_R) \quad := \quad Herm\left\{ \begin{bmatrix} AP(\Delta_R) - B_r \Delta_R C_{sr}(\Delta_R) \\ CP(\Delta_R) + \gamma C_s(\Delta_R) - D_r \Delta_R C_{sr}(\Delta_R) \end{bmatrix} \right.$$
$$\left. \begin{matrix} -B_r \Delta_R D_{mr}(\Delta_R) \\ \gamma D_m(\Delta_R) - D_r \Delta_R D_{mr}(\Delta_R) \end{matrix} \right] \right\} \tag{63}$$

$$C_s(\Delta_R) \quad = \quad \begin{bmatrix} C_{sr}(\Delta_R) \\ C_{sc}(\Delta_R) \end{bmatrix}$$

$$:= \quad \begin{bmatrix} C_{sr0} \\ C_{sc0} \end{bmatrix} + \sum_{i=1}^{l_r} r_i \begin{bmatrix} C_{sri} \\ C_{sci} \end{bmatrix} \tag{64}$$

$$D_m(\Delta_R) \quad = \quad \begin{bmatrix} D_{mr}(\Delta_R) \\ D_{mc}(\Delta_R) \end{bmatrix}$$

$$:= \quad \begin{bmatrix} D_{mr0} \\ D_{mc0} \end{bmatrix} + \sum_{i=1}^{l_r} r_i \begin{bmatrix} D_{mri} \\ D_{mci} \end{bmatrix} \tag{65}$$

$$P(\Delta_R) \quad = \quad P_0 + r_1 P_1 + \cdots + r_{l_r} P_{l_r} \tag{66}$$

$$Q(\Delta_R) \quad = \quad Q_0 + r_1 Q_1 + \cdots + r_{l_r} Q_{l_r} \tag{67}$$

$$J_i \quad := \quad diag\{0, \cdots, 0, I, 0, \cdots, 0\}. \tag{68}$$
$$[\begin{array}{cc} B_r & B_c \end{array}] \quad := \quad B \tag{69}$$
$$[\begin{array}{cc} D_r & D_c \end{array}] \quad := \quad D \tag{70}$$

Matrix partitions are correspond to the sizes of Δ_R and Δ_C.

Corollary 1 introduces some conservativeness. However the condition is solved by the finite sets of LMI's corresponding to all vertices of real repeated scalar uncertainty Δ_R.

5   State FB μ Synthesis

In this section, we consider the static state feedback μ synthesis problem for the system shown in Fig. 4. The plant $G(s)$ and the feedback law are given

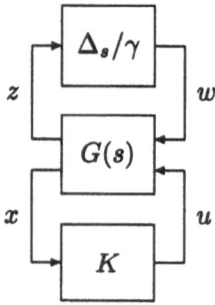

Figure 4: State feedback system for μ-synthesis.

by

$$\dot{x} \quad = \quad Ax + Bw + B_u u \tag{71}$$
$$z \quad = \quad Cx + Dw + D_u u \tag{72}$$
$$u \quad = \quad Kx \tag{73}$$

The coefficient matrices of the closed loop system $z = G_{cl}(s)w$ are given as follows:
$$\left[\begin{array}{cc} A_{cl} & B_{cl} \\ C_{cl} & D_{cl} \end{array} \right] = \left[\begin{array}{cc} A + B_u K & B \\ C + D_u K & D \end{array} \right] \tag{74}$$

By restricting $P(\Delta_s)$ to be constant P and using the variable transformation $W := KP$, we obtain the following result from Theorem 4, immediately.

Theorem 4 *There exists a state feedback law $u = Kx$ such that $\sup_\omega \mu(G_{cl}(j\omega)) < \gamma$ holds, if there exist $C_s(\Delta_s)$, $D_m(\Delta_s)$, $P > 0$, W such* *that*

$$X_k(\Delta_s) := Herm\left\{\begin{bmatrix} AP + B_u W - B\Delta_s C_s(\Delta_s) \\ CP + D_u W + (\gamma I - D\Delta_s)C_s(\Delta_s) \end{bmatrix}\right.$$
$$\left.\begin{bmatrix} -B\Delta_s D_m(\Delta_s) \\ (\gamma I - D\Delta_s)D_m(\Delta_s) \end{bmatrix}\right\} < 0, \forall \Delta_s \in \mathcal{U} \tag{75}$$

holds. Furthermore, if this condition is satisfied, such a state feedback gain is given by $K = WP^{-1}$.

Since we eliminate the A matrix of the multiplier in Theorem 3, we can reduce the state feedback μ synthesis problem to LMI. If we adopt the results by [8] or [10], the problem will end up with BMI.

In addition, similar to the previous sections, we can obtain K which satisfies $\sup_\omega \mu(G(j\omega)) < \gamma$ through the finite sets of LMI's, by using Lemmas 3 and 4.

[**Example 3**] Suppose the state space coefficients of $G(s)$ and the uncertainty set $\mathcal{U}$ be given by

$$A = \text{diag}\{-100, -100, -1, -1\} \tag{76}$$

$$B = \text{diag}\{100, 100, 1, -1\} \tag{77}$$

$$C = \begin{bmatrix} 0 & 1 & 0 & 1 \\ 0.5 & 0 & 0.5 & 0 \\ 2 & 0 & 1 & 0 \\ 0 & 2 & 0 & 1 \end{bmatrix} \tag{78}$$

$$D = \begin{bmatrix} 0 & 1 & 0 & 1 \\ 0.5 & 0 & 0.5 & 0 \\ 2 & 0 & 1 & 0 \\ 0 & -2 & 0 & -1 \end{bmatrix} \tag{79}$$

$$B_u = \begin{bmatrix} 0 & 0 & 1 & 0 \\ 0 & 0 & 0 & 1 \end{bmatrix}^T \tag{80}$$

$$D_u = \begin{bmatrix} 0 & 0 & 1 & 0 \\ 0 & 0 & 0 & 1 \end{bmatrix}^T \tag{81}$$

$$\mathcal{U} = \{\text{diag}\{r_1 I_2, r_2 I_2\} \mid r_i \in [-1, 1]\} \tag{82}$$

We try to minimize γ (the upper bound of μ of the closed loop) by state feedback $u = Kx$ based on Theorem 4. As a result, we obtain a γ as 1.719 (Fig. 5:dashed line). We compute the upper bound of μ of the resultant closed loop by two different methods. The dashed-and-dotted line shows the upper bound proposed by [1] and the solid line shows the upper bound proposed obtained by Theorem 1. The peak value of the upper bound by [1] is 2.383, while Theorem 1 gives us 1.718. This figure shows the effectiveness of the proposed method.

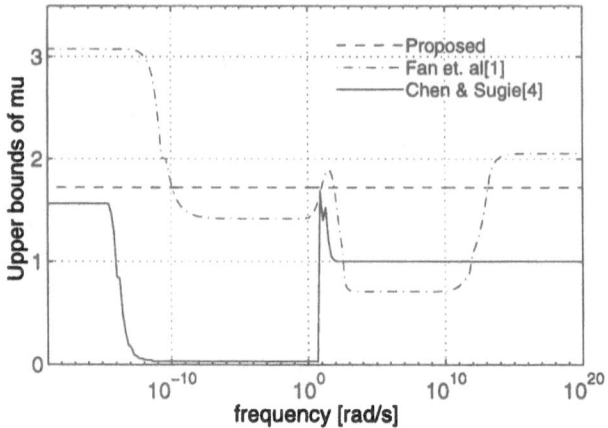

Figure 5: Upper bounds of $\mu(G_{cl})$

6   Conclusion

In this paper, we have discussed new methods of μ analysis and synthesis. First, we have given a computation method of new upper bounds of μ based on the parameter dependent multipliers, which are guaranteed to be less conservative than former ones. Second, using this type of multipliers, we have derived a necessary and sufficient condition for dynamical systems to have μ less than a specified value γ for every frequency, which requires neither frequency sweep nor higher order multipliers. Third, based on this result, we have solved a state feedback μ synthesis problem via LMI's. Numerical examples demonstrate the effectiveness of our method. Major demerit of our method is heavy computation burden. However, it is quite effective, especially for real repeated uncertainty case, our result provides an attractive alternative μ analysis/synthesis method.

References

[1] M.K.H. Fan, A.L. Tits and J.C. Doyle: Robustness in the Presence of Mixed Parametric Uncertainty and Unmodeled Dynamics, IEEE Trans. Automatic Control, **36**-1, 25/38(1991)

[2] M. Fu and N.E. Barabanov: Improved Upper Bounds of the Structured Singular Value, Proc. 34th IEEE CDC, 3115/3120(1995)

[3] T. Asai, S. Hara and T. Iwasaki: Simultaneous Modeling and Synthesis for Robust Control by LFT Scaling, Proc. IFAC World Congress, vol.G, 309/314(1996)

[4] G. Chen and T. Sugie: An Upper Bound of μ based on the Parameter Dependent Multipliers, Proc. 16th ACC, 2604/2608(1997)

[5] P. Gahinet, P. Apkarian, M. Chilali and E. Feron: Affine Parameter-Dependent Lyapunov Functions and Real Parameter Uncertainty, Proc. 3rd ECC, 2262/2267(1995)

[6] P. Gahinet, A. Nemirovski, A. J. Laub, M. Chilali: LMI Control Toolbox, MATH WORKS, 1995

[7] K. C. Goh, J. H. Ly, L. Turan and M. G. Safonov, μ/K_m-Synthesis via Bilinear Matrix Inequalities, Proc. 33rd IEEE CDC, 2032/2037(1994)

[8] M. Kawanishi and T. Sugie: Design of μ Suboptimal Controllers based on S-procedure and Generalized Strong Positive Real Lemma, 35th IEEE CDC, 3525/3526(1996)

[9] A. Packard and J. Doyle: The complex structured singular value, Automatica, **29**-1, 71/109(1993)

[10] M. G. Safonov, K. C. Goh and J. H. Ly: Control System Synthesis via Bilinear Matrix Inequalities, Proc. 14th ACC, 45/49(1994)

Optimal Controllers Which 'Work' in Presence of Plant & Controller Uncertainty

Edward J. Davison*

Abstract

It is shown that the design procedure proposed to solve the "perfect robust servomechanism" problem [1] has the property that it results in acceptable perturbation bounds with respect to both plant and controller uncertainty.

1 Introduction

There recently has been some concern expressed [2] that controllers which have been obtained using optimal control methods, for example, H_2, H_∞ ℓ_1 procedures, may have excessively poor robust properties, in the sense that the resulting closed loop system may be relatively intolerant to perturbations in the parameters of the resulting controller. It is the intent of this paper to show that there do exist optimal controller design approaches which do not present such a problem. In particular, it will be shown that the controller design approach proposed to solve the so-called "perfect servomechanism problem" [1], which is based on using a H_2 "cheap control" [3] approach, can result in controllers which have *excellent* tolerance to controller parameter perturbation.

2 Development

In the paper, the control problem to solved will be assumed to be the robust servomechanism problem (RSP) with constant disturbances and constant tracking signals [4].

In particular, the plant to be controlled is assumed to be described by the following nominal strictly proper LTI model, with input $u \in R^m$, output $y \in R^r$, disturbances $\omega \in R^\Omega$ and state $x \in R^n$:

$$\begin{aligned} \dot{x} &= Ax + Bu + E\omega \\ y &= Cx + Du + F\omega \end{aligned} \tag{1}$$

*Systems Control Group, Department of Electrical & Computer Engineering, University of Toronto, Toronto, Ontario, Canada M5S 3G4

where $D = 0$, and it is assumed that the following existence conditions for a solution to the RSP [4] are satisfied:

(a) (C, A, B) is stabilizable and detectable.

(b) $\text{rank}\begin{pmatrix} A & B \\ C & D \end{pmatrix} = n + r$

(c) y can be measured.

In this case, a controller with the following controller structure solves the robust servomechanism problem for (1) [4]:

$$\dot{\eta} = 0\eta + (y - y_{ref}) \tag{2a}$$

$$\begin{aligned} \dot{\xi} &= \bar{A}\xi + \bar{B}_1 y + \bar{B}_2 \eta \\ u &= \bar{C}\xi + \bar{D}_1 y + \bar{D}_2 \eta \end{aligned} \tag{2b}$$

where (2a) is the servo-compensator and (2b) is the stability compensator [4] for the controller. Assume now that the plant parameters are subject to the perturbation:

$$\begin{bmatrix} A & B \\ C & D \end{bmatrix} \rightarrow \begin{bmatrix} A & B \\ C & D \end{bmatrix} + \begin{bmatrix} \Delta A & \Delta B \\ \Delta C & \Delta D \end{bmatrix} \tag{3}$$

then the closed loop system is described by:

$$\begin{pmatrix} \dot{x} \\ \dot{\eta} \\ \dot{\xi} \end{pmatrix} = (\mathcal{A} + \mathcal{B}\Delta\mathcal{C}) \begin{pmatrix} x \\ \eta \\ \xi \end{pmatrix} \tag{4}$$

where

$$\mathcal{A} = \begin{bmatrix} A & B\bar{D}_2 & B\bar{C} \\ C & 0 & 0 \\ \bar{B}_1 C & \bar{B}_2 & \bar{A} \end{bmatrix}, \quad \mathcal{B} = \begin{bmatrix} I & 0 \\ 0 & I \\ 0 & \bar{B}_1 \end{bmatrix}, \quad \mathcal{C} = \begin{bmatrix} I & 0 & 0 \\ 0 & \bar{D}_2 & \bar{C} \end{bmatrix}$$

and

$$\Delta = \begin{bmatrix} \Delta A & \Delta B \\ \Delta C & \Delta D \end{bmatrix}$$

The following definitions of robustness are now made:

2.1 Plant Real Stability Margin (rstab_p)

Assume that the controller (2) has been designed so that the nominal closed loop system matrix $\mathcal{A}$ is asymptotically stable, and let rstab denote the *real stability radius* [5] of the perturbation matrix Δ with respect to the perturbed system $\mathcal{A} + \mathcal{B}\Delta\mathcal{C}$.

Then the *plant real stability margin* (rstab_p) is defined to be:

$$\text{rstab\_p} = \frac{\text{rstab}}{\bar{\sigma}\begin{pmatrix} A & B \\ C & D \end{pmatrix}} \times 100\% \tag{5}$$

where $\bar{\sigma}(\cdot)$ denotes the largest singular value of $(\cdot)$.

2.2 Controller Real Stability Margin (rstab_c)

Assume now that the controller parameters of the stabilizing controller (2b) are subject to the following perturbation:

$$\begin{bmatrix} \bar{A} & \bar{B}_1 & \bar{B}_2 \\ \bar{C} & \bar{D}_1 & \bar{D}_2 \end{bmatrix} \rightarrow \begin{bmatrix} \bar{A} & \bar{B}_1 & \bar{B}_2 \\ \bar{C} & \bar{D}_1 & \bar{D}_2 \end{bmatrix} + \begin{bmatrix} \Delta\bar{A} & \Delta\bar{B}_1 & \Delta\bar{B}_2 \\ \Delta\bar{C} & \Delta\bar{D}_1 & \Delta\bar{D}_2 \end{bmatrix} \tag{6}$$

then the closed loop system is described by:

$$\begin{pmatrix} \dot{x} \\ \dot{\eta} \\ \dot{\xi} \end{pmatrix} = (\mathcal{A} + \bar{\mathcal{B}}\bar{\Delta}\bar{\mathcal{C}}) \begin{pmatrix} x \\ \eta \\ \xi \end{pmatrix} \tag{7}$$

where

$$\bar{\mathcal{B}} = \begin{bmatrix} 0 & B \\ 0 & 0 \\ I & 0 \end{bmatrix}, \quad \bar{\mathcal{C}} = \begin{bmatrix} 0 & 0 & I \\ C & 0 & 0 \\ 0 & I & 0 \end{bmatrix}, \quad \bar{\Delta} = \begin{bmatrix} \Delta\bar{A} & \Delta\bar{B}_1 & \Delta\bar{B}_2 \\ \Delta\bar{C} & \Delta\bar{D}_1 & \Delta\bar{D}_2 \end{bmatrix}$$

Now let $\overline{\text{rstab}}$ denote the real stability radius of the perturbation matrix $\bar{\Delta}$ with respect to the perturbed system $\mathcal{A} + \bar{\mathcal{B}}\bar{\Delta}\bar{\mathcal{C}}$.

Then the *controller real stability margin* (rstab_c) is defined to be:

$$\text{rstab\_c} = \frac{\overline{\text{rstab}}}{\bar{\sigma}\begin{pmatrix} \bar{A} & \bar{B}_1 & \bar{B}_2 \\ \bar{C} & \bar{D}_1 & \bar{D}_2 \end{pmatrix}} \times 100\% \tag{8}$$

3   Results Obtained

The design procedure of [1] will now be used to design a controller (2) for a given plant model (1) by minimizing the cheap-control type of performance index J_ϵ:

$$J_\epsilon = \int_0^\infty (e'e + \epsilon\dot{u}'\dot{u})d\tau \tag{9}$$

where $e = (y - y_{ref})$ denotes the error in the system, and $\epsilon > 0$ is a scalar which penalizes rapid changes of control input signals in the system. The motivation for using this performance index is that 'perfect control' is achieved as $\epsilon \to 0$ iff the plant (1) is minimum phase with $m \geq r$, [1] i.e. the controller design approach will result in high performance control to be obtained (whenever it is possible to achieve); it is to be noted however that exactly the same design approach can also be applied when the plant is non-minimum phase.

3.1 Maximizing the Stability Margins

The question arises – what choice of $\epsilon > 0$ should be used in (9) in a given engineering problem? If one wishes to design controller (2) so as to maximize the stability margin rstab_p or rstab_c of the system, then it is observed that this can be simply achieved by carrying out a 1-dimensional search on $\epsilon > 0$ to determine:

$$
\begin{aligned}
\text{rstab\_p}^* &:= \sup_{\epsilon>0}\{\text{rstab\_p}(\epsilon)\} \\
\text{rstab\_c}^* &:= \sup_{\epsilon>0}\{\text{rstab\_c}(\epsilon)\}
\end{aligned}
\tag{10}
$$

These indices are respectively called the *maximum plant (controller) stability margin* of the closed loop system (1), (2) with respect to J_ϵ.

3.2 Minimizing $\epsilon > 0$ Subject to Specified Plant and Controller Stability Margins

Alternatively, one may wish to minimize $\epsilon > 0$ such that the resultant closed loop system has a plant stability margin $\geq 10^{-n_p}\%$ and a controller stability margin $\geq 10^{-n_c}\%$, say, where $n_p, n_c \in [-1, 0, 1, 2, 3, ...]$, corresponding to an accuracy specification that the plant data is accurate to at least $n_p + 2$ significant figure accuracy, and that the controller parameters can be implemented to at least $n_c + 2$ significant figure accuracy. This can be simply achieved by carrying out a minimization of $\epsilon > 0$, subject to the constraint that rstab_p$(\epsilon) \geq 10^{-n_p}$ and rstab_c$(\epsilon) \geq 10^{-n_c}$ [6]. The resulting controller obtained in this case has maximum performance (with respect to the performance index J_ϵ (9)), subject to the specified plant and controller stability margins.

4 Experimental Results Obtained

Three representative LTI SISO plant examples and one LTI MV plant example are considered:

Ex.1 A nonminimum phase, unstable system:

$$
\dot{x} = \begin{pmatrix} 1 & 2 \\ 1 & 0 \end{pmatrix} x + \begin{pmatrix} 1 \\ 0 \end{pmatrix} u
$$

$$y = (1 \quad -1)x$$

corresponding to:

$$G(s) = (s-1)/(s^2 - s - 2)$$

Ex.2 A minimum phase, unstable system:

$$\dot{x} = \begin{pmatrix} 0 & 1 \\ 0 & 0 \end{pmatrix} x + \begin{pmatrix} 0 \\ 1 \end{pmatrix} u$$

$$y = (1 \quad 0)x$$

corresponding to:

$$G(s) = 1/s^2$$

Ex.3 A minimum phase, stable system:

$$\dot{x} = \begin{pmatrix} -1 & -2 \\ 1 & 0 \end{pmatrix} x + \begin{pmatrix} 1 \\ 0 \end{pmatrix} u$$

$$y = (1 \quad 1)x$$

$$G(s) = (s+1)/(s^2 + s + 2)$$

Ex.4 A minimum phase, stable 2 input/2 output system which has large interaction (Rosenbrock's example):

$$\dot{x} = \begin{pmatrix} -1 & 1 & 0 \\ 0 & -1 & 0 \\ 0 & 0 & -1 \end{pmatrix} x + \begin{pmatrix} -0.1666 & 0 \\ 0.1666 & 1 \\ 0 & 0.5 \end{pmatrix} u,$$

$$y = \begin{pmatrix} 3 & -0.75 & -0.5 \\ 2 & 1 & 1 \end{pmatrix} x$$

4.1   Maximizing the Stability Margins

In this case, the following maximum plant and controller stability margins of the closed loop system (1), (2) with respect to J_ϵ are obtained (reported to 1.5 significant figure accuracy):

Ex.1	rstab_p* = 1.0%,	rstab_c* = 0.12%	with	$\epsilon = 1$
Ex.2	= 2.3%,	= 1.2%		$= 10^{-3}$
Ex.3	= 13%,	= 12%		$= 1$
Ex.4	= 2.8%,	= 2.0%		$= 1$

4.2 Finding Highest Performance Controllers Subject to Specified Plant and Controller Stability Margin

In this case, it is desired to minimize $\epsilon > 0$ in J_ϵ such that the resulting closed loop system has a plant stability margin $\geq 10^{-1}\%$ and a controller stability margin $\geq 10^{-4}\%$, corresponding to the assumption made that the plant data is accurate to at least 3 significant figures and that the controller parameters can be implemented to at least 6 significant figure accuracy. The following optimal values of ϵ are obtained in this case (reported to 1.5 significant figure accuracy).

$$
\begin{array}{llll}
\text{Ex.1} & \text{rstab\_p} = 1.0_{10-1}\%, & \text{rstab\_c} = 5.0_{10-4}\% & \text{with } \epsilon = 3_{10-8} \\
\text{Ex.2} & \quad\quad\quad = 1.0_{10-1}\%, & \quad\quad\quad = 2.1_{10-3}\% & \quad\quad \epsilon = 1_{10-9} \\
\text{Ex.3} & \quad\quad\quad = 1.0_{10-1}\%, & \quad\quad\quad = 1.3_{10-4}\% & \quad\quad \epsilon = 2_{10-11} \\
\text{Ex.4} & \quad\quad\quad = 1.1_{10-1}\%, & \quad\quad\quad = 8.3_{10-3}\% & \quad\quad \epsilon = 2_{10-7}
\end{array}
$$

4.3 Optimal Controllers Obtained

The controllers obtained have the structure:

$$
\begin{aligned}
\dot{\eta} &= 0\eta + (y - y_{ref}) \\
\dot{\xi} &= A_c \xi + (B_{c_1} \ \ B_{c_2}) \begin{pmatrix} y \\ \eta \end{pmatrix} \\
u &= C_c \xi + (D_{c_1} \ \ D_{c_2}) \begin{pmatrix} y \\ \eta \end{pmatrix}
\end{aligned}
$$

where, the controller parameters are given as follows for representative values of ϵ:

$$
\text{Ex.1} \begin{pmatrix} A_c & B_{c_1} & B_{c_2} \\ C_c & D_{c_1} & D_{c_2} \end{pmatrix}
$$

$$
= \left(\begin{array}{cc|cc} -11.50 & 3.55 & 8.00 & 1.00 \\ -3.00 & 4.00 & 4.00 & 0 \\ \hline -4.45 & -6.45 & 0 & 1.00 \end{array} \right) \quad \text{with } \epsilon = 1 \quad\quad (11a)
$$

$$
= \left(\begin{array}{cc|cc} -88.55 & -6396.1 & 8.00 & 3162.3 \\ -3.00 & 4.00 & 4.00 & 0 \\ \hline -81.55 & -6406.1 & 0 & 3162.3 \end{array} \right) \quad \text{with } \epsilon = 10^{-7} \quad\quad (11b)
$$

which gives rstab_p $= 1.0\%$, rstab_c $= 1.2_{10-1}\%$ for (11a) and rstab_p $= 1.4_{10-1}\%$, rstab_c $= 9.0_{10-4}\%$ for (11b).

Ex.2 $\begin{pmatrix} A_c & B_{c_1} & B_{c_2} \\ C_c & D_{c_1} & D_{c_2} \end{pmatrix}$

$$= \left(\begin{array}{cc|cc} -1.41 & 1.00 & 1.41 & 0 \\ -21.0 & -6.32 & 1.00 & -31.6 \\ \hline -20.0 & -6.32 & 0 & -31.6 \end{array} \right) \quad \text{with } \epsilon = 10^{-3} \qquad (12a)$$

$$= \left(\begin{array}{cc|cc} -44.72 & 1.00 & 44.72 & 0 \\ -1200.0 & -20.0 & 1000.0 & -1000.0 \\ \hline -200.0 & -20.0 & 0 & -1000.0 \end{array} \right) \quad \text{with } \epsilon = 10^{-6} \qquad (12b)$$

which gives rstab_p $= 2.3\%$, rstab_c $= 1.2\%$ for (12a) and rstab_p $= 1.6_{10-1}\%$, rstab_c $= 5.0_{10-2}\%$ for (12b).

Ex.3 $\begin{pmatrix} A_c & B_{c_1} & B_{c_2} \\ C_c & D_{c_1} & D_{c_2} \end{pmatrix}$

$$= \left(\begin{array}{cc|cc} -1.67 & -1.88 & 0 & -1.00 \\ 1.00 & 0 & 0 & 0 \\ \hline -0.665 & 0.114 & 0 & -1.00 \end{array} \right) \quad \text{with } \epsilon = 1 \qquad (13a)$$

$$= \left(\begin{array}{cc|cc} -448.2 & -447.2 & 0 & -1.00_{10^5} \\ 1.00 & 0 & 0 & 0 \\ \hline -447.2 & -445.2 & 0 & -1.00_{10^5} \end{array} \right) \quad \text{with } \epsilon = 10^{-10} \qquad (13b)$$

which gives rstab_p $= 13\%$, rstab_c $= 12\%$ for (13a) and rstab_p $= 1.6_{10-1}\%$, rstab_c $= 2.9_{10-4}\%$ for (13b).

Ex.4 $\begin{pmatrix} A_c & B_{c_1} & B_{c_2} \\ C_c & D_{c_1} & D_{c_2} \end{pmatrix}$

$$= \left(\begin{array}{ccc|cccc} -0.997 & 1.08 & -4.24_{10-2} & 0 & 0 & 7.66_{10-2} & -0.148 \\ -3.26 & -2.68 & 0.563 & 0 & 0 & -1.19 & 0.133 \\ -1.62 & -0.679 & -0.803 & 0 & 0 & -0.444 & -0.230 \\ \hline -2.04_{10-2} & -0.491 & 0.255 & 0 & 0 & -0.459 & 0.888 \\ -3.25 & -1.36 & 0.393 & 0 & 0 & -0.888 & -0.459 \end{array} \right)$$

$$\text{with } \epsilon = 1 \qquad (14a)$$

$$= \left(\begin{array}{ccc|cccc} -94.4 & -6.66 & 30.8 & 0 & 0 & -523.9 & -57.6 \\ 39.7 & -123.7 & 57.3 & 0 & 0 & 150 & 3374 \\ -166.9 & -76.6 & 89.3 & 0 & 0 & -172.8 & 1572 \\ \hline 560.2 & 45.9 & -185.0 & 0 & 0 & 3143 & 346 \\ -333.7 & -153.3 & 180.6 & 0 & 0 & -345.6 & 3143 \end{array} \right)$$

$$\text{with } \epsilon = 10^{-7} \qquad (14b)$$

which gives rstab_p $= 2.8\%$, rstab_c $= 2.0\%$ for (14a) and rstab_p $= 9.4_{10-2}\%$, rstab_c $= 8.3_{10-3}\%$ for (14b).

4.4   Destablizing Perturbation Matrices Obtained for Ex.1

The following are representative perturbation matrices obtained which destablize the resultant closed loop system. They are obtained for Ex.1 using controller (11a), (11b) respectively:

Plant perturbation (using controller (11a) with $\epsilon = 1$)

$$
\begin{pmatrix} \Delta A & \Delta B \\ \Delta C & \Delta D \end{pmatrix} = \left(\begin{array}{cc|c} -1.32_{10-2} & 3.01_{10-3} & 6.90_{10-3} \\ 1.11_{10-2} & 3.43_{10-3} & -1.89_{10-2} \\ \hline 1.61_{10-2} & -1.33_{10-2} & 1.25_{10-2} \end{array} \right)
$$

which gives rstab_p = 1.0%.

Controller perturbation (using controller (11a) with $\epsilon = 1$)

$$
\begin{pmatrix} \Delta A_c & \Delta B_{c_1} & \Delta B_{c_2} \\ \Delta C_c & \Delta D_{c_1} & \Delta D_{c_2} \end{pmatrix}
$$

$$
= \left(\begin{array}{cc|cc} -7.62_{10-4} & -5.43_{10-4} & 0 & -6.85_{10-3} \\ 1.84_{10-3} & 1.31_{10-3} & 0 & 1.65_{10-2} \\ \hline 7.22_{10-4} & 5.14_{10-4} & 0 & 6.48_{10-3} \end{array} \right)
$$

which gives rstab_c = 0.12%.

Plant perturbation (using controller (11b) with $\epsilon = 10^{-7}$)

$$
\begin{pmatrix} \Delta A & \Delta B \\ \Delta C & \Delta D \end{pmatrix} = \left(\begin{array}{cc|c} 2.44_{10-3} & -6.72_{10-7} & -7.09_{10-6} \\ -2.44_{10-3} & 6.72_{10-7} & 9.06_{10-6} \\ \hline -1.44_{10-5} & 9.85_{10-7} & -3.45_{10-3} \end{array} \right)
$$

which gives rstab_p = 0.14%.

Controller perturbation (using controller (11b) with $\epsilon = 10^{-7}$)

$$
\begin{pmatrix} \Delta A_c & \Delta B_{c_1} & \Delta B_{c_2} \\ \Delta C_c & \Delta D_{c_1} & \Delta D_{c_2} \end{pmatrix}
$$

$$
= \left(\begin{array}{cc|cc} -1.72_{10-2} & -1.22_{10-2} & 0 & -2.52_{10-2} \\ 4.08_{10-2} & 2.91_{10-2} & 0 & 6.00_{10-2} \\ \hline 1.72_{10-2} & 1.22_{10-2} & 0 & 2.52_{10-2} \end{array} \right)
$$

which gives rstab_c = 9.0_{10-4}%.

4.5   Closed Loop Step Response Simulations

The following are representative closed loop simulations of the plants Ex.1 to Ex.4 with their respective controllers, for the case of zero initial conditions with: (a) tracking signal $y_{ref} = 1$, unmeasurable disturbance signals $E\omega = [0, 0, ..., 0]$, $F\omega = 0$, (b) $y_{ref} = 0$, $E\omega = [1, 1, ..., 1]$, $F\omega = 0$; and (c) $y_{ref} = 0$, $E\omega = [0, 0, ..., 0]$, $F\omega = 1$. Figures 1a, 1b are for Ex.1, figures 2a, 2b are for Ex.2, figures 3a, 3b are for Ex.3 and figures 4a, 4b are for Ex.4. It is seen that in all cases, the transient response of the system is well-behaved, i.e. no excessive 'peaking' occurs, asymptotic tracking and disturbance rejection occur, and that as ϵ in J_ϵ becomes smaller, higher performance in control is obtained, e.g. the transient responses in the closed loop system become faster, the controller becomes more effective at attenuating disturbance effects, and smaller interaction effects occur in the case of Ex.4.

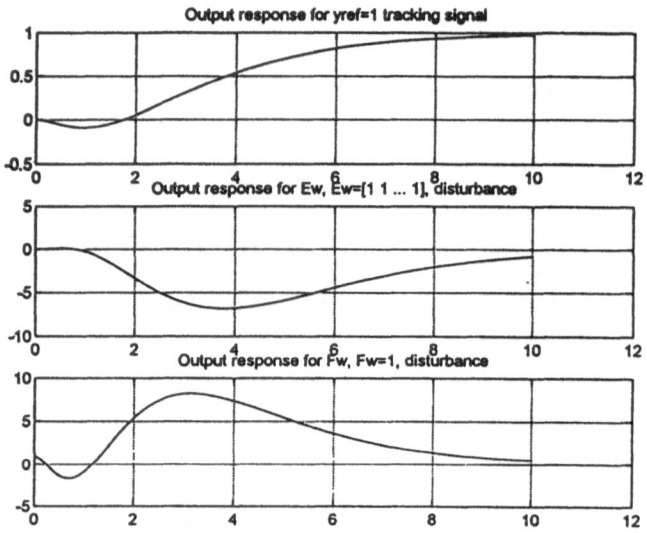

Figure 1a: Response of Ex.1 with controller (11a) corresponding to $\epsilon = 1$

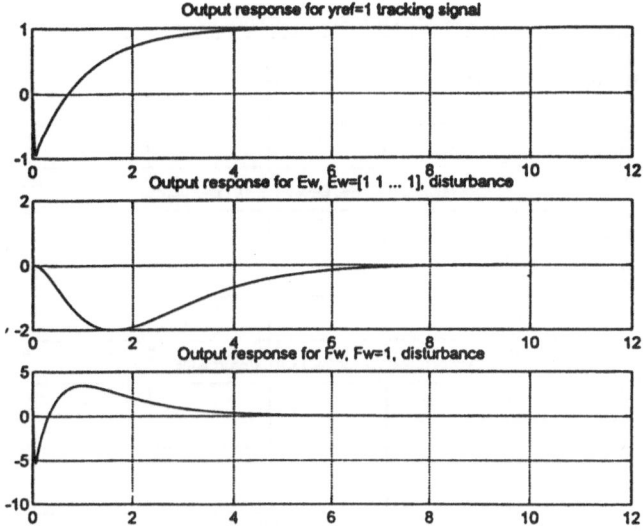

Figure 1b: Response of Ex.1 with controller (11b) corresponding to $\epsilon = 1e^{-7}$

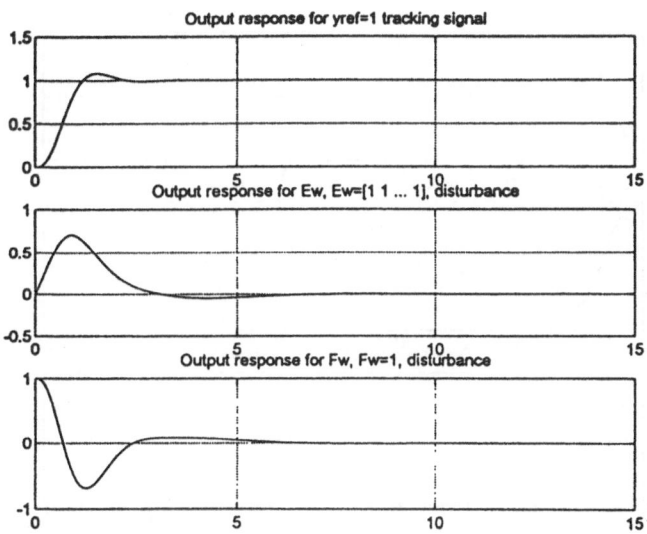

Figure 2a: Response of Ex.2 with controller (12a) corresponding to $\epsilon = 10^{-3}$

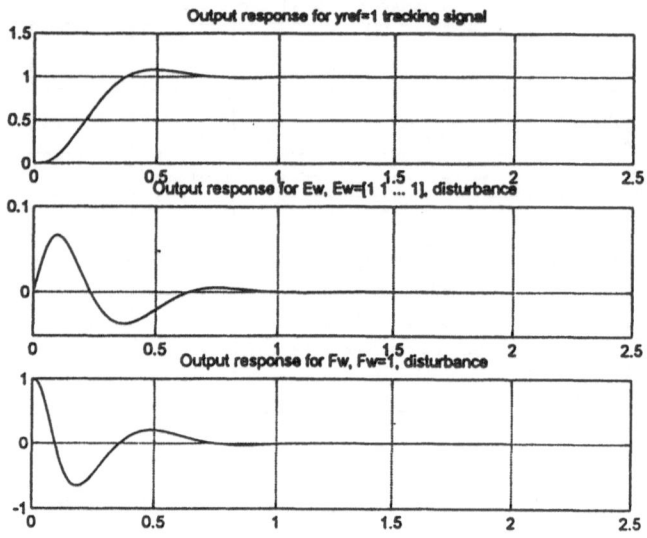

Figure 2b: Response of Ex.2 with controller (12b) corresponding to $\epsilon = 10^{-6}$

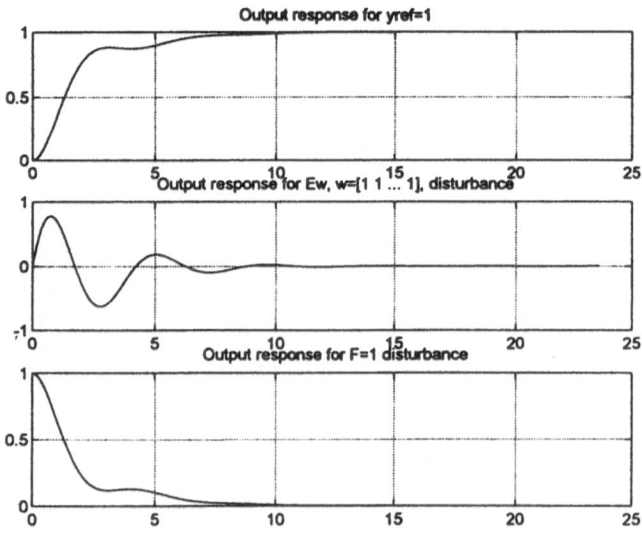

Figure 3a: Response of Ex.3 with controller (13a) corresponding to $\epsilon = 1$

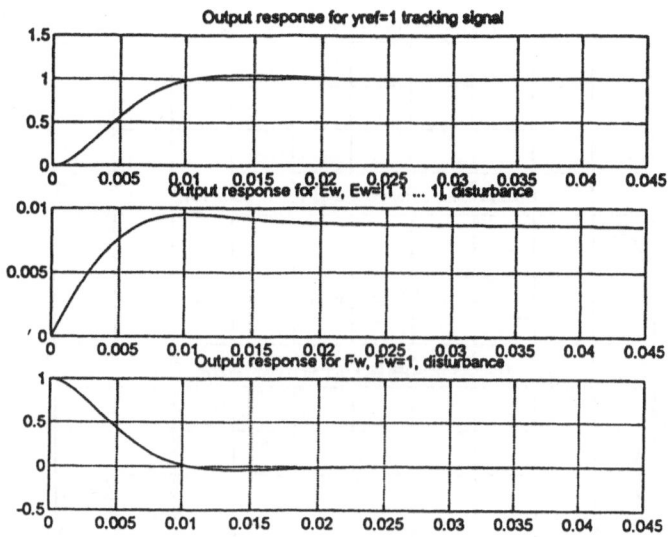

Figure 3b: Response of Ex.3 with controller (13b) corresponding to $\epsilon = 10^{-10}$

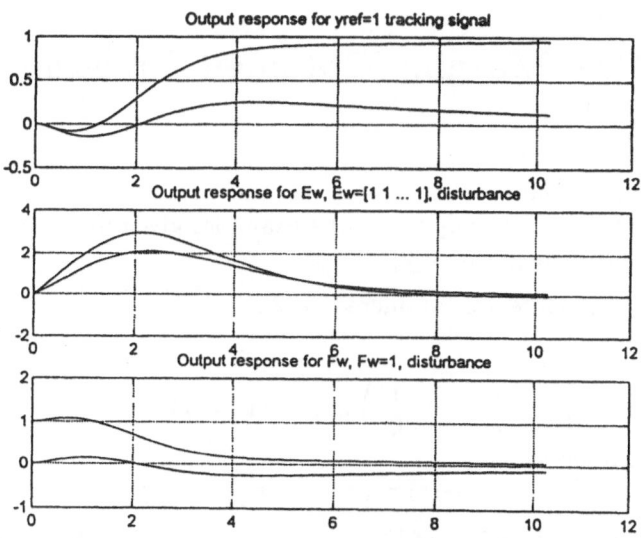

Figure 4a: Response of Ex.4 with controller (14a) corresponding to $\epsilon = 1$

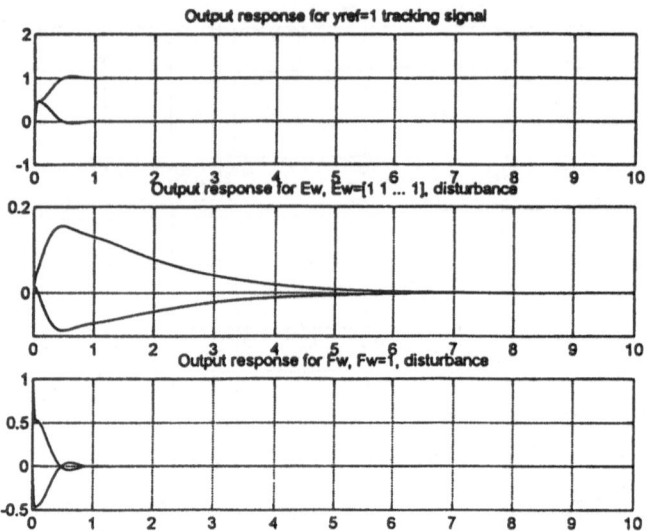

Figure 4b: Response of Ex.4 with controller (14b) corresponding to $\epsilon = 10^{-7}$

5   Stability Margins Obtained for Examples of [2]

It is of interest to compare the stability margins obtained, using the proposed design technique [1] when applied to the examples given in [2], with the corresponding controllers given in [2].

The following plants are considered in [2]:

Ex.5 $\dot{x} = \begin{pmatrix} 1 & 2 \\ 1 & 0 \end{pmatrix} x + \begin{pmatrix} 1 \\ 0 \end{pmatrix} u, y = (1 \quad -1)x$
corresponding to:
$$G(s) = (s - 1)/(s^2 - s - 2)$$

(This is example #1, 2 of [2] – in this case, a 6th order controller was found using H_∞ methods to maximize the upper gain margin of the system.)

Ex.6 $\dot{x} = \begin{pmatrix} -0.5 & 0.5 \\ 1 & 0 \end{pmatrix} x + \begin{pmatrix} 1 \\ 0 \end{pmatrix} u, y = (1 \quad -1)x$
corresponding to:
$$G(s) = (s - 1)/(s^2 + \frac{s}{2} - \frac{1}{2})$$

(This is example #3 of [2] – in this case, a 3rd order controller was found using H_∞ robust control methods to minimize a weighted complementary sensitivity function.)

Ex.7 $\quad \dot{x} = \begin{pmatrix} -1 & -2 \\ 1 & 0 \end{pmatrix} x + \begin{pmatrix} 1 \\ 0 \end{pmatrix} u, \, y = (-1 \ \ 1)x$

corresponding to:

$$G(s) = (-s+1)/(s^2 + s + 2)$$

(This is example #6 of [2] – in this case, a 6th order controller was found using H_∞ design to minimize a weighted H_2 norm of the disturbance transfer function.)

The following results are obtained, on determining the maximum plant and controller stability margin rstab_p*, rstab_c* for each of these examples using the proposed design technique (see Table 1). In addition, the stability margins rstab_p and rstab_c, for each of the corresponding controllers given in [2], are also determined.

Table 1: Computation of Stability Margins for Examples 5, 6, 7

	Using Proposed Design Method				Using Controllers of [2]	
	rstab_p*		rstab_c*		rstab_p	rstab_c
Ex.5	1.0%	$(\epsilon = 1)$	0.12%	$(\epsilon = 1)$	$1.2_{10-2}\%$	$1.8_{10-12}\%$
Ex.6	3.9%	$(\epsilon = 1)$	0.31%	$(\epsilon = 1)$	1.9%	$2.3_{10-5}\%$
Ex.7	8.6%	$(\epsilon = 10^{-3})$	0.17%	$(\epsilon = 10^{-3})$	$1.2_{10-2}\%$	$1.8_{10-12}\%$

It is to be noted that the results obtained in this case, i.e. the relatively small values of controller stability margin obtained for the controllers of [2], are consistent with the results of [2].

6 Results Obtained for 'Difficult-to-Control' Problems

There are some classes of problems which are inherently 'difficult-to-control', and it is of interest to determine the resultant stability margin obtained in this case, using the proposed design procedure [1]. An example of such a class of problems is the *multi-link inverted pendulum problem*. In this case, it is well known that one can 'easily' find a controller to experimentally stabilize an inverted pendulum consisting of a single link, and that, with care, one can also find a controller to experimentally stabilize a double-link inverted pendulum; however, it is most difficult, if not impossible, to experimentally stabilize an inverted pendulum consisting of, say, three or four links. For simplicity, a multi-link inverted pendulum system consisting of point masses $m_i = 1$ (kg), on a cascade of rigid massless links of length $\ell_i = 1$ (m), $i = 1, 2, 3, \ldots$ will be considered; the control input will be assumed to be the applied torque at the base of the bottom link, and the output will be assumed to be angle of the bottom link from the vertical. In this case, state space models of the linearized system are given as follows:

Ex.8 (1 link inverted pendulum)

$$A = \begin{pmatrix} 0 & 1 \\ 9.8 & 0 \end{pmatrix}, \quad B = \begin{pmatrix} 0 \\ 1 \end{pmatrix}, \quad C = (1 \ 0)$$

Ex.9 (2 link inverted pendulum)

$$A = \begin{pmatrix} 0 & 1 & 0 & 0 \\ 9.8 & 0 - 9.8 & 0 \\ 0 & 0 & 0 & 1 \\ -9.8 & 0 & 29.4 & 0 \end{pmatrix}, \quad B = \begin{pmatrix} 0 \\ 1 \\ 0 \\ -2 \end{pmatrix}, \quad C = (1 \ 0 \ 0 \ 0)$$

Ex.10 (3 link inverted pendulum)

$$A = \begin{bmatrix} 0 & 0 & 0 & 1 & 0 & 0 \\ 0 & 0 & 0 & 0 & 1 & 0 \\ 0 & 0 & 0 & 0 & 0 & 1 \\ 29.4 & -19.4 & 0 & 0 & 0 & 0 \\ -29.4 & 39.2 & -9.8 & 0 & 0 & 0 \\ 0 & -19.6 & 19.6 & 0 & 0 & 0 \end{bmatrix}, \quad B = \begin{pmatrix} 0 \\ 0 \\ 0 \\ 1.666 \\ -2.333 \\ 0.666 \end{pmatrix}$$

$$C = (1 \ 0 \ 0 \ 0 \ 0 \ 0 \ 0)$$

In this case, the following maximum plant and controller stability margins rstab_p*, rstab_c* are obtained using the proposed design technique [1] (see Table 2):

Table 2: Computation of Maximum Stability Margins for Examples No. 8, 9, 10 Inverted Pendulum System

		rstab_p*	rstab_c*
Ex.8	1-link	0.15%	$3.3_{10-2}\%$
Ex.9	2-link	$4.7_{10-4}\%$	$4.8_{10-6}\%$
Ex.10	3-link	$3.0_{10-6}\%$	$4.6_{10-10}\%$

It is to be noted in this case, that the *maximum* stability margins obtained rapidly become smaller as the number of links increase, and that for the case of the 3-link pendulum case, a plant data accuracy of 7-8 significant figures, and a controller data accuracy of 11-12 significant figures is required in order to guarantee closed loop stability! In contrast, with a 1-link pendulum, a plant data accuracy of only 2-3 significant figures and a controller data accuracy of only 3-4 significant figures is required. These results confirm that 3-link inverted pendulum systems are indeed difficult to control.

7 Conclusions

Given a LTI plant and corresponding LTI controller, a new definition of *plant and controller robust stability margin* is defined for the resultant perturbed closed loop system. These definitions are then used to (i) confirm the excessively poor controller stability margin results of certain types of optimal controllers presented in [2], and (ii) to show that alternate types of optimal controllers, which have been obtained to solve certain classes of control synthesis problems, may have acceptable perturbation bounds with respect to both plant and controller uncertainty. In particular, it is shown that the design procedure to solve the *"perfect robust servomechanism" problem* [1], which is based on using a H_2 "cheap controller" type of performance index, has this type of property, e.g. it will produce acceptable perturbation bounds with respect to both plant and controller uncertainty for the plants considered in [2].

In this case, on using the proposed design procedure [1], as the weighting of the control energy term in the performance index (9) becomes smaller, higher performance in the controller obtained is achieved, but at the expense of a smaller plant and controller robust stability margin. Thus, ignoring nonlinear effects, the limiting factor which dictates the performance of a controller for a plant is the position of the plant's unstable transmission zeros [7], the accuracy of plant data, and the accuracy of controller implementation which is possible to achieve.

References

[1] E.J. Davison and B.M. Scherzinger, "Perfect control of the robust servomechanism problem", *IEEE Trans on Automatic Control*, vol C-32, no 8, Aug 1987, pp 689-702.

[2] L.H. Keal and S.P. Bhattacharyya, "Robust, fragile, or optimal?", *IEEE Trans on Automatic Control*, vol 42, no 8, 1997, pp 1098-1105.

[3] B.M. Scherzinger and E.J. Davison, "The optimal LQ regulator with cheap control for not strictly proper systems", *Optimal Control Applications and Methods*, vol 6, 1985, pp 291-303.

[4] E.J. Davison and A. Goldenberg, "The robust control of a general servomechanism problem: The servo compensator", *Automatica*, vol 11, 1975, pp 461-471.

[5] Qiu Li, B. Bernhardsson, A. Rantzer, E.J. Davison, P.M. Young and J.C. Doyle, "A formulae for computation of the real stability radius", *Automatica*, vol 31, no 6, 1995, pp 879-890.

[6] E.J. Davison and I. Ferguson, "The design of controllers for the multivari-able servomechanism problem using parameter optimization methods", *IEEE Trans on Automatic Control*, vol AC-26, no 1, 1981, pp 93-110.

[7] Qiu Li and E.J. Davison, "Performance limitations for non-minimum phase systems in the servomechanism problem", *Automatica*, vol 29, no 2, 1993, pp 337-349.

Time Domain Characterizations of Performance Limitations of Feedback Control

Li Qiu*   Jie Chen[†]

Dedicated to B. A. Francis and M. Vidyasagar.

1 Introduction

The last 20 years have seen remarkable progress in the evolution of optimization-based control theory and design techniques. The new theories, apart from their theoretic elegance, have proven effective in various applications. However, the solutions of control optimization problems, in most cases in terms of numerical algorithms, do not provide a clear picture on the relationship between the optimal performance of the controlled system and the characteristics of the plant to be controlled, nor do they provide a clear idea on the effect on optimal performance attainable, due to changes in plant parameters, allocation of actuators and sensors, and choice of control structures.

On the other hand, control practice has long furnished heuristic as well as empirical understanding of the difficulty in feedback control due to plant characteristics, given in terms of rules-of-thumb largely applicable to scalar systems. For example, it is known that nonminimum phase systems are difficult to control, and that unstable poles close to nonminimum phase zeros pose additional difficulty. There has been effort to quantify such rules-of-thumb and to extend them to multivariable systems, and the subject itself has matured into a fruitful research area. Good results, mostly in the frequency domain, have been obtained to quantify, and to explain various design limitations and tradeoffs in multivariable feedback control. See [7, 15] and the references therein for the state-of-the-art.

In this paper, we survey some recent results which discuss fundamental limitations in achieving time-domain performance objectives. In particular, we consider the limitations in achieving small mean-square errors which are

---

*Department of Electrical & Electronic Engineering, Hong Kong University of Science & Technology, Clear Water Bay, Kowloon, Hong Kong, Tel: 852-2358-7067, Fax: 852-2358-1485, Email: eeqiu@ee.ust.hk.

[†]Department of Electrical Engineering, College of Engineering, University of California, Riverside, CA 92521-0425, Tel: 909-787-3688, Fax: 909-787-3188, Email: jchen@ee.ucr.edu.

to occur in tracking and in regulation. The materials are mainly based on
[3, 16] but we put them in a more unified framework. Our purpose is to show
a fundamental relationship between certain performance measures defined in
the time domain, and such simple plant characteristics as poles and zeros. The
results will complement those quantified in the frequency domain, obtained
elsewhere previously. We present these results for both continuous-time and
discrete-time systems, with an emphasis on the continuous-time case. The
discrete-time case is included since, apart from its own interest, it is essential
for the study of sampled-data systems.

Finally a note on the notation: A signal in the time domain is denoted by a
lower case letter, such as r, and a system, viewed as an input/output operator,
is denoted by a capital letter, such as F. The time domain to frequency domain
transform (Laplace transform in the continuous time case and λ-transform in
the discrete time case) is denoted by a hat " $\hat{\ }$ ", i.e., $\hat{r}$ is the Laplace or λ-
transform of r. If F is LTI, $\hat{F}$ represents the transfer function of F. For any
two nonzero vectors u and v with the same dimension, an angular measure is
provided by

$$\cos \angle(u,\ v) = \frac{|u^*v|}{\|u\|_2\|v\|_2}.$$

2 Preliminaries

Let $\hat{F}$ be the real rational matrix transfer function of a continuous time FDLTI
system F. Assume that $\hat{F}$ is right invertible. The poles and zeros of $\hat{F}$,
including multiplicity, are defined according to its Smith-McMillan form. $\hat{F}$ is
said to be minimum phase if all its zeros have a nonpositive real part; otherwise,
it is said to be nonminimum phase. Moreover, $\hat{F}$ is said to be semistable if all
its poles have a nonpositive real part, and otherwise strictly unstable. A pole
is said to be antistable if it has a positive real part.

Suppose that $\hat{F}$ is stable and z is a nonminimum phase zero of $\hat{F}$. Then,
there exists a unitary vector y such that

$$y^*\hat{F}(z) = 0.$$

We call y a (left or output) zero vector corresponding to the zero z. Let the
nonminimum phase zeros of $\hat{F}$ be ordered as as $z_1, z_2, \ldots, z_\nu$. Let also η_1 be a
zero vector corresponding to z_1. Define

$$\hat{F}_1(s) = V_1 \begin{bmatrix} \frac{j\omega_0 + z_1^*}{j\omega_0 - z_1}\frac{s - z_1}{s + z_1^*} & & & \\ & 1 & & \\ & & \ddots & \\ & & & 1 \end{bmatrix} V_1^*,$$

where $\omega_0 \in [0, \infty]$ and V_1 is a unitary matrix with the first column equal to η_1. Note that $\hat{F}_1$ is so constructed that it is inner, has only one zero at z_1 with η_1 as a zero vector, and additionally, $\hat{F}_1(j\omega_0) = I$. Note also that the choice of other columns in V_1 is immaterial. Now $\hat{F}_1^{-1}\hat{F}$ has zeros $z_2, z_3, \ldots, z_\nu$. Find a zero vector η_2 corresponding to the zero z_2 of $\hat{F}_1^{-1}\hat{F}$, and define

$$\hat{F}_2(s) = V_2 \begin{bmatrix} \dfrac{j\omega_0 + z_2^*}{j\omega_0 - z_2} \dfrac{s - z_2}{s + z_2^*} & & & \\ & 1 & & \\ & & \ddots & \\ & & & 1 \end{bmatrix} V_2^*,$$

where similarly, V_2 is a unitary matrix with the first column equal to η_2. It follows that $\hat{F}_2^{-1}\hat{F}_1^{-1}\hat{F}$ has zeros $z_3, z_4, \ldots, z_\nu$. Continue this process until $\eta_1, \ldots, \eta_\nu$ and $\hat{F}_1, \ldots, \hat{F}_\nu$ are obtained. Then we have one vector corresponding to each nonminimum zero, and the procedure yields a factorization of $\hat{F}$ in the form of

$$\hat{F} = \hat{F}_1 \cdots \hat{F}_\nu \hat{F}_0,$$

where $\hat{F}_0$ has no nonminimum phase zeros and

$$\hat{F}_i(s) = V_i \begin{bmatrix} \dfrac{j\omega_0 + z_i^*}{j\omega_0 - z_i} \dfrac{s - z_i}{s + z_i^*} & & & \\ & 1 & & \\ & & \ddots & \\ & & & 1 \end{bmatrix} V_i^*.$$

Since $\hat{F}_i$ is inner, has the only zero at z_i, and has η_i as a zero vector corresponding to z_i, it will be called a matrix Blaschke factor. Accordingly, the product

$$\hat{F}_z = \hat{F}_1 \cdots \hat{F}_\nu$$

will be called a matrix Blaschke product. The vectors $\eta_1, \ldots, \eta_\nu$ will be called Blaschke vectors of $\hat{F}$ at frequency ω_0. Keep in mind that these vectors depend on the order of the nonminimum zeros, and on ω_0. It can be shown that for a real rational $\hat{F}$ the Blaschke vectors corresponding to two complex conjugate zeros can be chosen as a complex conjugate pair provided that the two conjugate zeros are ordered consecutively.

For an unstable $\hat{F}$, there exist stable real rational matrix functions

$$\begin{bmatrix} \tilde{\hat{X}} & -\tilde{\hat{Y}} \\ -\tilde{\hat{N}} & \tilde{\hat{M}} \end{bmatrix}, \quad \begin{bmatrix} \hat{M} & \hat{Y} \\ \hat{N} & \hat{X} \end{bmatrix}$$

such that

$$\hat{F} = \hat{N}\hat{M}^{-1} = \tilde{\hat{M}}^{-1}\tilde{\hat{N}}$$

and

$$\begin{bmatrix} \hat{\tilde{X}} & -\hat{\tilde{Y}} \\ -\hat{\tilde{N}} & \hat{\tilde{M}} \end{bmatrix} \begin{bmatrix} \hat{M} & \hat{Y} \\ \hat{N} & \hat{X} \end{bmatrix} = I.$$

This is called a doubly coprime factorization of $\hat{F}$. Note that the nonminimum phase zeros of $\hat{F}$ are the nonminimum phase zeros of $\hat{N}$ and the antistable poles of $\hat{F}$ are the nonminimum phase zeros of $\hat{M}$. If we order the nonminimum phase zeros of F as $z_1, z_2, \ldots, z_\nu$ and the antistable poles of $\hat{F}$ as $p_1, p_2, \ldots, p_\mu$, then $\hat{N}$ and $\hat{M}$ can be factorized as

$$\begin{aligned} \hat{N} &= \hat{N}_1 \cdots \hat{N}_\nu \hat{N}_0, \\ \hat{M} &= \hat{M}_1 \cdots \hat{M}_\mu \hat{M}_0, \end{aligned}$$

with

$$\hat{N}_i(s) = V_i \begin{bmatrix} \dfrac{j\omega_z + z_i^*}{j\omega_z - z_i} \dfrac{s - z_i}{s + z_i^*} & & & \\ & 1 & & \\ & & \ddots & \\ & & & 1 \end{bmatrix} V_i^*,$$

$$\hat{M}_i(s) = U_i \begin{bmatrix} \dfrac{j\omega_p + p_i^*}{j\omega_p - p_i} \dfrac{s - p_i}{s + p_i^*} & & & \\ & 1 & & \\ & & \ddots & \\ & & & 1 \end{bmatrix} U_i^*.$$

Here $\hat{N}_0$ and $\hat{M}_0$ have no nonminimum phase zeros, and ω_z need not be equal to ω_p; i.e., $\hat{N}$ and $\hat{M}$ may be factorized at different frequencies. Consequently, for any real rational matrix $\hat{F}$ with nonminimum phase zeros $z_1, z_2, \ldots, z_\nu$ and strictly unstable poles $p_1, p_2, \ldots, p_\mu$, it can always be factorized to

$$\hat{F} = \hat{F}_z \hat{F}_0 \hat{F}_p^{-1},$$

where

$$\hat{F}_z(s) = \prod_{i=1}^{\nu} V_i \begin{bmatrix} \dfrac{j\omega_z + z_i^*}{j\omega_z - z_i} \dfrac{s - z_i}{s + z_i^*} & & & \\ & 1 & & \\ & & \ddots & \\ & & & 1 \end{bmatrix} V_i^*,$$

$$\hat{F}_p(s) = \prod_{i=1}^{\mu} U_i \begin{bmatrix} \dfrac{j\omega_p + p_i^*}{j\omega_p - p_i} \dfrac{s - p_i}{s + p_i^*} & & & \\ & 1 & & \\ & & \ddots & \\ & & & 1 \end{bmatrix} U_i^*,$$

and F_0 is a real rational matrix with neither nonminimum phase zero nor strictly unstable pole.

The relationship between zero vectors and zero Blaschke vectors is somewhat analogous to that between eigenvectors and Schur vectors of a square matrix. The eigenvectors are not in general completely defined in the sense that one may not find an eigenvector corresponding to each eigenvalue (with multiplicity counted) with desired property, say linear independence. However, a complete set of orthonormal Schur vectors exist as long as an order of eigenvalues is specified. Likewise, it is difficult to define a complete set of zero vectors corresponding to each nonminimum phase zero, and it is not clear what the desired property should be. Nevertheless, each Blaschke vector bears a natural correspondence to each nonminimum phase zero. The nice properties of the Blaschke vectors will become evident shortly.

The above factorization can be extended to transfer function matrices of discrete-time systems with much similarity and some differences. Consider a real rational transfer function matrix $\hat{F}$ of a discrete time FDLTI system under λ-transform ($\lambda = 1/z$). Let us assume that $\hat{F}$ is right invertible, and its poles and zeros with multiplicity included are defined according to its Smith-MacMillan form. Then $\hat{F}$ is said to be minimum phase if all its zeros have an absolute value no less than one, and otherwise nonminimum phase. It is said to be semistable if all its poles have an absolute value no less than one; otherwise, it is said to be strictly unstable. A pole is said to be antistable if it has an absolute value less than one.

At this point, we would like to emphasize that if z-transform is used for the transfer function instead, then the zeros at infinity should be considered nonminimum phase zeros. For example, transfer functions z^{-1} and $\frac{1}{z+0.5}$ represent nonminimum phase systems. This viewpoint is also more consistent with the definition of an outer function, i.e., a stable transfer function is outer iff it is minimum phase. Ambiguity often arises in this situation since in the continuous time case zeros at infinity are not considered nonminimum phase zeros. The reason is that in the continuous time case infinity is on the boundary of the stability region, whereas in the discrete time case when the z-transform is used, infinity is an interior point of the instability region and therefore should be considered the same as any other point in the same region. On the other hand, if λ-transform is used, the zeros at infinity are mapped to the origin, and so no confusion is likely to arise.

Based upon transfer functions under λ-transform, and using an analogous procedure, we may factorize $\hat{F}$ with nonminimum phase zeros $z_1, z_2, \ldots, z_\nu$ and antistable poles $p_1, p_2, \ldots, p_\mu$ as

$$\hat{F} = \hat{F}_z \hat{F}_0 \hat{F}_p^{-1},$$

where

$$\hat{F}_z(\lambda) = \prod_{i=1}^{\nu} V_i \begin{bmatrix} \dfrac{1 - z_i^* e^{j\omega_z}}{e^{j\omega_z} - z_i} \dfrac{\lambda - z_i}{1 - z_i^* \lambda} & & & \\ & 1 & & \\ & & \ddots & \\ & & & 1 \end{bmatrix} V_i^*,$$

$$\hat{F}_p(\lambda) = \prod_{i=1}^{\mu} U_i \begin{bmatrix} \dfrac{1 - p_i^* e^{j\omega_p}}{e^{j\omega_p} - p_i} \dfrac{\lambda - p_i}{1 - p_i^* \lambda} & & & \\ & 1 & & \\ & & \ddots & \\ & & & 1 \end{bmatrix} U_i^*,$$

and F_0 is a real rational matrix with no nonminimum phase zero or anti-stable pole. It thus becomes clear that any FDLTI system F, whether it is a continuous-time or discrete-time system, can be factorized into the cascade interconnection shown in Figure 1. In this factorization, F_0 is a minimum

Figure 1: Cascade factorization

phase and semistable system, $\hat{F}_{zi}$ and $\hat{F}_{pi}$ are matrix Blaschke factors with certain special properties.

3 Frequency domain characterizations

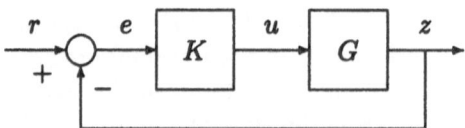

Figure 2: Unity feedback

Consider the unity feedback system shown in Figure 2. Assume that K and G are SISO LTI systems with real rational transfer functions $\hat{K}$ and $\hat{G}$ respectively. The loop gain is defined as $\hat{L} = \hat{G}\hat{K}$. The sensitivity and complementary sensitivity functions are defines as

$$\hat{S} = (1 + \hat{L})^{-1} \quad \text{and} \quad \hat{T} = \hat{L}(1 + \hat{L})^{-1}$$

respectively. Assume that $\hat{L}$ has antistable poles $p_1, p_2, \ldots, p_\mu$, and nonminimum phase zeros $z_1, z_2, \ldots, z_\nu$. Then, $\hat{L}$ has the factorization

$$\hat{L} = \hat{L}_z \hat{L}_0 \hat{L}_p^{-1},$$

with $\hat{L}_0$ being a real rational function of poles and zeros only in the closed left half plane, and $\hat{L}_z$ and $\hat{L}_p$ being the Blaschke products associated with the nonminimum phase zeros and the antistable poles, respectively:

$$\hat{L}_z(s) = \prod_{i=1}^{\nu} \frac{z_i - s}{z_i^* + s} \quad \text{and} \quad \hat{L}_p(s) = \prod_{i=1}^{\mu} \frac{p_i - s}{p_i^* + s}.$$

Suppose that $\hat{L}$ is proper ($\hat{L}(\infty)$ is finite), and that the feedback system is internally stable. Then we have

- Bode S-integral

$$\int_0^\infty \log \left| \frac{\hat{S}(j\omega)}{\hat{S}(\infty)} \right| d\omega = \frac{\pi}{2} \lim_{s \to \infty} \frac{s[\hat{S}(s) - \hat{S}(\infty)]}{\hat{S}(\infty)} + \pi \sum_{i=1}^{\mu} p_i, \qquad (1)$$

- Bode T-integral

$$\int_0^\infty \log \left| \frac{\hat{T}(j\omega)}{\hat{T}(0)} \right| \frac{d\omega}{\omega^2} = \frac{\pi}{2} \lim_{s \to 0} \frac{\hat{T}(s) - \hat{T}(0)}{s\hat{T}(0)} + \pi \sum_{i=1}^{\nu} \frac{1}{z_i}, \qquad (2)$$

- Poisson S-integrals

$$\int_{-\infty}^{\infty} \log |\hat{S}(j\omega)| \frac{\operatorname{Re} z_i}{|j\omega - z_i|^2} d\omega = \pi \log |\hat{L}_p^{-1}(z_i)|, \quad i = 1, 2, \ldots, \nu, \qquad (3)$$

- Poisson T-integrals

$$\int_{-\infty}^{\infty} \log |\hat{T}(j\omega)| \frac{\operatorname{Re} p_i}{|j\omega - p_i|^2} d\omega = \pi \log |\hat{L}_z^{-1}(p_i)|, \quad i = 1, 2, \ldots, \mu. \qquad (4)$$

In the discrete time case, $\hat{L}$ can be factorized similarly as

$$\hat{L} = \hat{L}_z \hat{L}_0 \hat{L}_p^{-1},$$

with $\hat{L}_0$ being a real rational function with no poles and zeros inside the unit circle, and $\hat{L}_z$ and $\hat{L}_p$ being the Blaschke products associated with the strictly nonminimum phase zeros and the strictly unstable poles, respectively:

$$\hat{L}_z(\lambda) = \prod_{i=1}^{\nu} \frac{\lambda - z_i}{z_i^* \lambda - 1} \quad \text{and} \quad \hat{L}_p(\lambda) = \prod_{i=1}^{\mu} \frac{\lambda - p_i}{p_i^* \lambda - 1}.$$

Under the condition that $\hat{L}$ is proper ($\hat{L}(0)$ is finite), and that the feedback system is internally stable, we have

- Bode S-integral

$$\int_0^\pi \log \left| \hat{S}(e^{j\omega}) \right| d\omega = \pi \lim_{\lambda \to 0} \log |\hat{S}(\lambda)| + \pi \sum_{i=1}^{\mu} \log \frac{1}{|p_i|}, \qquad (5)$$

- Bode T-integral

$$\int_0^\pi \log \left| \frac{\hat{T}(e^{j\omega})}{T(1)} \right| \frac{d\omega}{1 - \cos\omega} = \frac{\pi}{T(1)} \lim_{\lambda \to 1} \frac{\hat{T}(\lambda) - \hat{T}(1)}{\lambda - 1} + \pi \sum_{i=1}^{\nu} \frac{1 + z_i}{1 - z_i}, \qquad (6)$$

- Poisson S-integrals

$$\int_{-\pi}^\pi \log |\hat{S}(e^{j\omega})| \frac{|z_i|^2 - 1}{|e^{j\omega} - p_i|^2} d\omega = 2\pi \log |\hat{L}_p^{-1}(z_i)|, \quad i = 1, 2, \ldots, \nu, \qquad (7)$$

- Poisson T-integrals

$$\int_{-\pi}^\pi \log |\hat{T}(e^{j\omega})| \frac{|p_i|^2 - 1}{|e^{j\omega} - p_i|^2} d\omega = 2\pi \log |\hat{L}_z^{-1}(p_i)|, \quad i = 1, 2, \ldots, \mu. \qquad (8)$$

The performance limitations characterized by the above integral relations (except (5) and (6)) exhibit an interesting symmetry between sensitivity function and complementary sensitivity function, poles and zeros, etc.; see [9] for more details. These frequency domain characterizations have the following features, which may be undesirable in certain applications. The performance limitations characterized by the above integral relations (except (5) and (6)) exhibit an interesting symmetry between sensitivity function and complementary sensitivity function, poles and zeros, etc.; see [9] for more details. These frequency domain characterizations have the following features, which may be undesirable in certain applications.

- Sometimes it may not be desirable to characterize performance of a feedback system in terms of logarithmic integrals of $\hat{S}$ and/or $\hat{T}$, or pointwise in frequency; such is the case, for example, when the minimal $\mathcal{H}_\infty$ norm is sought after. Under this circumstance, the integral formulas give only indirect quantifications of the performance limitations and their interpretations must be carefully and delicately done. Nevertheless, one should note that the logarithmic integrals can be weakened to yield bounds on the performance.

- Since $\hat{L}$ contains both the plant $\hat{G}$ and the controller $\hat{K}$, the limitations expressed by the logarithmic integrals depend on both the plant and controller. While this may be advantageous in some cases, often one also desires to know the *a priori*, intrinsic performance achievable by designing the best controller possible. The latter, therefore, should depend on the plant only. Again, it should be pointed out that the integrals can also be weakened to lead to inequality versions depending upon solely on the plant.

- It is not clear from the integrals, even conceptually, what should be the time-varying and nonlinear generalizations of the limitations. (See [14] for an attempt.)

- The limitations are insensitive to controller used. For example, if the plant G is given with certain zero and pole pattern, then the integrals have the same values no matter what stabilizing controller K is used, as long as it does not introduce additional nonminimum phase zeros or antistable poles. Therefore, the Bode and Poisson integrals above may be more appropriately called performance invariances.

4   Minimum error tracking

We first consider a minimum error tracking problem. Let an FDLTI plant P be given with $\hat{P} = \begin{bmatrix} \hat{G} \\ \hat{H} \end{bmatrix}$, where G has output z and H has output y. Assume that we wish to design a feedback controller K in the structure shown in Figure 3 so that the closed loop system is internally stable (in any reasonable sense) and the output of the control system z tracks a vector step signal r with $r(t) = v$ when $t > 0$.

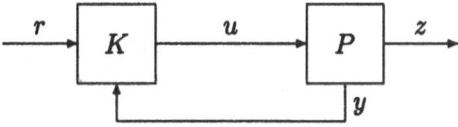

Figure 3: A general two-parameter control structure

In order for the problem to be solvable, we assume that $\hat{P}, \hat{G}, \hat{H}$ have the same unstable poles, and that $\hat{G}(0)$ has full row rank.

Let the tracking performance be measured by the energy of the tracking error

$$J(v) = \int_0^\infty \|r(t) - z(t)\|_2^2 dt.$$

This performance depends on v. A performance measure free of v can be obtained by averaging $J(v)$ over a reasonable set of v:

$$J_a = E\{J(v) : E(v) = 0, E(vv') = I\}.$$

Here E is the expectation operator. The best tracking performance achievable by designing K is then given, in the two cases, respectively by

$$J^*(v) = \inf_K J(v)$$

and

$$J_a^* = \inf_K J_a,$$

where K is chosen among all internally stabilizing (possibly nonlinear time-varying) controllers.

Let a doubly coprime factorization of $\hat{H}$ be

$$\hat{H} = \hat{N}\hat{M}^{-1} = \hat{\tilde{M}}^{-1}\hat{\tilde{N}}, \quad \begin{bmatrix} \hat{\tilde{X}} & -\hat{\tilde{Y}} \\ -\hat{\tilde{N}} & \hat{\tilde{M}} \end{bmatrix}\begin{bmatrix} \hat{M} & \hat{Y} \\ \hat{N} & \hat{X} \end{bmatrix} = I. \tag{9}$$

Then by the standard stabilization theory [19], the set of all stabilizing controllers is given by

$$K = \begin{bmatrix} Y + MQ & MR \end{bmatrix}\begin{bmatrix} 0 & I \\ X + NQ & NR \end{bmatrix}^{-1}, \tag{10}$$

where Q, R are arbitrary causal stable (possibly nonlinear time-varying) controllers. With this class of controllers applied to the system, the map from r to the error $e = r - z$ is given by $I - GMQ$. Let $\hat{G}$ have nonminimum phase zeros $z_1, z_2, \ldots, z_\nu$ with $\eta_1, \eta_2, \ldots, \eta_\nu$ being the corresponding Blaschke vectors at frequency 0. Since $\hat{P}, \hat{G}, \hat{H}$ have the same unstable poles, GM is stable and it has the factorization

$$\hat{G}\hat{M} = \hat{G}_1\hat{G}_2\cdots\hat{G}_\nu\hat{G}_0,$$

where

$$\hat{G}_i(s) = V_i\begin{bmatrix} -\dfrac{z_i^*}{z_i}\dfrac{s - z_i}{s + z_i^*} & & & \\ & 1 & & \\ & & \ddots & \\ & & & 1 \end{bmatrix} \quad V_i^* = I - \dfrac{2\,\mathrm{Re}\,z_i}{z_i}\dfrac{s}{s + z_i^*}\eta_i\eta_i^*,$$

and $\hat{G}_0$ is outer in $\mathcal{H}_\infty$. Using the Parseval's identity, we obtain

$$\begin{aligned} J(v) &= \|\hat{r} - \hat{G}_1\hat{G}_2\cdots\hat{G}_\nu\hat{G}_0\widehat{Qr}\|_2^2 \\ &= \|\hat{G}_\nu^{-1}\cdots\hat{G}_2^{-1}\hat{G}_1^{-1}\hat{r} - \hat{G}_0\widehat{Qr}\|_2^2 \\ &= \|(\hat{G}_\nu^{-1}\cdots\hat{G}_2^{-1}\hat{G}_1^{-1} - I)\hat{r} + (\hat{r} - \hat{G}_0\widehat{Qr})\|_2^2. \end{aligned}$$

Here the second equality follows from the fact that G_i are unitary operators in $\mathcal{L}_2$. When r is a step signal, $(\hat{G}_\nu^{-1}\cdots\hat{G}_2^{-1}\hat{G}_1^{-1} - I)\hat{r} \in \mathcal{H}_2^\perp$. Hence $J(v)$ is finite only if $\hat{r} - \hat{G}_0\widehat{Qr} \in \mathcal{L}_2$. Since Q is causal, we must have $\hat{r} - \hat{G}_0\widehat{Qr} \in \mathcal{H}_2$. Therefore,

$$J(v) = \|(\hat{G}_\nu^{-1}\cdots\hat{G}_2^{-1}\hat{G}_1^{-1} - I)\hat{r}\|_2^2 + \|\hat{r} - \hat{G}_0\widehat{Qr}\|_2^2 \geq \|(\hat{G}_\nu^{-1}\cdots\hat{G}_2^{-1}\hat{G}_1^{-1} - I)\hat{r}\|_2^2.$$

On the other hand, since G_0 has a dense image in $\mathcal{H}_2$, we can find an LTI system Q such that $\hat{G}(0)\hat{Q}(0) = I$ and $\|[I - \hat{G}_0(s)\hat{Q}(s)]\frac{1}{s}\|_2$ is arbitrarily small. Therefore,

$$J^*(v) = \|(\hat{G}_\nu^{-1} \cdots \hat{G}_2^{-1}\hat{G}_1^{-1} - I)\hat{r}\|_2^2.$$

Straightforward computation then shows that

$$
\begin{aligned}
J^*(v) &= \|(\hat{G}_\nu^{-1} \cdots \hat{G}_2^{-1}\hat{G}_1^{-1} - I)\hat{r}\|_2^2 = v^* \left(2 \sum_{i=1}^{\nu} \frac{1}{z_i} \eta_i \eta_i^* \right) v \\
&= 2\|v\|_2^2 \sum_{i=1}^{\nu} \frac{1}{z_i} \cos^2 \angle(v, \eta_i).
\end{aligned}
$$

Since a nearly optimal Q, i.e., a Q such that $\|[I - \hat{G}_0(s)\hat{Q}(s)]\frac{1}{s}\|_2$ is vanishingly small, can be chosen independently of v, we have

$$
\begin{aligned}
J_a^* &= \inf_K \{ EJ(v) : E(v) = 0, E(vv^*) = I \} \\
&= \{ EJ^*(v) : E(v) = 0, E(vv^*) = I \} \\
&= \mathrm{tr} \left(2 \sum_{i=1}^{\nu} \frac{1}{z_i} \eta_i \eta_i^* \right) \\
&= 2 \sum_{i=1}^{\nu} \frac{1}{\lambda_i}.
\end{aligned}
$$

We have thus established the following theorem.

Theorem 1 *Let $\hat{G}$ have nonminimum phase zeros $z_1, z_2, \ldots, z_\nu$ with $\eta_1, \eta_2, \ldots, \eta_\nu$ being the corresponding Blaschke vectors at frequency 0. Then*

$$J^*(v) = 2\|v\|_2^2 \sum_{i=1}^{\nu} \frac{1}{z_i} \cos^2 \angle(v, \eta_i)$$

and

$$J_a^* = 2 \sum_{i=1}^{\nu} \frac{1}{z_i}.$$

Remarks:

1. The limiting performance does not change if the controller is chosen from the set of LTI controllers or the set of nonlinear time-varying controllers,

2. The limiting performance does not depend on the poles of the plant,

3. The limiting performance does not depend on how the measurement is taken as long as the measurement does not introduce additional unstable poles and the stabilization can be accomplished by the measurement. This makes sense since the measurement does not provide any extra information on the behavior of the system when no uncertainty or disturbance is present.

The performance limitation exhibited in Theorem 1 is a fundamental one imposed by the plant. Since two-parameter control is the most general control structure, no other control scheme can do better. The use of other less general control structure can only introduce additional limitation. For example, robustness consideration motivates the use of error feedback in the tracking problem. If the one-parameter unity error feedback structure as in Figure 2 is used, it is shown in [3] that if the plant P is strictly unstable, then the best achievable performance will be worse than that given in Theorem 1. In this case, additional limitation on the tracking performance is introduced by the control structure. This provides further quantitative support to the observation made in [19] that one-parameter controller does not have enough freedom to accomplish both stabilization and tracking effectively. To take advantages of error feedback and two-parameter control, we may use the control structures shown in Figure 4 and Figure 5.

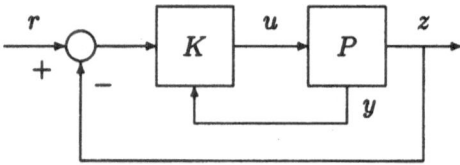

Figure 4: Separating stabilization and tracking error feedback

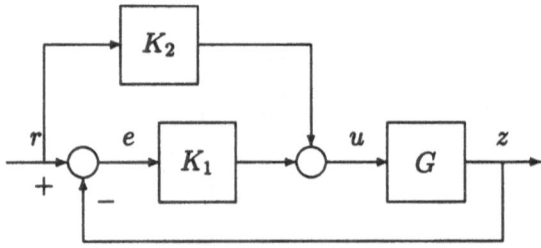

Figure 5: Feedback plus feedforward tracking

For discrete-time systems, analogous results can be obtained. Consider again the feedback controller structure in Figure 3 and a discrete-time FDLTI plant P, with $\hat{P} = \begin{bmatrix} \hat{G} \\ \hat{H} \end{bmatrix}$. Assume that we wish to design a feedback controller

K so that the closed loop system is internally stable (in any reasonable sense) and the output of the control system z tracks a vector step signal r with $r(k) = v$ when $k \geq 0$. Define the tracking error by

$$J(v) = \sum_{k=0}^{\infty} \|r(k) - z(k)\|_2^2$$

and the average tracking error by

$$J_a = E\{J(v) : E(v) = 0, E(vv') = I\}.$$

The best tracking performances achievable by designing K are then given by

$$J^*(v) = \inf_K J(v)$$

and

$$J_a^* = \inf_K J_a,$$

where K is chosen among all internally stabilizing (possibly time-varying, nonlinear) controllers. Similarly, we assume that $\hat{P}, \hat{G}, \hat{H}$ have the same unstable poles, and that $\hat{G}(1)$ has full row rank.

Theorem 2 *Let $\hat{G}$ has nonminimum phase zeros $z_1, z_2, \ldots, z_\nu$ with $\eta_1, \eta_2, \ldots, \eta_\nu$ being the corresponding zero Blaschke vectors at frequency 0. Then*

$$J^*(v) = \|v\|_2^2 \sum_{i=1}^{\nu} \frac{1 + z_i}{1 - z_i} \cos^2 \angle(v, \eta_i)$$

and

$$J_a^* = \sum_{i=1}^{\nu} \frac{1 + z_i}{1 - z_i}.$$

5 Minimum energy regulation

Next, we consider a minimum energy regulation problem. Let G be a given plant. Assume that we wish to design a feedback controller K in the structure shown in Figure 6 so that the closed loop system is stable. Assume that d is a vector impulse signal $d(t) = v\delta(t)$. The input energy is given by

$$E(v) = \int_0^{\infty} \|u(t)\|_2^2 dt.$$

A normalized average input energy independent of v can be obtained as

$$E_a = E\{E(v) : E(v) = 0, E(vv') = I\}.$$

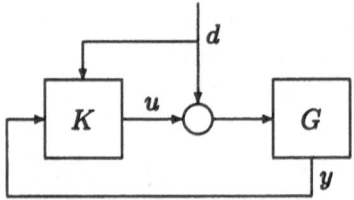

Figure 6: Feedback plus feedforward regulation

The minimum energy required in stabilizing the system is then determined as

$$E^*(v) = \inf_K E(v)$$

and

$$E_a^* = \inf_K E_a,$$

respectively, where K is chosen among all internally stabilizing (possibly non-linear time-varying) controllers.

Let a doubly coprime factorization of $\hat{G}$ be given as that of $\hat{H}$ in (9), and the set of all stabilizing controllers K be as in (10). The map from d to the input u is then given by

$$M(Y - R\tilde{N}) + MQ - I.$$

Write

$$Q = Q_0 - (Y - R\tilde{N}).$$

Then, according to the Parseval's identity, and in light of the fact that the map from Q_0 to Q is bijective over the set of all causal stable systems, we have

$$E^*(v) = \inf_{Q_0 \text{ stable}} \|\hat{M}\widehat{Q_0 d} - \hat{d}\|_2^2$$

Let $\hat{G}$ have antistable poles $p_1, p_2, \ldots, p_\mu$ with $\zeta_1, \zeta_2, \ldots, \zeta_\mu$ being the corresponding Blaschke vectors at the frequency ∞. Then $\hat{M}$ has the factorization

$$\hat{M} = \hat{M}_1 \hat{M}_2 \cdots \hat{M}_\mu \hat{M}_0,$$

where

$$\hat{M}_i(s) = U_i \begin{bmatrix} \dfrac{s - z_i}{s + z_i^*} & & & \\ & 1 & & \\ & & \ddots & \\ & & & 1 \end{bmatrix} \qquad U_i^* = I - \frac{2\,\mathrm{Re}\,p_i}{s + p_i^*}\eta_i\eta_i^*,$$

and $\hat{M}_0$ is outer in $\mathcal{H}_\infty$. It follows that

$$\begin{aligned} \|\hat{M}\widehat{Q_0 d} - \hat{d}\|_2^2 &= \|\hat{M}_0\widehat{Q_0 d} - \hat{M}_\mu^{-1}\cdots\hat{M}_2^{-1}\hat{M}_1^{-1}\hat{d}\| \\ &= \|(\hat{M}_0\widehat{Q_0 d} - \hat{d}) + (I - \hat{M}_\mu^{-1}\cdots\hat{M}_2^{-1}\hat{M}_1^{-1})\hat{d}\|_2^2 \end{aligned}$$

Since $(I - \hat{M}_\mu^{-1} \cdots \hat{M}_1^{-1})\hat{d} \in \mathcal{H}_2^{\perp}$, $\|\widehat{M\hat{Q}_0 d} - \hat{d}\|_2^2$ is finite only if $\hat{M}_0\widehat{\hat{Q}_0 d} - \hat{d} \in \mathcal{L}_2$. Since Q_0 is causal, we mush have $\hat{M}_0\widehat{\hat{Q}_0 d} - \hat{d} \in \mathcal{H}_2$. Therefore,

$$E(v) = \|\hat{M}_0\widehat{\hat{Q}_0 d} - \hat{d}\|_2^2 + \|(I - \hat{M}_\mu^{-1} \cdots \hat{M}_2^{-1}\hat{M}_1^{-1})\hat{d}\|_2^2 \geq \|(I - \hat{M}_\mu^{-1} \cdots \hat{M}_1^{-1})\hat{d}\|_2^2.$$

On the other hand, since M_0 has a dense image in $\mathcal{H}_2$, we can find an LTI system Q_0 such that $\|\hat{M}_0\hat{Q}_0 - I\|_2$ is arbitrarily small. Therefore,

$$E^*(v) = \|(I - \hat{M}_\mu^{-1} \cdots \hat{M}_2^{-1}\hat{M}_1^{-1})\hat{d}\|_2^2.$$

Straightforward computation shows that

$$E^*(v) = v^* \left(2 \sum_{i=1}^{\mu} p_i \zeta_i \zeta_i^* \right) v = 2\|v\|_2^2 \sum_{i=1}^{\mu} p_i \cos^2 \angle(v, \zeta_i).$$

Since a nearly optimal Q_0, i.e., a Q_0 such that $\|\hat{M}_0\hat{Q}_0 - I\|_2$ is vanishingly small, can be chosen independently of v, we have

$$E_a^* = 2 \sum_{i=1}^{\mu} p_i.$$

This proves the following theorem.

Theorem 3 *Let $\hat{G}$ have antistable poles $p_1, p_2, \ldots, p_\mu$ with $\zeta_1, \zeta_2, \ldots, \zeta_\mu$ being the corresponding pole Blaschke vectors at the frequency ∞. Then,*

$$E^*(v) = 2\|v\|_2^2 \sum_{i=1}^{\mu} p_i \cos^2 \angle(v, \zeta_i)$$

and

$$E_a^* = 2 \sum_{i=1}^{\mu} p_i.$$

Remarks:

1. The limiting input energy does not change if the controller is chosen from the set of LTI controllers or the set of nonlinear time-varying controllers,

2. The limiting input energy does not depend on the zeros of the plant,

3. The limiting input energy does not depend on how the measurement is taken as long as the stabilization can be accomplished by the measurement. This makes sense since the measurement does not provide any extra information on the behavior of the system when no uncertainty or disturbance is present.

Qiu and Chen

In practical situations, the excitation (considered as a disturbance) signal may not be accessible. A natural question is then if the measurement feedback structure in Figure 7 would lead to the same performance limitation? The answer is yes if and only G is minimum phase. Consequently, in order to achieve good performance in regulation, measurement variables should be selected, whenever possible, in such a way that the input to measurement transfer function is minimum phase. This is the case when the measurement vector contains all states.

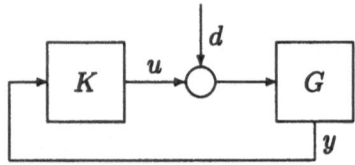

Figure 7: Measurement feedback regulation

Finally, for discrete time systems, the same problem can be studied but the result takes a different form. Indeed, assume that d is a vector impulse signal $d(k) = v\delta(k)$ and define the input energy measures similarly by

$$E(v) = \sum_{k=1}^{\infty} \|u(k)\|_2^2,$$

and

$$E_a = E\{E(v) : E(v) = 0, E(vv') = I\}.$$

Furthermore, define the optimal versions of $E(v)$ and E_a as

$$E^*(v) = \inf_K E(v)$$

and

$$E_a^* = \inf_K E,$$

respectively. Here K is, likewise, chosen among all internally stabilizing (possibly time-varying, nonlinear) controllers. It turns out a clean formula for $E^*(v)$ as in Theorems 1-3 is not available in this context.

Theorem 4 *Let $\hat{G}$ have antistable poles $p_1, p_2, \ldots, p_\mu$. Then,*

$$E_a^* = 1 - \prod_{i=1}^{\mu} p_i^2.$$

6   Concluding remarks

It is observed (first in [16]) that in the continuous-time case the sum of the reciprocal of the nonminimum phase zeros, interestingly, shows up in both the Bode T-integral and the expression of the minimum tracking error, and that the sum of the antistable poles shows up in both the Bode S-integral and the expression of the minimum regulation energy. Notice that in the special case when G is stable and the unity feedback in Figure 2 is used, the minimum tracking error can also be defined, via the Parseval's identity, in the frequency domain as

$$J_a^* = \inf_{K\,\text{stabilizing}} \frac{1}{\pi} \int_0^\infty \|S(\mathrm{j}\omega)\|_F^2 \frac{\mathrm{d}\omega}{\omega^2}. \tag{11}$$

Furthermore, notice that in the special case when G is minimum phase and the feedback control structure in Figure 7 is used, the minimum regulation energy can be defined, via the Parseval's identity, in the frequency domain as

$$E_a^* = \inf_{K\,\text{stabilizing}} \frac{1}{\pi} \int_0^\infty \|T(\mathrm{j}\omega)\|_F^2 \mathrm{d}\omega. \tag{12}$$

This leads to the speculation that there may be a deep connection between the square integrals (11-12) and the Bode type logarithmic integrals. Investigation is being undertaken to clarify this issue.

In the continuous-time case, we again observe a nice symmetry between the tracking problem and the regulation problem, which complements the symmetry between the Bode type sensitivity and complementary sensitivity integrals.

In the discrete time case, the asymmetry between the tracking problem and the regulation problem (cf. Theorem 2 and Theorem 3) is not an isolated phenomenon. A similar asymmetry occurs between (5) and (6).

Several other extensions of the problems and results in this paper have been studied recently or are currently under study, including:

1. Minimum tracking error for sinusoidal signals [4].

2. Minimum tracking error in systems with delays [3].

3. Tracking and regulation performance limitation of sampled-data systems.

4. Time domain performance limitation in filtering and estimation problems [17].

References

[1] J. Chen, "Sensitivity integral relations and design trade-offs in linear multivariable feedback systems", *IEEE Trans. Autom. Contr.*, vol. 40, pp. 1700-1716, 1995.

[2] J. Chen, "On logarithmic integrals and performance bounds for MIMO system, part I and part II", *Preprints*, 1997.

[3] J. Chen, L. Qiu, and O. Toker, "Limitations on maximal tracking accuracy", *Proc. 35th IEEE Conf. on Decision and Control*, pp. 726–731, 1996, submitted to *IEEE Trans. Autom. Contr.*.

[4] J. Chen, L. Qiu, and O. Toker, "Limitation on maximal tracking accuracy, part 2: tracking sinusoidal and ramp signals", *Proc. 1997 American Control Conf.*, pp. 1757-1761, 1997.

[5] E. J. Davison and B. M. Scherzinger, "Perfect control of the robust servomechanism problem", *IEEE Trans. Autom. Contr.*, vol. 32, pp. 689-702, 1987.

[6] B. A. Francis, *A Course in $\mathcal{H}_\infty$ Control Theory*, Springer-Verlag, 1987.

[7] J. Freudenberg and D. Looze, *Frequency Domain Properties of Scalar and Multivariable Feedback Systems*, Springer-Verlag, 1988.

[8] G. C. Goodwin and M. M. Seron, "Fundamental design tradeoffs in filtering, prediction, and smoothing", *IEEE Trans. Autom. Contr.*, vol. 42, pp. 1240-1251, 1997.

[9] H. Kwakernaak, "Symmetries in control system design", *Trends in Control: A European Perspective*, A. Isidori, editor, pp. 17–51, Springer, 1995.

[10] H. Kwakernaak and R. Sivan, "The maximal achievable accuracy of linear optimal regulators and linear optimal filters", *IEEE Trans. Autom. Contr.*, vol. 17, pp. 79-86, 1972.

[11] H. Kwakernaak and R. Sivan, *Linear Optimal Control Systems*, Wiley Interscience, New York, 1972.

[12] M. Morari and E. Zafiriou, *Robust Process Control*, Prentice Hall, Englewood Cliffs, NJ, 1989.

[13] L. Qiu and E. J. Davison, "Performance limitations of non-minimum phase systems in the servomechanism problem", *Automatica*, vol. 29, pp. 337-349, 1993.

[14] J. Shamma, "Performance limitations in sensitivity reduction for nonlinear plants", *System & Control Letters*, vol. 27, pp. 249-254, 1991.

[15] M. M. Seron, J. H. Braslavsky, and G. C. Goodwin, *Fundamental Limitations in Filtering and Control*, Springer, London, 1997.

[16] M. M. Seron, J. H. Braslavsky, P. V. Kokotović, and D. Q. Mayne, "Feedback limitations in nonlinear systems: from Bode integrals to cheap control", *IEEE Trans. on Automatic Control*, 1998, to appear.

[17] M. M. Seron, J. H. Braslavsky, D. G. Mayne, and P. V. Kokotović, "Limiting performance of optimal linear filters", *Preprints*, 1997.

[18] O. Toker, J. Chen, and L. Qiu, "Tracking performance limitations for multivariable discrete time systems", *Proc. 1997 American Control Conf.*, pp. 3887-3891, 1997. Also submitted to *Automatica*.

[19] M. Vidyasagar, *Control System Synthesis: A Factorization Approach*, The MIT Press, 1985.

Loopshaping by Quadratic Constraints on Open-Loop Transfer Functions

Ichijyo Hodaka   Masayuki Suzuki   Noboru Sakamoto*

Abstract

The purpose of this note is to propose to execute loopshaping by imposing quadratic constraints on open-loop transfer functions. Quadratic constraints are defined to be consistent with feedback stable systems. Then, a robust stability analysis and a loopshaping method under quadratic constraints on open-loop transfer functions are given. Our results provide a general robustness of feedback systems in terms of open-loop transfer functions, including robustness based on gain and phase margins.

1 Introduction

It is a basic requirement to guarantee robust stability in feedback system design, because no mathematical model can exactly represent a physical system and a stabilizing controller for a mathematical model might cause instability of the feedback system. To reward this requirement, numerous design methods of robust controllers have been developed. In particular, H_∞ control theory (e.g. [6, 8]) provides a powerful tool for robust controller design and is widely recognized to be a practical controller design method. The aim of H_∞ control is to guarantee robustness of feedback systems by optimizing the H_∞-norm of appropriate closed-loop transfer functions, such as sensitivity and complementary sensitivity functions. On the other hand, as more classical indicators of robustness of systems, gain and phase margins have been used for a long time. In H_∞ control theory, however, these stability margins are not explicitly taken into consideration and this causes a controller design including trial and error in practical control engineering, where those margins are often employed. Furthermore, it is pointed out in [1] that some controllers designed by several modern robust control theories including H_∞ control theory lead extremely poor gain or phase margins. The reason for this fact is that H_∞ control focuses on closed-loop transfer functions, while gain and phase margins are, essentially, characteristics of open-loop transfer functions. Therefore, it

---

*Department of Aerospace Engineering, School of Engineering, Nagoya University, Furo-cho, Chikusaku, Nagoya, Japan

is necessary to discuss robustness of feedback systems in terms of open-loop transfer functions.

In this note, we consider robustness of feedback systems from a modern viewpoint of a classical open-loop shaping and propose to adopt quadratic constraints on open-loop transfer functions for the shaping. Quadratic constraints are generalizations of the concepts of bounded realness and positive realness [2, 3], and are suitable for specifications of unstable open-loop transfer functions. We present a robust stability analysis and a controller synthesis method under quadratic constraints on open-loop transfer functions. This results enable us to guarantee a general robustness of feedback systems in terms of open-loop transfer functions, including direct specifications of gain and phase margins for control systems.

2 Preliminaries

Our notation follows the book [8]. In addition, for an $n \times m$ matrix B with $\mathrm{rank}B = r$, let $B^\perp((n-r) \times n)$ be any matrix such that $B^\perp B = 0$ and $\mathrm{rank}B^\perp = n - r$.

Also, we often use the following lemma.

Lemma 1 *Let matrices $B(n \times m)$, $C(q \times n)$ and $P = P^*(n \times n)$ be given. Suppose $\mathrm{rank}B < n$ and $\mathrm{rank}C < n$. Then, (i)*

$$\exists \mu > 0 : P - \mu BB^* < 0 \Leftrightarrow B^\perp PB^{\perp *} < 0$$

and (ii)

$$\exists K(m \times q) : BKC + (BKC)^T + P < 0 \Leftrightarrow B^\perp PB^{\perp T} < 0 \text{ and } C^{T\perp} PC^{T\perp T} < 0.$$

For simplicity, a pair of proper transfer functions (P, K) is said to be feedback stable if the feedback system described by Figure 1 is internally stable, i.e. $I + K(\infty)P(\infty)$ is nonsingular and four transfer functions $(I + KP)^{-1}$, $K(I + PK)^{-1}$, $P(I + KP)^{-1}$, $(I + PK)^{-1}$ belong to $\mathbf{RH}_\infty$.

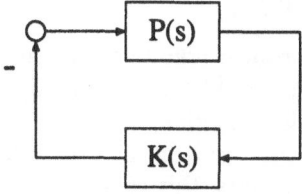

Figure 1: Feedback system

3   Open-loop quadratic constraints

In this section, we define quadratic constraints on open-loop transfer functions for loopshaping. Before, we state the definition, let us illustrate several conventional constraints with respect to robustness of feedback systems and show that these constraints can be indeed expressed as quadratic constraints.

Consider a feedback system consisting of a plant $P(s)$ and controller $K(s)$, described by Figure 1. An open-loop transfer function $L(s) := K(s)P(s)$ is assumed to be scalar and have no pole on the imaginary axis, temporarily.

The first requirement for the system is stability. This implies from Nyquist criterion that the Nyquist diagram of $L(s)$ does not path through the point $-1 + j0$ and encircles it p times in the counterclockwise, where p is the number of unstable poles of L. Therefore, for robust stability, the diagram must be away from -1 to some distance; that is, there exists $\epsilon \geq 0$ such that

$$|L(j\omega) + 1| > \epsilon, \ \forall \omega \in \mathbf{R} \cup \{\infty\},$$

or equivalently,

$$\begin{bmatrix} L(-j\omega) & 1 \end{bmatrix} \begin{bmatrix} 1 & 1 \\ 1 & 1-\epsilon \end{bmatrix} \begin{bmatrix} L(j\omega) \\ 1 \end{bmatrix} > 0, \ \forall \omega \in \mathbf{R} \cup \{\infty\}.$$

The circle criterion [2] in absolute stability theory is a more definite robust stability result. Consider a feedback system consisting of a transfer function L and a nonlinear element ϕ which belongs to sector $[a, b]$, i.e.

$$\phi(0) = 0, \ a \leq \frac{\phi(y)}{y} \leq b, \ \forall y \neq 0.$$

In the case when $0 < a < 1 < b$, the criterion is that if the Nyquist diagram of L does not enter the closed disk whose center is at $-(a + b)/(2ab)$ and radius is $(b - a)/(2ab)$ and encircles it p times in the counterclockwise, then the system is asymptotically stable for any ϕ belonging to sector $[a, b]$. This yields a quadratic constraint:

$$\begin{bmatrix} L(-j\omega) & 1 \end{bmatrix} \begin{bmatrix} 1 & \frac{a+b}{2ab} \\ \frac{a+b}{2ab} & \frac{1}{ab} \end{bmatrix} \begin{bmatrix} L(j\omega) \\ 1 \end{bmatrix} > 0, \ \forall \omega \in \mathbf{R} \cup \{\infty\}$$

with the encirclement condition.

Next, let us consider a robust stability with respect to H_∞-norm bounded multiplicative uncertainties Δs [8, 6]. By small gain theorem, if a complementary sensitivity transfer function $T := L/(1 + L)$ is in $\mathbf{RH_\infty}$ and if $||T||_\infty < 1/||\Delta||_\infty$, then the feedback system is robustly stable for any Δ. In terms of quadratic constraints, this requires

$$\begin{bmatrix} L(-j\omega) & 1 \end{bmatrix} \begin{bmatrix} \epsilon - 1 & \epsilon \\ \epsilon & 1 \end{bmatrix} \begin{bmatrix} L(j\omega) \\ 1 \end{bmatrix} > 0, \ \forall \omega \in \mathbf{R} \cup \{\infty\}.$$

Each condition mentioned above assures some robustness of feedback systems by quadratic constraints on open-loop transfer functions, in addition to stability of nominal feedback systems. In order to generalize these constraints and to discuss open-loop loopshaping by them, let us define quadratic constraints consistent with feedback stable but unstable open-loop transfer functions Ls.

Definition 1 *Let an $m \times m$ proper transfer function $L(s)$ and $m \times m$ real matrices $Q_1 = Q_1^T$, Q_2, $Q_3 = Q_3^T$ be given. Assume that a pair (L, I) is feedback stable and that*

$$J := [\, I \quad -I \,] \begin{bmatrix} Q_1 & Q_2 \\ Q_2^T & Q_3 \end{bmatrix} \begin{bmatrix} I \\ -I \end{bmatrix} \leq 0, \tag{1}$$

and introduce

$$\Phi(s) \; := \; [\, L^T(-s) \quad I \,] \begin{bmatrix} Q_1 & Q_2 \\ Q_2^T & Q_3 \end{bmatrix} \begin{bmatrix} L(s) \\ I \end{bmatrix} \tag{2}$$

$$\Omega_L \; := \; \{ \omega \in \mathbf{R} \cup \{\infty\} \mid L(j\omega) \in \mathbf{C}^{m \times m} \}. \tag{3}$$

Then, L is said to satisfy a quadratic constraint with parameter matrices (Q_1, Q_2, Q_3) ((Q_1, Q_2, Q_3)-quadratic constraint) if

$$\Phi(j\omega) > 0, \forall \omega \in \Omega_L. \tag{4}$$

It is clear that we can describe the previously-mentioned robust stability constraints can be described in the form of (4) under feedback stability of (L, I) and (1) in every examples. In the definition, the quadratic constraint (4) is assumed to hold for all $j\omega$ except for $j\omega$-poles of L, accepting that L has $j\omega$-poles. The assumption (1) with respect to (Q_1, Q_2, Q_3) plays an important role, as follows. Applying Lemma 1–(i) to (4), we see that

$$[\, L^*(j\omega) \quad I \,] \begin{bmatrix} Q_1 & Q_2 \\ Q_2^T & Q_3 \end{bmatrix} \begin{bmatrix} L(j\omega) \\ I \end{bmatrix} > 0$$

$$\Leftrightarrow \quad \exists \mu > 0 : \begin{bmatrix} Q_1 & Q_2 \\ Q_2^T & Q_3 \end{bmatrix} + \mu \begin{bmatrix} I \\ -L^*(j\omega) \end{bmatrix} [\, I \quad -L(j\omega) \,] > 0.$$

Then, under the assumption (1),

$$(I + L(j\omega))^*(I + L(j\omega))$$

$$= \; [\, I \quad -I \,] \begin{bmatrix} I \\ -L^*(j\omega) \end{bmatrix} [\, I \quad -L(j\omega) \,] \begin{bmatrix} I \\ -I \end{bmatrix}$$

$$> \; -\frac{1}{\mu} [\, I \quad -I \,] \begin{bmatrix} Q_1 & Q_2 \\ Q_2^T & Q_3 \end{bmatrix} \begin{bmatrix} I \\ -I \end{bmatrix}$$

$$= \; -\frac{1}{\mu} J \geq 0$$

follows. Therefore, we obtain

$$\exists \epsilon > 0 : |\det(I + L(j\omega))| > \epsilon, \forall \omega \in \Omega,$$

which leads to a robust stability in the sense that the Nyquist diagram of $\det(I + L(j\omega))$ is away from the origin at least ϵ.

Conversely, any L so that (L, I) is feedback stable satisfies (Q_1, Q_2, Q_3)-quadratic constraint for some (Q_1, Q_2, Q_3) satisfying (1). This follows from the feedback stability constraint

$$(I + L(j\omega))^* \, (I + L(j\omega)) = \left[\begin{array}{cc} L^T(-j\omega) & I \end{array} \right] \left[\begin{array}{cc} I & I \\ I & I \end{array} \right] \left[\begin{array}{c} L(j\omega) \\ I \end{array} \right] > 0, \forall \omega \in \Omega,$$

which is exactly (I, I, I) – quadratic constraint, and then (1) is satisfied as well.

Thus, we only need to consider quadratic constraints with the assumption (1) when we refer to feedback stable Ls.

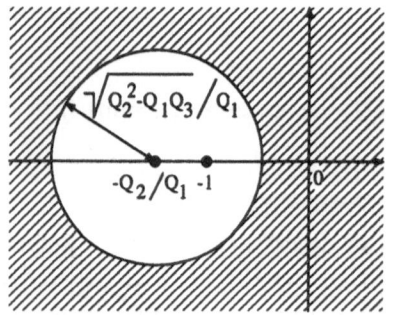

Figure 2: Quadratic constraint when $m = 1$ and $Q_1 > 0$

To interpret quadratic constraints graphically, let us assume single-input single-output ($m = 1$) L and $L(\infty) = 0$. Then we see that

$$(4) \Leftrightarrow \begin{cases} \left(L + \frac{Q_2}{Q_1}\right)^* \left(L + \frac{Q_2}{Q_1}\right) > \frac{Q_2^2 - Q_1 Q_3}{Q_1^2} & (Q_1 > 0) \\ \Re(L) > -\frac{Q_3}{2Q_2} & (Q_1 = 0) \\ \left(L + \frac{Q_2}{Q_1}\right)^* \left(L + \frac{Q_2}{Q_1}\right) < \frac{Q_2^2 - Q_1 Q_3}{Q_1^2} & (Q_1 < 0) \end{cases} .$$

From these inequalities, it is seen by noting $Q_3 > 0$ from $L(\infty) = 0$ and (1) that $L(j\omega)$ lies in the region, which does not contain the critical point $-1 + j0$, described by circles or lines. For example, Figure 2 describes the region when $Q_1 > 0$.

An important application of quadratic constraints is a specification of gain and phase margins of feedback systems. Let real numbers g_m and ϕ_m be desired gain and phase margins, respectively. Calculate a circle or line passing through the points $-1/g_m$ and $\exp(\pm j(\pi - \phi_m))$ in the complex plane. If $L(j\omega)$ lies in the region, which does not contain the point $-1 + j0$, decided by the circle or line and if (L, I) is feedback stable, then the feedback system has gain margin g_m and phase margin ϕ_m degree. Then, it is easily seen that the region can be expressed by a quadratic constraint with parameters, e.g.,

$$
\begin{aligned}
Q_1 &= 2g_m(g_m \cos\phi_m - 1) \\
Q_2 &= g_m^2 - 1 \\
Q_3 &= 2(g_m - \cos\phi_m).
\end{aligned}
$$

Note that these parameter matrices satisfy the assumption (1). Thus, we can specify gain and phase margins of feedback systems through loopshaping by quadratic constraints together with nominal feedback stability.

4 Stability analysis by quadratic constraints

In this section, we will analyze robustness of feedback systems under quadratic constraints on open-loop transfer functions. Here, let us consider a feedback system which has an uncertain transfer function Δ in the feedback loop. In usual models of uncertainties, only a plant is assumed to perturb. On the other hand, the feature of our uncertainty models is that the open-loop transfer function $L = KP$ is assumed to perturb, which is consistent with investigation of robustness by open-loop loopshaping.

Theorem 1 *Consider $m \times m$ proper transfer functions L and Δ. Assume that (L, I) and $(\Delta, L(\infty))$ are both feedback stable. Let $m \times m$ real matrices $Q_1 = Q_1^T$, Q_2, $Q_3 = Q_3^T$ satisfying (1) be given. If L satisfies (Q_1, Q_2, Q_3)-quadratic constraint and Δ satisfies*

$$
\Psi(j\omega) \leq 0, \quad \forall \omega \in \Omega_\Delta, \tag{5}
$$

then the pair (L, Δ) is feedback stable, where

$$
\Psi(s) := \begin{bmatrix} I & -\Delta^T(-s) \end{bmatrix} \begin{bmatrix} Q_1 & Q_2 \\ Q_2^T & Q_3 \end{bmatrix} \begin{bmatrix} I \\ -\Delta(s) \end{bmatrix} \tag{6}
$$

$$
\Omega_\Delta := \left\{ \omega \in \mathbf{R} \cup \{\infty\} \mid \Delta(j\omega) \in \mathbf{C}^{m \times m} \right\}. \tag{7}
$$

Proof of Theorem 1.

In the following, we will assume $L, \Delta \in \mathbf{RL}_\infty^{m \times m}$. Otherwise, a similar proof is possible by taking the Nyquist contour avoiding to the left around the $j\omega$-poles of L and Δ. Let us first prove the theorem when $D := L(\infty) = 0$. Then, $\Delta \in \mathbf{RH}_\infty^{m \times m}$ follows from feedback stability of (Δ, D). The quadratic constraint (4) implies $Q_3 > 0$ and

$$
\exists \mu > 0 : \begin{bmatrix} Q_1 & Q_2 \\ Q_2^T & Q_3 \end{bmatrix} + \mu \begin{bmatrix} I \\ -L^*(j\omega) \end{bmatrix} \begin{bmatrix} I & -L(j\omega) \end{bmatrix} > 0
$$

by Lemma 1–(i). Combining the above inequality and (5) gives

$$(I + L(j\omega)\Delta(j\omega))^*(I + L(j\omega)\Delta(j\omega))$$
$$= \begin{bmatrix} I & -\Delta^*(j\omega) \end{bmatrix} \begin{bmatrix} I \\ -L^*(j\omega) \end{bmatrix} \begin{bmatrix} I & -L(j\omega) \end{bmatrix} \begin{bmatrix} I \\ -\Delta(j\omega) \end{bmatrix}$$
$$> -\frac{1}{\mu} \begin{bmatrix} I & -\Delta^*(j\omega) \end{bmatrix} \begin{bmatrix} Q_1 & Q_2 \\ Q_2^T & Q_3 \end{bmatrix} \begin{bmatrix} I \\ -\Delta(j\omega) \end{bmatrix} \geq 0.$$

Therefore,

$$\det(I + L(j\omega)\Delta(j\omega)) = \det(I + \Delta(j\omega)L(j\omega)) \neq 0, \ \forall \omega \in \mathbf{R} \cup \{\infty\}.$$

Next, we prove that the number of encirclements of the Nyquist diagram of $\det(I + \Delta L)$ around the origin is equal to that of $\det(I + L)$. Let

$$\mathcal{M} := \left\{ \Delta \in \mathbf{RL}_\infty^{m \times m} | \Psi(j\omega) \geq 0, \forall \omega \in \mathbf{R} \cup \{\infty\} \right\}.$$

It is easily seen from $Q_3 > 0$ that $\mathcal{M}$ is a convex set. Then, by observing $\Delta, I \in \mathcal{M}$,

$$\Delta_\epsilon := \epsilon\Delta + (1 - \epsilon)I \in \mathcal{M}, \ \forall \epsilon \in [0, 1].$$

Thus, we obtain

$$f(\omega, \epsilon) := \det(I + \Delta_\epsilon(j\omega)L(j\omega)) \neq 0, \forall \omega \in \mathbf{R} \cup \{\infty\}, \forall \epsilon \in [0, 1].$$

By noting that the function $f(\omega, \epsilon)$ is continuous on ω, ϵ and

$$f(\omega, 0) = \det(I + L(j\omega))$$
$$f(\omega, 1) = \det(I + \Delta(j\omega)L(j\omega)),$$

we can ensure that the number of encirclements of $\det(I + \Delta L)$ and $\det(I + L)$ are the same.[5] Consequently, we can conclude the theorem is true when $D = 0$. Consider next the case of $D \neq 0$. From feedback stability of (L, I), if we set $L_0 := L - D$, then $(L_0, (I + D)^{-1})$ is also feedback stable. Moreover if we let $\hat{L} := (I + D)^{-1}L_0$, it is straightforward that $(\hat{L}, I)$ is feedback stable, $\hat{L}(\infty) = 0$ and $\hat{L}$ satisfies $(\hat{Q}_1, \hat{Q}_2, \hat{Q}_3)$–quadratic constraint. Here, the triplet $(\hat{Q}_1, \hat{Q}_2, \hat{Q}_3)$ is defined as

$$\begin{bmatrix} \hat{Q}_1 & \hat{Q}_2 \\ \hat{Q}_2^T & \hat{Q}_3 \end{bmatrix} := \begin{bmatrix} I + D^T & 0 \\ D^T & I \end{bmatrix} \begin{bmatrix} Q_1 & Q_2 \\ Q_2^T & Q_3 \end{bmatrix} \begin{bmatrix} I + D & D \\ 0 & I \end{bmatrix}.$$

Note that $(\hat{Q}_1, \hat{Q}_2, \hat{Q}_3)$ satisfies

$$\hat{J} := \begin{bmatrix} I & -I \end{bmatrix} \begin{bmatrix} \hat{Q}_1 & \hat{Q}_2 \\ \hat{Q}_2^T & \hat{Q}_3 \end{bmatrix} \begin{bmatrix} I \\ -I \end{bmatrix} = J \leq 0.$$

On the other hand, $\hat{\Delta} := \Delta(I + D\Delta)^{-1}(I + D) \in \mathbf{RH}_\infty^{m \times m}$ follows from the feedback stability of (Δ, D). Also, it is readily verified that

$$\hat{\Psi}(j\omega) := \begin{bmatrix} I & -\hat{\Delta}^T(-s) \end{bmatrix} \begin{bmatrix} \hat{Q}_1 & \hat{Q}_2 \\ \hat{Q}_2^T & \hat{Q}_3 \end{bmatrix} \begin{bmatrix} I \\ -\hat{\Delta}(s) \end{bmatrix} = \Psi(j\omega)$$

and accordingly, $\hat{\Delta}$ satisfies

$$\hat{\Psi}(j\omega) \le 0, \forall \omega \in \mathbf{R} \cup \{\infty\}.$$

Finally, applying the theorem for $D = 0$ to the pair $(\hat{L}, \hat{\Delta})$ defined above completes the proof.   Q.E.D.

5 Controller design under quadratic constraints

The purpose of this section is to derive a stabilizing controller design method under quadratic constraints. At first, we will express quadratic constraints in the frequency-domain as some matrix inequalities in the time-domain. This step is a similar one of derivations of H_∞ controllers through the bounded real lemma.

Theorem 2 *Let real matrices $A(n \times n)$, $B(n \times m)$, $C(m \times n)$, $D(m \times m)$, and $m \times m$ real matrices $Q_1 = Q_1^T$, Q_2, $Q_3 = Q_3^T$ satisfying (1) be given. Then, the following statements (i), (ii) and (iii) are equivalent.*

(i) Define

$$L(s) := \left[\begin{array}{c|c} A & B \\ \hline C & D \end{array} \right] \tag{8}$$

$$N(s) := \left[\begin{array}{cc} A - sI & 0 \\ -C^T Q_1 C & -A^T - sI \\ (Q_1 D + Q_2)^T C & B^T \end{array} \right]. \tag{9}$$

The triplet (A, B, C) is stabilizable and detectable, (L, I) is feedback stable, L satisfies (Q_1, Q_2, Q_3)-quadratic constraint (4), and moreover,

$$\mathrm{rank} N(j\omega) = 2n, \forall \omega \in \mathbf{R}. \tag{10}$$

(ii) If we set $E := I + D$, then $\det E \ne 0$ and there exists a real symmetric positive definite matrix $Z(n \times n)$ such that

$$\left[\begin{array}{cc} Z\tilde{A} + \tilde{A}^T Z & Z\tilde{B} \\ \tilde{B}^T Z & 0 \end{array} \right] - \left[\begin{array}{cc} \tilde{C}^T & 0 \\ \tilde{D}^T & I \end{array} \right] \left[\begin{array}{cc} \tilde{Q}_1 & \tilde{Q}_2 \\ \tilde{Q}_2^T & \tilde{Q}_3 \end{array} \right] \left[\begin{array}{cc} \tilde{C} & \tilde{D} \\ 0 & I \end{array} \right] < 0, \tag{11}$$

where

$$\left[\begin{array}{cc} \tilde{A} & \tilde{B} \\ \tilde{C} & \tilde{D} \end{array} \right] := \left[\begin{array}{cc} A - BE^{-1}C & BE^{-1} \\ E^{-1}C & DE^{-1} \end{array} \right] \tag{12}$$

$$\left[\begin{array}{cc} \tilde{Q}_1 & \tilde{Q}_2 \\ \tilde{Q}_2^T & \tilde{Q}_3 \end{array} \right] := \left[\begin{array}{cc} I & -I \\ 0 & I \end{array} \right] \left[\begin{array}{cc} Q_1 & Q_2 \\ Q_2^T & Q_3 \end{array} \right] \left[\begin{array}{cc} I & 0 \\ -I & I \end{array} \right]. \tag{13}$$

(iii) There exists a real symmetric positive definite matrix $Z(n \times n)$ such that

$$\begin{bmatrix} ZA + A^T Z & ZB \\ B^T Z & 0 \end{bmatrix} - \begin{bmatrix} C^T & 0 \\ D^T & I \end{bmatrix} \begin{bmatrix} Q_1 & Q_2 \\ Q_2^T & Q_3 \end{bmatrix} \begin{bmatrix} C & D \\ 0 & I \end{bmatrix} < 0. \quad (14)$$

To prove the theorem, we need the next lemma, which is fairly standard and can be easily obtained by using a spectral factorization result [8].

Lemma 2 *Let real matrices $A(n \times n)$, $B(n \times m)$, $C(m \times n)$, $D(m \times m)$, and $m \times m$ real matrices $Q_1 = Q_1^T$, Q_2, $Q_3 = Q_3^T$ satisfying $Q_1 \leq 0$ be given. Then, the following statements (i) and (ii) are equivalent.*

(i) $\det(sI - A) \neq 0$, $\forall \mathrm{Re}(s) \geq 0$ *and L defined by (8) satisfies (4).*

(ii) *There exists a real symmetric positive definite matrix $Z(n \times n)$ such that (14) holds.*

Proof of Theorem 2.

((iii)$\Rightarrow$(ii)) The (2,2) – block of (14) is

$$\begin{bmatrix} D^T & I \end{bmatrix} \begin{bmatrix} Q_1 & Q_2 \\ Q_2^T & Q_3 \end{bmatrix} \begin{bmatrix} D \\ I \end{bmatrix} > 0,$$

which is equivalent to

$$\exists \mu > 0 : \begin{bmatrix} I \\ -D^T \end{bmatrix} \begin{bmatrix} I & -D \end{bmatrix} > -\mu \begin{bmatrix} Q_1 & Q_2 \\ Q_2^T & Q_3 \end{bmatrix} \quad (15)$$

by Lemma 1–(i). Then,

$$\begin{aligned} E^T E &= \begin{bmatrix} I & -I \end{bmatrix} \begin{bmatrix} I \\ -D^T \end{bmatrix} \begin{bmatrix} I & -D \end{bmatrix} \begin{bmatrix} I \\ -I \end{bmatrix} \\ &> -\mu \begin{bmatrix} I & -I \end{bmatrix} \begin{bmatrix} Q_1 & Q_2 \\ Q_2^T & Q_3 \end{bmatrix} \begin{bmatrix} I \\ -I \end{bmatrix} = -\mu J \geq 0. \end{aligned}$$

This yields $\det E \neq 0$. Pre- and post-multiplying (14) by matrices

$$\begin{bmatrix} I & -C^T E^{-T} \\ 0 & E^{-T} \end{bmatrix}, \begin{bmatrix} I & 0 \\ -E^{-1}C & E^{-1} \end{bmatrix}$$

respectively, we obtain (11).

((ii)$\Rightarrow$(iii)) Pre- and post-multiplying (11) by

$$\begin{bmatrix} I & C^T \\ 0 & E^T \end{bmatrix}, \begin{bmatrix} I & 0 \\ C & E \end{bmatrix}$$

respectively, we obtain (14).

((ii)⇒(i)) Noting $\tilde{Q}_1 = J \leq 0$ and defining

$$\tilde{L}(s) := \left[\begin{array}{c|c} \tilde{A} & \tilde{B} \\ \hline \tilde{C} & D \end{array} \right] \tag{16}$$

together with Lemma 2 give that $\tilde{A}$ is stable. Therefore, (L, Δ) is feedback stable, (A, B) is stabilizable, (C, A) is detectable and

$$\tilde{\Phi}(j\omega) > 0, \forall \omega \in \mathbf{R} \cup \{\infty\}, \tag{17}$$

where

$$\tilde{\Phi}(s) := [\ \tilde{L}^T(-s) \quad I\ ] \left[\begin{array}{cc} \tilde{Q}_1 & \tilde{Q}_2 \\ \tilde{Q}_2^T & \tilde{Q}_3 \end{array} \right] \left[\begin{array}{c} \tilde{L}(s) \\ I \end{array} \right]. \tag{18}$$

Also, by noting the identity

$$\tilde{\Phi}(s) = \left(I + L^T(-s) \right)^{-1} \Phi(s) \left(I + L(s) \right)^{-1}, \tag{19}$$

we know

$$\Phi(j\omega) = (I + L(j\omega))^* \, \tilde{\Phi}(j\omega) \, (I + L(j\omega)) \geq 0, \forall \omega \in \Omega. \tag{20}$$

Since $\det (I + L(j\omega)) \neq 0, \forall \omega \in \mathbf{R} \cup \{\infty\}$ from feedback stability, $\det \Phi(j\omega) \neq 0, \forall \omega \in \mathbf{R} \cup \{\infty\}$ follows and leads to (4). For convenience, let us define

$$\left[\begin{array}{cc} M_1 & M_2 \\ M_2^T & M_3 \end{array} \right] := \left[\begin{array}{cc} C^T & 0 \\ D^T & I \end{array} \right] \left[\begin{array}{cc} Q_1 & Q_2 \\ Q_2^T & Q_3 \end{array} \right] \left[\begin{array}{cc} C & D \\ 0 & I \end{array} \right]$$

$$\left[\begin{array}{cc} \tilde{M}_1 & \tilde{M}_2 \\ \tilde{M}_2^T & \tilde{M}_3 \end{array} \right] := \left[\begin{array}{cc} \tilde{C}^T & 0 \\ \tilde{D}^T & I \end{array} \right] \left[\begin{array}{cc} \tilde{Q}_1 & \tilde{Q}_2 \\ \tilde{Q}_2^T & \tilde{Q}_3 \end{array} \right] \left[\begin{array}{cc} \tilde{C} & \tilde{D} \\ 0 & I \end{array} \right].$$

One of realizations of $\tilde{\Phi}(s)$ is given by

$$\tilde{\Phi}(s) = \left[\begin{array}{cc|c} \tilde{A} & 0 & \tilde{B} \\ -\tilde{M}_1 & -\tilde{A}^T & -\tilde{M}_2 \\ \hline \tilde{M}_2^T & \tilde{B}^T & \tilde{M}_3 \end{array} \right]. \tag{21}$$

Because $\det \tilde{\Phi}(j\omega) \neq 0$ and $\tilde{A}$ is stable,

$$\det \left[\begin{array}{ccc} \tilde{A} - j\omega I & 0 & \tilde{B} \\ -\tilde{M}_1 & -\tilde{A}^T - j\omega I & -\tilde{M}_2 \\ \tilde{M}_2^T & \tilde{B}^T & \tilde{M}_3 \end{array} \right] \neq 0, \forall \omega \in \mathbf{R}. \tag{22}$$

By noting the identity:

$$\left[\begin{array}{ccc} \tilde{A} - j\omega I & 0 & \tilde{B} \\ -\tilde{M}_1 & -\tilde{A}^T - j\omega I & -\tilde{M}_2 \\ \tilde{M}_2^T & \tilde{B}^T & \tilde{M}_3 \end{array} \right]$$

$$= \left[\begin{array}{ccc} I & 0 & 0 \\ 0 & I & C^T E^{-T} \\ 0 & 0 & E^{-T} \end{array} \right] \left[\begin{array}{ccc} A - j\omega I & 0 & B \\ -M_1 & -A^T - j\omega I & -M_2 \\ M_2^T & B^T & M_3 \end{array} \right]$$

$$\times \left[\begin{array}{ccc} I & 0 & 0 \\ 0 & I & 0 \\ -E^{-1}C & 0 & E^{-1} \end{array} \right],$$

we have

$$\det \begin{bmatrix} A - j\omega I & 0 & B \\ -M_1 & -A^T - j\omega I & -M_2 \\ M_2^T & B^T & M_3 \end{bmatrix} \neq 0, \forall \omega \in \mathbf{R} \qquad (23)$$

and this implies (10).

((i)$\Rightarrow$(ii)) Notice that (10) implies a realization of $\Phi(s)$

$$\Phi(s) = \left[\begin{array}{cc|c} A & 0 & B \\ -M_1 & -A^T & -M_2 \\ \hline M_2^T & B^T & M_3 \end{array} \right] \qquad (24)$$

has neither uncontrollable nor unobservable mode on the imaginary axis. Then, a realization of $\Phi^{-1}(s)$:

$$\Phi^{-1}(s) = \left[\begin{array}{cc|c} A - BM_3^{-1}M_2^T & -BM_3^{-1}B^T & -BM_3^{-1} \\ -M_1 + M_2M_3^{-1}M_2^T & -A^T + M_2M_3^{-1}B^T & M_2M_3^{-1} \\ \hline M_3^{-1}M_2^T & M_3^{-1}B^T & M_3^{-1} \end{array} \right] \qquad (25)$$

has also neither uncontrollable nor unobservable mode on the imaginary axis, since $M_3 = \Phi(\infty) > 0$. On the other hand, with respect to (21), because $\tilde{A}$ is stable, $\tilde{\Phi}(j\omega)$ has no pole on the imaginary axis. Therefore, from (4) and (19), we can ensure that

$$\tilde{\Phi}(j\omega) \geq 0, \forall \omega \in \mathbf{R} \cup \{\infty\}.$$

Now let $\det \tilde{\Phi}(j\omega_0) = 0$ for some $\omega_0 \in \mathbf{R}$. Then $\tilde{\Phi}^{-1}(s)$ has a pole at $s = j\omega_0$. If we notice the following identities:

$$\begin{aligned} \tilde{\Phi}^{-1}(s) &= \left[\begin{array}{cc|c} \tilde{A} - \tilde{B}\tilde{M}_3^{-1}\tilde{M}_2^T & -\tilde{B}\tilde{M}_3^{-1}B^T & -\tilde{B}\tilde{M}_3^{-1} \\ -\tilde{M}_1 + \tilde{M}_2\tilde{M}_3^{-1}\tilde{M}_2^T & -\tilde{A}^T + \tilde{M}_2\tilde{M}_3^{-1}\tilde{B}^T & \tilde{M}_2\tilde{M}_3^{-1} \\ \hline \tilde{M}_3^{-1}\tilde{M}_2^T & \tilde{M}_3^{-1}\tilde{B}^T & \tilde{M}_3^{-1} \end{array} \right] \\[2mm] &= \left[\begin{array}{cc|c} A - BM_3^{-1}M_2^T & -BM_3^{-1}B^T & -BM_3^{-1}E^T \\ -M_1 + M_2M_3^{-1}M_2^T & -A^T + M_2M_3^{-1}B^T & M_2M_3^{-1}E^T - C^T \\ \hline EM_3^{-1}M_2^T - C & EM_3^{-1}B^T & EM_3^{-1}E^T \end{array} \right], \end{aligned}$$

then the A - matrix, which is exactly the same as the A - matrix of (25), must have an eigenvalue $j\omega_0$. However, because

$$\det \Phi(j\omega) \neq 0, \forall \omega \in \mathbf{R} \cup \{\infty\},$$

$\Phi^{-1}(s)$ has no pole on the imaginary axis and consequently, $j\omega_0$ is either an uncontrollable pole or an unobservable pole of (25). This leads to a contradiction. Finally, we see that $\tilde{A}$ is stable, $\tilde{Q}_1 \leq 0$ and

$$\tilde{\Phi}(j\omega) > 0, \forall \omega \in \mathbf{R} \cup \{\infty\}.$$

By Lemma 2, (11) must be satisfied.   Q.E.D.

In Theorem 2, it should be noted that, while the statements (i) and (iii) are both conditions with respect to open-loop transfer functions, the statement (ii) is a condition with respect to the closed-loop transfer function $\tilde{L}$. Indeed, we see that $\tilde{L} = L(I + L)^{-1}$ from (12), and moreover, by noting that (14) and (11) have the same form, we can conclude that (Q_1, Q_2, Q_3) - quadratic constraint on open-loop transfer functions Ls with feedback stability is equivalent to $(\tilde{Q}_1, \tilde{Q}_2, \tilde{Q}_3)$ - quadratic constraint on closed-loop transfer functions $\tilde{L}$s with stability of $\tilde{L}$s. This fact will be used for a controller synthesis satisfying quadratic constraints on open loop transfer functions.

Before we state the controller synthesis, let us investigate the relation between quadratic constraints and dissipativity[4], by using Theorem 2. Introduce a state space representation of L:

$$\Sigma_L : \begin{cases} \dot{x} = Ax + Bu \\ y = Cx + Du \end{cases}.$$

Setting P as the left-hand side of (14), we can write

$$\begin{bmatrix} x^T & u^T \end{bmatrix} \begin{bmatrix} ZA + A^T Z & ZB \\ B^T Z & 0 \end{bmatrix} \begin{bmatrix} x \\ u \end{bmatrix} - \begin{bmatrix} y^T & u^T \end{bmatrix} \begin{bmatrix} Q_1 & Q_2 \\ Q_2^T & Q_3 \end{bmatrix} \begin{bmatrix} y \\ u \end{bmatrix}$$

$$= - \begin{bmatrix} x^T & u^T \end{bmatrix} P \begin{bmatrix} x \\ u \end{bmatrix} < 0.$$

If we define a function V by $V(x) := x^T Z x$, V is positive definite. Let $x(t)$ be a solution of the system Σ_L corresponding to an initial condition $x(t) = x(0)$ and an input u. Then, we obtain a dissipation inequality:

$$V(x(t)) - V(x(0)) = \int_0^t (y^T Q_1 y + 2y^T Q_2 u + u^T Q_3 u) dt$$

$$- \int_0^t \begin{bmatrix} x^T & u^T \end{bmatrix} P \begin{bmatrix} x \\ u \end{bmatrix} dt, \forall x(0), \forall u, \forall t \geq 0.$$

Therefore, the system Σ_L is dissipative with respect to supply rate $y^T Q_1 y + 2y^T Q_2 u + u^T Q_3 u$. Moreover, under the unity-feedback $u = -y$,

$$\dot{V}(x) = u^T J u - \begin{bmatrix} x^T & u^T \end{bmatrix} P \begin{bmatrix} x \\ u \end{bmatrix}.$$

In this equation, $J \leq 0$ implies that $\dot{V}(x) < 0, \forall x \neq 0$ and consequently, (L, I) is feedback stable. Therefore, under the assumption (1), L satisfies quadratic constraints with feedback stability if and only if Σ_L is dissipative in the above sense.

Now, let us proceed to discuss a synthesis problem of a stabilizing controller which achieves (Q_1, Q_2, Q_3) – quadratic constraints. Consider a feedback system of Figure 1, consisting of a strictly proper plant:

$$P(s) := \left[\begin{array}{c|c} A_p & B_p \\ \hline C_p & 0 \end{array} \right], A_p(n_p \times n_p), B_p(n_p \times m), C_p(q \times n_p) \qquad (26)$$

and a proper controller:

$$K(s) := \left[\begin{array}{c|c} A_k & B_k \\ \hline C_k & D_k \end{array}\right], A_k(n_k \times n_k), B_k(n_k \times q), C_k(m \times n_k), D_k(m \times q).$$

Define an open-loop transfer function $L(s) := K(s)P(s)$ and a realization of L:

$$L(s) = \left[\begin{array}{c|c} A & B \\ \hline C & D \end{array}\right] := \left[\begin{array}{cc|c} A_p & 0 & B_p \\ B_k C_p & A_k & 0 \\ \hline D_k C_p & C_k & 0 \end{array}\right]. \tag{27}$$

Again, the triplet (Q_1, Q_2, Q_3) is assumed to satisfy (1) and additionally, we assume (10). In the light of Theorem 2, the existence condition of our controllers is equivalent to the condition (ii) in the theorem. Then, a derivation of the existence condition of the controllers can be done in the same way as that of H_∞ controllers by an LMI-based approach [7].

To begin with, decompose $\tilde{Q}_1 = J \leq 0$ as

$$\tilde{Q}_1 = J = U_1 \Lambda_1^{-1} U_1^T, \Lambda_1 < 0,$$

and by using this, equivalently express (11) as follows.

$$\left[\begin{array}{ccc} Z\tilde{A} + \tilde{A}^T Z & Z\tilde{B} - \tilde{C}^T \tilde{Q}_2 & \tilde{C}^T U_1 \\ \tilde{B}^T Z - \tilde{Q}_2^T \tilde{C} & -\tilde{Q}_3 & 0 \\ U_1^T \tilde{C} & 0 & \Lambda_1 \end{array}\right] < 0 \tag{28}$$

We use the following compact expressions, taking care of (27) and (12):

$$\tilde{A} = A_a + B_a K C_a, \tilde{B} = -B_a M, \tilde{C} = M^T K C_a$$

where

$$A_a := \left[\begin{array}{cc} A_p & 0 \\ 0 & 0 \end{array}\right], B_a := \left[\begin{array}{cc} -B_p & 0 \\ 0 & I \end{array}\right], C_a := \left[\begin{array}{cc} C_p & 0 \\ 0 & I \end{array}\right],$$

$$M := \left[\begin{array}{c} I \\ 0 \end{array}\right], K := \left[\begin{array}{cc} D_k & C_k \\ B_k & A_k \end{array}\right].$$

Substituting these expressions to (28), we have

$$\left[\begin{array}{ccc} ZA_a + A_a^T Z & -ZB_a M & 0 \\ -M^T B_a^T Z & -\tilde{Q}_3 & 0 \\ 0 & 0 & \Lambda_1 \end{array}\right] + \left[\begin{array}{c} ZB_a \\ -\tilde{Q}_2^T M^T \\ U_1^T M^T \end{array}\right] K \left[\begin{array}{ccc} C_a & 0 & 0 \end{array}\right]$$

$$+ \left(\left[\begin{array}{c} ZB_a \\ -\tilde{Q}_2^T M^T \\ U_1^T M^T \end{array}\right] K \left[\begin{array}{ccc} C_a & 0 & 0 \end{array}\right]\right)^T < 0.$$

By Lemma 1-(ii), there exists a matrix K satisfying the above inequality if and only if the next two inequalities hold.

$$
\begin{bmatrix} B_a \\ -\tilde{Q}_2^T M^T \\ U_1^T M^T \end{bmatrix}^{\perp}
\begin{bmatrix} A_a Z^{-1} + Z^{-1} A_a^T & -B_a M & 0 \\ -M^T B_a^T & -\tilde{Q}_3 & 0 \\ 0 & 0 & \Lambda_1 \end{bmatrix}
$$
$$
\times \begin{bmatrix} B_a \\ -\tilde{Q}_2^T M^T \\ U_1^T M^T \end{bmatrix}^{\perp T} < 0, \tag{29}
$$

$$
\begin{bmatrix} C_a^T \\ 0 \\ 0 \end{bmatrix}^{\perp}
\begin{bmatrix} Z A_a + A_a^T Z & -Z B_a M & 0 \\ -M^T B_a^T Z & -\tilde{Q}_3 & 0 \\ 0 & 0 & \Lambda_1 \end{bmatrix}
\begin{bmatrix} C_a^T \\ 0 \\ 0 \end{bmatrix}^{\perp T} < 0. \tag{30}
$$

To reduce a redundancy of these inequalities, partition Z such as

$$
Z =: \begin{bmatrix} Y & Y_2 \\ Y_2^T & Y_3 \end{bmatrix}, Z^{-1} =: \begin{bmatrix} X & X_2 \\ X_2^T & X_3 \end{bmatrix}, X, Y(n_p \times n_p), X_3, Y_3(n_k \times n_k).
$$

Then, $Z > 0$ leads to $X - Y^{-1} = X_2 X_3^{-1} X_2^T \geq 0$ and then

$$
X > 0, Y > 0, \begin{bmatrix} X & I \\ I & Y \end{bmatrix} \geq 0. \tag{31}
$$

Also, by noting

$$
\begin{bmatrix} B_a \\ -\tilde{Q}_2^T M^T \\ U_1^T M^T \end{bmatrix}^{\perp}
= \begin{bmatrix} \begin{bmatrix} -B_p \\ -\tilde{Q}_2^T \\ U_1^T \end{bmatrix}^{\perp} & 0, \end{bmatrix}
\begin{bmatrix} I & 0 & 0 & 0 \\ 0 & 0 & I & 0 \\ 0 & 0 & 0 & I \\ 0 & I & 0 & 0 \end{bmatrix}
$$

$$
\begin{bmatrix} C_a^T \\ 0 \\ 0 \end{bmatrix}^{\perp}
= \begin{bmatrix} C_p^{T\perp} & 0 & 0 & 0 \\ 0 & 0 & I & 0 \\ 0 & 0 & 0 & I \end{bmatrix},
$$

we see that (30) and (30) are respectively equivalent to

$$
\begin{bmatrix} -B_p \\ -Q_2^T + Q_3 \\ U_1^T \end{bmatrix}^{\perp}
\begin{bmatrix} A_p X + X A_p^T & B_p & 0 \\ B_p^T & -Q_3 & 0 \\ 0 & 0 & \Lambda_1 \end{bmatrix}
\begin{bmatrix} -B_p \\ -Q_2^T + Q_3 \\ U_1^T \end{bmatrix}^{\perp T} < 0 \tag{32}
$$

$$
\begin{bmatrix} C_p^{T\perp} & 0 \\ 0 & I \end{bmatrix}
\begin{bmatrix} Y A_p + A_p^T Y & Y B_p \\ B_p^T Y & -Q_3 \end{bmatrix}
\begin{bmatrix} C_p^{T\perp T} & 0 \\ 0 & I \end{bmatrix} < 0. \tag{33}
$$

Conversely, (31), (32) and (33) imply that there exist $X_2(n_p \times n_p)$ and $X_3(n_k \times n_k) > 0$ such that $X - Y^{-1} = X_2 X_3^{-1} X_2^T \geq 0$. Thus, $Z^{-1} > 0$, (30) and (30) can be obtained.

Proposition 1 *Let a plant (26) and $m \times m$ matrices $Q_1 = Q_1^T$, Q_2, $Q_3 = Q_3^T$ with (1) be given. Then, the next statements (i) and (ii) are equivalent.*

(i) There exists a controller $K(s)$ such that (P, K) is feedback stable, L defined by (27) satisfies (Q_1, Q_2, Q_3) – quadratic constraint and (10) holds.

(ii) There exist $m \times m$ matrices X and Y such that the inequalities (31),(32) and (33) hold.

The inequalities (31),(32) and (33) are called LMIs and numerically reliable algorithms to solve them are available on PCs. Notice also that a parameterization of all controllers stated in Proposition 1 can be obtained by direct substitutions to Theorem 1 in [7].

6 Conclusion

Many modern robust control theories including H_∞ control theory focus on specifications of some closed-loop transfer functions, and resultant feedback systems have occasionally small gain and phase margins [1]. Therefore, we should discuss robustness in terms of open-loop transfer functions, since these stability margins are characteristics of open-loop transfer functions. On the other hand, although quadratic constraints on transfer functions are known as generalizations of the concepts of bounded realness and positive realness, there are few applications of quadratic constraints to controller synthesis.

In this note, we have proposed to execute loopshaping by imposing quadratic constraints on open-loop transfer functions. We have defined quadratic constraints consistent with loopshaping method under feedback stability. Then, we have proved a robust stability theorem with respect to feedback systems under quadratic constraints on open-loop transfer functions, and furthermore, we have presented a loopshaping method by imposing quadratic constraints on open-loop transfer functions. Thus, we have provided a general aspect of a classical loopshaping by adopting a modern tool of quadratic constraints.

Acknowledgement

The authors acknowledge Professor K. Sugimoto of Nagoya University for helpful discussions.

References

[1] L. H. Keel and S. P. Bhattacharyya: Robust, Fragile, or Optimal?, IEEE Trans. on Automatic Control; Vol. 42, No. 8, pp. 1098–1105 (1997)

[2] M. Vidyasagar: Nonlinear Systems Analysis 2nd Ed., Prentice Hall (1993)

[3] C. A. Desoer and M. Vidyasagar: Feedback Systems: Input-Output Properties; Academic Press, New York (1975)

[4] J. C. Willems: Dissipative Dynamical Systems, Part I and Part II; Arch. Rational Mech. Anal., Vol. 45, pp. 321–393 (1972)

[5] M. J. Chen and C. A. Desoer: Necessary and sufficient condition for robust stability of linear distributed feedback systems; Int. J. Control, Vol. 35, No. 2, pp. 255–267 (1982)

[6] J. C. Doyle, B. A. Francis and A. R. Tannenbaum: Feedback Control Theory; Macmillan Publishing Company (1992)

[7] T. Iwasaki and R. E. Skelton: All Controllers for the General H_∞ Control Problem: LMI Existence Conditions and State Space Formulas, Automatica, Vol. 30, No. 8, pp. 1307–1317 (1994)

[8] K. Zhou with J. C. Doyle and K. Glover: Robust and Optimal Control, Prentice Hall (1996)

Control of Hysteretic Systems: A State-Space Approach

R.B. Gorbet[*]   K.A. Morris[†]   D.W.L. Wang[‡]

1 Introduction

Hysteresis occurs in a number of applications, and has many causes. A very important class of hysteretic systems are "smart materials" such as shape memory alloys (SMA), magnetostrictive materials and piezoceramics. When a shape memory alloy undergoes a temperature-induced phase transformation, its deformation characteristics are altered. This change is significant enough that an SMA wire can be used as an actuator by cycling the temperature through the transformation range of the alloy. SMA actuators have been employed successfully in robotics research applications for many years[8, e.g.], and are beginning to be used in more commercial applications such as valves and dampers[14].

The hysteresis in a typical SMA actuator is illustrated in Figure 1. The highly non-linear behaviour and poor identification of hysteretic systems can make design of reliable controllers for these systems difficult. We are interested in control of SMA actuators, and hysteretic systems in general.

Several experimental results suggest that a model first developed to describe the dynamics in magnetic materials, the Preisach model, also describes the behaviour of SMAs and other phase transformation problems[6, 9, e.g.]. In this paper we describe the Preisach model and show that it can be formulated as a classical dynamical system. Using a purely abstract analysis of the Preisach operator, it is shown that these systems are dissipative, in a generalization of the sense first described in [13].

An initial difficulty in studying hysteresis is to arrive at a precise definition. The word comes from the Greek word meaning "to lag behind". It is tempting to characterize hysteresis in terms of this lag, or perhaps by the fact that hysteretic systems have "memory", or that they generate "loops"[10]. While these are all characteristics of hysteretic systems, many linear systems also

---

[*]Systems Control Group, University of Toronto, Toronto, Ontario M5S 3G4

[†]Dept. of Applied Mathematics, University of Waterloo, Waterloo, Ontario N2L 3G1

[‡]Dept. of Electrical & Computer Engineering, University of Waterloo, Waterloo, Ontario N2L 3G1

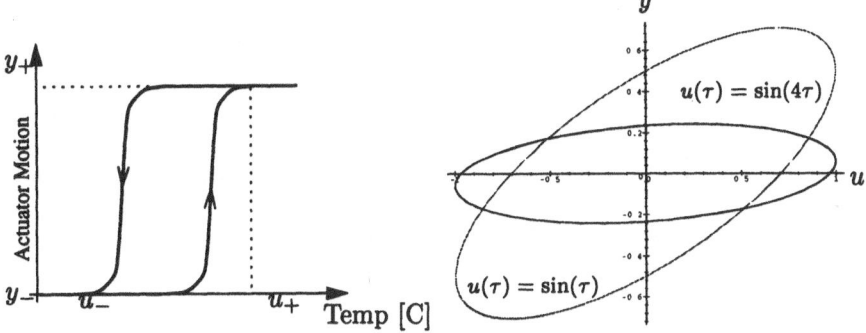

Figure 1: SMA Actuator Hysteresis   Figure 2: Output of LTI system:
$$y(t) = \int_0^t e^{-(t-\tau)} u(\tau) d\tau, \quad t = 4..12$$

have delays or memory. As well, for certain inputs, some linear systems can display looping behaviour (cf. Figure 2).

The most useful definition for our purposes is obtained by restricting ourselves to *rate independent* or *static* hysteresis. *Rate independence* means that the curves described in R^2 by the couple (u, y) are invariant for changes in the input rate, such as changes of the frequency. Linear systems are not rate-independent, as evidenced by the different outputs obtained in Figure 2. While this restriction excludes some types of hysteresis, it does include many commonly encountered, such as those occurring in phase transitions[3].

Many static hysteretic systems of practical interest, including piezoceramics and shape memory alloys, can be modelled by a Preisach operator. Essentially, the behaviour of the system is described by a weighted sum over a number of relays, each of which has output $+1$ or -1. In magnetic materials these relays have an interpretation as representing individual magnetic dipoles. A physical description is more elusive for other systems; however, the validity of these models has been demonstrated by a number of experiments[6, 9, e.g.].

An important consequence of the rate independence of these systems is that the output is dependent only on the past input extrema, not on all input values. This property will be significant in obtaining a state-space description of Preisach systems.

2   Preisach Operators

The basic building block of the Preisach model is the hysteresis relay γ. A relay is characterized by its half-width $r > 0$ and the input offset s, and is denoted by $\gamma_{r,s}$. The behaviour of the relay is described schematically in Figure 3. The structure of the model is illustrated in Figure 4. The output is computed as the weighted sum of relay outputs; the value $\mu(r, s)$ represents the weighting of

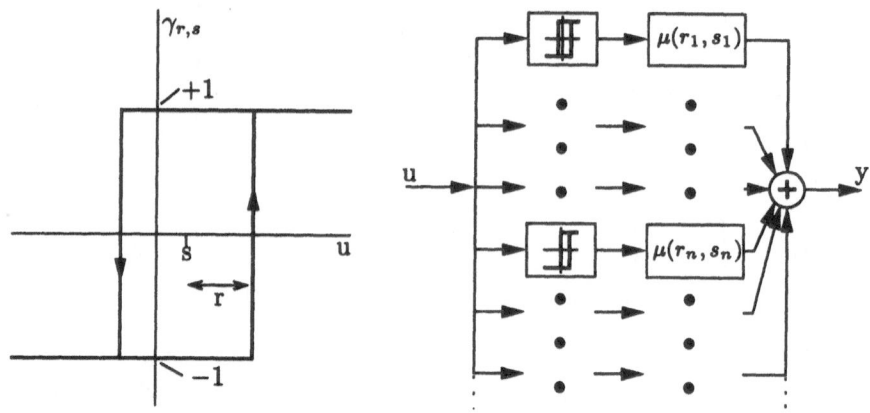

Figure 3: Hysteresis Relay   Figure 4: Preisach Model Structure

the relay $\gamma_{r,s}$. The relay output, and hence the Preisach model, is only defined for continuous inputs u. As this input varies with time, each individual relay adjusts its output according to the current input value, and the weighted sum of all relay outputs provides the overall system output (cf. Figure 4)

$$y(t) = \int_0^\infty \int_{-\infty}^\infty \mu(r,s) \gamma_{r,s}[u](t)\,ds\,dr. \tag{1}$$

The notation Γ will be used to denote the operator which maps a continuous input function to some output function according to (1): $y = \Gamma u$.

The identification of a relay $\gamma_{r,s}$ with the point (r,s) allows each relay to be uniquely represented as a point in $\mathbf{R}_+ \times \mathbf{R}$. This half-plane plays an important role in understanding the Preisach model, and is often referred to as the *Preisach plane*, $\mathcal{P}$. The collection of weights $\mu(r,s)$ forms the *Preisach weighting function* $\mu : \mathcal{P} \mapsto \mathbf{R}$. This weighting function is experimentally determined for a given system; see [2, 10] for two different approaches to the identification problem.

In any physical setup, there are limitations which can be interpreted as a restriction on the support of μ. For instance, control input saturation, say at $\hat{u}$, means that some relays in $\mathcal{P}$ can never be exercised and cannot contribute to a change in output. This effectively restricts the domain of μ to a triangle in $\mathcal{P}$ defined by $\mathcal{P}_r = \{(r,s) \in \mathcal{P} | |s| \le \hat{u} - r\}$, illustrated in Figure 5. In this case, μ effectively has compact support $\mathcal{P}_r$. We will henceforth assume that μ is only non-zero in some region $\mathcal{P}_r$. We will also assume that μ is bounded, piece-wise continuous, and non-negative inside $\mathcal{P}_r$; the set of such μ will be called $\mathcal{M}_p$. These are common assumptions when dealing with Preisach models for physical systems[3, 10, e.g.], and a model of an SMA actuator identified in [6] satisfied $\mu \in \mathcal{M}_p$.

In [5], it was shown that if μ is bounded and piece-wise continuous, then

$\Gamma : C^0 \mapsto C^0$. If furthermore μ is non-negative, then[1] $\Gamma : W_1^2 \mapsto W_1^2$.

One ambiguity remains in this definition of the model, and that is the question of the initial state of the relays γ. Obviously, the output y depends not only on u but on the initial configuration of the relays of $\mathcal{P}_r$. It is common to assume an initial relay output of -1 if $s > 0$ and $+1$ otherwise[3, e.g.]. There is some physical justification for this choice. Since magnetic materials have weighting functions μ which are symmetric about $s = 0$, the model output corresponding to this assumed initial state is zero. It represents the *de-magnetized* state of a magnetic material: the state in which no remnant magnetization is present[12].

The Preisach plane can be used to track individual relay states by observing the evolution of the *Preisach plane boundary*, ψ, in $\mathcal{P}_r$. First, the relays are divided into two time-varying sets, represented by the regions $\mathcal{P}_-$ and $\mathcal{P}_+$ defined as follows:

$$\mathcal{P}_\pm(t) = \{(r, s) \in \mathcal{P}_r \mid \text{output of } \gamma_{r,s} \text{ at } t \text{ is } \pm 1\}. \tag{2}$$

It will become clear that each set is connected; ψ is the line separating $\mathcal{P}_+$ from $\mathcal{P}_-$. The boundary corresponding to the assumed initial relay configuration (the line $s = 0$) will be denoted ψ^*.

As an example, suppose the input u starts at $u = 0$ and increases monotonically to $u = 3$. Initially, ψ is the line $s = 0$ in $\mathcal{P}_r$. As u increases, relay outputs switch from -1 to $+1$ when $u = s + r$. Hence, $\mathcal{P}_+$ grows at the expense of $\mathcal{P}_-$, and the moving boundary defining this growth is the line $s = u - r$. In Figure 5a, the thick line represents ψ for some input value $0 < u < 3$; the arrow indicates that the sloped segment moves upwards as u increases. The line in Figure 5b shows the state of $\mathcal{P}_r$ when $u = 3$. Similarly, if the input reverses direction at $u = 3$ and decreases monotonically to $u = -1$, relays in $\mathcal{P}_+$ switch over to $\mathcal{P}_-$ when $u = s - r$. A new segment is generated on ψ, corresponding to the line $s = u + r$ (cf. Figure 5c). Subsequent input reversals generate further segments of ψ.

Note that the boundary ψ always intersects the axis $r = 0$ at the current input value. With ψ written as a map $\mathbf{R} \times \mathbf{R}_+ \mapsto \mathbf{R}$, then $\psi(t, 0) = u(t)$. If the boundary at time t is $\psi(t, r)$, applying an input for which $u(t) \neq \psi(t, 0)$ amounts to applying an input with a discontinuity at t. In this case, the output is not defined. This observation leads to the following definition.

Definition 2.1 (Admissible Inputs) *An input $u \in \mathcal{U}$ is said to be admissible to state ψ at time t if $u(t) = \psi(t, 0)$.*

The Preisach plane boundary is the memory of the Preisach model. When an arbitrary input is applied, monotonically increasing input segments gener-

---

[1]W_1^2 is the Sobolev space: the space of real-valued functions satisfying $\int_{-\infty}^{\infty}(\dot{u}^2 + u^2)dt < \infty$.

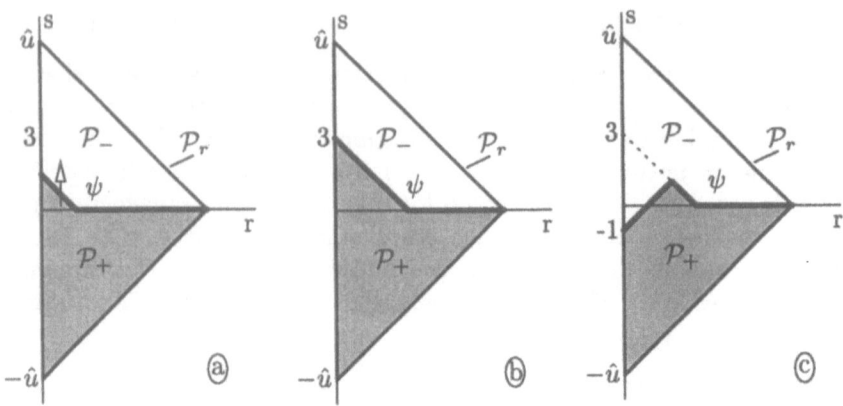

Figure 5: Preisach Boundary Behaviour

ate boundary segments of slope -1, while monotonically decreasing input segments generate boundary segments of slope +1. Input reversals cause corners
in the boundary. The history of past input reversals—and hence of hysteresis
branching behaviour—is stored in the corners of the boundary.

We have seen that input extrema generate the corners of the Preisach
boundary, and that this boundary represents the memory of the Preisach
model. The *wiping out property* states that some input extrema can remove
the effects of previous extrema, essentially "wiping out" the memory of the
model. That a system display this behaviour has been shown to be one of two
necessary and sufficient conditions for existence of a Preisach model. For more
on representation conditions, see [10]. The wiping out behaviour is sketched
in Figure 6, and explained as follows. Consider once more the input of Figure
5. Continuing from Figure 5c, suppose that u reverses and increases monotonically until it reaches $u = 5$. A segment of slope -1 sweeps upward through $\mathcal{P}_r$,
switching relays from $\mathcal{P}_-$ to $\mathcal{P}_+$ (cf. Figure 6a). As the input reaches $u = 3$ and
continues to increase, the two corners which had been generated by previous
reversals at $u = 3$ and $u = -1$ are eliminated (cf. Figure 6b). For $u > 3$, the
influence of those two previous input extrema has been completely removed
(cf. Figure 6c). So input extrema which exceed previous extrema in magnitude
can "wipe out" part of the memory. Those extrema which remain in memory
at any time form a *reduced memory sequence*, which will be discussed further
in Section 3.1.

3   State Space Representation

The previous section described the traditional input-output representation of
the Preisach model. Sector conditions, such as passivity, form a very important part of non-linear input-output stability theory. Unfortunately some

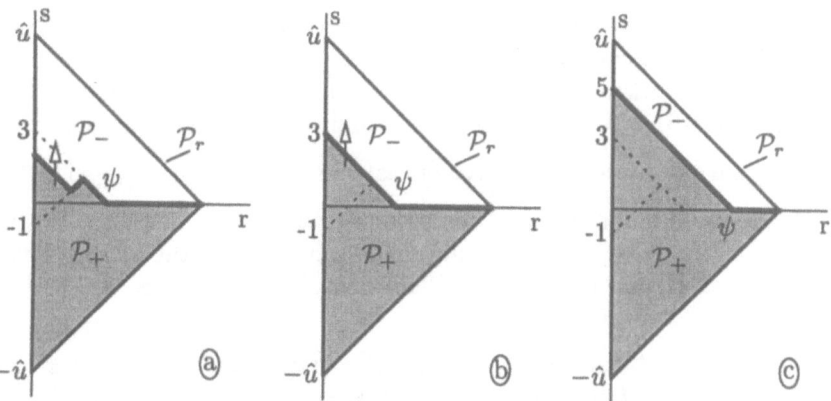

Figure 6: Wiping Out Property

hystereses, such as those found in magnetic materials, do not satisfy a sector condition. This section is concerned with developing a state-space representation for Preisach models for which $\mu \in \mathcal{M}_p$. By placing the model in a state-space framework, more general stability techniques such as Lyapunov and dissipativity theory may be applied.

The dynamical system framework used is that of [13], where a complete definition can be found. The system is defined through the input, output and state spaces $\mathcal{U}$, $\mathcal{Y}$ and $\mathcal{X}$, as well as the state transition operator ϕ and the read-out operator r. $\phi : \mathbf{R}^2 \times \mathcal{X} \times \mathcal{U} \mapsto \mathcal{X}$ must satisfy the standard axioms:

consistency: $\phi(t_o, t_o, x_o, u) = x_o$ for all $t_o \in \mathbf{R}$, $x_o \in \mathcal{X}$, $u \in \mathcal{U}$;

determinism: $\phi(t_1, t_o, x_o, u_1) = \phi(t_1, t_o, x_o, u_2)$ for all $t_o, t_1 \in \mathbf{R}$, $t_1 \geq t_o$, $x_o \in \mathcal{X}$, and all $u_1, u_2 \in \mathcal{U}$ satisfying $u_1(t) = u_2(t)$ for all $t_o \leq t \leq t_1$;

semi-group: $\phi(t_2, t_o, x_o, u) = \phi(t_2, t_1, \phi(t_1, t_o, x_o, u), u)$ for all $t_o \leq t_1 \leq t_2$, $x_o \in \mathcal{X}$, $u \in \mathcal{U}$;

stationarity: $\phi(t_1 + T, t_o + T, x_o, \sigma_T u) = \phi(t_1, t_o, x_o, u)$ for all $t_o \in \mathbf{R}$, $t_1 \geq t_o$, $T \in \mathbf{R}$, $x_o \in \mathcal{X}$, and $u \in \mathcal{U}$. σ_T is the shift operator: $\sigma_T u(t) = u(t + T)$.

The input space $\mathcal{U}$ is defined, for some system-dependent $\hat{u} > 0$, as

$$\mathcal{U} = \{u \in C^0(-\infty, \infty)| \, \|u\|_\infty \leq \hat{u} \text{ and } \lim_{t \to -\infty} u(t) = 0\}.$$

For any interval $[t_0, t_1]$ in $\mathbf{R}$, the notation $u_{[t_0, t_1]}$ denotes the restriction of u to $[t_0, t_1]$. The notation $\mathcal{U}[t_0, t_1]$ denotes the set obtained when every element of $\mathcal{U}$ is restricted to $[t_0, t_1]$.

The input restriction $\|u\|_\infty \leq \hat{u}$ allows the Preisach plane to be bounded, and arises naturally in systems where input signals are subject to saturation.

In Section 2, it was seen that in order to be admissible to state ψ at time t, an input must satisfy $u(t) = \psi(t, 0)$ (cf. Definition 2.1). Since the initial boundary ψ^* is assumed to be the line $s = 0$, this implies that $\lim_{t \to -\infty} u(t) = \lim_{t \to -\infty} \psi(t, 0) = 0$. Hence the second limitation on the input space.

The output space $\mathcal{Y}$ is the set of real-valued functions $C^0(-\infty, \infty)$.

Since the boundary ψ embodies the memory of the model, it is a natural choice for the state. The following definition captures the salient features of the boundaries, and fits Willems' definition of a state space.

Definition 3.1 (The State Space) *The state space $\mathcal{X}$ is defined to be the set of continuous functions $\psi : [0, \hat{u}] \mapsto \mathbf{R}$ which satisfy the following properties:*

(BP1) *Lipschitz condition:* $|\psi(r_1) - \psi(r_2)| \leq |r_1 - r_2|, \ \forall r_1, r_2 \in [0, \hat{u}];$

(BP2) *initial condition:* $\psi(\hat{u}) = 0.$

The Lipschitz property is more general than required, since boundaries may only be composed of segments of slope ± 1. However, including all functions with Lipschitz constant 1 leads to a complete state space, as will be seen shortly. The second property, along with (BP1), ensures that elements of $\mathcal{X}$ are within the triangle defined by $\mathcal{P}_r$.

Before proceeding to define the operators ϕ and r, we first examine some properties of the state space: boundedness, completeness and reachability. The distance between two states $\psi_1, \psi_2 \in \mathcal{X}$ is defined as the area between the boundaries,

$$d(\psi_1, \psi_2) = \int_0^{\hat{u}} |\psi_1(r) - \psi_2(r)| \, dr.$$

It is clear that $d(\cdot, \cdot)$ is a metric. From the definition of $\mathcal{X}$, $d(\psi_1, \psi_2) \leq \hat{u}^2$ for all $\psi_1, \psi_2 \in \mathcal{X}$, and the state space is bounded.

Definition 3.2 (Reachability) *We say that $\psi_o \in \mathcal{X}$ is reachable' if there exists finite T and $u \in \mathcal{U}(-\infty, T]$ so that $\psi(T) = \psi_o$.*

As the input to the Preisach model evolves, it generates boundaries composed only of alternating segments of slope ± 1, plus a segment on the line $s = 0$ if any memory of the initial condition remains. However, boundaries in $\mathcal{X}$ may be continuous curves. Inputs which decrease in amplitude and increase in frequency as they approach T may generate a countably infinite number of boundary segments, accumulating at the $r = 0$ axis in $\mathcal{P}_r$. For finite time T, however, there is no input in $\mathcal{U}(-\infty, T]$ which can generate a smooth curve over all of $[0, \hat{u}]$, so the entire state space $\mathcal{X}$ is not reachable.

Theorem 3.1 (Reachable Boundary) *A boundary $\psi \in \mathcal{X}$ is reachable from ψ^* if there exists $c \in [0, \hat{u}]$ such that ψ is composed of a finite or countably infinite number of segments of slope exactly ± 1 over $[0, c]$, and $\psi = 0$ over $[c, \hat{u}]$.*

Proof. The proof is clear from the boundary evolution rules described in the previous section. ∎

The set of all reachable boundaries will be denoted $\mathcal{X}_r$.

The following concept of *approximate reachability* is similar to that of approximate controllability for infinite-dimensional systems, proposed in [4].

Definition 3.3 (Approximate Reachability) *The state space $(\mathcal{X}, d)$ of a dynamical system is said to be "approximately reachable from x_o" if, for any $x \in \mathcal{X}$ and $\varepsilon > 0$, there exists a state $x_r \in \mathcal{X}$, reachable from x_o, such that $d(x_r, x) < \varepsilon$.*

Theorem 3.2 (Approximate Reachability) *The state space $\mathcal{X}$ is approximately reachable from the initial state ψ^*.*

Proof. We need to show that $\overline{\mathcal{X}_r} = \mathcal{X}$. Consider any $\psi \in \mathcal{X}$ and $\varepsilon > 0$. For each $n = 1, 2, \ldots$ let $\{r_i\}_{i=0\ldots n}$ be a uniform partition of $[0, \hat{u}]$, and define a second partition $\{R_i\}_{i=1\ldots n}$ by

$$R_i = \frac{\psi(r_i) - \psi(r_{i-1})}{2} + \frac{r_i + r_{i-1}}{2}.$$

Because ψ has Lipschitz constant 1, we have $r_{i-1} \leq R_i \leq r_i$. For every n (and every corresponding partition), define the function ψ_n by

$$\psi_n(r) = \begin{cases} \psi(r_{i-1}) + (r - r_{i-1}), & r_{i-1} \leq r \leq R_i \\ \psi(r_i) - (r - r_i). & R_i < r < r_i \end{cases}$$

By construction, every ψ_n is composed of alternating segments of slope ± 1, and satisfies $\psi_n(\hat{u}) = \psi(\hat{u}) = 0$, so $\{\psi_n\} \subset \mathcal{X}_r$.

Also, $\psi_n(r_i) = \psi(r_i)$ for $i = 0 \ldots n$. Since ψ and all ψ_n are uniformly continuous then for any $\varepsilon > 0$ there exists a finite N such that $d(\psi_n, \psi) < \varepsilon$ for all $n > N$, and $\overline{\mathcal{X}_r} = \mathcal{X}$. ∎

Theorem 3.3 (Completeness) *The metric space $(\mathcal{X}, d)$ is complete.*

Proof. First, we show that $\mathcal{X}$ forms an equicontinuous family of curves. Choose any $\varepsilon > 0$. For any $\psi \in \mathcal{X}$ and any $r_o \in [0, \hat{u}]$, if $|r - r_o| < \varepsilon$ then

$$\begin{aligned} |\psi(r) - \psi(r_o)| &\leq |r - r_o| \\ &< \varepsilon. \end{aligned}$$

Since ε is arbitrary, the family of curves $\mathcal{X}$ is equicontinuous.

Choose any Cauchy sequence $\{\psi_n\}$ in $\mathcal{X}$. It is required to show that there exists $\psi \in \mathcal{X}$ with $\psi_n \xrightarrow{d} \psi$. For every n, $\psi_n(\hat{u}) = 0$ and ψ is Lipschitz continuous, so for every point $r \in [0, \hat{u}]$,

$$
\begin{aligned}
|\psi_n(r)| &= |\psi_n(r) - \psi_n(\hat{u})| \\
&\leq |r - \hat{u}|,
\end{aligned}
$$

and the sequence $\{\psi_n\}$ is point-wise bounded over X. By the Arzela-Ascoli theorem, $\{\psi_n\}$ has a subsequence $\{\psi_{n_k}\}$ which converges both uniformly and point-wise to a continuous function ψ. This implies that

$$
\int_0^{\hat{u}} |\psi_{n_k}(r) - \psi(r)|\, dr \longrightarrow 0
$$

and so $\psi_{n_k} \xrightarrow{d} \psi$.

Also, d is simply the L_1 norm on $C^0[0, \hat{u}]$, and L_1 is the completion of C^0. Since $\{\psi_n\} \subset C^0[0, \hat{u}]$ is Cauchy, $\{\psi_n\}$ converges in d to some limit $\psi_o \in L_1[0, \hat{u}]$. Since the limit of a sequence which converges in a metric space is unique, $\psi_o = \psi$, and $\psi_n \xrightarrow{d} \psi$.

We now have convergence of the Cauchy sequence $\psi_n \xrightarrow{d} \psi$. To prove completeness, it remains to show that this limit is in $\mathcal{X}$. We know already that ψ is continuous. Choose any $r_1, r_2 \in [0, \hat{u}]$. Recall that $\{\psi_n(r)\}$ is a bounded sequence of real numbers. Then

$$
\begin{aligned}
|\psi(r_2) - \psi(r_1)| &= \left| \lim_{n \to \infty} \psi_n(r_2) - \lim_{n \to \infty} \psi_n(r_1) \right| \\
&= \left| \lim_{n \to \infty} (\psi_n(r_2) - \psi_n(r_1)) \right| \\
&= \lim_{n \to \infty} |(\psi_n(r_2) - \psi_n(r_1))| \\
&\leq \lim_{n \to \infty} |r_2 - r_1| \\
&= |r_2 - r_1|,
\end{aligned}
$$

so ψ is Lipschitz continuous with constant 1.

The point-wise convergence $\psi_{n_k} \longrightarrow \psi$ implies that $\psi(\hat{u}) = 0$, since $\psi_{n_k}(\hat{u}) = 0$ for every n_k. So ψ also satisfies (BP2), and $\psi \in \mathcal{X}$. Hence, $(\mathcal{X}, d)$ is complete. ∎

3.1   Reduced Memory Sequences

In this section, we introduce an intermediate space $\mathcal{S}$, which will be used in the construction of the state transition operator ϕ.

The wiping out property of the Preisach model was described in Section 2. In essence, any input maximum which exceeds previous maxima will wipe out the memory of those maxima; minima can be similarly "wiped out". At a given time t, only certain past extrema are retained and affect the output. They form an alternating set of input maxima and minima, in which each maximum is smaller in amplitude than the previous one, and each minimum is larger than the previous one. The two series converge to $u(t)$. This alternating sequence is known as the *reduced memory sequence*, and the following mathematical construction is based on that of [12]. A different but equivalent approach to memory sequences in hysteresis operators can be found in [3].

For any input $u \in \mathcal{U}(-\infty, T]$ and any $\tau \leq T$, set $s_0 = 0$ and $\eta = \max_{t \in (-\infty, \tau]} |u(t)|$. This is well-defined, since $\lim_{t \to -\infty} u(t) = 0$. Let $t_1 = \max\{t \in (-\infty, \tau] \mid |u(t)| = \eta\}$, and define the elements $s_i|_{i=1,2,\ldots}$ of the reduced memory sequence $s(u, \tau)$ as follows:

$$
\begin{aligned}
i = 1: \quad & s_1 = u(t_1), \\
s_{i-1} < s_{i-2}: \quad & s_i = \max_{t \in (t_{i-1}, \tau]} u(t), \quad \text{and} \quad t_i = \max\{t \in (t_{i-1}, \tau] \mid u(t) = s_i\}, \\
s_{i-1} > s_{i-2}: \quad & s_i = \min_{t \in (t_{i-1}, \tau]} u(t), \quad \text{and} \quad t_i = \max\{t \in (t_{i-1}, \tau] \mid u(t) = s_i\},
\end{aligned}
\tag{3}
$$

terminating the sequence if $t_i = \tau$.

Note that the values s_i are well-defined: by definition of s_{i-1} in (3), $u(t) > s_{i-1}$ (or $u(t) < s_{i-1}$) over $(t_{i-1}, \tau]$. Since u is continuous, the required maximum (or minimum) is well-defined. The times t_i are similarly well-defined, since the maximum is being taken over a non-empty set and τ is finite. The sequence $\{t_i\}$ is merely used to construct $\{s_i\}$ and then discarded: the time at which extrema occur is of no significance in the Preisach model due to its static nature.

If the input u has a finite number of extrema in $(-\infty, \tau]$, the above sequence has finite length N, $t_N = \tau$ and $u(t_N) = u(\tau)$. In this case, the tail of the sequence is formed by setting $s_i = s_N$ for $i > N$. If the sequence is infinite, then setting $t^* = \sup\{t_i\}$, the input u must be constant over $[t^*, \tau]$. Note that in both cases, $\lim_{i \to \infty} s_i = u(\tau)$.

This section will make use of the notation $N(s) = \sup\{i \mid s_{i-1} \neq s_i\}$. For any reduced memory sequence $s(u, \tau)$, this is the index beyond which s_i is constant and equal to $u(\tau)$. If u has a finite number of extrema in $(-\infty, \tau]$, then $N(s(u, \tau))$ is finite; otherwise, $N(s)$ may be infinite. Also, for any sequence $s = \{s_i\}$, let s^e and s^o denote the even and odd subsequences $s^e = \{s_i\}_{i=2,4,\ldots} = \{s_i^e\}$ and $s^o = \{s_i\}_{i=1,3,\ldots} = \{s_i^o\}$.

Definition 3.4 (Space of Reduced Memory Sequences) *The space of reduced memory sequences, $S \subset l_\infty$, is composed of all sequences s with $\|s\|_\infty \leq \hat{u}$, and for which the even subsequence s^e and odd subsequence s^o satisfy*

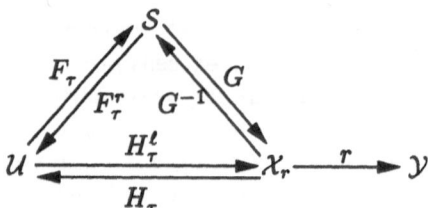

Figure 7: Relationship Between System Spaces

1. s^e *is strictly decreasing (strictly increasing) up to* $N(s)$ *and, if* $N(s) <$ ∞, *constant thereafter;*

2. s^o *is strictly increasing (strictly decreasing) up to* $N(s)$ *and, if* $N(s) <$ ∞, *constant thereafter;*

3. $\lim_{i \to \infty} s_i^e = \lim_{i \to \infty} s_i^o$.

Figure 7 shows the spaces defined thus far. We now define mappings between these spaces, and note some of their properties.

$F_\tau : \mathcal{U} \mapsto \mathcal{S}$
For any time $\tau < \infty$ and any input $u \in \mathcal{U}(-\infty, \tau]$, the reduced memory sequence $F_\tau u = s(u, \tau)$ is defined as in (3).

$F_\tau^r : \mathcal{S} \mapsto \mathcal{U}$
The reduced memory sequence $s(u, \tau)$ captures only information regarding dominant extrema of $u_{(-\infty, \tau]}$. There are therefore an infinite number of inputs $u_i \neq u$ which are equivalent, in the sense that they have the same reduced memory sequence: $F_\tau u_i = F_\tau u$. Hence, no inverse of F_τ exists. A right-inverse $F_\tau^r : \mathcal{S} \mapsto \mathcal{U}(-\infty, \tau]$ is defined below.

For any $s(\cdot, \tau) \in \mathcal{S}$, it is required to construct an input $u \in \mathcal{U}(-\infty, \tau]$ with extrema equal to the elements of s and satisfying $u(\tau) = \lim_{i \to \infty} s_i$. Choose any $t_0 < \tau$, and let $\{t_i\}$ be a partition of $[t_0, \tau]$ defined for all $i \geq 1$ by

$$t_i = t_o + \frac{\tau - t_0}{2} \sum_{k=0}^{i-1} \frac{1}{2^k}.$$

Note that $\lim_{i \to \infty} t_i = \tau$. Set $s_0 = 0$ and define $u(t)$ on $(-\infty, \tau]$ by straight-line interpolation between the points (t_i, s_i):

$$u(t) = \begin{cases} 0 & t \leq t_0, \\ s_i & t = t_i, \\ (t - t_{i-1})\frac{s_i - s_{i-1}}{t_i - t_{i-1}} + s_{i-1} & t_{i-1} < t < t_i. \end{cases}$$

The resulting output $u \in \mathcal{U}(-\infty, \tau]$ has extrema corresponding to elements of $s(\cdot, \tau)$, and

$$u(\tau) = u(\lim_{i \to \infty} t_i) = \lim_{i \to \infty} s_i. \tag{4}$$

We now give some properties of these operators. The concatenation operator $\diamond : C^0 \times C^0 \mapsto C^0$ is defined by

$$u_{(t_o, t_1]} \diamond v_{(t_1, t_2]} = \begin{cases} u, & t_o < t \le t_1 \\ v, & t_1 < t \le t_2. \end{cases}$$

Lemma 3.1 *The operators F_τ, F_τ^r satisfy the following properties, for any $\tau < \infty$.*

1. *F_τ^r is a right-inverse of F_τ: for any $s \in \mathcal{S}$, $F_\tau F_\tau^r s(\cdot, \tau) = s(\cdot, \tau)$.*

2. *F_τ is deterministic; that is, for any $\tau < \infty$ and $u_1, u_2 \in \mathcal{U}(-\infty, \tau]$ satisfying $u_1(t) = u_2(t)$ over $(-\infty, \tau]$, then $F_\tau u_1 = F_\tau u_2$.*

3. *For any $u \in \mathcal{U}(-\infty, T]$ the composition $F_\tau^r F_\tau$ preserves the continuity of the input for every $\tau < T$; that is,*

$$\left(F_\tau^r F_\tau u_{(-\infty, \tau]} \right) \diamond u_{(\tau, T]} \in \mathcal{U}(-\infty, T].$$

4. *For any $T > \tau$ and $u \in \mathcal{U}(-\infty, T]$, the operators satisfy the identity*

$$F_T \left[\left(F_\tau^r F_\tau u_{(-\infty, \tau]} \right) \diamond u_{(\tau, T]} \right] = F_T u. \tag{5}$$

Proof. The first two properties follow from the definitions of F_τ and F_τ^r. To prove the other two, choose any $u \in \mathcal{U}(-\infty, T]$, $\tau < T$. Let the reduced memory sequence at τ be $s(u, \tau) = F_\tau u$, and the input which is reconstructed from $s(u, \tau)$ be $\tilde{u} = F_\tau^r s$. From (4), $\tilde{u}(\tau) = \lim_{i \to \infty} s_i$. But from the definition of F_τ, $\lim_{i \to \infty} s_i = u(\tau)$. So $\tilde{u}(\tau) = u(\tau)$ and $F_\tau^r F_\tau u \diamond u_{(\tau, T]} \in \mathcal{U}(-\infty, T]$, proving Property 3.

By construction, $\tilde{u}$ above contains all of the same extremum information as $u_{(-\infty, \tau]}$. Therefore, $\tilde{u} \diamond u_{(\tau, T]}$ has extrema identical to those of $u_{(-\infty, \tau]}$. Since $\tilde{u} \diamond u_{(\tau, T]} \in \mathcal{U}(-\infty, T]$, $F_T(\tilde{u} \diamond u_{(\tau, T]})$ is defined and the identity (5) holds. ∎

$G : \mathcal{S} \mapsto \mathcal{X}_\tau$

Any reduced memory sequence $s(u, \tau) \in \mathcal{S}$, defines a corresponding boundary $\psi = G(s)$. The elements of s correspond to the corners of the boundary curve, as defined below. For all $i < \infty$, define the set of points $p_i \in \mathbf{R}^2$

$$\begin{aligned} p_0 &= (\hat{u}, 0) \\ p_1 &= (|s_1|, 0) \\ p_i &= \begin{cases} \left(\frac{s_{i-1} - s_i}{2}, \frac{s_{i-1} + s_i}{2} \right), & s_i < s_{i-1} \\ \left(\frac{s_i - s_{i-1}}{2}, \frac{s_i + s_{i-1}}{2} \right), & s_i > s_{i-1} \end{cases} \end{aligned} \tag{6}$$

and $G(s)$ to be the linear interpolate between the points p_i. Note that if $N(s) < \infty$ then for all $i > N(s)$, $s_i = s_{i-1} = s_{N(s)}$ and $p_i = (0, s_{N(s)}) = (0, u(\tau))$. If $N(s)$ is infinite, then the boundary $G(s)$ has an infinite number of corners p_i. In this case, since $\lim_{i \to \infty} s_i^e = \lim_{i \to \infty} s_i^o = u(\tau)$, then $\lim_{i \to \infty} p_i = (0, u(\tau))$. In both cases, the result is as expected for the Preisach model: the boundary at time τ intersects the axis $r = 0$ at the point $(0, u(\tau))$: $\psi(\tau, 0) = u(\tau)$.

Note that the range of G, $Ra(G)$, is the set of all curves $\psi \in \mathcal{X}$ which have a finite or countably infinite number of alternating segments of slope ± 1. This set is not quite $\mathcal{X}$, since $\mathcal{X}$ also contains continuous curves. However, $Ra(G) = \mathcal{X}_r$.

$G^{-1} : \mathcal{X}_r \mapsto \mathcal{S}$

For every sequence $s \in \mathcal{S}$, the boundary $G(s)$ is unique, by definition of G. Since $Ra(G) = \mathcal{X}_r$, the inverse mapping $G^{-1} : \mathcal{X}_r \mapsto \mathcal{S}$ exists. The construction of a sequence $s \in \mathcal{S}$ from any boundary in $\psi \in \mathcal{X}_r$ is done by extracting the coordinates of the corners and calculating the corresponding extrema from (6). If the number of corners N is finite, the reduced memory sequence is completed by setting the tail to $s_i = s_N$ for all $i > N$.

$H_\tau^\ell : \mathcal{U} \mapsto \mathcal{X}_r$
$H_\tau : \mathcal{X}_r \mapsto \mathcal{U}$

The mappings $H_\tau^\ell : \mathcal{U} \mapsto \mathcal{X}_r$ and $H_\tau : \mathcal{X}_r \mapsto \mathcal{U}$ are defined as the compositions $H_\tau = F_\tau^r G^{-1}$, $H_\tau^\ell = G F_\tau$.

Lemma 3.2 *The operators H_τ, H_τ^ℓ satisfy the following properties, for any $\tau < \infty$.*

1. *H_τ^ℓ is a left-inverse of H_τ: for any $\psi \in \mathcal{X}_r$, $H_\tau^\ell H_\tau \psi = \psi$.*

2. *H_τ^ℓ is deterministic; that is, for any $\tau < \infty$ and $u_1, u_2 \in \mathcal{U}(-\infty, \tau]$ satisfying $u_1(t) = u_2(t)$ over $(-\infty, \tau]$, then $H_\tau^\ell u_1 = H_\tau^\ell u_2$.*

3. *For any $T > \tau$ and $u \in \mathcal{U}(-\infty, T]$, the operators satisfy the identity*

$$H_T^\ell \left[(H_\tau H_\tau^\ell u_{(-\infty, \tau]}) \Diamond u_{(\tau, T]} \right] = H_T^\ell u. \tag{7}$$

Proof. The proof follows from the definitions of H_τ and H_τ^ℓ, and Lemma 3.1. ∎

3.2 State Transition and Read-Out Operators

The state transition operator ϕ determines the state $\psi = \phi(t_1, t_o, \psi_o, u)$ which results at time t_1 from applying an input $u \in \mathcal{U}[t_o, t_1]$ to a system starting in state ψ_o at time t_o. For this operation to be well-posed, the state ψ_o must be reachable and u must be admissible to ψ_o at t_o; that is, $u(t_o) = \psi(t_o, 0)$ (cf. Definition 2.1).

The state transition operator ϕ is defined using the mappings introduced in the previous section. Given some interval $[t_0, t_1]$, some initial state $\psi_o \in \mathcal{X}_r$, and some input $u \in \mathcal{U}[t_0, t_1]$ admissible to ψ_o at t_o, $\phi(t_1, t_o, \psi_o, u)$ is defined as follows:

1. determine the memory sequence for the initial state: $s(\cdot, t_0) = G^{-1}\psi_o$,

2. construct an input $u_o \in \mathcal{U}(-\infty, t_0]$ which generates $s(\cdot, t_0)$: $u_o = F_{t_0}^r s(\cdot, t_0)$,

3. concatenate the inputs u_o and u to form $\tilde{u} = u_o \Diamond u \in \mathcal{U}(-\infty, t_1]$,

4. determine the corresponding boundary ψ_1 at time t_1: $\psi_1 = GF_{t_1}\tilde{u}$.

Recalling that $H_r = F_r^r G^{-1}$ and $H_r^\ell = GF_r$, the state transition function ϕ is given by

$$\phi(t_1, t_o, \psi_o, u) = H_{t_1}^\ell \left[H_{t_o} \psi_o \Diamond u_{(t_0, t_1]} \right]. \tag{8}$$

We now prove that the state transition operator as defined in (8) satisfies the four required axioms.

consistency: Choose any $t_o \in \mathbf{R}$, $\psi_o \in \mathcal{X}_r$, and admissible $u \in \mathcal{U}$. Then

$$\begin{aligned}
\phi(t_o, t_o, \psi_o, u) &= H_{t_o}^\ell \left[H_{t_o} \psi_o \Diamond u_{(t_o, t_o]} \right] \\
&= H_{t_o}^\ell \left[H_{t_o} \psi_o \right] \\
&= \psi_o.
\end{aligned}$$

determinism: Choose any $t_1 \geq t_o$, $\psi_o \in \mathcal{X}_r$, and admissible $u_1, u_2 \in \mathcal{U}$ such that $u_1(t) = u_2(t)$ over $[t_o, t_1]$. Let $u_o = H_{t_o}\psi_o$. Then

$$\begin{aligned}
\phi(t_1, t_o, \psi_o, u_1) &= H_{t_1}^\ell \left(u_o \Diamond u_{1(t_o, t_1]} \right), \text{ and} \\
\phi(t_1, t_o, \psi_o, u_2) &= H_{t_1}^\ell \left(u_o \Diamond u_{2(t_o, t_1]} \right).
\end{aligned}$$

But $u_{1(t_o, t_1]} = u_{2(t_o, t_1]}$, and from Lemma 3.2 H_r^ℓ is deterministic, so we have $\phi(t_1, t_o, \psi_o, u_1) = \phi(t_1, t_o, \psi_o, u_2)$.

semi-group: Choose any $t_o \in \mathbf{R}$, $t_2 \geq t_1 \geq t_o$, $\psi_o \in \mathcal{X}_r$ and admissible $u \in \mathcal{U}$. Let $u_o = H_{t_o}\psi_o$, $\tilde{u} = u_o \Diamond u_{(t_o, t_2]}$, and note that $u_o \Diamond u_{(t_o, t_1]} = \tilde{u}_{(-\infty, t_1]}$ and $u_{(t_1, t_2]} = \tilde{u}_{(t_1, t_2]}$. Then

$$\begin{aligned}
\phi(t_2, t_1, \phi(t_1, t_o, \psi_o, u), u) &= H_{t_2}^\ell \left[H_{t_1} \left\{ \phi(t_1, t_o, \psi_o, u) \right\} \Diamond u_{(t_1, t_2]} \right] \\
&= H_{t_2}^\ell \left[H_{t_1} \left\{ H_{t_1}^\ell \left[H_{t_o} \psi_o \Diamond u_{(t_o, t_1]} \right] \right\} \Diamond u_{(t_1, t_2]} \right] \\
&= H_{t_2}^\ell \left[H_{t_1} \left\{ H_{t_1}^\ell \left[u_o \Diamond u_{(t_o, t_1]} \right] \right\} \Diamond u_{(t_1, t_2]} \right] \\
&= H_{t_2}^\ell \left[H_{t_1} H_{t_1}^\ell \tilde{u}_{(-\infty, t_1]} \Diamond \tilde{u}_{(t_1, t_2]} \right] \\
&= H_{t_2}^\ell \tilde{u}
\end{aligned}$$

by Property 3 of Lemma 3.2. But

$$
\begin{aligned}
\phi(t_2, t_o, \psi_o, u) &= H_{t_2}^\ell \left[H_{t_o} \psi_o \Diamond u_{(t_o, t_2]} \right] \\
&= H_{t_2}^\ell \left[u_o \Diamond u_{(t_o, t_2]} \right] \\
&= H_{t_2}^\ell \tilde{u},
\end{aligned}
$$

and so ϕ satisfies the semi-group property.

stationarity: Choose any $t_o \in \mathbf{R}$, $t_1 \geq t_o$, $\psi_o \in \mathcal{X}_r$, admissible $u \in \mathcal{U}$ and $T < \infty$. It is required to show that

$$
H_{t_1}^\ell \left[H_{t_o} \psi_o \Diamond u_{(t_o, t_1]} \right] = H_{t_1 + T}^\ell \left[H_{t_o + T} \psi_o \Diamond \sigma_T u_{(t_o, t_1]} \right]. \tag{9}
$$

Recall that σ_T is the shift operator. Set

$$
\begin{aligned}
u_1 &= H_{t_o} \psi_o \in \mathcal{U}(-\infty, t_o], \\
u_2 &= H_{t_o + T} \psi_o \in \mathcal{U}(-\infty, t_o + T].
\end{aligned}
$$

Then

$$
s(u_1, t_o) = F_{t_o} u_1 = F_{t_o} F_{t_o}^r G^{-1} \psi_o = G^{-1} \psi_o,
$$

and similarly, $s(u_2, t_o + T) = G^{-1} \psi_o$. Now define

$$
\begin{aligned}
\tilde{u}_1 &= u_1 \Diamond u_{(t_o, t_1]}, \\
\tilde{u}_2 &= u_2 \Diamond \sigma_T u_{(t_o, t_1]}.
\end{aligned}
$$

Using these definitions in (9), it must be shown that $H_{t_1}^\ell \tilde{u}_1 = H_{t_1 + T}^\ell \tilde{u}_2$. Since $s(u_1, t_o) = s(u_2, t_o + T)$, then $s(\tilde{u}_1, t_1) = F_{t_1} \tilde{u}_1$ is the same reduced memory sequence as $s(\tilde{u}_2, t_1 + T) = F_{t_1 + T} \tilde{u}_2$. Thus

$$
\begin{aligned}
H_{t_1}^\ell \tilde{u}_1 &= G F_{t_1} \tilde{u}_1 \\
&= G F_{t_1 + T} \tilde{u}_2 \\
&= H_{t_1 + T}^\ell \tilde{u}_2,
\end{aligned}
$$

and ϕ is stationary.

The read-out function r gives the system output which corresponds to a particular state ψ. Recall that the Preisach model output is (1):

$$
y(t) = \int_0^\infty \int_{-\infty}^\infty \mu(r, s) \gamma_{r,s}[u](t) \, ds \, dr.
$$

Since relays below ψ have output $+1$ and relays above, -1, r can be defined as a function of ψ:

$$
y(t) = r(\psi(t)) = \int_0^\infty \int_{-\infty}^{\psi(t)} \mu(r, s) \, ds \, dr - \int_0^\infty \int_{\psi(t)}^\infty \mu(r, s) \, ds \, dr.
$$

4 Dissipativity of the Preisach Model

In his pioneering work on dissipative dynamical systems[13], Willems shows that the major input-output stability results can all be cast as special cases of dissipativity theory. Dissipativity is defined in terms of the relationship between two functions known as the *supply rate* and the *storage function*.

Definition 4.1 (Dissipativity [13]) *A dynamical system is said to be dissipative with respect to the (locally integrable) supply rate* $w : U \times Y \mapsto \mathbf{R}$ *if there exists a non-negative function* $S : \mathcal{X} \mapsto \mathbf{R}^+$, *called the* storage function, *such that for all* $t_1 \geq t_o$, $x_o \in \mathcal{X}$, *and* $u \in \mathcal{U}$,

$$S(x_o) + \int_{t_o}^{t_1} w(u(t), y(t))dt \geq S(\phi(t_1, t_o, x_o, u)). \qquad (10)$$

In this section, it is shown that the Preisach model is "dissipative" in a more general sense, since the supply rate will include the derivative of the output. This type of supply rate has been investigated in [11].

Essentially, for a system to be dissipative, the sum of the storage in the initial state and the supply generated by the input must not be less than the storage in the final state. In other words, there is no internal generation of storage. The word "energy" is conspicuously absent from this description: while dissipativity theory is based on energy concepts, the supply rate and storage function are generalizations of the physical concepts of "rate of energy supply" and "amount of stored energy". There need not be any physical energy interpretation in order for the definition or related results to hold.

4.1 Energy Storage in the Preisach Model

In general, storage functions for physical systems are not unique. However, it is often the case that the actual energy stored in a system is a storage function with some related supply rate.

In [5], a formula was derived for Q, the energy stored in the Preisach model, by looking at energy transfer and storage in individual relays. It was assumed that the input and output variables were related such that the units of the product $u\dot{y}$ are power. This is often the case in actuators, where u is some form of electrical or mechanical force, and y is displacement. It was shown that the energy transfer to a relay in a switch from -1 to +1 was $q_+ = 2\mu(r, s)(s+r)$; the energy transfer in switching from +1 to -1 is $q_- = -2\mu(r, s)(s - r)$. Note that the energy transferred to any relay during a complete cycle is $q_+ + q_- = 4r\mu(r, s)$, which is the area of the weighted relay. This agrees with the well-known result that the hysteretic energy loss in each cycle of a magnetic circuit is equal to the area of the hysteresis loop[1, e.g.].

Defining the regions

$$
\begin{aligned}
\mathcal{Q}_1 &= \{(r,s) \in \mathcal{P}_r | \, s > r\} \\
\mathcal{Q}_2 &= \{(r,s) \in \mathcal{P}_r | \, |s| \le r\} \\
\mathcal{Q}_3 &= \{(r,s) \in \mathcal{P}_r | \, s < -r\},
\end{aligned}
$$

we observe that if $\mu(r,s) \ge 0$, $q_+ \le 0$ for relays in $\mathcal{Q}_3$ and $q_- \le 0$ for relays in $\mathcal{Q}_1$. Energy transfer is positive for all other switches. Negative energy transfer represents energy being recovered from the system: relays whose next switch will result in negative energy transfer are *storing energy*. The formula for total stored energy is

$$
Q(\psi(t)) = 2 \int_{\mathcal{Q}_1 \cap \mathcal{P}_+(t)} \mu(r,s)(s-r)dsdr - 2 \int_{\mathcal{Q}_3 \cap \mathcal{P}_-(t)} \mu(r,s)(s+r)dsdr. \quad (11)
$$

Recall from their definition in equation (2) that the regions $\mathcal{P}_+(t)$ and $\mathcal{P}_-(t)$ are entirely defined by the boundary $\psi(t)$. If $\mu \in \mathcal{M}_p$ then $Q(\psi) \ge 0$, since $s > r$ in $\mathcal{Q}_1$ and $s < -r$ in $\mathcal{Q}_3$.

Proposition 4.1 (Minimum Energy) *If $\mu \in \mathcal{M}_p$, then whenever $u(t) = 0$, the Preisach model is in a state of minimum stored energy.*

Proof. If $u(t) = 0$, then $\psi(t,0) = 0$. Since boundaries have Lipschitz constant 1, ψ must be entirely contained in the region $\mathcal{Q}_2$. Thus $\mathcal{Q}_1 \cap \mathcal{P}_+ = \emptyset$ and similarly, $\mathcal{Q}_3 \cap \mathcal{P}_- = \emptyset$. So from (11), $Q(\psi) = 0$ since the areas of integration are empty. ∎

It will be shown that the Preisach model satisfies the dissipation inequality (10) with the generalized supply rate $w = u\dot{y}$. We do this by demonstrating that the recoverable stored energy Q is a storage function for this supply rate. For a more abstract approach to the topic of dissipation in hysteresis operators, see the work on *hysteresis potentials* in [3].

Theorem 4.1 *If $\mu \in \mathcal{M}_p$, the Preisach model satisfies the generalized dissipation inequality*

$$
Q(\psi_o) + \int_{t_o}^{t_1} u\dot{y}dt \ge Q(\psi_1) \quad (12)
$$

for any $\psi_o \in \mathcal{X}$, $t_1 \ge t_o$ and $u \in \mathcal{U}[t_o,t_1]$ such that $\phi(t_1,t_o,\psi_o,u) = \psi_1$.

Proof. Recall that if $\mu \in \mathcal{M}_p$ then $Q(\psi) \ge 0$. Also, μ and $\mathcal{P}_r$ are bounded, so $Q(\psi) < \infty$ for all ψ, and $Q : \mathcal{X} \mapsto \mathbf{R}^+$. It remains to show that for any initial state ψ_o and $u \in \mathcal{U}[t_o,t_1]$ such that $\psi_1 = \phi(t_1,t_o,\psi_o,u)$, the inequality (12) is satisfied.

The remainder of the proof is outlined below; for a detailed proof, see [5]. The idea is to consider an arbitrary input u and write out the expression for the energy transferred by u to the system on a relay by relay basis. This gives the expression

$$\int_{t_o}^{t_1} u\dot{y}dt = \int_{\mathcal{P}_r} n_+(r,s)q_+dsdr + \int_{\mathcal{P}_r} n_-(r,s)q_-dsdr$$

where $n_+(r,s)$ is the number of times the relay $\gamma_{r,s}$ is switched from -1 to +1, and $n_-(r,s)$ is the number of times it is switched from +1 to -1. We then remove from the right hand side quantities which are positive: energy transferred when a relay is fully cycled; positive energy transfer during partial cycling, depending on the region of $\mathcal{P}_r$. After a non-trivial step in regrouping terms on the right-hand side, this results in the inequality

$$\int_{t_o}^{t_1} u\dot{y}dt$$

$$\geq 2\int_{\mathcal{Q}_1 \cap \mathcal{P}_+(t_1)} \mu(r,s)(s-r)dsdr - 2\int_{\mathcal{Q}_3 \cap \mathcal{P}_-(t_1)} \mu(r,s)(s+r)dsdr$$

$$-2\int_{\mathcal{Q}_1 \cap \mathcal{P}_+(t_o)} \mu(r,s)(s-r)dsdr + 2\int_{\mathcal{Q}_3 \cap \mathcal{P}_-(t_o)} \mu(r,s)(s+r)dsdr$$

$$= Q(\psi(t_1)) - Q(\psi(t_o)),$$

which shows that the dissipation inequality (12) is satisfied. ∎

The Passivity Theorem allows the determination of classes of feedback stabilizing controllers for systems which are *passive*.

Definition 4.2 (Passivity) *An operator G is said to be passive if there exists $\delta \geq 0$ such that, for all $T < \infty$ and $u \in Do(G)$,*

$$\int_{-\infty}^{T} u \cdot Gu \, dt \geq \delta \int_{-\infty}^{T} u^2 dt. \tag{13}$$

G is said to be strictly passive if equation (13) holds for $\delta > 0$.

If the product $u \cdot Gu$ has units of power, this says that the net energy input over time is positive for every possible input. In other words, the system does not generate any power internally. The underlying assumption in the above definition is that the system starts in a state of minimum stored energy. Otherwise, an input could be generated which recovers any stored energy, causing $\int u \cdot Gu \, dt$ to be negative.

The definition of passivity (with $\delta = 0$) is a special case of dissipativity, in which the supply rate is $w(u,y) = uy$ and the storage function is zero. Like dissipativity, the theory is motivated by the study of energy storage, where the input-output pair represents instantaneous power. Again, the theory and related results continue to hold when such an interpretation is not available.

Corollary 4.1 (Preisach Model Passivity) *If $\mu \in \mathcal{M}_p$, the composite operator $\frac{d}{dt}\Gamma : W_1^2 \mapsto L_2$ is passive.*

Proof. The assumption on μ guarantees that the Preisach model satisfies the inequality (12). Suppose the system starts in a state of minimum energy storage, i.e. $u(t_o) = 0$. Then $Q(\psi_o) = 0$ (cf. Proposition 4.1) and

$$\int_{t_o}^{t_1} u\dot{y}dt \geq Q(\phi(t_1, t_o, \psi_o, u)) \geq 0.$$

But since $\lim_{t \to -\infty} u(t) = 0$, then for any finite T and $u \in Do(\Gamma)$, we have

$$\int_{-\infty}^{T} u\dot{y}dt \geq 0,$$

which completes the proof.   ∎

This passivity result has been reported in [7], where a Passivity Theorem was applied to obtain a stability result for rate feedback of Preisach models for which $\mu \in \mathcal{M}_p$. The derivation presented here takes a more general approach, emphasizing that passivity is simply a special case of dissipativity.

5 Conclusion

We have shown that the Preisach hysteresis operator can be placed in the standard dynamical system framework. Although the system has memory, it is entirely contained in the state, allowing treatment of these non-linearities in the more familiar state-space setting. What distinguishes these static hysteretic systems from more common dynamical systems is that the time scale is irrelevant. This framework is used to show a generalized form of dissipativity, and the passivity of the relationship from input to output derivative. Current research goals involve the extension of the rate feedback result of [7] to position feedback. Research directions specific to SMA actuators include the incorporation of time-varying actuator stresses in the dissipativity and stability results.

References

[1] J.C. Anderson, *Magnetism and Magnetic Materials*, Chapman and Hall, 1968.

[2] H.T. Banks, A.J. Kurdilla and G. Webb, "Identification of Hysteretic Control Influence Operators Representing Smart Actuators: Convergent Approximations", *CRSC-TR97-7*, Center for Research in Scientific Computation, NCSU, 1997.

[3] M. Brokate and J. Sprekels, *Hysteresis and Phase Transitions*, Springer, 1996.

[4] R.F. Curtain and A.J. Pritchard, *Infinite Dimensional Linear Systems Theory*, Springer-Verlag, 1978.

[5] R.B. Gorbet, *Control of Hysteretic Systems with Preisach Representations*, Ph.D. Thesis, University of Waterloo, Waterloo, Canada, 1997.

[6] R.B. Gorbet, D.W.L. Wang and K.A. Morris, "Preisach Model Identification of a Two-Wire SMA Actuator", *IEEE ICRA 1998*, submitted, October, 1997.

[7] R.B. Gorbet, K.A. Morris and D.W.L. Wang, "Stability of Control Systems for the Preisach Hysteresis Model", *Journal of Engineering Design and Automation*, to appear.

[8] M. Hashimoto, M. Takeda, H. Sagawa, I. Chiba and K. Sato, "Application of shape memory alloy to robotic actuators", *Journal of Robotic Systems*, Vol. 2, No. 1, pp. 3-25, 1985.

[9] D. Hughes and J.T. Wen, "Preisach modeling of piezoceramic and shape memory alloy hysteresis", *1995 IEEE Control Conference on Applications*, Albany, New York, September 1995.

[10] I.D. Mayergoyz, *Mathematical Models of Hysteresis*, Springer-Verlag, 1991.

[11] K.A. Morris and J.N. Juang, "Dissipative Controller Designs for Second-Order Dynamic Systems", *IEEE Transactions on Automatic Control*, Vol. 39, No. 5, pp. 1056-1063, 1994.

[12] A. Visintin, *Differential Models of Hysteresis*, Springer-Verlag, 1994.

[13] J.C. Willems, "Dissipative Dynamical Systems, Part I: General Theory", *Archives for Rational Mechanics and Analysis*, Vol. 45, pp. 321-351, 1972.

[14] S. Yokota, K. Yoshida, K. Bandoh and M. Suhara, "Response of proportional valve using shape-memory-alloy array actuators", *13th IFAC World Congress*, pp. 505-510, IFAC, 1996.

Lecture Notes in Control and Information Sciences

Edited by M. Thoma

1993–1998 Published Titles:

Vol. 186: Sreenath, N.
Systems Representation of Global Climate
Change Models. Foundation for a Systems
Science Approach.
288 pp. 1993 [3-540-19824-5]

Vol. 187: Morecki, A.; Bianchi, G.;
Jaworeck, K. (Eds)
RoManSy 9: Proceedings of the Ninth
CISM-IFToMM Symposium on Theory and
Practice of Robots and Manipulators.
476 pp. 1993 [3-540-19834-2]

Vol. 188: Naidu, D. Subbaram
Aeroassisted Orbital Transfer: Guidance
and Control Strategies
192 pp. 1993 [3-540-19819-9]

Vol. 189: Ilchmann, A.
Non-Identifier-Based High-Gain Adaptive
Control
220 pp. 1993 [3-540-19845-8]

Vol. 190: Chatila, R.; Hirzinger, G. (Eds)
Experimental Robotics II: The 2nd
International Symposium, Toulouse,
France, June 25-27 1991
580 pp. 1993 [3-540-19851-2]

Vol. 191: Blondel, V.
Simultaneous Stabilization of Linear
Systems
212 pp. 1993 [3-540-19862-8]

Vol. 192: Smith, R.S.; Dahleh, M. (Eds)
The Modeling of Uncertainty in Control
Systems
412 pp. 1993 [3-540-19870-9]

Vol. 193: Zinober, A.S.I. (Ed.)
Variable Structure and Lyapunov Control
428 pp. 1993 [3-540-19869-5]

Vol. 194: Cao, Xi-Ren
Realization Probabilities: The Dynamics of
Queuing Systems
336 pp. 1993 [3-540-19872-5]

Vol. 195: Liu, D.; Michel, A.N.
Dynamical Systems with Saturation
Nonlinearities: Analysis and Design
212 pp. 1994 [3-540-19888-1]

Vol. 196: Battilotti, S.
Noninteracting Control with Stability for
Nonlinear Systems
196 pp. 1994 [3-540-19891-1]

Vol. 197: Henry, J.; Yvon, J.P. (Eds)
System Modelling and Optimization
975 pp approx. 1994 [3-540-19893-8]

Vol. 198: Winter, H.; Nüßer, H.-G. (Eds)
Advanced Technologies for Air Traffic Flow
Management
225 pp approx. 1994 [3-540-19895-4]

Vol. 199: Cohen, G.; Quadrat, J.-P. (Eds)
11th International Conference on
Analysis and Optimization of Systems –
Discrete Event Systems: Sophia-Antipolis,
June 15–16–17, 1994
648 pp. 1994 [3-540-19896-2]

Vol. 200: Yoshikawa, T.; Miyazaki, F. (Eds)
Experimental Robotics III: The 3rd
International Symposium, Kyoto, Japan,
October 28-30, 1993
624 pp. 1994 [3-540-19905-5]

Vol. 201: Kogan, J.
Robust Stability and Convexity
192 pp. 1994 [3-540-19919-5]

Vol. 202: Francis, B.A.; Tannenbaum, A.R.
(Eds)
Feedback Control, Nonlinear Systems,
and Complexity
288 pp. 1995 [3-540-19943-8]

Vol. 203: Popkov, Y.S.
Macrosystems Theory and its Applications:
Equilibrium Models
344 pp. 1995 [3-540-19955-1]

Vol. 204: Takahashi, S.; Takahara, Y.
Logical Approach to Systems Theory
192 pp. 1995 [3-540-19956-X]

Vol. 205: Kotta, U.
Inversion Method in the Discrete-time
Nonlinear Control Systems Synthesis
Problems
168 pp. 1995 [3-540-19966-7]

Vol. 206: Aganovic, Z.; Gajic, Z.
Linear Optimal Control of Bilinear Systems
with Applications to Singular Perturbations
and Weak Coupling
133 pp. 1995 [3-540-19976-4]

Vol. 207: Gabasov, R.; Kirillova, F.M.;
Prischepova, S.V.
Optimal Feedback Control
224 pp. 1995 [3-540-19991-8]

Vol. 208: Khalil, H.K.; Chow, J.H.;
Ioannou, P.A. (Eds)
Proceedings of Workshop on Advances
inControl and its Applications
300 pp. 1995 [3-540-19993-4]

Vol. 209: Foias, C.; Özbay, H.;
Tannenbaum, A.
Robust Control of Infinite Dimensional
Systems: Frequency Domain Methods
230 pp. 1995 [3-540-19994-2]

Vol. 210: De Wilde, P.
Neural Network Models: An Analysis
164 pp. 1996 [3-540-19995-0]

Vol. 211: Gawronski, W.
Balanced Control of Flexible Structures
280 pp. 1996 [3-540-76017-2]

Vol. 212: Sanchez, A.
Formal Specification and Synthesis of
Procedural Controllers for Process Systems
248 pp. 1996 [3-540-76021-0]

Vol. 213: Patra, A.; Rao, G.P.
General Hybrid Orthogonal Functions and
their Applications in Systems and Control
144 pp. 1996 [3-540-76039-3]

Vol. 214: Yin, G.; Zhang, Q. (Eds)
Recent Advances in Control and Optimization
of Manufacturing Systems
240 pp. 1996 [3-540-76055-5]

Vol. 215: Bonivento, C.; Marro, G.;
Zanasi, R. (Eds)
Colloquium on Automatic Control
240 pp. 1996 [3-540-76060-1]

Vol. 216: Kulhavý, R.
Recursive Nonlinear Estimation: A Geometric
Approach
244 pp. 1996 [3-540-76063-6]

Vol. 217: Garofalo, F.; Glielmo, L. (Eds)
Robust Control via Variable Structure and
Lyapunov Techniques
336 pp. 1996 [3-540-76067-9]

Vol. 218: van der Schaft, A.
L_2 Gain and Passivity Techniques in Nonlinear
Control
176 pp. 1996 [3-540-76074-1]

Vol. 219: Berger, M.-O.; Deriche, R.;
Herlin, I.; Jaffré, J.; Morel, J.-M. (Eds)
ICAOS '96: 12th International Conference on
Analysis and Optimization of Systems -
Images, Wavelets and PDEs:
Paris, June 26-28 1996
378 pp. 1996 [3-540-76076-8]

Vol. 220: Brogliato, B.
Nonsmooth Impact Mechanics: Models,
Dynamics and Control
420 pp. 1996 [3-540-76079-2]

Vol. 221: Kelkar, A.; Joshi, S.
Control of Nonlinear Multibody Flexible Space
Structures
160 pp. 1996 [3-540-76093-8]

Vol. 222: Morse, A.S.
Control Using Logic-Based Switching
288 pp. 1997 [3-540-76097-0]

Vol. 223: Khatib, O.; Salisbury, J.K.
Experimental Robotics IV: The 4th International Symposium, Stanford, California,
June 30 - July 2, 1995
596 pp. 1997 [3-540-76133-0]

Vol. 224: Magni, J.-F.; Bennani, S.;
Terlouw, J. (Eds)
Robust Flight Control: A Design Challenge
664 pp. 1997 [3-540-76151-9]

Vol. 225: Poznyak, A.S.; Najim, K.
Learning Automata and Stochastic
Optimization
219 pp. 1997 [3-540-76154-3]

Vol. 226: Cooperman, G.; Michler, G.;
Vinck, H. (Eds)
Workshop on High Performance Computing
and Gigabit Local Area Networks
248 pp. 1997 [3-540-76169-1]

Vol. 227: Tarbouriech, S.; Garcia, G. (Eds)
Control of Uncertain Systems with Bounded
Inputs
203 pp. 1997 [3-540-76183-7]

Vol. 228: Dugard, L.; Verriest, E.I. (Eds)
Stability and Control of Time-delay Systems
344 pp. 1998 [3-540-76193-4]

Vol. 229: Laumond, J.-P. (Ed.)
Robot Motion Planning and Control
360 pp. 1998 [3-540-76219-1]

Vol. 230: Siciliano, B.; Valavanis, K.P. (Eds)
Control Problems in Robotics and Automation
328 pp. 1998 [3-540-76220-5]

Vol. 231: Emel'yanov, S.V.; Burovoi, I.A.;
Levada, F.Yu.
Control of Indefinite Nonlinear Dynamic
Systems
196 pp. 1998 [3-540-76245-0]

Vol. 232: Casals, A.; de Almeida, A.T. (Eds)
Experimental Robotics V: The Fifth
International Symposium Barcelona, Catalonia,
June 15-18, 1997
190 pp. 1998 [3-540-76218-3]

Vol. 233: Chiacchio, P.; Chiaverini, S. (Eds)
Complex Robotic Systems
189 pp. 1998 [3-540-76265-5]

Vol. 234: Arena, P.; Fortuna, L.; Muscato, G.;
Xibilia, M.G.
Neural Networks in Multidimensional Domains:
Fundamentals and New Trends in Modelling
and Control
179 pp. 1998 [1-85233-006-6]

Vol. 235: Chen, B.M.
H∞ Control and Its Applications
361 pp. 1998 [1-85233-026-0]

Vol. 236: de Almeida, A.T.; Khatib, O. (Eds)
Autonomous Robotic Systems
283 pp. 1998 [1-85233-036-8]

Vol. 237: Kreigman, D.J.; Hagar, G.D.;
Morse, A.S. (Eds)
The Confluence of Vision and Control
304 pp. 1998 [1-85233-025-2]

Vol. 238: Elia, N. ; Dahleh, M.A.
Computational Methods for Controller Design
200 pp. 1998 [1-85233-075-9]

Vol. 239: Wang, Q.G.; Lee, T.H.; Tan, K.K.
Finite Spectrum Assignment for Time-Delay
Systems
200 pp. 1998 [1-85233-065-1]

Vol. 240: Lin, Z.
Low Gain Feedback
376 pp. 1999 [1-85233-081-3]